Proceedings of the 5th International Conference "Advanced Composite Materials Engineering" COMAT 2014

Honoring Professor Ioan Goia at his 80th Birthday

Transilvania University of Braşov

16 - 17 October 2014, Braşov, Romania

György SZEIDL, Chairman
Sorin VLASE, President
Michael M. DEDIU, Editor

DERC Publishing House
Tewksbury (Boston), Massachusetts, U. S. A.

Published and printed in 2014 in the United States of America

American Mathematical Society
2010 Mathematical Subject Classification: 65-xx, 70-xx, 74-xx, 76-xx, 82-xx

Library of Congress Cataloging in Publication Data

Advanced Composite Materials Engineering, the 5th International Conference
Research & Innovation in Engineering, the 3rd International Conference, 16 – 17 October 2014, Braşov, Romania / György SZEIDL, Chairman, Sorin Vlase, President, Michael M. Dediu, Editor

p. cm. – (Proceedings of the 5th International Conference "Advanced Composite" Materials Engineering 3rd International Conference, "Research & Innovation in Engineering", COMAT 2014)

Includes bibliographical references

ISBN-13: 978-1-939757234

PREFACE

After receiving many favorable comments and having interesting discussions regarding the Proceedings of the 4th International Conference "Advanced Composite Materials Engineering" COMAT 2012, and the Proceedings of the 5th International Conference "Computational Mechanics and Virtual Engineering" COMEC 2013, we are pleased to present these Proceedings of the 5th International Conference "Advanced Composite Materials Engineering" and the 3rd International Conference "Research & Innovation in Engineering" COMAT 2014. This conference was organized with the important support of Transilvania University of Brasov, Romanian Academy of Technical Sciences and Romanian Society of Theoretical and Applied Mechanics, to whom we thank very much.

This conference was dedicated to Professor Ioan Goia, at his 80th birthday, because he was a great Professor at Transilvania University of Brasov for many years, and he supported these conferences from the very beginning. He is an extraordinary Professor, who had astonishing accomplishments, kindness, patience, dedication and optimism. Professor Ioan Goia is a great example for the younger generations, and he is always remembered for his great devotion to teaching and helping students.

The research and innovation in engineering in general, and in particular advanced composite materials engineering, are ever more important in everyday activities, and they include, between many other techniques, geometric models, analysis, simulation, optimization and decision making tools, within a computer-generated environment, that accelerates product development, which is the main purpose of any engineering effort. The applications are very diverse and include manufacturing, aerospace, civil engineering structures, communication, geotechnics, flow problems, automotive engineering, predictive surgery, transportation, geo-environmental modeling, biomechanics, electromagnetism, medicine, prediction of natural physical events, and metal forming, just to mention a few.

These Proceedings of COMAT 2014 include 50 papers, which analyze many important practical applications. The topics range from energy release rate evaluation in sandwich composite structures to structural synthesis, analysis and design of the modular anthropomorphic grippers for industrial robots, from in-plane vibrations of pinned-fixed heterogeneous curved beams under a concentrated force to modeling and simulation of an acoustic cloak, from the finite elements analysis vs. experimental buckling behavior of the thinwalled plane gussets to the mathematical model of a special vehicle clutch servomechanism.

We want to thank very much Professor Sorin Vlase for his remarkable effort and dedication for the organization of this international conference, Professor György Szeidl for his distinguished chairmanship, and the organizing committee for their continuous assistance.

We also thank all the conference participants their interesting presentations, and Mrs. Sophia Dediu for her assistance in preparing this volume.

There is, certainly, much more that can be said about these engineering research than we presented here. We hope that the papers included here will provide ideas for our audience, and will stimulate more research, development and applications.

We look forward to receiving comments and suggestions from our readers.

Michael M. Dediu

Boston, U.S.A., December 8, 2014

Previously published in this series:

1. Ioan Goia, *Mechanics of Materials*
2. Ionel Staretu, *Gripping Systems*
3. Proceedings of the 4th International Conference "*Advanced Composite Materials Engineering*" COMAT 2012,
4. Proceedings of the 5th International Conference "*Computational Mechanics and Virtual Engineering*" COMEC 2013

The 5th International Conference
"Advanced Composite Materials Engineering"
and the 3rd International Conference
"Research & Innovation in Engineering"
COMAT 2014
16-17 October 2014, Braşov, Romania

ENERGY RELEASE RATE EVALUATION IN SANDWICH COMPOSITE STRUCTURES BY USING THE DIC

Octavian POP[1], Ioan GOIA[2], Dorin ROSU[2], Sorin VLASE[2]
[1]University of Limoges, France, ion-octavian.pop@unilim.fr
[2]Transilvania University of Braşov, Romania, i.goia@unitbv.ro

Abstract: *In this paper a new formalism based on the coupling between the optical full field techniques and the integral invariants is proposed in order to evaluate the fracture parameters in cracked sandwich composite structures. The formalism allows identifying the fracture parameters in terms of energy release rate. From the experimental tests the displacement field is obtained by digital image correlation measurements. In this case the experimental displacement fields are employed to calculate the strain and stress fields by a numerical approach. Then using the mechanical fields defined on the surface of specimen the integral invariants can be used in order to characterize the fracture process. The present study is limited to the identification of the mechanical fields using the experimental displacement fields measured by the optical methods.*

1. INTRODUCTION

Since some years the optical methods find more and more, their applications in the mechanical characterisation of materials and structures. Associated to the full fields techniques, the optical methods can be easily correlated with the energetically approaches such as the integral invariants. Among the optical methods, the digital image correlation seem to be the best to characterise the mechanical behaviour in the case of composite. Another particularity of this optical full fields' technique is the possibility to coupling this one with the numerical approach such as the Finite Element Method (FEM). Using the optical methods, the zone of interest (ZOI) can be discretized either by the subsets similar to the finite elements of the mesh, in the case of DIC. In this paper a feasibility study in order to characterize the fracture behavior in sandwich composite structures [1,2] by digital image correlation technique, is proposed.

OPTICAL FULL-FIELD TECHNIQUES [3, 4, 5, 6]

Related with the optical full field methods, digital image correlation is an optical method allows to measure in-plane displacement.

The basic principle of method is based on the comparison between two images of sample plane surface acquired at different states one before deformation and the other one after. Then, the displacements are estimated by comparing the degree of resemblance between these two images subsets. As is shown in Fig. 1, in DIC the displacements are calculated into Zone Of Interest (ZOI) discretized by small areas with multiple pixel called subsets (where, m is the subset numbers). Note, that each subset is characterized by a unique light intensity distribution (i.e. gray level). Assuming that during the test the light intensity distribution on the plane surface of sample does not change, the displacements are estimated by searching the subsets changes (translation + rotation + rigid body motion) between the undeformed and the deformed images. Finally the displacement field is build from the displacement vectors (u1, u2) corresponding to displacements of all subsets centers.

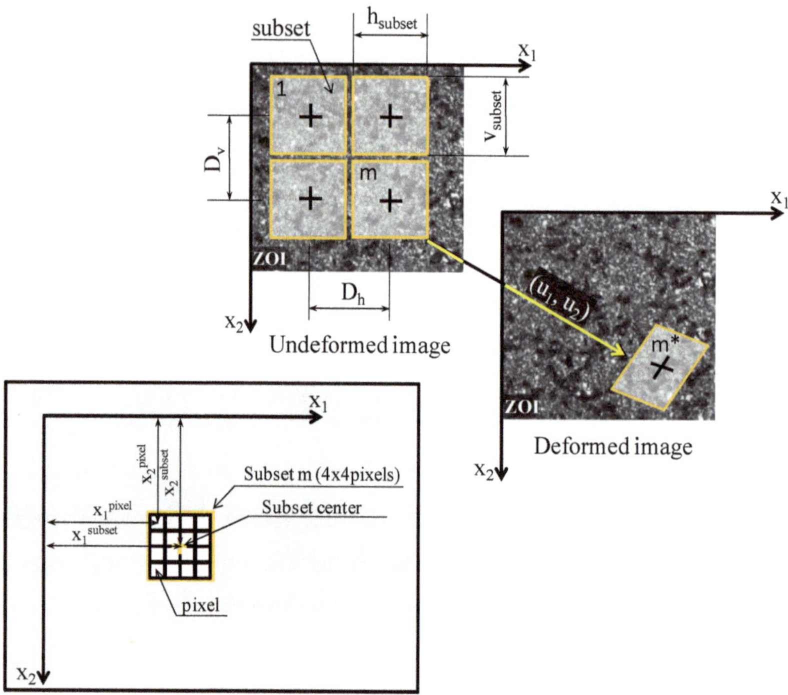

Fig. 1: Principle of Digital Image Correlation

Another important aspect in DIC is the specimen plane surface preparation. So, in order to obtain a gray level distribution, it is necessary to create a characteristic speckle pattern a black speckle is deposited on the white background by spraying the black and white paints.

2. EXPERIMENTAL SETUP

The testing machine is an electromechanical press with a load capacity of 50kN. The test is run under displacement control and the velocity of the cross-head is fixed at 0.01 mm/s. As shown in Fig. 2 the sandwich composite specimen is loaded in opening mode using a system which includes a steel pyramid and two rollers. The loading system allows perfect lips displacement symmetry.

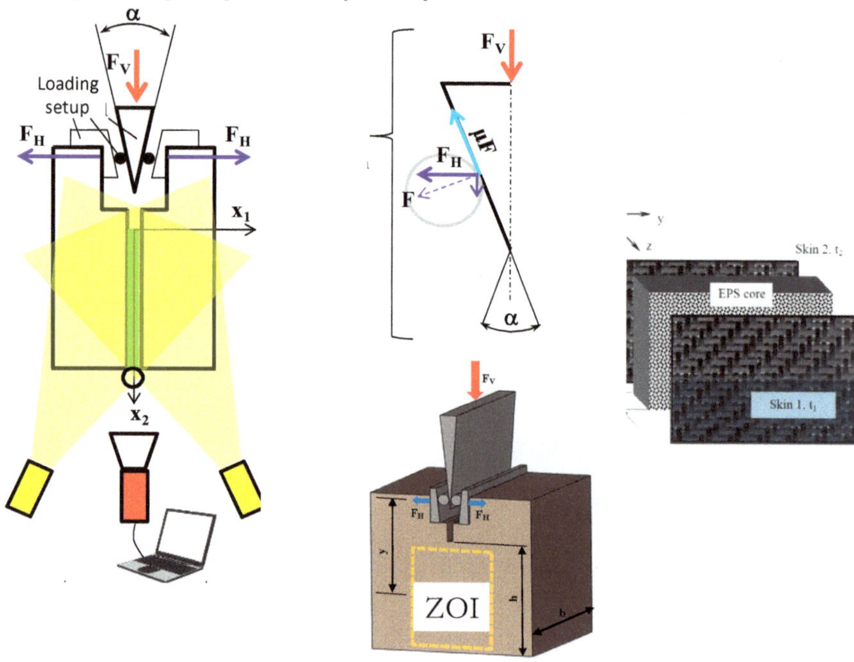

Fig. 2: Experimental setup

Besides, Fig. 3 puts in evidence the crack lips displacement symmetry during the test. In fact the plotted displacements correspond to the rollers displacement measured by mark tracking method.

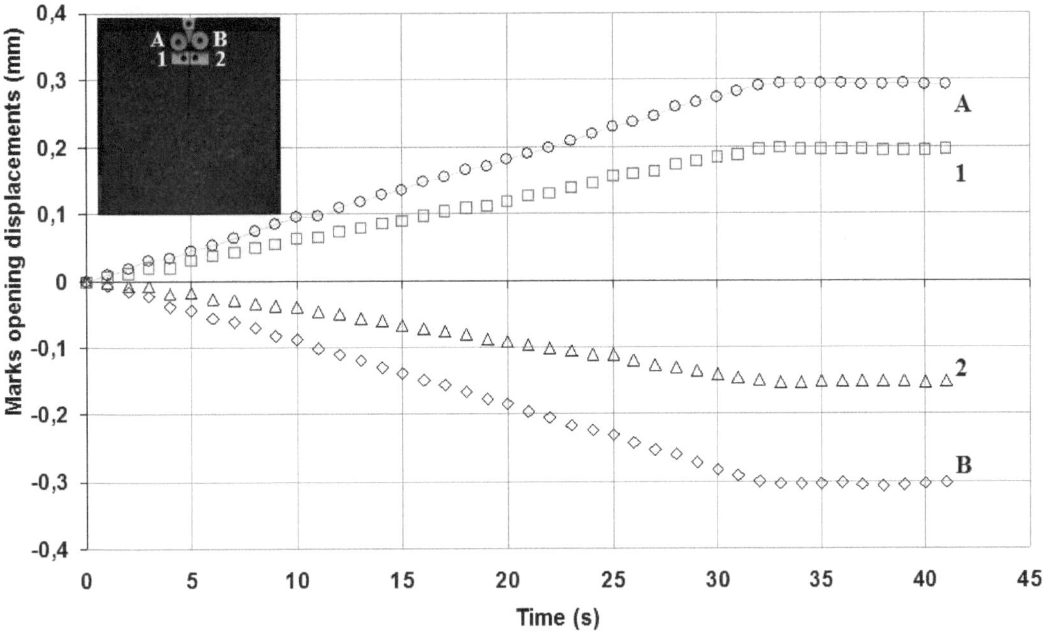

Fig. 3: Experimental displacement of the loading rollers

As is shown in Fig.2, the specimen is a sandwich composite. The sandwich structure taken into account to accomplish the damping analysis presents two carbon/epoxy skins reinforced with a 0.3 kg/m2 twill weave fabric and an expanded polystyrene (EPS) 9 mm thick core with a density of 30 kg/m3 [1], [2]. The final structure's thickness is 10 mm. The carbon-fibre fabric used in this structure is a high rigidity one, that presents so called twill weave. The main feature of this weave is that the warp and the weft threads are crossed in a programmed order and frequency, to obtain a flat appearance. The skins were impregnated under vacuum with epoxy resin and sticked to the core.

During the test, the load-displacement evolution is recorded using a LVDT sensor and a load cell. Furthermore, an 8-bit Charge Coupled Device (CCD) camera (1392x1040 pixels^2) is synchronized with the testing machine to measure the displacement fields by DIC.

3. INTEGRAL INVARIANT J-INTEGRAL

In fracture mechanics the J-integral is associated with strain energy release rate or the work per unit of crack area. The theoretical concept of the J-integral was developed by Cherepanov [7, 8] and Rice [9, 10] who showed that the J-integral is independent of the path defined around the crack tip (Fig. 4). The crack is oriented in x_1-direction, and the crack tip represents the origin of the coordinate system.

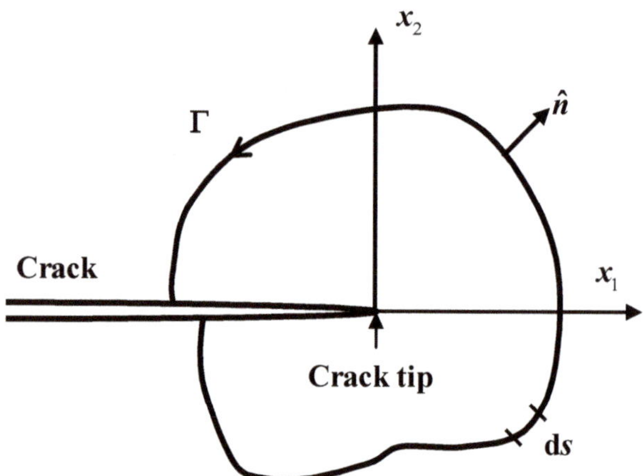

Fig. 4: Domain of integration of J

The J-integral can be expressed in the following form:

$$J=\int_{\Gamma}\left(W\cdot n_1-T_i\cdot\frac{\partial u_i}{\partial x_1}\right)\cdot ds \tag{4}$$

$$W=\frac{1}{2}\cdot\sigma_{ij}\cdot\varepsilon_{ij} \tag{5}$$

$$T_i=\sigma_{ij}\cdot n_j \tag{6}$$

Where: W is the elastic strain energy density, T_i is the traction along the contour Γ, n the unit normal of the contour path, u is the displacement vector, σ_{ij} and ε_{ij} are the stress and strain tensors, respectively.

By introducing Equations (5) and (6) in (4), we can see that the J-integral evaluation is based on the knowledge of the mechanical fields in terms of displacement, strain and stress. Therefore, the displacement vector (u) will be evaluated experimentally by DIC method, while the strain (ε_{ij}) and stress (σ_{ij}) tensors will be calculated from numerical approach.

EXPERIMENTAL RESULTS

As illustrated in Fig. 5 the domain of integration can be assimilated by crowns defined by the centers of the correlation windows. By using the finite elements mesh generated from the experimental data, the J-integral is evaluated numerically via Equation (4) along different crowns (Γ_i) surrounding the crack tip.

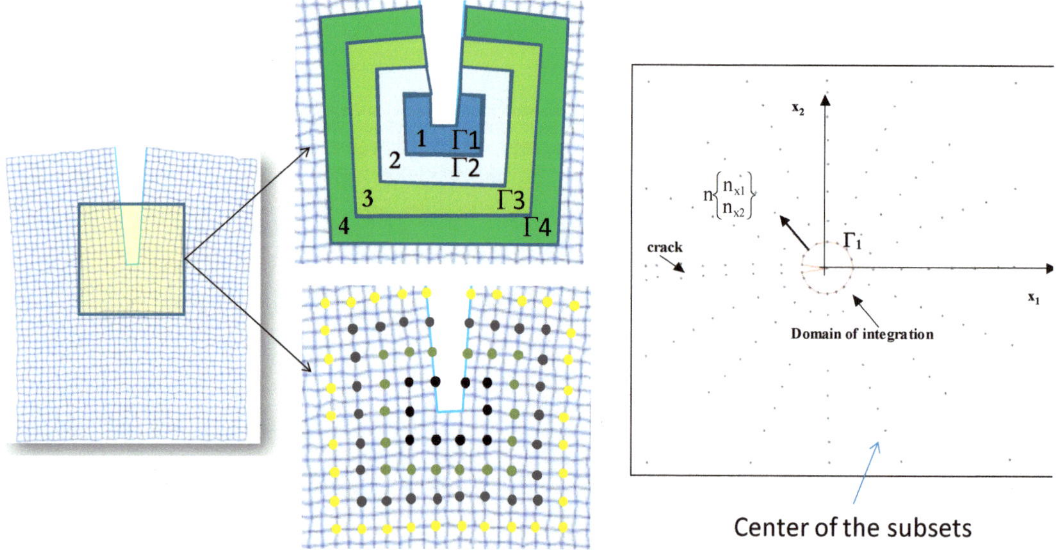

Fig. 5: Domain of integration using the experimental results

The J-integral values versus loading amplitude and size of the domain of integration are presented in Table 1.

Load (N)	J –integral (J/m²)			
	Γ_1	Γ_2	Γ_3	Γ_4
280	25.33	25.01	24.99	25.42
560	55.23	55.03	55.02	54.98

The results presented in Table 1 show a low variation of the energy release according to integral path. These slight differences can be explained by experimental displacement field noises. In this case the displacement field optimization may be considered in order to increase the accuracy of the experimental fields.

CONCLUSION

In this study, a new experimental technique is proposed in order to characterize the fracture parameters in sandwich composites. The DIC is employed in order to measure the displacements fields on the specimen surface. Based on the experimental optical full field measurements the energetic approach is used to evaluate the energy release rate. Using the experimental displacement and the strain fields the stress tensor is calculated via a constitutive law. The stress fields are evaluated using the finite elements method. In fact the approach of this study consists of a combination of experimental and numerical techniques in order to evaluate the different mechanical fields employed in the energetic approach such as J or G-integral. For this purpose the experimental data are implemented in a finite element code in order to generate a finite element mesh. In this case the stress fields are calculated using the experimental data by imposing real boundary conditions. Now, using the mechanical fields measured or evaluated by coupling experimental with numerical approaches the J-integral is calculated for several loading steps and for different crowns defined around the crack tip. The J-integral evolution versus crown size shows the energy release rate invariance. The slight differences can be explained by the displacement field noises.

REFERENCES

[1] D.H. Teodorescu, S. Vlase, "Dynamic analysis of an ultra-lightweight sandwich structure for multiple applications". 3rd WSEAS International Conference on Dynamical Systems and Control, Arcachon, France, 13 – 15 October 2007, pp. 229-234.

[2] D.H. Teodorescu, S. Vlase, D.L. Motoc, I. Popa, R. Dorin, F. Teodorescu, "Mechanical behaviour of an advanced sandwich composite structure". WSEAS International Conference on ENGINEERING MECHANICS, STRUCTURES, ENGINEERING GEOLOGY (EMESEG '08), Heraklion, Greece, July 22-24, 2008, pp. 280-285.

[3] Sutton M.A., McNeill S.R., Helm J.D., Chao Y.J., (2000) Advances in two-dimensional and three-dimensional computer vision, In: Rastogi P.K. (ed) Photomechanics, Springer, Berlin Heidelberg New York, pp 323–372.

[4] Sutton, M. A., Turner, J. L., Bruck, H. A. and Chae, T. A. (1991) Full-field Representation of Discretely Sampled Surface Deformation for Displacement and Strain Analysis. Experimental Mechanics, 31(2), 168-177.

[5] Sutton, M. A., McNeill, S. R., Jang, J. and Babai, M., (1988) Effects of Subpixel Image Restoration on Digital Correlation Error. Journal of Optical Engineering, 27(10), 870-877.

[6] Sutton, M. A., Cheng, M. Q., Peters, W. H., Chao Y. J. and McNeill, S. R., (1986) Application of an Optimized Digital Correlation Method to Planar Deformation Analysis. Image and Vision Computing, 4(3), 143-151.

[7] Cherepanov GP (1962) The stress in a heterogeneous plate with slits, in Russian, Izvestia AN SSSR, OTN, Mekhan. I Mashin. 1:131-137.

[8] Cherepanov GP (1979) Mechanics of brittle Fracture, McGraw-Hill, New York.

[9] Budiansky B, Rice JR (1973) Conservation laws and energy-release rates, ASME, J. Appl. Mech. 40: 201-203.

[10] Rice JR (1968) A path independent integral and the approximate analysis of strain concentration by notches and cracks, Journal of Applied Mechanics, 35:379-386.

\

LOW – COST CNC DIMENSIONAL INSPECTION GAUGES IMPROVING

Braun B.[1], Drugă C.[2]
[1] TRANSILVANIA University of Brasov, ROMANIA, braun@unitbv.ro
[2] TRANSILVANIA University of Brasov, ROMANIA, druga@unitbv.ro

Abstract (TNR 9 pt Bold): *The paper describes the way in which a low cost and efficient measuring and scanning system was obtained, starting from existing equipment in the Faculty of Product Design from Transilvania University of Brasov. It refers to a YAMAHA robotic 1-D axis, for translation entraining, which was disposed to another 1-D entraining axis (which was manufactured with recycled components and controlled by an Arduino microcontroller). For dimensional inspection and scanning, a portable and flexible support as third axis was adapted, may include any type of measuring device. In this way, a low - cost three axes ordering system the coordinate measuring was possible to be obtained. The advantages of the proposed solution are: simplicity, low-cost, high accuracy and precision. The main disadvantage is that it does not allow a quick displaying and processing of the results. This is the subject of future research on improving the proposed solution.*
Keywords (TNR 9 pt Bold): *entraining axis, microcontroller, measuring, scanning*

1. COORDINATE MEASURING SYSTEMS – KEY ROLE FOR QUALITY INSURANCE

Due to the fact that allow very accurate and efficient dimensional measurement and scans of parts components with applications in all fields, in the last years the use of the coordinate measuring machines has experienced unprecedented development. Equipped with complex software environment, these are very user friendly, allowing access without the need for excessive baggage of knowledge on their use. The only disadvantage on their widespread use is the high purchase price.
Therefore, one of the main goals of our research is to create and develop a low-cost and effective type CNC coordinate machine for testing and scanning of small complex components with applications in industry.

2. TWO AXES COORDINATE MEASURING SYSTEM – STEPS TO ACHIEVE

The starting point for manufacturing and implementing the system consisted into a 1-D robotic entraining axis, manufactured by YAMAHA (figure 1), which technical characteristics are presented in the table below [1], [2]:

Table 1: First robotic entraining axis, technical characteristics

Range	600 mm
Ordering	DC motor coupled with ball screw
Accuracy	0.005 mm

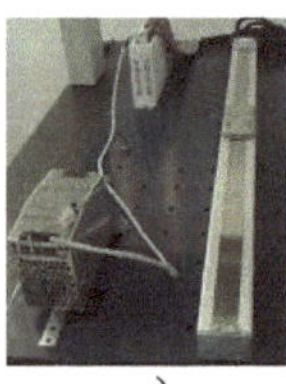
a)

b)

c)

Figure 1: The first 1-D entraining axis: a) axis arrangement; b) power supply module; c) hardware interface for PC connecting

The axis is equipped with a conveyor for different components transport, technological processing or manufacturing quality inspection.

To increase the applications field in terms of quality inspection, the research aim was to extend the possibilities to measure and to scan different type of components. For this reason, it was took the question to dispose a second axis, which orientation is at 90° to the first one. In this way a 2D translation coordinate system could be obtained. Mechanical standpoint, the second entraining axis (figure 2) was composed by several recycled components, providing from a scrapped printer. Over the conveyor was then ordered first axis.

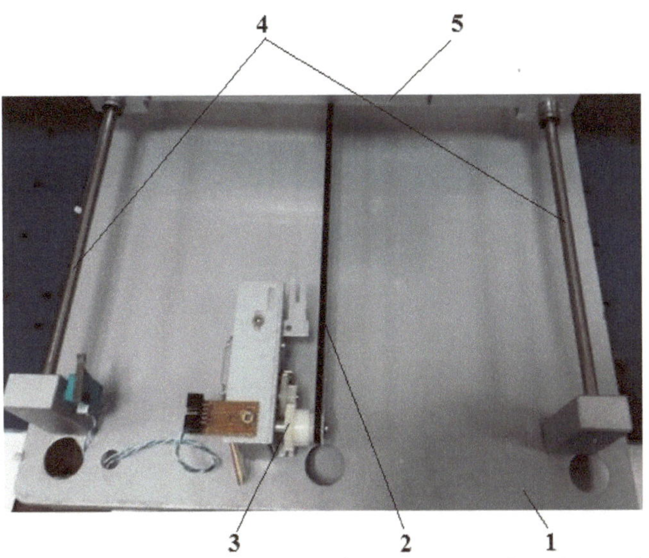

Figure 2: The second entraining axis: 1 – support plate, 2 – toothed belt; 3 – engine with reduction; 4 – guide bars; 5 - conveyor

Hardware standpoint, a base plate was used, on which some electronic components were fixed (figure 3) (an *Arduino* microcontroller, a display, connection sockets etc.) [3]. Another plate serve (figure 3) as support for some command buttons [4].

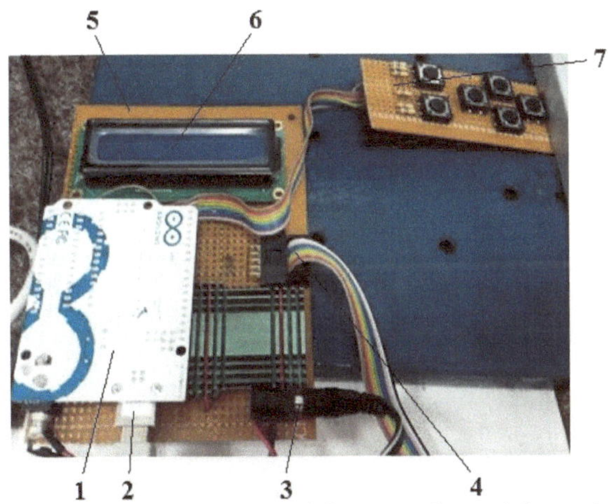

Figure 3: Hardware system to command the second entraining axis: 1 – Arduino microcontroller; 2 – USB plug-in connection; 3 – supply cable socket; 4 – plug-in connection for the engine; 5 – base support; 6 – display; 7 – second plate with push buttons for command

Composing both axes, a two axes drive system was obtained, as can be seen in figure 4. Besides, a measuring and scanning system was disposed to complete the 3-D CNC dimensional inspection system (figure 6). The measuring system, as the third axis was obtained using a magnetic flexible support and a measuring device (figure 5). The system allows disposing the measuring support in any position, the latter one could be in contact or non-contact principle. Thus means that different type of measuring devices can be used, including also optical non contact displacement transducer with LASER ray.

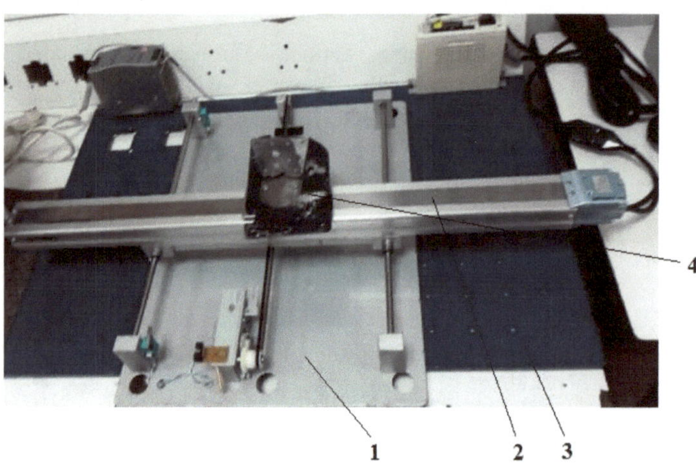

Figure 4: Two axes drive system: 1 – the first axis; 2 – the second axis; 3 – base support; 4 – conveyor with fixing system of the tested probes

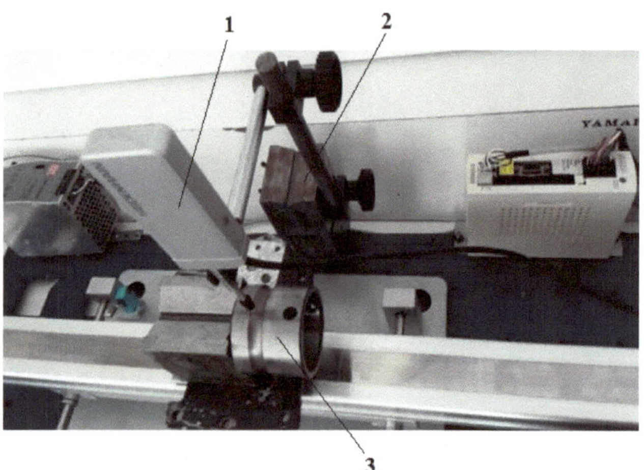

Figure 5: Measuring system support disposing: 1 – magnetic flexible support; 2 – measuring device; 3 – tested probe

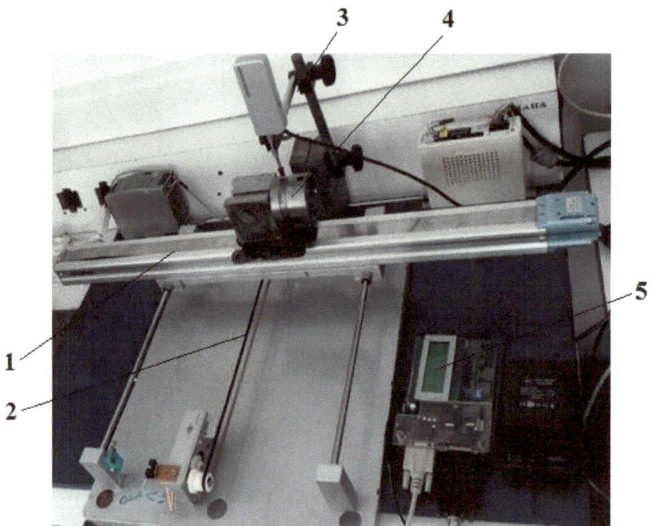

Figure 6: The 3-D CNC dimensional inspection system: 1 – the first axis; 2 – the second axis; 3 – the measuring system; 4 – the tested probe; 5 – the hardware device corresponding to the measuring system

Via a hardware interface, it was possible to connect to the PC the measuring system. Thus means that dimensional measurements and/or scans aided by PC, in real time, of the tested components, can be performed. The hardware interface consists into a signal decoder (providing from the displacement measuring transducer), a microcontroller, a display, push buttons a socket for transducer connection, a plug-in socket for supplying and an RS-232 interface protocol for data transfer to the PC.

3. DIMENSIONAL INSPECTION/SCANNING EXAMPLES, USING THE IMPROVED SYSTEM

Due to the hardware and software interfaces, the YAMAHA robotic axis was programmed for different dimensional measuring cycles, using the YAMAHA's axis software environment [5]. An example is presented in figure 2, representing measuring cycle command programming routine, in several successive points of the tested component.

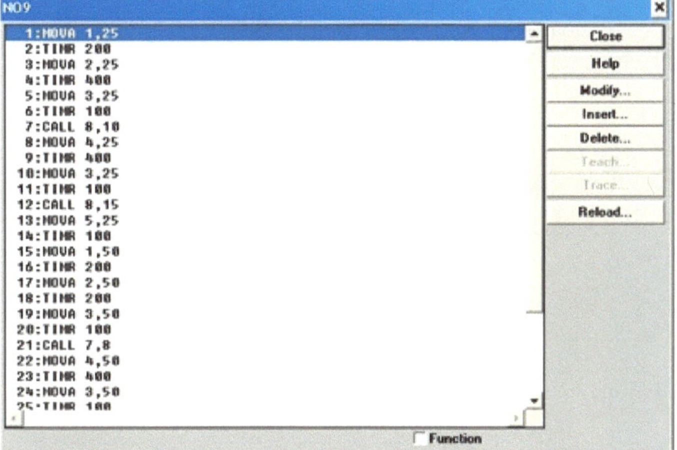

Figure 2: Programming routine for a step by step dimensional measuring cycle, using the first 1-entraining axis

On the second drive axis, the Arduino microcontroller allows the command programming routine, a case being presented in figure 3. It represents an example, in which the tested probe could be displaced step by step (which increment was predefined at 1 mm), in order to ensure a measuring cycle, along the second axis [4], [6].

```
lcd.setCursor(0, 1);
lcd.print("     ");
// position value incrementing by 1 mm
position = position + 24; }}
if (position < 7080) {
if (digitalRead(inc10) == LOW) {
lcd.setCursor(0, 1);
lcd.print("     ");
// position value decrementing by 1 mm
position = position - 24; }}
if (position > 120) {
if (digitalRead(dec10) == LOW) {
lcd.setCursor(0, 1);
lcd.print("     ");
```

Figure 3: Programming routine for tested probe's positioning displacement with 1 mm increment, for a measuring cycle along the second axis

Composing both algorithms, a measuring cycle for probes with complex geometry could be done, due to the fact that a composed displacement of the probe (fixed by the conveyor) along both axis could be performed.

For the probes testing, as devices can be used displacement transducers for dimensional measuring, having different principles. Until now, an incremental displacement transducer, optical principle (figure 4) could be adapted to the system, which technical characteristics are presented in the table below [7]:

Table 2: Technical characteristics for the implemented dimensional measuring device

Measuring range	25 mm
Accuracy	0.2 μm
Output signal type	digital signal, type TTL

Figure 4: Example of dimensional measuring device arrangement for 3-D scanning and dimensional measuring of components with applications in automobiles field

For signal decoding in order to, a PIC18F452 microcontroller was adapted as hardware interface, in order to connect the measuring device to the PC. The hardware interface is presented in figure 5 [8].

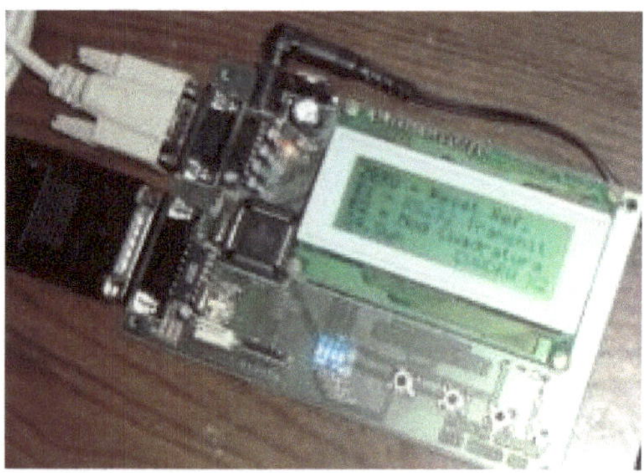

Figure 5: The hardware interface for signals decoding and measured values displaying, associated to the incremental displacement transducer

4. CONCLUSION

The research led to obtain a 3-D coordinate dimensional measuring and/or scanning system, CNC type, with performances close to the Coordinate Measuring Machines (CMM). The main advantage of the proposed system is that the cost is estimated to be 10 times less than those of CMMs. Besides, the system can be successfully used for

dimensional inspection or scanning of different components with complex geometry. Another advantage refers to the measuring devices very quick and simple interchanging.
However, the main disadvantage is that currently the measuring system cannot allow synchronizing both axes, thus means that each axis must be programmed and commanded separately, meaning an inefficient testing process. Besides, the communication hardware protocol needs also serial devices, meaning to use PC old generation. For this reason, in the future, the research will be focused to improve the hardware and software system to solve the problem with the drive system synchronization. Another goal is to use only USB or wireless hardware communication protocols.

REFERENCES

[1] Yamaha, Single-axis robot, http://www.yamaharobotics.com/Catalog/PDF/CurrentManuals/ Discontinued/ T4-T5_E_V2.01.pdf , 2013.
[2] Yamaha, Robot user's manual, YAMAHA, 2008.
[3] Arduino, Functional characteristic and use, http://en.wikipedia.org/wiki/Arduino, 2014.
[4] Pirvulescu, I., Studies and research on optimizing training in translational applications in industry, Diploma Project, Transilvania University of Brasov, 2014.
[5] POPCOM, Yamaha drive axis programming instructions, http://www.yamaharobotics.com/ Catalog/PDF/POPMANE.pdf, 2012.
[6] Visual Basic, Programming instructions, http://ro.wikipedia.org/wiki/Visual_Basic, 2013.
[7] HEIDENHAIN, Langen messen. Messgerate fur das Messen von Langen, www.heidenhain.de, Deutschland 2006.
[8] Stefanescu, O., Integrating incremental transducers in measuring systems, Diploma Project, Transilvania University of Brasov, 2006.

VIBRATION OF CIRCULAR PLATES SUBJECTED TO CONSTANTRADIAL LOAD IN THEIR PLANE

György Szeidl[1],Nóra Szücs[2]

[1]Institute of Applied Mechanics, Universityof Miskolc,Miskolc, HUNGARY
gyorgy.szeidl@uni-miskolc.hu
[2]Robert BoschEnergy and Body Systems KFT.,RobertBoschPark3, H-3526Miskolc,HUNGARY
noraszucs@gmail.com

Abstract: *The present paper deals with the vibration of circular plates provided that the plates are prestressed in such a way that the stresses due to the in-plane load are constants. We have determined the Green functions of the governing equations. With the knowledge of the Green function the self adjoint eigenvalue problems giving the natural frequencies are replaced by homogenous Fredholm integral equations for which the symmetric Green functions constitute the kernel. According to the numerical solution the square of the natural frequencies are linear, or approximately linear functions of the constant in plane load.*

Keywords: *curved beams, heterogeneous material, natural frequency as a function of the load, Green function matrices*

1. INTRODUCTION

It is well known [1, 2006] that the natural frequency of a simply supported beam subjected to a compressive axial force – the beam is in a prestressed state – satisfies the equation

$$\frac{\alpha^2}{\alpha_1^2} = 1 - \frac{f}{f_1} \tag{1}$$

in which α and α_1 are the first natural frequencies of the loaded and unloaded beam while f and f_1 are the compressive force and its smallest critical value.

Paper by [1, 2006] Lawther, which we have already cited, attacks the problem how a prestressed state of the body affects the natural frequencies in a more general form. He studies finite dimensional multiparameter eigenvalue problems and comes to the conclusion that for multiparameter problems, the eigenvalue part of the solution is described by interaction curves in an eigenvalue space, and every such eigenvalue solution has an associated eigenvector. If all points on a curve have the same eigenvector then the curve is necessarily a straight line, but the converse is far more complex.

Boundary element solutions for the plate buckling problem have been published in the papers [2, 1994], [3, 1996], [4, 1999], [5, 2000], [6, 2005]. The authors of the papers cited investigate the buckling phenomenon under various assumptions and use different mechanical models. However it is a common feature of each paper that the fundamental solutions utilized do not involve the effect of the in-plane stresses directly. The reason for attacking the problem in this way is simple: it makes possible that the model is applicable for any in-plane load exerted on the boundary. The price one has to pay for the generality of the model concerning the in-plane load is the presence of a domain integral which should be handled in a way. However there are various cases when the stress state due to the in-plane load is a constant one. We also remark that the papers cited above are not concerned with the issue how the in-plane stresses affect the vibrations of the plate.

Paper [7, 1965] is devoted to the issue what vibration characteristics of centrally clamped, variable thickness disks have if they are subjected to rotational and thermal in-plane stresses. This paper contains useful references for the most important earlier results which are not cited here. Pardoen [8, 1974] devotes his attention to the vibration of prestressed circular orthotropic plates and uses a finite element procedure for solving the problem. In her master thesis [9, 1980] D. Chotova investigated the effect of a uniformly distributed in-plane compressive force system – exerted on the outer boundary of the plate – on the natural frequencies of circular plates assuming axisymmetric behavior. Numerical solutions were found by determining the zeros of the corresponding frequency determinant which is regarded as a function of the load. It is also important to cite papers [10, 1983] [11, 1984] by Chen and his co-authors who have dealt among others with large amplitude vibration of an initially stressed hick circular plate. They have applied energetic methods in order to clarify how the load influences the vibrations. The results achieved are presented graphically. Paper [12, 2003] is devoted to the stability of parametric vibrations of circular plates subjected to in-plane forces by using the Liapunov method. Younesian [13, 2012] studies the forced vibrations of annular plates under the action of a transverse load rotating on the outer boundary.

The main objective of the present paper is to clarify mathematically – by setting up polynomial relationships – how the constant in plane load influences the vibrations of clamped and simply supported circular plates. We attack the problem by determining the Green functions of the prestressed plate which might also be useful for solving various boundary value problems in a closed form if the plate is subjected to a transverse load. With the knowledge of the Green function the two parameter eigenvalue problem for the natural frequencies is reduced to an eigenvalue problem governed by homogenous Fredholm integral equations with a symmetric kernel. This formulation is basically the same as that of the boundary element method, however the kernel is not singular. After having set up the eigenvalue problem this way we solve it by the boundary element method.

The paper is organized into seven sections. Section 2 is a collection of the most important notations and notational conventions. Section 3 presents the governing equations of the problem raised. Section 4 outlines the determination of the Green functions. It is also shown that the third Green function contains the other two at the limit that is if the spring constant tends to zero or infinity. The algorithm of the numerical solutions is considered in Section 5. Computational and analytical (closed form approximate) solutions are presented in Section 6. The last section is a conclusion. We remark that some preliminary results were published earlier in Hungarian [14, 2007].

2. Notations and Notational Conventions

The table below presents the most important notations and notational conventions.

$\mathcal{A}=\sqrt{\frac{\gamma}{g}\frac{\alpha^2 R_o^4}{I_1 E_1}}$	dimensionless natural frequency
$2b$	thickness of the plate
f	radial load in the mid-plane
$I_1=8b^3/12$	moment of inertia
E	Young modulus
$E_1=E/(1-\nu^2)$	modified Young modulus
$\mathcal{F}=R_o^2\frac{f}{I_1 E_1}$	dimensionless in-plane load
$\mathcal{F}_\nu=\frac{\mathcal{F}}{1-\nu}$	modified dimensionless in-plane load
g	gravitational acceleration
$G(\rho,\xi)=G(\xi,\rho)$	Green function
$J_n, Y_n, I_n, K_n, n=0,1,2,\dots$	Bessel functions
k_γ	spring constant
$\mathcal{K}=\frac{R_o k_\gamma}{I_1 E_1}$	dimensionless spring constant
$\mathcal{K}_\nu=1-\frac{\mathcal{K}}{1-\nu}$	modified dimensionless spring constant
$p_z(\rho)$	distributed load in the z direction
$Q_R(\rho)$	shear force
R	radius
R_o	outer radius of the plate
$w(\rho)$	deflection
$\tilde{\Delta}=\frac{\mathrm{d}^2}{\mathrm{d}\rho^2}+\frac{1}{\rho}$	differential operator
α	natural frequency
γ	density
ν	Poisson number
$\rho=R/R_o$	dimensionless independent variable
ξ	dimensionless independent variable

The text contains further notations defined at the place of their first appearance.

3. Governing Equations

If a circular plate is subjected to a constant radial load f in its plane then the deflection w due to the load $p_z(\rho)$ acting perpendicularly to the middle plane of the plate should meet the differential equation

$$\tilde{\Delta}\tilde{\Delta}w \pm \mathcal{F}\tilde{\Delta}w = \frac{R_o^4}{I_1 E_1}p_z, \qquad \tilde{\Delta}=\frac{\mathrm{d}^2}{\mathrm{d}\rho^2}+\frac{1}{\rho}\frac{\mathrm{d}}{\mathrm{d}\rho} \tag{2}$$

if axisymmetric deformations are assumed – the sign preceding $\mathcal{F}$ is positive for compression and negative for tension – Figure 1 shows a compressive load [15, 2003]. Depending on what the supports are equation (2) should be associated with appropriate boundary conditions. As regards the outer boundary it is clear from Figure 1 that for a clamped plate

$$w|_{\rho=1} = 0, \qquad \left.\frac{\mathrm{d}w}{\mathrm{d}\rho}\right|_{\rho=1} = 0, \tag{3}$$

are the boundary conditions. If the plate is simply supported then the boundary conditions are of the form

$$w|_{\rho=1} = 0, \qquad \left.\left(\frac{\mathrm{d}^2 w}{\mathrm{d}\rho^2} + \frac{\nu}{\rho}\frac{\mathrm{d}w}{\mathrm{d}\rho}\right)\right|_{\rho=1} = 0. \tag{4}$$

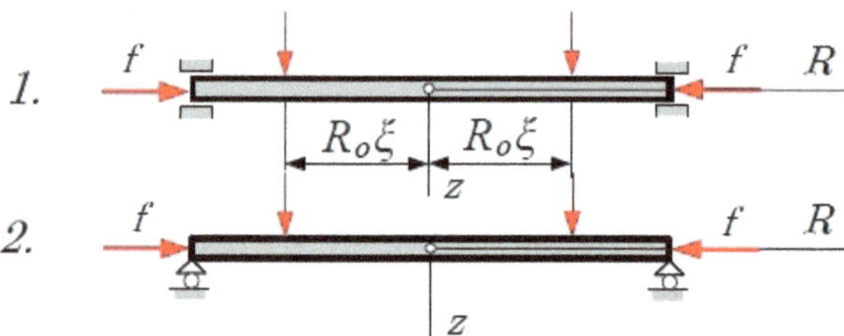

Figure 1.

It is also clear that the deflection at the center of the plate should meet the conditions

$$w|_{\rho=0} = \text{finite}, \qquad \left.\frac{\mathrm{d}w}{\mathrm{d}\rho}\right|_{\rho=0} = 0 \tag{5}$$

as well. For our later considerations we remark that the shear force Q_R is related to the deflection via equation

$$Q_R \frac{R_o^3}{I_1 E_1} = \frac{\mathrm{d}}{\mathrm{d}\rho}\left[\frac{\mathrm{d}^2}{\mathrm{d}\rho^2} + \frac{1}{\rho}\frac{\mathrm{d}}{\mathrm{d}\rho} \pm \mathcal{F}\right] w(\rho). \tag{6}$$

We remark that the boundary conditions and the formula for the shear force are all taken from book [16, 1967] by I. Kozák.

4. Green functions

4.1. **Definition of the Green function.** Assume that the plate is subjected to a uniform load distributed on the circle with radius $R_o\xi$ (ξ is also a dimensionless coordinate) – see Figure 1. The resultant of the total load is assumed to be 1. The deflection due to the load at ρ is denoted by $G(\rho,\xi)$ and is referred to as the Green function. Observe that the Green function should satisfy the homogenous equation in (2) if $0 \le \rho < \xi$ and $\xi < \rho \le 1$.

4.2. **Green functions for a compressive f.**

4.2.1. *General solution* As is well known the general solution of the homogenous equation in (2) assumes the form

$$w(\rho) = c_1 + c_2 \ln\rho + c_3 J_o(\sqrt{\mathcal{F}}\rho) + c_4 Y_o(\sqrt{\mathcal{F}}\rho) \tag{7}$$

where c_1, c_2, c_3 and c_4 are undetermined constants of integration.

Since the Green function should meet conditions (5) it assumes the form

$$G(\rho,\xi) = A_1 + A_2 J_0(\sqrt{\mathcal{F}}\rho), \qquad 0 \le \rho \le \xi, \tag{8a}$$

$$G(\rho,\xi) = B_1 + B_2 \ln\rho + B_3 J_0(\sqrt{\mathcal{F}}\rho) + B_4 Y_0(\sqrt{\mathcal{F}}\rho), \qquad \xi \le \rho \le 1 \tag{8b}$$

where the constants of integration A_1, A_2 and B_1, B_2, B_3, B_4 are to be determined from the continuity and discontinuity conditions prescribed at $\rho = \xi$ and the boundary conditions which are imposed on the boundary $\rho = 1$.

Observe that the continuity conditions

$$G(\xi-0,\xi) = G(\xi+0,\xi) \qquad G'(\xi-0,\xi) = G'(\xi+0,\xi) \qquad G''(\xi-0,\xi) = G''(\xi+0,\xi) \tag{9}$$

and the discontinuity condition

$$2\pi R_o\xi\left[Q_R(\xi+0) - Q_R(\xi-0)\right] = 2\pi R_o\xi Q_R(\xi+0) = 1, \tag{10}$$

in which $Q_R(\xi-0)=0$ from the vertical equilibrium, are all independent of the supports. Here (a) derivatives with respect to ρ are denoted by primes and (b) it follows from (6) that

$$Q_R\frac{R_o^3}{I_1E_1}=\frac{\mathrm{d}}{\mathrm{d}\rho}\left[\frac{\mathrm{d}^2}{\mathrm{d}\rho^2}+\frac{1}{\rho}\frac{\mathrm{d}}{\mathrm{d}\rho}+\mathcal{F}\right]G(\xi,\rho)\,. \tag{11}$$

4.2.2. *Green function for the clamped plate.* If the plate clamped the continuity and discontinuity conditions are associated with the following boundary conditions

$$G(1,\xi)=0\,,\qquad G'(1,\xi)=0\,. \tag{12}$$

In what follows we shall need the derivatives

$$G'(\rho,\xi)=-A_2\sqrt{\mathcal{F}}J_1(\sqrt{\mathcal{F}}\rho)\qquad 0\le\rho\le\xi\,, \tag{13a}$$

$$G'(\rho,\xi)=B_2\frac{1}{\rho}-B_3\sqrt{\mathcal{F}}J_1(\sqrt{\mathcal{F}}\rho)-B_4\sqrt{\mathcal{F}}Y_1(\sqrt{\mathcal{F}}\rho)\qquad \xi\le\rho\le1\,, \tag{13b}$$

$$G''(\rho,\xi)=A_2\mathcal{F}\left[\frac{J_1\left(\sqrt{\mathcal{F}}\rho\right)}{\sqrt{\mathcal{F}}\rho}-J_0\left(\sqrt{\mathcal{F}}\rho\right)\right]\qquad 0\le\rho\le\xi\,, \tag{14a}$$

$$G''(\rho,\xi)=-B_2\frac{1}{\rho^2}+B_3\mathcal{F}\left[\frac{J_1\left(\sqrt{\mathcal{F}}\rho\right)}{\sqrt{\mathcal{F}}\rho}-J_0\left(\sqrt{\mathcal{F}}\rho\right)\right]+B_4\mathcal{F}\left(\frac{Y_1\left(\sqrt{\mathcal{F}}\rho\right)}{\sqrt{\mathcal{F}}\rho}-Y_0\left(\sqrt{\mathcal{F}}\rho\right)\right)\quad \xi\le\rho\le1\,, \tag{14b}$$

$$G'''(\rho,\xi)=-A_2\frac{\mathcal{F}^{3/2}}{4}\left[J_3(\sqrt{\mathcal{F}}\rho)-3J_1(\sqrt{\mathcal{F}}\rho)\right]=A_2\frac{\mathcal{F}^{3/2}}{4}\left[\frac{1}{\sqrt{\mathcal{F}}\rho}J_2(\sqrt{\mathcal{F}}\rho)-J_1\left(\sqrt{\mathcal{F}}\rho\right)\right]\qquad 0\le\rho\le\xi \tag{15a}$$

and

$$\begin{aligned}G'''(\rho,\xi)&=\frac{2}{\rho^3}B_2-B_3\frac{\mathcal{F}^{3/2}}{4}\left[J_3(\sqrt{\mathcal{F}}\rho)-3J_1(\sqrt{\mathcal{F}}\rho)\right]-B_4\frac{\mathcal{F}^{3/2}}{4}\left[Y_3(\sqrt{\mathcal{F}}\rho)-3Y_1(\sqrt{\mathcal{F}}\rho)\right]=\\&=\frac{2}{\rho^3}B_2+B_3\frac{\mathcal{F}^{3/2}}{4}\left[\frac{1}{\sqrt{\mathcal{F}}\rho}J_2(\sqrt{\mathcal{F}}\rho)-J_1\left(\sqrt{\mathcal{F}}\rho\right)\right]+B_4\frac{\mathcal{F}^{3/2}}{4}\left[\frac{1}{\sqrt{\mathcal{F}}\rho}Y_2(\sqrt{\mathcal{F}}\rho)-Y_1\left(\sqrt{\mathcal{F}}\rho\right)\right]\quad \xi\le\rho\le1\end{aligned} \tag{15b}$$

obtained by using relations (44).

Substituting the Green function and its derivatives into the continuity conditions $(9)_1$ and combining then the continuity conditions $(9)_2$ and $(9)_3$ we have

$$A_1+A_3J_0(\sqrt{\mathcal{F}}\xi)=B_1+B_2\ln\xi+B_3J_0(\sqrt{\mathcal{F}}\xi)+B_4Y_0(\sqrt{\mathcal{F}}\xi)\,, \tag{16a}$$

$$-A_3\sqrt{\mathcal{F}}J_1\left(\sqrt{\mathcal{F}}\xi\right)=B_2\frac{1}{\xi}-B_3\sqrt{\mathcal{F}}J_1\left(\sqrt{\mathcal{F}}\xi\right)-B_4\sqrt{\mathcal{F}}Y_1\left(\sqrt{\mathcal{F}}\xi\right)\,, \tag{16b}$$

$$A_2J_0(\sqrt{\mathcal{F}}\xi)=B_3J_0(\sqrt{\mathcal{F}}\xi)+B_4Y_0(\sqrt{\mathcal{F}}\xi)\,. \tag{16c}$$

It follows from (11) that

$$\begin{aligned}\frac{R_K^3}{I_1E_1}Q_R(\xi+0)&=B_2\frac{d}{d\rho}\left\{-\frac{1}{\rho^2}+\frac{1}{\rho^2}+\mathcal{F}\ln\rho\right\}\bigg|_{\rho=\xi}+\\&+B_3\frac{d}{d\rho}\left\{\left[\frac{\mathcal{F}}{2}\left[J_2(\sqrt{\mathcal{F}}\rho)-J_0(\sqrt{\mathcal{F}}\rho)\right]-\mathcal{F}\frac{J_1(\sqrt{\mathcal{F}}\rho)}{\sqrt{\mathcal{F}}\rho}\right]+\mathcal{F}J_0(\sqrt{\mathcal{F}}\rho)\right\}\bigg|_{\rho=\xi}\\&+B_4\frac{d}{d\rho}\left\{\left[\frac{\mathcal{F}}{2}\left[Y_2(\sqrt{\mathcal{F}}\rho)-Y_0(\sqrt{\mathcal{F}}\rho)\right]-\mathcal{F}\frac{Y_1(\sqrt{\mathcal{F}}\rho)}{\sqrt{\mathcal{F}}\rho}\right]+\mathcal{F}Y_0(\sqrt{\mathcal{F}}\rho)\right\}\bigg|_{\rho=\xi}=\end{aligned}$$

$$B_2\frac{d}{d\rho}\left\{-\frac{1}{\rho^2}+\frac{1}{\rho^2}+\mathcal{F}\ln\rho\right\}\bigg|_{\rho=\xi}+\underbrace{B_3\frac{d}{d\rho}\left\{\left[\frac{\mathcal{F}}{2}\left[J_2(\sqrt{\mathcal{F}}\rho)+J_0(\sqrt{\mathcal{F}}\rho)\right]-\mathcal{F}\frac{J_1(\sqrt{\mathcal{F}}\rho)}{\sqrt{\mathcal{F}}\rho}\right]\right\}}_{=zero}\Bigg|_{\rho=\xi}$$

$$+ B_4 \frac{d}{d\rho} \underbrace{\left\{ \left[\frac{\mathcal{F}}{2} \left[Y_2(\sqrt{\mathcal{F}}\rho) + Y_0(\sqrt{\mathcal{F}}\rho) \right] - \mathcal{F} \frac{Y_1(\sqrt{\mathcal{F}}\rho)}{\sqrt{\mathcal{F}}\rho} \right] \right\}}_{=zero} \Bigg|_{\rho=\xi} = B_2 \mathcal{F} \frac{1}{\xi} \,.$$

Consequently discontinuity condition (10) leads to the equation

$$B_2 = \frac{R_o^3}{I_1 E_1} \frac{1}{2\pi\xi R_o} \frac{1}{\mathcal{F}} \xi = \frac{R_o^2}{I_1 E_1 \frac{f R_o^2}{I_1 E_1}} \frac{1}{2\pi\xi} \xi = \frac{1}{2\pi f} \,. \tag{16d}$$

The last two equations for the integration constants are obtained from the boundary conditions (12)

$$B_1 + B_3 J_0(\sqrt{\mathcal{F}}) + B_4 Y_0(\sqrt{\mathcal{F}}) = 0 \,, \tag{16e}$$

$$B_2 - B_3 \sqrt{\mathcal{F}} J_1(\sqrt{\mathcal{F}}) - B_4 \sqrt{\mathcal{F}} Y_1(\sqrt{\mathcal{F}}) = 0 \,. \tag{16f}$$

Introducing the notations $J_{n\rho} = J_n(\sqrt{\mathcal{F}}\rho)$, $J_{n\xi} = J_n(\sqrt{\mathcal{F}}\xi)$, $J_{n1} = J_n(\sqrt{\mathcal{F}})$ and $Y_{n\rho} = Y_n(\sqrt{\mathcal{F}}\rho)$, $Y_{n\xi} = Y_n(\sqrt{\mathcal{F}}\xi)$, $Y_{n1} = Y_n(\sqrt{\mathcal{F}})$ we can rewrite equations (16) in the following form:

$$\begin{bmatrix} 1 & J_{0\xi} & -1 & -\ln\xi & -J_{0\xi} & -Y_{0\xi} \\ 0 & -\sqrt{\mathcal{F}} J_{1\xi} & 0 & -\frac{1}{\xi} & \sqrt{\mathcal{F}} J_{1\xi} & \sqrt{\mathcal{F}} Y_{1\xi} \\ 0 & J_{0\xi} & 0 & 0 & -J_{0\xi} & -Y_{0\xi} \\ 0 & 0 & 0 & 1 & 0 & 0 \\ 0 & 0 & 1 & 0 & J_{01} & Y_{01} \\ 0 & 0 & 0 & 1 & -\sqrt{\mathcal{F}} J_{11} & -\sqrt{\mathcal{F}} Y_{11} \end{bmatrix} \begin{bmatrix} A_1 \\ A_2 \\ B_1 \\ B_2 \\ B_3 \\ B_4 \end{bmatrix} = \begin{bmatrix} 0 \\ 0 \\ 0 \\ \frac{1}{2\pi f} \\ 0 \\ 0 \end{bmatrix} \tag{17}$$

After solving equation system (17) and substituting solutions (46) given in the Appendix into (8) we have

$$G(\rho,\xi) = \frac{J_{11}\ln\xi + \frac{1}{\sqrt{\mathcal{F}}}(J_{0\rho} + J_{0\xi} - J_{01}) + \frac{\pi}{2} J_{0\rho}(J_{0\xi} Y_{11} - Y_{0\xi} J_{11})}{2\pi f J_{11}} \qquad 0 \le \rho \le \xi \,, \tag{18a}$$

$$G(\rho,\xi) = \frac{J_{11}\ln\rho + \frac{1}{\sqrt{\mathcal{F}}}(J_{0\xi} + J_{0\rho} - J_{01}) + \frac{\pi}{2} J_{0\xi}(J_{0\rho} Y_{11} - Y_{0\rho} J_{11})}{2\pi f J_{11}} \qquad \xi \le \rho \le 1 \,, \tag{18b}$$

which is the Green function for the clamped plate. It can be checked with ease that the Green function is symmetric, i.e., $G(\rho,\xi) = G(\xi,\rho)$.

4.2.3. *Green function for the simply supported plate.* If the plate is simply supported the continuity and discontinuity conditions are associated with the following boundary conditions:

$$G(1,\xi) = 0 \,, \qquad G''(1,\xi) + \frac{\nu}{\rho} G'(1,\xi) = 0 \tag{19}$$

Observe that only one equation differs from those we used earlier – compare $(19)_2$ and $(12)_2$. For the sake of brevity we omit the paper and pencil calculations and present the final result only:

$$G(\rho,\xi) = \frac{J_{01} - J_{0\xi} - J_{0\rho} + \frac{\pi}{2} J_{0\xi} J_{0\rho}(\mathcal{F}_\nu Y_{01} - \sqrt{\mathcal{F}} Y_{11})}{2\pi f(\mathcal{F}_\nu J_{01} - \sqrt{\mathcal{F}} J_{11})} + \frac{\ln\xi - \frac{\pi}{2} J_{0\rho} Y_{0\xi}}{2\pi f} \qquad 0 \le \rho \le \xi \,, \tag{20a}$$

$$G(\rho,\xi) = \frac{J_{01} - J_{0\rho} - J_{0\xi} + \frac{\pi}{2} J_{0\rho} J_{0\xi}\left(\mathcal{F}_\nu Y_{01} - \sqrt{\mathcal{F}} Y_{11}\right)}{4\pi f(\mathcal{F}_\nu J_{01} - \sqrt{\mathcal{F}} J_{11})} + \frac{\ln\rho - \frac{\pi}{2} J_{0\xi} Y_{0\rho}}{2\pi f} \qquad \xi \le \rho \le 1 \,. \tag{20b}$$

This Green function is also symmetric, i.e., $G(\rho,\xi) = G(\xi,\rho)$.

4.3. **Green functions for a tensile f.**

4.3.1. *General solution* If the in-plane load is a tensile one

$$w(\rho) = c_1 + c_2 \ln\rho + c_3 I_0\left(\sqrt{\mathcal{F}}\rho\right) + c_4 K_0\left(\sqrt{\mathcal{F}}\rho\right) \tag{21}$$

is the general solution we need when determining the Green function – c_1, c_2, c_3 and c_4 are again undetermined constants of integration. With the knowledge of the general solution we assume that the Green function, which satisfies conditions (5), can be given in the following form

$$G(\rho,\xi) = A_1 + A_3 I_0(\sqrt{\mathcal{F}}\rho) \,, \qquad \rho < \xi \,, \tag{22a}$$

$$G(\rho,\xi) = B_1 + B_2 \ln\rho + B_3 I_0(\sqrt{\mathcal{F}}\rho) + B_4 K_0(\sqrt{\mathcal{F}}\rho) \,, \qquad \rho > \xi \tag{22b}$$

in which the integration constants are denoted in the same was as for the case of compressive f – however this may not cause any misunderstanding.

4.3.2. *Green function for the clamped plate.* When determining the Green function we shall need the following derivatives

$$G'(\rho,\xi) = A_3\sqrt{\mathcal{F}}I_1(\sqrt{\mathcal{F}}\rho) \qquad \rho<\xi\,, \tag{23a}$$

$$G'(\rho,\xi) = B_2\frac{1}{\rho} + B_3\sqrt{\mathcal{F}}I_1(\sqrt{\mathcal{F}}\rho) - B_4\sqrt{\mathcal{F}}K_1(\sqrt{\mathcal{F}}\rho) \qquad \rho>\xi\,, \tag{23b}$$

$$G''(\rho,\xi) = A_3\mathcal{F}\left(I_0\left(\sqrt{\mathcal{F}}\rho\right) - \frac{1}{\sqrt{\mathcal{F}}\rho}I_1\left(\sqrt{\mathcal{F}}\rho\right)\right) \qquad \rho<\xi\,, \tag{24a}$$

$$G''(\rho,\xi) = -B_2\frac{1}{\rho^2} + B_3\mathcal{F}\left(I_0\left(\sqrt{\mathcal{F}}\rho\right) - \frac{1}{\sqrt{\mathcal{F}}\rho}I_1\left(\sqrt{\mathcal{F}}\rho\right)\right) + B_4\mathcal{F}\left(K_0\left(\sqrt{\mathcal{F}}\rho\right) + \frac{1}{\sqrt{\mathcal{F}}\rho}K_1\left(\sqrt{\mathcal{F}}\rho\right)\right) \quad \rho>\xi\,, \tag{24b}$$

$$G'''(\rho,\xi) = A_3\frac{\mathcal{F}^{3/2}}{4}\left[3I_1(\sqrt{\mathcal{F}}\rho) + I_3(\sqrt{\mathcal{F}}\rho)\right] = -A_3\mathcal{F}^{3/2}\left(\frac{1}{\sqrt{\mathcal{F}}x}I_2\left(\sqrt{\mathcal{F}}x\right) - I_1\left(\sqrt{\mathcal{F}}x\right)\right) \quad \rho<\xi\,. \tag{25a}$$

$$\begin{aligned} G'''(\rho,\xi) &= -2B_2\frac{1}{\rho^3} + B_3\frac{\mathcal{F}^{3/2}}{4}\left[3I_1(\sqrt{\mathcal{F}}\rho) + I_3(\sqrt{\mathcal{F}}\rho)\right] - B_4\frac{\mathcal{F}^{3/2}}{4}\left[3K_1(\sqrt{\mathcal{F}}\rho) + K_3(\sqrt{\mathcal{F}}\rho)\right] = \\ &= -2B_2\frac{1}{\rho^3} + B_3\mathcal{F}^{3/2}\left(\frac{1}{\sqrt{\mathcal{F}}x}I_2\left(\sqrt{\mathcal{F}}x\right) - I_1\left(\sqrt{\mathcal{F}}x\right)\right) - B_4\mathcal{F}^{3/2}\left[\frac{1}{\sqrt{\mathcal{F}}x}K_2\left(\sqrt{\mathcal{F}}x\right) + K_1\left(\sqrt{\mathcal{F}}x\right)\right], \quad \rho>\xi\,. \end{aligned} \tag{25b}$$

obtained by utilizing (45). Substituting the Green function and its derivatives into the continuity conditions $(9)_1$ and combining $(9)_2$ and $(9)_3$ we have

$$A_1 + A_3I_0(\sqrt{\mathcal{F}}\xi) = B_1 + B_2\ln\xi + B_3I_0(\sqrt{\mathcal{F}}\xi) + B_4K_0(\sqrt{\mathcal{F}}\xi)\,, \tag{26a}$$

$$A_3\sqrt{\mathcal{F}}I_1(\sqrt{\mathcal{F}}\xi) = B_2\frac{1}{\xi} + B_3\sqrt{\mathcal{F}}I_1(\sqrt{\mathcal{F}}\xi) - B_4\sqrt{\mathcal{F}}K_1(\sqrt{\mathcal{F}}\xi)\,, \tag{26b}$$

$$A_3I_0(\sqrt{\mathcal{F}}\xi) = B_3I_0(\sqrt{\mathcal{F}}\xi) + B_4K_0(\sqrt{\mathcal{F}}\xi)\,. \tag{26c}$$

By repeating the line of thought resulting in equation (16d) discontinuity condition (10) leads to the equation

$$B_2 = -1/2\pi f\,. \tag{26d}$$

The last two equations for the integration constants are provided again by the boundary conditions (12)

$$\begin{aligned} B_1 + B_3I_0(\sqrt{\mathcal{F}}) + B_4K_0(\sqrt{\mathcal{F}}) &= 0\,, \qquad (26e)\\ B_2 + B_3\sqrt{\mathcal{F}}I_1(\sqrt{\mathcal{F}}) - B_4\sqrt{\mathcal{F}}K_1(\sqrt{\mathcal{F}}) &= 0\,. \qquad (26f) \end{aligned}$$

Let $I_{n\rho} = I_n(\sqrt{\mathcal{F}}\rho)$, $I_{n\xi} = I_n(\sqrt{\mathcal{F}}\xi)$, $I_{n1} = I_n(\sqrt{\mathcal{F}})$, $K_{n\rho} = K_n(\sqrt{\mathcal{F}}\rho)$, $K_{n\xi} = K_n(\sqrt{\mathcal{F}}\xi)$ and $K_{n1} = K_n(\sqrt{\mathcal{F}})$. Making use of the notations introduced equation system (26) can be cast into the following form

$$\begin{bmatrix} 1 & I_{0\xi} & -1 & -\ln\xi & -I_{0\xi} & -K_{0\xi} \\ 0 & \sqrt{\mathcal{F}}J_{1\xi} & 0 & -\frac{1}{\xi} & -\sqrt{\mathcal{F}}I_{1\xi} & \sqrt{\mathcal{F}}K_{1\xi} \\ 0 & I_{0\xi} & 0 & 0 & -I_{0\xi} & -K_{0\xi} \\ 0 & 0 & 0 & 1 & 0 & 0 \\ 0 & 0 & 1 & 0 & I_{01} & K_{01} \\ 0 & 0 & 0 & 1 & \sqrt{\mathcal{F}}I_{11} & -\sqrt{\mathcal{F}}K_{11} \end{bmatrix} \begin{bmatrix} A_1 \\ A_3 \\ B_1 \\ B_2 \\ B_3 \\ B_4 \end{bmatrix} = \begin{bmatrix} 0 \\ 0 \\ 0 \\ -\frac{1}{2\pi f} \\ 0 \\ 0 \end{bmatrix} \tag{27}$$

With the knowledge of the integration constants $A_1,\ A_3,\ B_1,\dots,B_4$ equation (22) yields the Green function as

$$G(\rho,\xi) = \frac{I_{11}\ln\xi + \frac{1}{\sqrt{\mathcal{F}}}\left(I_{01} - I_{0\xi} - I_{0\rho}\right) + I_{0\rho}\left(I_{0\xi}K_{11} + K_{0\xi}I_{11}\right)}{2I_{11}\pi f} \qquad 0<\rho\le\xi\,, \tag{28a}$$

$$G(\xi,\rho) = \frac{I_{11}\ln\rho + \frac{1}{\sqrt{\mathcal{F}}}\left(I_{01} - I_{0\rho} - I_{0\xi}\right) + I_{0\xi}\left(I_{0\rho}K_{11} + K_{0\rho}I_{11}\right)}{2I_{11}\pi f} \qquad \xi\le\rho\le 1\,. \tag{28b}$$

4.3.3. *Green function for the simply supported plate.* We present the Green function without providing details concerning the hand made calculations

$$G(\rho,\xi)=\frac{I_{0\xi}+I_{0\rho}-I_{01}-I_{0\rho}I_{0\xi}\left(\sqrt{\mathcal{F}}K_{11}+\mathcal{F}_\nu K_{01}\right)}{2\pi f(\mathcal{F}_\nu I_{01}-\sqrt{\mathcal{F}}I_{11})}+\frac{\ln\xi+I_{0\rho}K_{0\xi}}{2\pi f}\qquad 0<\rho\le\xi\,, \tag{29a}$$

$$G(\rho,\xi)=\frac{I_{0\rho}+I_{0\xi}-I_{01}-I_{0\xi}I_{0\rho}\left(\sqrt{\mathcal{F}}K_{11}-\mathcal{F}_\nu K_{01}\right)}{2\pi f(\mathcal{F}_\nu I_{01}-\sqrt{\mathcal{F}}I_{11})}+\frac{\ln\rho+I_{0\xi}K_{0\rho}}{2\pi f}\qquad \xi\le\rho\le1\,. \tag{29b}$$

4.4. **Solutions of statical boundary value problems** Given the Green functions the deflection due to an axisymmetric load $p_z(\rho)$ can always be calculated as

$$w(\rho)=2\pi R_o^2\int_0^1 G(\rho,\xi)p_z(\xi)\,\xi\,\mathrm{d}\xi\,. \tag{30}$$

5. Integral equation for the natural frequencies

5.1. **Integral equation of the problem** Under the assumption of harmonic vibrations the amplitude $W(\rho)$ of the vibrations $w(\rho,t)$ should satisfy the differential equation

$$\left(\frac{\mathrm{d}^2}{\mathrm{d}\rho^2}+\frac{1}{\rho}\frac{\mathrm{d}}{\mathrm{d}\rho}\right)\left[\left(\frac{\mathrm{d}^2}{\mathrm{d}\rho^2}+\frac{1}{\rho}\frac{\mathrm{d}}{\mathrm{d}\rho}\right)W+\mathcal{F}W\right]=\frac{R_o^4}{I_1E_1}\frac{\gamma}{g}\alpha^2W \tag{31}$$

where γ is the plate weight for the unit area of the middle surface and g is the gravitational acceleration. The eigenvalue problems defined by equation (31) and boundary conditions (3), (4) and (??) are all self adjoint. Since $\alpha^2\gamma W/g$ corresponds to p_z in equation (2) it follows from equation (30) that the amplitude $W(\rho)$ should satisfy the integral equation

$$W(\rho)=\lambda\int_0^1\hat{G}(\rho,\xi)\,W(\xi)\,\xi\,\mathrm{d}\xi\quad\text{where}\quad\lambda=\frac{\gamma}{g}\alpha^2\,,\qquad\hat{G}(\rho,\xi)=2\pi R_o^2G(\rho,\xi)\,, \tag{32}$$

which can be manipulated into the following form:

$$W(\rho)=\frac{\gamma}{g}\alpha^2\int_0^1\frac{R_o^2}{f}\tilde{G}(\rho,\xi)\,W(\xi)\,\xi\,\mathrm{d}\xi=\underbrace{\frac{\gamma}{g}\frac{\alpha^2R_o^4}{I_1E_1}}_{\mathcal{A}}\int_0^1\frac{1}{\underbrace{\frac{fR_o^2}{I_1E_1}}_{\mathcal{F}}}\tilde{G}(\rho,\xi)\,W(\xi)\,\xi\,\mathrm{d}\xi\,. \tag{33}$$

Here $\mathcal{A}$ and $\mathcal{F}$ are dimensionless quantities: $\mathcal{A}$ is proportional to the square of a natural frequency, $\mathcal{F}$ is proportional to the load. If we introduce a new unknown function

$$y(\rho)=\sqrt{\rho}W(\rho) \tag{34}$$

then we have

$$\underbrace{\sqrt{\rho}W(\rho)}_{y(\rho)}=\mathcal{A}\int_0^1\underbrace{\sqrt{\rho}\frac{\tilde{G}(\rho,\xi)}{\mathcal{F}}\sqrt{(\xi)}}_{\mathcal{G}(\rho,\xi)}\underbrace{\sqrt{\xi}W(\xi)}_{y(\xi)}\mathrm{d}\xi\qquad\text{that is}\qquad y(\rho)=\mathcal{A}\int_0^1\mathcal{G}(\rho,\xi)y(\xi)\mathrm{d}\xi\,. \tag{35}$$

The above equation is a homogenous Fredholm integral equation with a symmetric kernel. At the same time this equation is an eigenvalue problem with $\mathcal{A}$ as an eigenvalue which is a function of the dimensionless in-plane load $\mathcal{F}$.

5.2. **Computational algorithm** Numerical solution to the eigenvalue problem (35) can be sought by quadrature methods [17]. Consider the integral formula

$$J(\phi)=\int_0^1\phi(\xi)\mathrm{d}\xi\equiv\sum_{j=0}^n w_j\phi(\xi_j)\,,\qquad\xi_j\in[0,1]\,,\quad j=0,1,\dots,n \tag{36}$$

where $\phi(\xi)$ is a scalar and the weights w_j are known. Making use of the above equation we obtain from (??) that

$$\sum_{j=0}^n w_j\mathcal{G}(\rho,\xi_j)\,y(\xi_j)=\chi y(\rho)\,,\qquad\chi=1/\mathcal{A}\,,\quad x\in[0,1]\,,\quad j=0,1,\dots,n \tag{37}$$

the solution of which yields an approximate eigenvalue $\mathcal{A}=1/\chi$ and an approximate eigenfunction $y(\rho)$. After setting ρ to ρ_i $(i=0,1,2,\dots,n)$ we have

$$\sum_{j=0}^n w_j\mathcal{G}(\rho_i,\xi_j)\,y(\xi_j)=\chi y(\rho_i)\quad\chi=1/\mathcal{A}\qquad x\in[0,1] \tag{38}$$

or

$$\mathcal{G}\mathcal{D}\tilde{y} = \chi\,\tilde{y} \tag{39a}$$

where $\mathcal{G} = [G(\rho_i, \xi_j)]$ is symmetric,

$$\mathcal{D} = \text{diag}(w_0|\dots|w_k|\dots|w_n|) \quad \text{and} \quad \tilde{y}^T = [y(x_0)|y(x_1)|\dots|y(x_n)]. \tag{39b}$$

After solving the algebraic eigenvalue problem (39) we have the approximate eigenvalues χ_r and eigenvectors $\tilde{y}_r$ $(r = 0, 1, \dots, n)$. The corresponding eigenfunction is obtained by substituting back into equation (37):

$$y(\rho) = \frac{1}{\chi_r}\sum_{j=0}^{n} w_j \mathcal{G}(\rho, \xi_j)\, y(\xi_j)\ . \tag{40}$$

Divide the interval $[0, 1]$ into equidistant subintervals of length h and apply the integration formula to each subinterval. By repeating the line of thought leading to equation (39) one can show that the algebraic eigenvalue problem obtained is of the same structure as that of equation (39).

6. Computational results

6.1. Clamped plate Let us introduce the following dimensionless quantities

$$\mathcal{A}_{oi} = \frac{\gamma}{g}\frac{R_o^2}{I_1 E_1}\alpha_{oi}\ , \qquad \mathcal{A}_i = \frac{\gamma}{g}\frac{R_o^2}{I_1 E_1}\alpha_i \quad \text{and} \quad \mathcal{F}_{o1} = \frac{\gamma}{g}\frac{R_o^2}{I_1 E_1} f_1\ . \tag{41}$$

where α_{oi} and α_i are the i-th $(i = 1, 2, \dots)$ natural frequencies of the unloaded and loaded plates, while f_1 is the first critical load. The computational results for $i = 1$ are presented in Table 1.

TABLE 1

$\mathcal{F}/\mathcal{F}_{o1}$	0.068	0.136	0.204	0.272	0.341	0.409	0.477
$\mathcal{A}/\mathcal{A}_{o1}$ – compression	0.929	0.863	0.796	0.730	0.663	0.595	0.528
$\mathcal{A}/\mathcal{A}_{o1}$ – tension	1.061	1.127	1.193	1.258	1.324	1.389	1.454
$\mathcal{F}/\mathcal{F}_{o1}$	0.545	0.613	0.681	0.749	0.817	0.886	0.954
$\mathcal{A}/\mathcal{A}_{o1}$ – compression	0.460	0.392	0.324	0.255	0.186	0.117	0.048
$\mathcal{A}/\mathcal{A}_{o1}$ – tension	1.5189	1.584	1.648	1.713	1.777	1.841	1.906

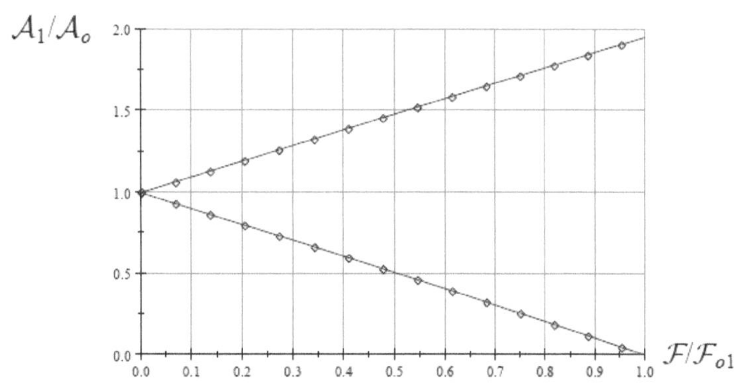

Figure 2.

With the knowledge of computational results we have fitted the following curves onto the discrete points – these are denoted by diamonds in Figure 2 ($\mathcal{A}_{o1} = 104.85$, $\mathcal{F}_{o1} = 14.68$)

$$\frac{\mathcal{A}_1}{\mathcal{A}_{o1}} = 0.994\,99 - 0.965\,94\frac{\mathcal{F}}{\mathcal{F}_{o1}} - 2.851\,4\times 10^{-2}\left(\frac{\mathcal{F}}{\mathcal{F}_{o1}}\right)^2 \tag{42a}$$

$$\frac{\mathcal{A}_1}{\mathcal{A}_{o1}} = 0.995\,61 + 0.968\,14\frac{\mathcal{F}}{\mathcal{F}_{o1}} - 1.474\,2\times 10^{-2}\left(\frac{\mathcal{F}}{\mathcal{F}_{o1}}\right)^2 \tag{42b}$$

Observe that the approximate solution (42) is practically linear in the interval $\mathcal{F}/\mathcal{F}_{o1} \in [0, 1]$.

6.2. **Simply supported plate** For a simply supported plate the computational results presented in Table 2 under the assumption that $i = 1$. These are denoted by diamonds in Figure 3.

TABLE 2

$\mathcal{F}/\mathcal{F}_{o1}$	0.070	0.140	0.210	0.280	0.350	0.420	0.490
$\mathcal{A}/\mathcal{A}_{o1}$ – compression	0.930	0.860	0.790	0.720	0.650	0.580	0.510
$\mathcal{A}/\mathcal{A}_{o1}$ – tension	1.071	1.141	1.211	1.281	1.351	1.421	1.491
$\mathcal{F}/\mathcal{F}_{o1}$	0.560	0.630	0.700	0.770	0.817	0.840	0.910
$\mathcal{A}/\mathcal{A}_{o1}$ – compression	0.440	0.370	0.300	0.230	0.159	0.089	0.019
$\mathcal{A}/\mathcal{A}_{o1}$ – tension	1.561	1.631	1.701	1.771	1.841	1.911	1.981

The approximate solutions obtained again by fitting a curve onto to the computational results are practically linear functions. These are are also shown in Figure 3. ($\mathcal{A}_{o1} = 24.838, \mathcal{F}_{o1} = 4.285$)

$$\frac{\mathcal{A}_1}{\mathcal{A}_{o1}} =\simeq 1.000 - 1.000\frac{\mathcal{F}}{\mathcal{F}_{o1}}, \qquad \frac{\mathcal{A}_1}{\mathcal{A}_{o1}} \simeq 1.000 + 1.000\frac{\mathcal{F}}{\mathcal{F}_{o1}} \tag{43}$$

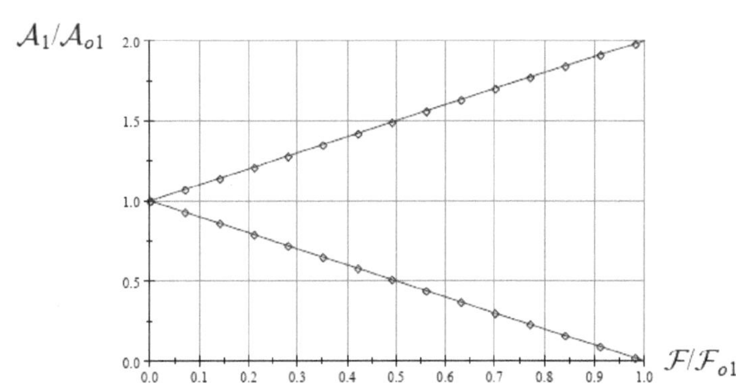

Figure 3.

The square of the first natural frequency of the unloaded plate and the first critical load can be calculated by equation (41).

7. Concluding remarks

We have dealt with the vibrations of circular plates subjected to a constant radial load in its plane on the outer boundary. The load can be either a compressive one or a tensile one. When solving the problem we have assumed that the deformations due to the load are also axisymmetric.

(a) We have determined the Green functions for two support arrangements – clamped plate, simply supported plate – and for tensile and compressive in-plane loads as well. With the knowledge of the Green functions, the self adjoint eigenvalue problems giving the natural frequencies of the vibrations for the circular plates and loads considered have been replaced by four eigenvalue problems each of which is governed by a Fredholm integral equation.

(b) If the plate is subjected to a load perpendicular to the middle plane the deflections can be determined by integration – see equation (30).

(c) The eigenvalue problems governed by the Fredholm integral equations are reduced to algebraic eigenvalue problems and the eigenvalues as functions of the load are computed using the boundary element method and the QZ algorithm. According to the results the square of the first natural frequency can be approximated with a good accuracy (or accurately – simply supported plate) by linear functions of the load for the parameters (material constants and geometrical data) considered.

Finally we remark that investigations for annular plates are in progress. Some preliminary results for compressive loads were presented earlier [18, 2005], [19, 2005].

Appendix A. Bessel functions – fundamental relations

Let us denote $J_n(x)$ and $Y_n(x)$ uniformly by $H_n(x)$. The following relations hold

$$\frac{dH_0(x)}{dx} = -H_1(x)\,, \qquad \frac{dH_1(x)}{dx} = \frac{1}{x}\left(H_0(x)\,x - H_1(x)\right) = H_0(x) - \frac{1}{x}H_1(x) = \frac{1}{2}\left(H_0(x) - H_2(x)\right) \tag{44a}$$

$$\frac{dH_n(x)}{dx} = \frac{1}{2}\left(H_{n-1}(x) - H_{n+1}(x)\right)\,, \qquad \frac{2n}{x}H_n(x) = H_{n-1}(x) + H_{n+1}(x) \tag{44b}$$

and

$$Y_n(x)J_{n+1}(x) - Y_{n+1}(x)J_n(x) = \frac{2}{\pi x} \ . \tag{44c}$$

See [20, 1977a] for more details.

As regards the Bessel functions $I_n(x)$ and $K_n(x)$ use has been made of the following relations

$$\frac{dI_0(x)}{dx} = I_1(x)\,, \quad \frac{dK_0(x)}{dx} = -K_1(x)\,, \quad \frac{2n}{x}I_n(x) = I_{n-1}(x) - I_{n+1}(x)\,, \tag{45a}$$

$$\frac{dI_1(x)}{dx} = \frac{1}{x}\left(xI_0(x) - I_1(x)\right) = I_0(x) - \frac{1}{x}I_1(x) = \frac{1}{2}\left(I_0(x) + I_2(x)\right)\,, \tag{45b}$$

$$\frac{dI_n(x)}{dx} = \frac{1}{2}\left(I_{n-1}(x) + I_{n+1}(x)\right)\,, \quad -\frac{2n}{x}I_n(x) = K_{n-1}(x) - K_{n+1}(x) \tag{45c}$$

$$\frac{dK_1(x)}{dx} = -\frac{1}{x}\left(xK_0(x) + K_1(x)\right) = -K_0(x) - \frac{1}{x}K_1(x) = -\frac{1}{2}\left(K_0(x) + K_2(x)\right)\,, \tag{45d}$$

$$\frac{dK_n(x)}{dx} = -\frac{1}{2}\left(K_{n-1}(x) + K_{n+1}(x)\right)\,, \tag{45e}$$

and

$$I_n(x)K_{n+1}(x) + I_{n+1}(x)K_n(x) = \frac{1}{x} \ . \tag{45f}$$

See [20, 1977b] for more details.

APPENDIX B. CLAMPED PLATE – CALCULATION OF THE INTEGRATION CONSTANTS

After solving equation system (16) we have

$$A_1 = \frac{1}{2\pi f}\left(\ln\xi - \frac{J_{01} - J_{0\xi}}{\sqrt{\mathcal{F}}J_{11}}\right)\,, \qquad A_2 = \frac{1}{4f}\left(\frac{2}{\pi\sqrt{\mathcal{F}}J_{11}} - Y_{0\xi} + \frac{Y_{11}J_{0\xi}}{J_{11}}\right)\,, \tag{46a}$$

$$B_1 = \frac{1}{2\pi f}\frac{J_{0\xi} - J_{01}}{\sqrt{\mathcal{F}}J_{11}}\,,\,, \qquad B_2 = \frac{1}{2\pi f}\,, \qquad B_3 = \frac{1}{2\pi f}\left(\frac{1}{\sqrt{\mathcal{F}}J_{11}} + \frac{\pi Y_{11}J_{0\xi}}{2J_{11}}\right)\,, \qquad B_4 = -\frac{1}{2f\pi}\frac{\pi J_{0\xi}}{2} \ . \tag{46b}$$

Substitution of the solutions above into (8) yields the Green function for the clamped plate.

REFERENCES

[1] Ray Lawther. On the straightness of eigenvalue iterations. *Computational Mechanics*, 37:362–368, 2005.

[2] S. Syngellakis and A. Elzein. Plate buckling loads by the the boundary element method. *International Journal for Numerical Methods in Engineering*, 37:1763–1778, 1994.

[3] M. S. Nerantzaki and J. T. Katsikadelis. Buckling of plates with variable thickness – an analog solution. *Engineering Analysis with Boundary elements*, 18:149–154, 1996.

[4] J. Lin, R. C. Duffield, , and H. Shih. Buckling analysis of elastic plates by boundary element method. *Engineering Analysis with Boundary Elements*, 23:131–137, 1999.

[5] P. H. Wen, M. H. Aliabadi, and A. Young. Application of dual reciprocity method to plates and shells. *Engineering Analysis with Boundary elements*, 24:583–590, 2000.

[6] J. Purbolaksono and M. H. Aliabadi. Buckling analysis of shear deformable plates by boundary element method. *International Journal for Numerical Methods in Engineering*, 62:537–563, 2005.

[7] C.D. Mote Jr. Free vibrations of initially stressed circular disks. *ASME J. Engng. for Industry*, 87:258–364, 1965.

[8] G. C. Pardoen. Vibration and buckling analysis of axisymmetric polar orthotropic circular plates. *Computers & Structures*, 4:951–960, 1974.

[9] D. Csotova. Vibrations of circular plates subjected to an in plane load. Master thesis (in hungarian), Department of Mecahnics, University of Miskolc, Hungary, 1980.

[10] Lien-When Chen & Ji-Liang Dong. Large amplitude vibration of an initially stressed hick circular plate. *AIAA*, 21:1317–1324, 1983.

[11] Lien-When Chen & Ji-Liang Dong. Vibrations of an initially stressed transversely isotropic circular thick plate. *International Journal of Mechanical Sciences*, 26(4):253–263, 1984.

[12] A. Tylikowskia and K. Frischmuthb. Stability and stabilization of circular plate parametric vibrations. *International Journal of Solids and Structures*, 40:5187–5196, 2003.

[13] D. Younesian and M. H. Aleghafourian. A direct formulation and numerical solution of the general transient elastodynamic problem. *International Journal of Civil Engineering and Building Materials*, 2(4), 2012.

[14] Szűcs Nóra. Vibrations of circular plates subjected to an in-plane load. *GÉP*, LVIII.(5-6):41–47, 2007 (in Hungarian).

[15] Zdeněk P. Bažant and Luigi Cedolin. *Stability of structures, Elastic, Inelastic, Fracture, and Damage Theories*. Number ISBN 0-486-42568-1. Dover Publications, Inc. Mineola, New York, 2nd edition, 2003, p. 427.

[16] I. Kozák. *Strength of materials V. – Thin walled structures and the theory of plates and shells*. Tankönyvkiadó (Publisher of Textbooks), Budapest, Hungary, 1967, 287-291.o. (in Hungarian).

[17] C. T. H. Baker. *The Numerical Treatment of Integral Equations – Monographs on Numerical Analysis*. Clarendon Press, Oxford, 1977.

[18] G. Szeidl, D. Georgieva, and N. Szűcs. Vibration of a circular plate with a hole subjected to a radial uniform load in its plane. In *Section G: Applied Mechanics, Modern Numerical Methods*, pages 149–154, microCAD 2005, University of Miskolc, March 10-11, 2005.

[19] G. Szeidl, N. Szűcs, and B. Tóth. Vibration of circular plates subjected to uniform loads in their plane. In V. Kompis, editor, *Numerical Methods on Continuum Mechanics and 4th Workshop on Trefftz Methods*, Slovakia, August 23-26, 2005., Extended six page abstract on the conference CD.

[20] E. Janke, F. Emde, and F. Lösch. *Tafelen Höheren Funktionen*. Nauka, Moscow (in Russian), 1977, (a) p. 241-242 and (b) 245-246.

IN-PLANE VIBRATIONS OF PINNED-FIXED HETEROGENEOUS CURVED BEAMS UNDER A CONCENTRATED FORCE

György Szeidl[1], László Kiss[1]

[1]Institute of Applied Mechanics, Universityof Miskolc,Miskolc, HUNGARY
gyorgy.szeidl@uni-miskolc.hu, mechkiss@uni-miskolc.hu

Abstract: *The paper deals with the vibrations of loaded heterogeneous curved beams when a central load (a constant force) is exerted at the crown point of the beam. The effect of the loading is accounted by the strain it causes. It is assumed that the radius of curvature is constant and the Young's modulus and the Poisson's number are functions of the cross-sectional coordinates. The paper presents the determination of the Green function matrices for loads directed upwards and downwards. An appropriate numerical model is also provided, which makes possible to determine how the natural frequencies are related to the load. It is also shown that when the strain is zero, the corresponding formulae yield results valid for the free vibrations of curved beams.*

Keywords: *curved beams, heterogeneous material, natural frequency as a function of the load, Green function matrices*

1. INTRODUCTION

Curved beams are widely used in numerous engineering applications – let us consider, for instance, arch bridges, roof structures, or stiffeners in aerospace applications. Research into the mechanical behavior of such structural elements started in the 19th century – see book [1] by Love for further details. The free vibrations of curved beams have been under extensive investigation: survey papers were published by Markus and Nanasi [2], Laura and Maurizi [3] as well as Chidampram and Lessia [4]. The PhD thesis by Szeidl [5] clarifies how the extensibility of the centerline affects the free vibrations and stability of circular beams subjected to a constant radial load (dead load) within the frames of the linear theory. The natural frequencies were computed by utilizing different numerical models. One of these relies on the Green function matrix of the corresponding boundary value problem. Unfortunately, the results of this work have not been published in English language. Paper [6] by Huang et al. takes shear deformations into account provided that the beam vibrates under the action of a constant vertical distributed load.

Lawther [7] investigates how a pre-stressed state of a body affects the natural frequencies. He studies finite dimensional multiparameter eigenvalue problems and finds that for multiparameter problems, the eigenvalue part of the solution is described by interaction curves in an eigenvalue space, and every such eigenvalue solution has a corresponding eigenvector. If all points on a curve have the same eigenvector, then the curve is necessarily a straight line, but the converse problem is far more complex. In the light of Lawther's results, there arises the question: how the frequencies change when a curved beam is subjected to a vertical force at the crown point. We assume that the curved beam is made of heterogeneous, isotropic and linearly elastic material. The cross-section is uniform in terms of both the geometry and the material composition. The material parameters are functions of the cross-sectional coordinates. Our main objectives: (1) derivation of those boundary value problems, which make possible to clarify how the load affects the natural frequencies; (2) determination of the Green function matrices, which can be used to reduce the eigenvalue problems set up for the natural frequencies to eigenvalue problems governed by systems of Fredholm integral equations; (3) reduction of the eigenvalue problems to algebraic ones and (4) the numerical solution of these.

2. THE PROBLEM FORMULATION

Figure 1 (a) shows a portion of the beam with the applied curvilinear coordinate system ($\xi = s, \eta, \zeta$) and (b) presents a pinned-fixed beam subjected to a load, directed downwards. By assumption, the uniform cross-section is symmetric with respect to the axis ζ. The Young's modulus and the Poisson number depend on the cross sectional coordinates in such a way that $E(\eta, \zeta) = E(-\eta, \zeta)$ and $\nu(\eta, \zeta) = \nu(-\eta, \zeta)$. Observe that the coordinate line $\xi = s$ coincides with the so-called

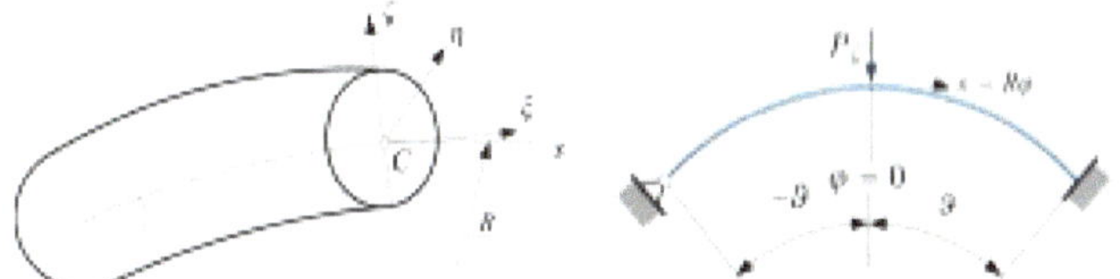

Figure 1. (a) Coordinate system, (b) Pinned-fixed beam

(E-weighted) centerline. This centerline intersects the cross-section at the point C. The location of the latter is obtained from the condition that the E-weighted first moment of the cross section with respect to the axis η is zero there:

$$Q_{e\eta} = \int_A E(\eta, \zeta)\,\zeta\,\mathrm{d}A = 0\,. \tag{1}$$

Let us now separate the load-induced, and otherwise time-independent mechanical quantities from those, which belong to the vibrations of the loaded beam. The latter ones are the time-dependent increments and are uniformly denoted by a subscript $_b$. Let u_o, w_o and R be the tangential and radial displacements and the radius of the centerline, respectively. Since this radius is constant, the coordinate line s and the angle coordinate φ are related to each other by $s = R\varphi$. The axial strain $\varepsilon_{o\xi}$ and the rigid body rotation $\psi_{o\eta}$ on the centerline can be expressed in terms of the displacements as

$$\varepsilon_{o\xi} = \frac{\mathrm{d}u_o}{\mathrm{d}s} + \frac{w_o}{R}, \qquad \psi_{o\eta} = \frac{u_o}{R} - \frac{\mathrm{d}w_o}{\mathrm{d}s}\,. \tag{2}$$

The principle of virtual work for a beam under distributed loading yields that the equilibrium equations

$$\frac{\mathrm{d}N}{\mathrm{d}s} + \frac{1}{R}\left[\frac{\mathrm{d}M}{\mathrm{d}s} - \left(N + \frac{M}{R}\right)\psi_{o\eta}\right] + f_t = 0, \quad \frac{\mathrm{d}}{\mathrm{d}s}\left[\frac{\mathrm{d}M}{\mathrm{d}s} - \left(N + \frac{M}{R}\right)\psi_{o\eta}\right] - \frac{N}{R} + f_n = 0 \tag{3a}$$

should be fulfilled by the axial force N and the bending moment M. Here f_t and f_n denote the intensity of the distributed loads on the centerline in the tangential and normal directions.

Hooke's law expresses the relation between the inner forces and the deformations [8] in such a way that

$$N = \frac{I_{e\eta}}{R^2}\varepsilon_{o\xi} - \frac{M}{R}, \qquad M = -I_{e\eta}\left(\frac{\mathrm{d}^2 w_o}{\mathrm{d}s^2} + \frac{w_o}{R^2}\right), \quad N + \frac{M}{R} = \frac{I_{e\eta}}{R^2}\varepsilon_{o\xi}, \quad \text{where} \tag{4}$$

$$A_e = \int_A E(\eta,\zeta)\mathrm{d}A\,, \qquad I_{e\eta} = \int_A E(\eta,\zeta)\zeta^2\mathrm{d}A\,, \qquad m = \frac{A_e R^2}{I_{e\eta}} - 1 \tag{5}$$

A_e is referred to as the E-weighted area, $I_{e\eta}$ is the E-weighted moment of inertia with respect to the bending axis and m is a parameter. For the sake of brevity, we introduce dimensionless displacements and a notation for the derivatives:

$$U_o = \frac{u_o}{R}, \qquad W_o = \frac{w_o}{R}; \qquad (\ldots)^{(n)} = \frac{\mathrm{d}^n(\ldots)}{\mathrm{d}\varphi^n}, \quad n \in \mathbb{Z}\,. \tag{6}$$

Upon substitution of (4) and (2) into equilibrium equations (3), we obtain the following system of differential equations (DEs):

$$\begin{bmatrix} 0 & 0 \\ 0 & 1 \end{bmatrix}\begin{bmatrix} U_o \\ W_o \end{bmatrix}^{(4)} + \begin{bmatrix} -m & 0 \\ 0 & 2 - m\varepsilon_{o\xi} \end{bmatrix}\begin{bmatrix} U_o \\ W_o \end{bmatrix}^{(2)} + $$
$$+ \begin{bmatrix} 0 & -m \\ m & 0 \end{bmatrix}\begin{bmatrix} U_o \\ W_o \end{bmatrix}^{(1)} + \begin{bmatrix} 0 & 0 \\ 0 & 1 + m(1 - \varepsilon_{o\xi}) \end{bmatrix}\begin{bmatrix} U_o \\ W_o \end{bmatrix} = \frac{R^3}{I_{e\eta}}\begin{bmatrix} f_t \\ f_n \end{bmatrix}. \tag{7}$$

If we neglect the effects of the deformations to the equilibrium (so we set $\varepsilon_{o\xi}$ to zero), then we have

$$\begin{bmatrix} 0 & 0 \\ 0 & 1 \end{bmatrix}\begin{bmatrix} U_o \\ W_o \end{bmatrix}^{(4)} + \begin{bmatrix} -m & 0 \\ 0 & 2 \end{bmatrix}\begin{bmatrix} U_o \\ W_o \end{bmatrix}^{(2)} + \begin{bmatrix} 0 & -m \\ m & 0 \end{bmatrix}\begin{bmatrix} U_o \\ W_o \end{bmatrix}^{(1)} + \begin{bmatrix} 0 & 0 \\ 0 & 1+m \end{bmatrix}\begin{bmatrix} U_o \\ W_o \end{bmatrix} = \frac{R^3}{I_{e\eta}}\begin{bmatrix} f_t \\ f_n \end{bmatrix}. \tag{8}$$

With our notational conventions, quantities like the total tangential displacement is equal to the sum $u_o + u_{ob}$. It turns out that the increments in the axial strain and in the rotation have a very similar structure to equations (2):

$$\varepsilon_{mb} = \varepsilon_{o\xi\,b} + \psi_{o\eta}\psi_{o\eta\,b}\,, \qquad \psi_{o\eta\,b} = \frac{u_{ob}}{R} - \frac{\mathrm{d}w_{ob}}{\mathrm{d}s}\,, \qquad \varepsilon_{o\xi\,b} = \frac{\mathrm{d}u_{ob}}{\mathrm{d}s} + \frac{w_{ob}}{R}\,. \tag{9}$$

Based on the principle of virtual work, it can be shown that the equations of equilibrium with the increments are

$$\frac{\mathrm{d}}{\mathrm{d}s}\left(N_b + \frac{M_b}{R}\right) - \frac{1}{R}\left(N + \frac{M}{R}\right)\psi_{o\eta\,b} + f_{tb} = 0\,, \tag{10a}$$

$$\frac{\mathrm{d}^2M_b}{\mathrm{d}s^2}-\frac{N_b}{R}-\frac{\mathrm{d}}{\mathrm{d}s}\left[\left(N+\frac{M}{R}\right)\psi_{o\eta\,b}+\left(N_b+\frac{M_b}{R}\right)\psi_{o\eta}\right]+f_{nb}=0\,. \tag{10b}$$

Due to the dynamical nature of the problem, the increments f_{tb} and f_{nb} are forces of inertia:

$$f_{tb}=-\rho_a A\frac{\partial^2 u_{ob}}{\partial t^2},\qquad f_{nb}=-\rho_a A\frac{\partial^2 w_{ob}}{\partial t^2}\,. \tag{11}$$

Here A is the area of the cross-section, and ρ_a is the averaged density of the cross-section. Application of Hooke's law for the increments in the inner forces yields

$$N_b=\frac{I_{e\eta}}{R^2}m\varepsilon_{o\xi\,b}-\frac{M_b}{R},\qquad M_b=-I_{e\eta}\left(\frac{\mathrm{d}^2w_{ob}}{\mathrm{d}s^2}+\frac{w_{ob}}{R^2}\right),\qquad N_b+\frac{M_b}{R}=\frac{I_{e\eta}}{R^2}m\varepsilon_{o\xi\,b}\,. \tag{12a}$$

A comparison of equations (9), (10) and (12) result in the equations of motion:

$$\begin{bmatrix}0&0\\0&1\end{bmatrix}\begin{bmatrix}U_{ob}\\W_{ob}\end{bmatrix}^{(4)}+\begin{bmatrix}-m&0\\0&2-m\varepsilon_{o\xi}\end{bmatrix}\begin{bmatrix}U_{ob}\\W_{ob}\end{bmatrix}^{(2)}+$$
$$+\begin{bmatrix}0&-m\\m&0\end{bmatrix}\begin{bmatrix}U_{ob}\\W_{ob}\end{bmatrix}^{(1)}+\begin{bmatrix}0&0\\0&1+m\,(1-\varepsilon_{o\xi})\end{bmatrix}\begin{bmatrix}U_{ob}\\W_{ob}\end{bmatrix}=\frac{R^3}{I_{e\eta}}\begin{bmatrix}f_{tb}\\f_{nb}\end{bmatrix}. \tag{13}$$

We remark that during the formal derivations we have neglected the quadratic term $\varepsilon_{o\xi}\varepsilon_{o\xi\,b}$ in (10a) and we have used the inequalities $\varepsilon_{o\xi\,b}\gg(\varepsilon_{o\xi\,b}\psi_{o\eta})^{(1)}$ and $1\gg\varepsilon_{o\xi}$ in (10b), when utilizing Hooke's law. If the vibrations are assumed to be harmonic with the dimensionless displacement amplitudes $\hat{U}_{ob}$ and $\hat{W}_{ob}$, then we have the following system of DEs

$$\begin{bmatrix}0&0\\0&1\end{bmatrix}\begin{bmatrix}\hat{U}_{ob}\\\hat{W}_{ob}\end{bmatrix}^{(4)}+\begin{bmatrix}-m&0\\0&2-m\varepsilon_{o\xi}\end{bmatrix}\begin{bmatrix}\hat{U}_{ob}\\\hat{W}_{ob}\end{bmatrix}^{(2)}+$$
$$+\begin{bmatrix}0&-m\\m&0\end{bmatrix}\begin{bmatrix}\hat{U}_{ob}\\\hat{W}_{ob}\end{bmatrix}^{(1)}+\begin{bmatrix}0&0\\0&1+m\,(1-\varepsilon_{o\xi})\end{bmatrix}\begin{bmatrix}\hat{U}_{ob}\\\hat{W}_{ob}\end{bmatrix}=\lambda\begin{bmatrix}\hat{U}_{ob}\\\hat{W}_{ob}\end{bmatrix};\quad\lambda=\rho_a A\frac{R^3}{I_{e\eta}}\alpha^2\,, \tag{14}$$

where λ and α denote the eigenvalues and eigenfrequencies.

For an unloaded beam ($\varepsilon_{o\xi}=0$), we get back the equations, which govern the free vibrations – compare equation

$$\begin{bmatrix}0&0\\0&1\end{bmatrix}\begin{bmatrix}\hat{U}_{ob}\\\hat{W}_{ob}\end{bmatrix}^{(4)}+\begin{bmatrix}-m&0\\0&2\end{bmatrix}\begin{bmatrix}\hat{U}_{ob}\\\hat{W}_{ob}\end{bmatrix}^{(2)}+\begin{bmatrix}0&-m\\m&0\end{bmatrix}\begin{bmatrix}\hat{U}_{ob}\\\hat{W}_{ob}\end{bmatrix}^{(1)}+\begin{bmatrix}0&0\\0&m+1\end{bmatrix}\begin{bmatrix}\hat{U}_{ob}\\\hat{W}_{ob}\end{bmatrix}=\lambda\begin{bmatrix}\hat{U}_{ob}\\\hat{W}_{ob}\end{bmatrix}$$

to equation (11) in [9]. This system is associated with the boundary conditions valid for pinned-fixed beams and together they constitute an eigenvalue problem. The left side of equation (14) can be rewritten in the brief form

$$\mathbf{K}\left[\mathbf{y}\left(\varphi\right),\varepsilon_{o\xi}\right]=\overset{4}{\mathbf{P}}\mathbf{y}^{(4)}+\overset{2}{\mathbf{P}}\mathbf{y}^{(2)}+\overset{1}{\mathbf{P}}\mathbf{y}^{(1)}+\overset{0}{\mathbf{P}}\mathbf{y}^{(0)},\qquad\mathbf{y}^T=\left[\hat{U}_{ob}\,|\,\hat{W}_{ob}\right]\,. \tag{15}$$

Observe that the i-th eigenfrequency α_i in the eigenvalue problem depends on the heterogeneity parameters m and ρ_a; and also, on the magnitude and the direction of the concentrated force. The effects of the latter one are accounted through the axial strain: $\varepsilon_{o\xi}=\varepsilon_{o\xi}(\mathcal{P})$. Here $\mathcal{P}$ is a dimensionless load, defined by $\mathcal{P}=P_\zeta R^2\vartheta/(2I_{e\eta})$.

3. THE GREEN FUNCTION MATRIX

Differential equations (15) are degenerated, since the matrix $\overset{4}{\mathbf{P}}$ has no inverse. Let $\mathbf{r}(\varphi)$ be a prescribed inhomogeneity. Consider the boundary value problems defined by

$$\mathbf{K}(\mathbf{y})=\sum_{\nu=0}^{4}\overset{\nu}{\mathbf{P}}(\varphi)\mathbf{y}^{(\nu)}(\varphi)=\mathbf{r}(\varphi),\qquad\overset{3}{\mathbf{P}}(\varphi)=\mathbf{0} \tag{16}$$

and the boundary conditions valid for pinned-fixed beams:

$$\hat{U}_{ob}(-\vartheta)=0\quad\hat{W}_{ob}(-\vartheta)=0\quad\hat{W}^{(2)}_{ob}(-\vartheta)=0\quad|\quad\hat{U}_{ob}(\vartheta)=0\quad\hat{W}_{ob}(\vartheta)=0\quad\hat{W}^{(1)}_{ob}(\vartheta)=0\,. \tag{17}$$

Solution to the homogeneous part of differential equation (15) depends on whether the axial strain $\varepsilon_{o\xi}$ is positive or negative – i.e.: whether the concentrated force is directed upwards or downwards. Let

$$\chi^2=\begin{cases}1-m\varepsilon_{o\xi}&\text{if}\quad\varepsilon_{o\xi}<0\\m\varepsilon_{o\xi}-1&\quad\ \ \varepsilon_{o\xi}>0\text{ and }m\varepsilon_{o\xi}>1\,.\end{cases} \tag{18}$$

This solution is of the form

$$\mathbf{y}=\left[\sum_{i=1}^{4}\underset{(2\times2)}{\mathbf{Y}_i}\underset{(2\times2)}{\mathbf{C}_i}\right]\underset{(2\times1)}{\mathbf{e}},\quad\text{where} \tag{19a}$$

$$\mathbf{Y}_1 = \begin{bmatrix} \cos\varphi & 0 \\ \sin\varphi & 0 \end{bmatrix}, \mathbf{Y}_2 = \begin{bmatrix} -\sin\varphi & 0 \\ \cos\varphi & 0 \end{bmatrix}, \mathbf{Y}_3 = \begin{bmatrix} \cos\chi\varphi & \mathcal{M}\varphi \\ \chi\sin\chi\varphi & -1 \end{bmatrix}, \mathbf{Y}_4 = \begin{bmatrix} -\sin\chi\varphi & 1 \\ \chi\cos\chi\varphi & 0 \end{bmatrix} \tag{19b}$$

when $\varepsilon_{o\xi} < 0$. However, $\mathbf{Y}_3$ and $\mathbf{Y}_4$ are different when $\varepsilon_{o\xi} > 0$ and $m\varepsilon_{o\xi} > 1$:

$$\mathbf{Y}_3 = \begin{bmatrix} \cosh\chi\varphi & \mathcal{M}\varphi \\ \chi\sinh\chi\varphi & -1 \end{bmatrix}, \mathbf{Y}_4 = \begin{bmatrix} -\sinh\chi\varphi & 1 \\ \chi\cosh\chi\varphi & 0 \end{bmatrix}. \tag{19c}$$

In the above relations $\mathbf{C}_i$ are arbitrary constant matrices, $\mathbf{e}$ are arbitrary column matrices and $\mathcal{M} = (m+1)/[m(1+\varepsilon_{o\xi})]$. Solutions to the boundary value problems (16) and (17) are sought in the form

$$\mathbf{y}(\varphi) = \int_a^b \mathbf{G}(\varphi,\psi)\mathbf{r}(\psi)\mathrm{d}\psi, \quad \mathbf{G}(\varphi,\psi) = \begin{bmatrix} G_{11}(\varphi,\psi) & G_{12}(\varphi,\psi) \\ G_{21}(\varphi,\psi) & G_{22}(\varphi,\psi) \end{bmatrix}, \tag{20}$$

where $\mathbf{G}(\varphi,\psi)$ is the Green function matrix, defined by the following properties [5]:

(1) The Green function matrix is continuous in φ and ψ in each of the triangular ranges $-\vartheta \le \varphi \le \psi \le \vartheta$ and $-\vartheta \le \xi \le \varphi \le \vartheta$. The elements $(G_{11}(\varphi,\psi), G_{12}(\varphi,\psi))$ $[G_{21}(\varphi,\psi), G_{22}(\varphi,\psi)]$ are (2 times) [4 times] differentiable with respect to φ, and the following derivatives are continuous in φ and ψ:

$$\frac{\partial^\nu \mathbf{G}(\varphi,\psi)}{\partial x^\nu} = \mathbf{G}^{(\nu)}(\varphi,\psi), \quad \nu = 1,2, \quad \frac{\partial^\nu G_{2i}(\varphi,\psi)}{\partial x^\nu} = G_{2i}^{(\nu)}(\varphi,\psi), \quad \nu = 1,\dots,4;\ i = 1,2. \tag{21}$$

(2) Let ψ be fixed in $[-\vartheta,\vartheta]$. Although the functions $G_{11}(\varphi,\psi)$, $G_{12}^{(1)}(\varphi,\psi)$, $G_{21}^{(\nu)}(\varphi,\psi)$ $\nu = 1,2,3$; $G_{22}^{(\nu)}(\varphi,\psi)$ $\nu = 1,2$ are continuous everywhere, the derivatives $G_{11}^{(1)}(\varphi,\psi)$, $G_{22}^{(3)}(\varphi,\psi)$ have a jump at $\varphi = \psi$:

$$\lim_{\varepsilon\to 0}\left[G_{11}^{(1)}(\varphi+\varepsilon,\varphi) - G_{11}^{(1)}(\varphi-\varepsilon,\varphi)\right] = 1/\overset{1}{P}_{11}(\varphi), \quad \lim_{\varepsilon\to 0}\left[G_{22}^{(3)}(\varphi+\varepsilon,\varphi) - G_{22}^{(3)}(\varphi-\varepsilon,\varphi)\right] = 1/\overset{4}{P}_{22}(\varphi). \tag{22}$$

(3) Let $\boldsymbol{\alpha}$ denote an arbitrary constant vector. For a given $\psi \in [-\vartheta,\vartheta]$, the vector $\mathbf{G}(\varphi,\psi)\boldsymbol{\alpha}$, as a function of φ $(\varphi \ne \psi)$ should satisfy the homogeneous differential equation $\mathbf{K}[\mathbf{G}(\varphi,\psi)\boldsymbol{\alpha}] = \mathbf{0}$.

(4) The vector $\mathbf{G}(\varphi,\psi)\boldsymbol{\alpha}$, as a function of φ, should satisfy the boundary conditions (17). Moreover, there exists only one Green function matrix to each of the boundary value problems.

If the Green function matrix exists then (20) satisfies the differential equations (16) and the boundary conditions (17).

Consider the system of differential equations in the form of

$$\mathbf{K}[\mathbf{y}] = \lambda\mathbf{y}, \tag{23}$$

where $\mathbf{K}[\mathbf{y}]$ is given by (15) and λ is the unknown eigenvalue. The system of ordinary DEs (23) and the homogeneous boundary conditions (17) constitute a boundary value problem, which is, in fact, an eigenvalue problem with λ as the eigenvalue.

Vectors $\mathbf{u}^T = [u_1|u_2]$ and $\mathbf{v}^T = [v_1|v_2]$ are comparison vectors, if they are different from zero, satisfy the boundary conditions and are differentiable as many times as required. The eigenvalue problems (23), (17) are self-adjoint if the product

$$(\mathbf{u},\mathbf{v})_M = \int_{-\vartheta}^{\vartheta} \mathbf{u}^T\mathbf{K}\mathbf{v}\,\mathrm{d}\varphi \tag{24}$$

is commutative, i.e., $(\mathbf{u},\mathbf{v})_M = (\mathbf{v},\mathbf{u})_M$ over the set of comparison vectors and it is positive definite for any comparison vector $\mathbf{u}$, if $(\mathbf{u},\mathbf{u})_M > 0$. If the eigenvalue problems (23), (17) are self-adjoint, then the related Green function matrices are cross-symmetric: $\mathbf{G}(\varphi,\psi) = \mathbf{G}^T(\psi,\varphi)$.

4. Numerical Solution to the Eigenvalue Problems

Making use of (20), the eigenvalue problems (23), (17) can be replaced by homogeneous integral equation systems:

$$\mathbf{y}(\varphi) = \lambda\int_{-\vartheta}^{\vartheta} \mathbf{G}(\varphi,\psi)\mathbf{y}(\psi)\mathrm{d}\psi. \tag{25}$$

Numerical solution to such problems can be sought e.g., by quadrature methods [10]. Consider the integral formula

$$J(\phi) = \int_{-\vartheta}^{\vartheta} \phi(\psi)\,\mathrm{d}\psi \equiv \sum_{j=0}^{n} w_j\phi(\psi_j), \qquad \psi_j \in [-\vartheta,\vartheta], \tag{26}$$

where $\psi_j(\varphi)$ is a vector and w_j are the known weights. Having utilized the latter equation, we obtain from (25) that

$$\sum_{j=0}^{n} w_j\mathbf{G}(\varphi,\psi_j)\tilde{\mathbf{y}}(\psi_j) = \tilde{\kappa}\tilde{\mathbf{y}}(\varphi) \qquad \tilde{\kappa} = 1/\tilde{\lambda} \qquad \in [-\vartheta,\vartheta] \tag{27}$$

is the solution, which yields an approximate eigenvalue $\bar{\lambda} = 1/\bar{\kappa}$ and the corresponding approximate eigenfunction $\bar{\mathbf{y}}(\varphi)$. After setting φ to ψ_i $(i = 0, 1, 2, \ldots, n)$, we have

$$\sum_{j=0}^{n} w_j \mathbf{G}(\psi_i, \psi_j)\bar{\mathbf{y}}(\psi_j) = \bar{\kappa}\bar{\mathbf{y}}(\psi_i) \qquad \bar{\kappa} = 1/\bar{\lambda} \qquad \psi_i, \psi_j \in [-\vartheta, \vartheta], \quad \text{or} \quad \mathcal{G}\mathcal{D}\bar{\mathcal{Y}} = \bar{\kappa}\bar{\mathcal{Y}}, \tag{28}$$

where $\mathcal{G} = [\mathbf{G}(\psi_i, \psi_j)]$ for self-adjoint problems, while $\mathcal{D} = \text{diag}(w_0, \ldots, w_0|\ldots|w_n, \ldots, w_n)$ and $\bar{\mathcal{Y}}^T = [\bar{\mathbf{y}}^T(\psi_0)|\bar{\mathbf{y}}^T(\psi_1)|\ldots|\bar{\mathbf{y}}^T(\psi_n)]$. After solving the generalized algebraic eigenvalue problem (28), we have the approximate eigenvalues $\bar{\lambda}_r$ and eigenvectors $\mathcal{Y}_r$, while the corresponding eigenfunction is obtained via substituting into (27):

$$\bar{\mathbf{y}}_r(\varphi) = \bar{\lambda}_r \sum_{j=0}^{n} w_j \mathbf{G}(\varphi, \psi_j)\bar{\mathbf{y}}_r(\psi_j) \qquad r = 0, 1, 2, \ldots, n\,. \tag{29}$$

Divide the interval $[-\vartheta, \vartheta]$ into equidistant subintervals of length h and apply the integration formula to each subinterval. By repeating the line of thought leading to (29), the algebraic eigenvalue problem obtained has the same structure as (29).

It is also possible to consider the integral equations (25) as if they were boundary integral equations and apply isoparametric approximation on the subintervals (elements). If this is the case, one can approximate the eigenfunction on the e-th element (the e-th subinterval which is mapped onto the interval $\gamma \in [-1, 1]$ and is denoted by $\mathfrak{L}_e$) by

$$\overset{e}{\mathbf{y}} = \mathbf{N}_1(\gamma)\overset{e}{\mathbf{y}}_1 + \mathbf{N}_2(\gamma)\overset{e}{\mathbf{y}}_2 + \mathbf{N}_3(\gamma)\overset{e}{\mathbf{y}}_3\,, \tag{30}$$

where quadratic local approximation is assumed: $\mathbf{N}_i = \text{diag}(N_i)$, $N_1 = 0.5\gamma(\gamma - 1)$, $N_2 = 1 - \gamma^2$, $N_3 = 0.5\gamma(\gamma + 1)$, $\overset{e}{\mathbf{y}}_i$ is the value of the eigenfunction $\mathbf{y}(\varphi)$ at the left endpoint, the midpoint and the right endpoint of the element, respectively. Upon substitution of approximation (30) into (25), we have

$$\bar{\mathbf{y}}(\varphi) = \bar{\lambda}\sum_{e=1}^{n_{be}} \int_{\mathfrak{L}_e} \mathbf{G}(x, \gamma)[\mathbf{N}_1(\eta)|\mathbf{N}_2(\gamma)|\mathbf{N}_3(\gamma)]d\gamma \left[\overset{e}{\bar{\mathbf{y}}}_1|\overset{e}{\bar{\mathbf{y}}}_2|\overset{e}{\bar{\mathbf{y}}}_3\right]^T\,, \tag{31}$$

in which, n_{be} is the number of elements. Using equation (31) as a point of departure, and repeating the line of thought leading to (28), we again get an algebraic eigenvalue problem.

5. COMPUTATION OF THE GREEN FUNCTION MATRICES

Based on the definition presented in Section 3, here we show the calculation of the corresponding Green function matrices for the two loading cases of pinned-fixed beams. With regard to property 3, the Green function can be given in the form

$$\underbrace{\mathbf{G}(\varphi, \psi)}_{(2\times 2)} = \sum_{j=1}^{4} \mathbf{Y}_j(\varphi)\left[\mathbf{A}_j(\psi) \pm \mathbf{B}_j(\psi)\right]\,, \tag{32}$$

where (a) the sign is [positive](negative) if $[\varphi \le \psi](\varphi \ge \psi)$; (b) the matrices $\mathbf{A}_j$ and $\mathbf{B}_j$ have the following structure

$$\mathbf{A}_j = \begin{bmatrix} \overset{j}{A}_{11} & \overset{j}{A}_{12} \\ \overset{j}{A}_{21} & \overset{j}{A}_{22} \end{bmatrix} = \begin{bmatrix} \mathbf{A}_{j1} & \mathbf{A}_{j2} \end{bmatrix}, \qquad \mathbf{B}_j = \begin{bmatrix} \overset{j}{B}_{11} & \overset{j}{B}_{12} \\ \overset{j}{B}_{21} & \overset{j}{B}_{22} \end{bmatrix} = \begin{bmatrix} \mathbf{B}_{j1} & \mathbf{B}_{j2} \end{bmatrix} \quad j = 1, \ldots, 4; \tag{33}$$

(c) the coefficients in $\mathbf{B}_j$ are independent of the boundary conditions. As the matrices $\mathbf{Y}_3$ and $\mathbf{Y}_4$ are different for $\varepsilon_{o\xi} < 0$ and for $\varepsilon_{o\xi} > 0$, when $m\varepsilon_{o\xi} > 1$, we deal with the two possibilities separately.

The Green functions matrix if $\varepsilon_{o\xi} < 0$. Let us now introduce the following notational conventions

$$a = \overset{1}{B}_{11},\ b = \overset{2}{B}_{11},\ c = \overset{3}{B}_{11},\ d = \overset{3}{B}_{21},\ e = \overset{4}{B}_{11},\ f = \overset{4}{B}_{21}\,. \tag{34}$$

We note that $\overset{1}{B}_{21} = \overset{2}{B}_{21} = \overset{1}{B}_{22} = \overset{2}{B}_{22} = 0$ – see Section 3. The systems of equations for the unknowns $a, \ldots, f$ can be set up by fulfilling the continuity and discontinuity conditions mentioned in property 1 and 2 for the Green function matrix if $\varphi = \psi$. Therefore, if $i = 1$, we have

$$\begin{bmatrix} \cos\psi & -\sin\psi & \cos(\chi\psi) & \mathcal{M}\psi & -\sin(\chi\psi) & 1 \\ \sin\psi & \cos\psi & \chi\sin(\chi\psi) & -1 & \chi\cos(\chi\psi) & 0 \\ -\sin\psi & -\cos\psi & -\chi\sin(\chi\psi) & \mathcal{M} & -\chi\cos(\chi\psi) & 0 \\ \cos\psi & -\sin\psi & \chi^2\cos(\chi\psi) & 0 & -\chi^2\sin(\chi\psi) & 0 \\ -\sin\psi & -\cos\psi & -\chi^3\sin(\chi\psi) & 0 & -\chi^3\cos(\chi\psi) & 0 \\ -\cos\psi & \sin\psi & -\chi^4\cos(\chi\psi) & 0 & \chi^4\sin(\chi\psi) & 0 \end{bmatrix} \begin{bmatrix} a \\ b \\ c \\ d \\ e \\ f \end{bmatrix} = \begin{bmatrix} 0 \\ 0 \\ \frac{1}{2m} \\ 0 \\ 0 \\ 0 \end{bmatrix}\,. \tag{35}$$

from where we get the constants as

$$\begin{aligned}
a &= \overset{1}{B}_{11} = \frac{\chi^2}{(1-\chi^2)(1-\mathcal{M})m}\frac{\sin\psi}{2}\,, & b &= \overset{2}{B}_{11} = \frac{\chi^2}{(1-\chi^2)(1-\mathcal{M})m}\frac{\cos\psi}{2}\,,\\
c &= \overset{3}{B}_{11} = -\frac{\chi^2}{(1-\chi^2)(1-\mathcal{M})m}\frac{\sin\chi\psi}{2\chi^3}\,, & d &= \overset{3}{B}_{21} = -\frac{1}{2(1-\mathcal{M})m}\,,\\
e &= \overset{4}{B}_{11} = -\frac{1}{\chi(1-\chi^2)(1-\mathcal{M})m}\frac{\cos\chi\psi}{2}\,, & f &= \overset{4}{B}_{21} = \frac{1}{2}\mathcal{M}\frac{\psi}{m(1-\mathcal{M})}\,.
\end{aligned} \tag{36}$$

If $i=2$, then

$$\begin{bmatrix}
\cos\psi & -\sin\psi & \cos(\chi\psi) & \mathcal{M}\psi & -\sin(\chi\psi) & 1\\
\sin\psi & \cos\psi & \chi\sin(\chi\psi) & -1 & \chi\cos(\chi\psi) & 0\\
-\sin\psi & -\cos\psi & -\chi\sin(\chi\psi) & \mathcal{M} & -\chi\cos(\chi\psi) & 0\\
\cos\psi & -\sin\psi & \chi^2\cos(\chi\psi) & 0 & -\chi^2\sin(\chi\psi) & 0\\
-\sin\psi & -\cos\psi & -\chi^3\sin(\chi\psi) & 0 & -\chi^3\cos(\chi\psi) & 0\\
-\cos\psi & \sin\psi & -\chi^4\cos(\chi\psi) & 0 & \chi^4\sin(\chi\psi) & 0
\end{bmatrix}
\begin{bmatrix} a\\ b\\ c\\ d\\ e\\ f\end{bmatrix}
=
\begin{bmatrix} 0\\ 0\\ 0\\ 0\\ 0\\ -\frac{1}{2}\end{bmatrix} \tag{37}$$

is the equation system, the solution of which assumes the form

$$\begin{aligned}
a &= \overset{1}{B}_{12} = \frac{1}{2}\frac{\cos\psi}{(1-\chi^2)}\,, & b &= \overset{2}{B}_{12} = -\frac{1}{2}\frac{\sin\psi}{(1-\chi^2)}\,, & c &= \overset{3}{B}_{12} = -\frac{1}{2}\frac{\cos\chi\psi}{(1-\chi^2)\chi^2}\,,\\
d &= \overset{3}{B}_{22} = 0\,, & e &= \overset{4}{B}_{12} = \frac{1}{2}\frac{\sin\chi\psi}{(1-\chi^2)\chi^2}\,, & f &= \overset{4}{B}_{22} = \frac{1}{2\chi^2}\,.
\end{aligned} \tag{38}$$

Regarding the unknown scalars $\overset{1}{A}_{1i}(\psi)$, $\overset{2}{A}_{1i}(\psi)$, $\overset{3}{A}_{1i}(\psi)$, $\overset{3}{A}_{2i}(\psi)$, $\overset{4}{A}_{1i}(\psi)$, $\overset{4}{A}_{2i}(\psi)$, $i=1,2;\psi\in[-\vartheta,\vartheta]$ in the matrices $\mathbf{A}_j$, property (4), that is the boundary conditions (17) yield

$$\begin{bmatrix}
\cos\vartheta & \sin\vartheta & \cos(\chi\vartheta) & -\mathcal{M}\vartheta & \sin(\chi\vartheta) & 1\\
\cos\vartheta & -\sin\vartheta & \cos(\chi\vartheta) & \mathcal{M}\vartheta & -\sin(\chi\vartheta) & 1\\
-\sin\vartheta & \cos\vartheta & -\chi\sin(\chi\vartheta) & -1 & \chi\cos(\chi\vartheta) & 0\\
\sin\vartheta & \cos\vartheta & \chi\sin(\chi\vartheta) & -1 & \chi\cos(\chi\vartheta) & 0\\
\cos\vartheta & \sin\vartheta & \chi^2\cos(\chi\vartheta) & 0 & \chi^2\sin(\chi\vartheta) & 0\\
-\sin\vartheta & -\cos\vartheta & -\chi^3\sin(\chi\vartheta) & 0 & -\chi^3\cos(\chi\vartheta) & 0
\end{bmatrix}
\begin{bmatrix} \overset{1}{A}_{1i}\\ \overset{2}{A}_{1i}\\ \overset{3}{A}_{1i}\\ \overset{3}{A}_{2i}\\ \overset{4}{A}_{1i}\\ \overset{4}{A}_{2i}\end{bmatrix}
=$$

$$=\begin{bmatrix}
-a\cos\vartheta - b\sin\vartheta - c\cos(\chi\vartheta) + d\mathcal{M}\vartheta - e\sin(\chi\vartheta) - f\\
a\cos\vartheta - b\sin\vartheta + c\cos(\chi\vartheta) + d\mathcal{M}\vartheta - e\sin(\chi\vartheta) + f\\
a\sin\vartheta - b\cos\vartheta + c\chi\sin(\chi\vartheta) + d - e\chi\cos(\chi\vartheta)\\
a\sin\vartheta + b\cos\vartheta + c\chi\sin(\chi\vartheta) - d + e\chi\cos(\chi\vartheta)\\
-a\cos\vartheta - b\sin\vartheta - c\chi^2\cos(\chi\vartheta) - e\chi^2\sin(\chi\vartheta)\\
-a\sin\vartheta - b\cos\vartheta - c\chi^3\sin(\chi\vartheta) - e\chi^3\cos(\chi\vartheta)
\end{bmatrix}. \tag{39}$$

Closed form solution to this system was generated using Maple 15. The unknowns are not detailed here.

Calculation of the Green functions matrix if $\varepsilon_{o\xi}>0$ and $m\varepsilon_{o\xi}>1$. For the other loading case, repeating a procedure similar to the procedure leading to (35), we get the following equations if $i=1$

$$\begin{bmatrix}
\cos\psi & -\sin\psi & \cosh(\chi\psi) & \mathcal{M}\psi & \sinh(\chi\psi) & 1\\
\sin\psi & \cos\psi & -\chi\sinh(\chi\psi) & -1 & -\chi\cosh(\chi\psi) & 0\\
-\sin\psi & -\cos\psi & \chi\sinh(\chi\psi) & \mathcal{M} & \chi\cosh(\chi\psi) & 0\\
\cos\psi & -\sin\psi & -\chi^2\cosh(\chi\psi) & 0 & -\chi^2\sinh(\chi\psi) & 0\\
-\sin\psi & -\cos\psi & -\chi^3\sinh(\chi\psi) & 0 & -\chi^3\cosh(\chi\psi) & 0\\
-\cos\psi & \sin\psi & -\chi^4\cosh(\chi\psi) & 0 & -\chi^4\sinh(\chi\psi) & 0
\end{bmatrix}
\begin{bmatrix} a\\ b\\ c\\ d\\ e\\ f\end{bmatrix}
=
\begin{bmatrix} 0\\ 0\\ \frac{1}{2m}\\ 0\\ 0\\ 0\end{bmatrix}. \tag{40}$$

The solutions are as follows

$$\begin{aligned}
a &= \overset{1}{B}_{11} = -\frac{\chi^2}{(1+\chi^2)(1-\mathcal{M})m}\frac{\sin\psi}{2}\,, & b &= \overset{2}{B}_{11} = -\frac{\chi^2}{(1+\chi^2)(1-\mathcal{M})m}\frac{\cos\psi}{2}\,,\\
c &= \overset{3}{B}_{11} = -\frac{1}{\chi(1+\chi^2)(1-\mathcal{M})m}\frac{\sinh\chi\psi}{2}\,, & d &= \overset{3}{B}_{21} = -\frac{1}{2(1-\mathcal{M})m}\,,\\
e &= \overset{4}{B}_{11} = \frac{1}{\chi(1+\chi^2)(1-\mathcal{M})m}\frac{\cosh\chi\psi}{2}\,, & f &= \overset{4}{B}_{21} = \frac{1}{2(1-\mathcal{M})m}\mathcal{M}\psi\,.
\end{aligned} \tag{41}$$

If $i = 2$

$$\begin{bmatrix} \cos\psi & -\sin\psi & \cosh(\chi\psi) & \mathcal{M}\psi & \sinh(\chi\psi) & 1 \\ \sin\psi & \cos\psi & -\chi\sinh(\chi\psi) & -1 & -\chi\cosh(\chi\psi) & 0 \\ -\sin\psi & -\cos\psi & \chi\sinh(\chi\psi) & \mathcal{M} & \chi\cosh(\chi\psi) & 0 \\ \cos\psi & -\sin\psi & -\chi^2\cosh(\chi\psi) & 0 & -\chi^2\sinh(\chi\psi) & 0 \\ -\sin\psi & -\cos\psi & -\chi^3\sinh(\chi\psi) & 0 & -\chi^3\cosh(\chi\psi) & 0 \\ -\cos\psi & \sin\psi & -\chi^4\cosh(\chi\psi) & 0 & -\chi^4\sinh(\chi\psi) & 0 \end{bmatrix} \begin{bmatrix} a \\ b \\ c \\ d \\ e \\ f \end{bmatrix} = \begin{bmatrix} 0 \\ 0 \\ 0 \\ 0 \\ 0 \\ -\frac{1}{2} \end{bmatrix} \tag{42}$$

is the equation system to be solved – compare it to (37) – and the solutions we have obtained are

$$\begin{aligned} &a = \overset{1}{B}_{12} = \frac{1}{2}\frac{\cos\psi}{(1+\chi^2)}, \quad b = \overset{2}{B}_{12} = -\frac{1}{2}\frac{\sin\psi}{(1+\chi^2)}, \qquad c = \overset{3}{B}_{12} = \frac{1}{2}\frac{\cosh\chi\psi}{(1+\chi^2)\chi^2} \\ &d = \overset{3}{B}_{22} = 0, \qquad\qquad e = \overset{4}{B}_{12} = -\frac{1}{2}\frac{\sinh\chi\psi}{\chi^2(1+\chi^2)}, \quad f = \overset{4}{B}_{22} = -\frac{1}{2\chi^2}. \end{aligned} \tag{43}$$

For the elements of the matrices $\mathbf{A}_j$, boundary conditions (17) yield the equation system

$$\begin{bmatrix} \cos\vartheta & \sin\vartheta & \cosh(\chi\vartheta) & -\mathcal{M}\vartheta & -\sinh(\chi\vartheta) & 1 \\ \cos\vartheta & -\sin\vartheta & \cosh(\chi\vartheta) & \mathcal{M}\vartheta & \sinh(\chi\vartheta) & 1 \\ -\sin\vartheta & \cos\vartheta & \chi\sinh(\chi\vartheta) & -1 & -\chi\cosh(\chi\vartheta) & 0 \\ \sin\vartheta & \cos\vartheta & -\chi\sinh(\chi\vartheta) & -1 & -\chi\cosh(\chi\vartheta) & 0 \\ \cos\vartheta & \sin\vartheta & -\chi^2\cosh(\chi\vartheta) & 0 & \chi^2\sinh(\chi\vartheta) & 0 \\ -\sin\vartheta & -\cos\vartheta & -\chi^3\sinh(\chi\vartheta) & 0 & -\chi^3\cosh(\chi\vartheta) & 0 \end{bmatrix} \begin{bmatrix} \overset{1}{A}_{11} \\ \overset{2}{A}_{11} \\ \overset{3}{A}_{11} \\ \overset{3}{A}_{21} \\ \overset{4}{A}_{11} \\ \overset{4}{A}_{21} \end{bmatrix} =$$

$$= \begin{bmatrix} -a\cos\vartheta - b\sin\vartheta - c\cosh(\chi\vartheta) + d\mathcal{M}\vartheta + e\sinh(\chi\vartheta) - f \\ a\cos\vartheta - b\sin\vartheta + c\cosh(\chi\vartheta) + d\mathcal{M}\vartheta + e\sinh(\chi\vartheta) + f \\ a\sin\vartheta - b\cos\vartheta - c\chi\sinh(\chi\vartheta) + d + e\chi\cosh(\chi\vartheta) \\ a\sin\vartheta + b\cos\vartheta - c\chi\sinh(\chi\vartheta) - d - e\chi\cosh(\chi\vartheta) \\ -a\cos\vartheta - b\sin\vartheta + c\chi^2\cosh(\chi\vartheta) - e\chi^2\sinh(\chi\vartheta) \\ -a\sin\vartheta - b\cos\vartheta - c\chi^3\sinh(\chi\vartheta) - e\chi^3\cosh(\chi\vartheta) \end{bmatrix}. \tag{44}$$

the solutions of which are omitted here.

6. THE LOAD-STRAIN RELATION AND THE CRITICAL STRAIN

In practise, the loading is generally the known quantity. However, the formulation has the axial strain $\varepsilon_{o\xi}$ as a parameter. For a first, linearized model, the effect the deformations have on the equilibrium is neglected. We can establish the load-strain relationship $\varepsilon_{o\xi} = \varepsilon_{o\xi}(\mathcal{P})$ on the basis of differential equations (8) if $f_t = f_n = 0$, and by applying the

$$\begin{aligned} &U_o|_{\pm\vartheta} = W_o|_{\pm\vartheta} = M|_{-\vartheta} = \psi_{o\eta}|_{+\vartheta} = 0, \\ &U_o|_{\varphi=-0} = U_o|_{\varphi=+0}, \qquad W_o|_{\varphi=-0} = W_o|_{\varphi=+0}, \qquad \psi_{o\eta}|_{\varphi=-0} = \psi_{o\eta}|_{\varphi=+0}, \end{aligned} \tag{45a}$$

$$N|_{\varphi=-0} = N|_{\varphi=+0}, \qquad M|_{\varphi=-0} = M|_{\varphi=+0}, \qquad \mathrm{d}M/\mathrm{d}s|_{\varphi=+0} - \mathrm{d}M\,\mathrm{d}s|_{\varphi=-0} - P_\zeta = 0 \tag{45b}$$

boundary and continuity (discontinuity) conditions prescribed at the crown point. Here the physical quantities can all be given in terms of the dimensionless displacements U_o and W_o as

$$\psi_{o\eta} = U_o - W_o^{(1)}, \quad N = A_e\varepsilon_{o\xi} - \frac{M}{\rho_o} \approx A_e\varepsilon_{o\xi}, \quad M = -\frac{I_{e\eta}}{\rho_o^2}\left(W_o^{(2)} + W_o\right). \tag{46}$$

After solving the boundary value problem defined by the ODEs (8) ($f_t = f_n = 0$) and the above continuity and discontinuity conditions we get the axial strain in the following form

$$\varepsilon(\mathcal{P}, m, \vartheta) = -\frac{\mathcal{P}}{\vartheta}\left[(3\vartheta\cos\vartheta - 4\cos\vartheta\sin\vartheta + 4\sin\vartheta)\cos^2\vartheta + (\vartheta^2 - 2 + 2\cos\vartheta)\sin\vartheta + \vartheta(2 - 5\cos\vartheta)\right] \times$$

$$\times \frac{1}{m\left[(-2\vartheta\sin\vartheta + 11\cos\vartheta - 4\vartheta\cos^2\vartheta\sin\vartheta - 7\cos^3\vartheta)\cos\vartheta - 4 + 3\vartheta^2\right] + 2\vartheta(\cos\vartheta\sin\vartheta - 2\cos^3\vartheta\sin\vartheta + \vartheta)}. \tag{47}$$

If $\mathcal{P}$ is [negative] (positive), then $\varepsilon_{o\xi}$ is [negative] (positive). The critical strain, at which, curved beams lose their stability, can be obtained by solving the eigenvalue problem governed by equations (13) – with the right side set to zero – and the corresponding boundary conditions. The eigenvalue sought is $\chi = \sqrt{1 - m\varepsilon_{o\xi}}$.

The general solution for the displacement increments and the boundary conditions to be satisfied are:

$$W_{ob} = -E_2 - E_3\cos\varphi + E_4\sin\varphi - \chi E_5\cos\chi\varphi + \chi E_6\sin\chi\varphi, \tag{48}$$

$$U_{ob} = E_1 + E_2\mathcal{M}\varphi + E_3\sin\varphi + E_4\cos\varphi + E_5\sin\chi\varphi + E_6\cos\chi\varphi, \tag{49}$$

$$U_{ob}|_{\pm\vartheta} = W_{ob}|_{\pm\vartheta} = W_{ob}^{(2)}\Big|_{-\vartheta} = W_{ob}^{(1)}\Big|_{+\vartheta} = 0\,. \tag{50}$$

Here E_i $(i = 1, \ldots, 6)$ are undetermined constants of integration. The boundary conditions lead to the homogenous equation system

$$\begin{bmatrix} 1 & -\mathcal{M}\vartheta & -\sin\vartheta & \cos\vartheta & -\sin\chi\vartheta & \cos\chi\vartheta \\ 1 & \mathcal{M}\vartheta & \sin\vartheta & \cos\vartheta & \sin\chi\vartheta & \cos\chi\vartheta \\ 0 & 1 & \cos\vartheta & \sin\vartheta & \chi\cos\chi\vartheta & \chi\sin\chi\vartheta \\ 0 & 1 & \cos\vartheta & -\sin\vartheta & \chi\cos\chi\vartheta & -\chi\sin\chi\vartheta \\ 0 & 0 & \cos\vartheta & \sin\vartheta & \chi^3\cos\chi\vartheta & \chi^3\sin\chi\vartheta \\ 0 & 0 & \sin\vartheta & \cos\vartheta & \chi^2\sin\chi\vartheta & \chi^2\cos\chi\vartheta \end{bmatrix} \begin{bmatrix} E_1 \\ E_2 \\ E_3 \\ E_4 \\ E_5 \\ E_6 \end{bmatrix} = \begin{bmatrix} 0 \\ 0 \\ 0 \\ 0 \\ 0 \\ 0 \end{bmatrix}. \tag{51}$$

The vanishing of the determinant results in the following non-linear equation:

$$\left[\chi^4\left(1-2\cos^2\chi\vartheta\right) + \left(1-2\chi^2\right)\sin^2\chi\vartheta + \chi\vartheta\left(\chi^2-1\right)\cos\chi\vartheta\sin\chi\vartheta\right]\sin^2\vartheta + \chi^2\vartheta\left(\chi^2-1\right)\cos\vartheta\sin\vartheta\cos^2\chi\vartheta =$$
$$= \left[\left(\chi\vartheta\left(\chi^2-1\right)\cos\chi\vartheta + \sin\chi\vartheta\right)\cos\vartheta + \chi^2\vartheta\left(\chi^2-1\right)\sin\vartheta\sin\chi\vartheta + \chi\left(-\chi^2-1\right)\sin\vartheta\cos\chi\vartheta\right]\cos\vartheta\sin\chi\vartheta\,. \tag{52}$$

The lowest reasonable critical value is then approximated by the polynomial

$$\chi\vartheta = g_{pf}(\vartheta) = 0.014\,875,\vartheta^5 - 0.078\,701\,\vartheta^4 + 0.168\,958\,\vartheta^3 - 0.119\,606\,\vartheta^2 + 0.057\,002\,\vartheta + 3.749\,293\,, \tag{53}$$

and the critical strain is

$$\varepsilon_{o\xi\,\mathrm{crit}} = -\frac{1}{m}\left(\chi^2-1\right) = -\frac{1}{m}\left[\left(\frac{g_{pf}}{\vartheta}\right)^2 - 1\right]. \tag{54}$$

7. Computational Results

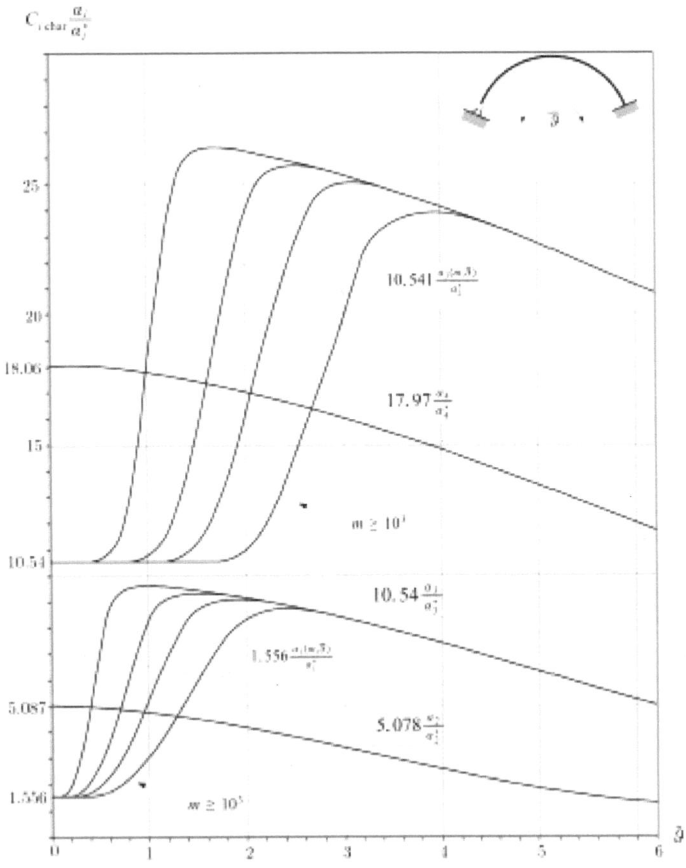

Figure 2. Results for pinned-fixed beams when $\varepsilon_{o\xi} \simeq 0$

We have solved the eigenvalue problems governing the vibrations of curved beams using a Fortran90 code. The numerical results have been compared to those valid for the free vibrations of curved beams with the same geometry and material. For more details about the natural frequencies of planar curved beams see [5].

When we set the strain to a very small value, i.e., to $|\varepsilon_{o\xi}| = |\varepsilon_{o\xi\,\mathrm{crit}} \cdot 10^{-6}|$ for both loading cases, we get back the results valid for the free vibrations of curved beams – see [5] and [11] for details.

It is known that – see e.g.: [9] – the i-th eigenfrequency for the free transverse vibrations of heterogeneous straight beams is obtained from the relation

$$\alpha_i^* = \frac{C_{i,\,\mathrm{char}}\pi^2}{\sqrt{\frac{\rho_a A}{I_{e\eta}}}\ell_b^2}\,, \tag{55}$$

where the constant $C_{i,\,\mathrm{char}}$ depends on the supports and the ordinal number of the frequency sought. This time $C_{1,\,\mathrm{char}} = 1.556$, $C_{2,\,\mathrm{char}} = 5.078$, $C_{3,\,\mathrm{char}} = 10.541$, $C_{4,\,\mathrm{char}} = 17.97$ and ℓ_b is the length of the beam. If we recall Eq.(14)$_2$ which, for such a small strain considered, expresses the relation between the eigenvalues λ_i and the eigenfrequencies $\alpha_i = \alpha_{i\,\mathrm{free}}$ for the free vibrations of curved beams we may write

$$C_{i,\mathrm{char}}\frac{\alpha_i}{\alpha_i^*} = \frac{\frac{\sqrt{\lambda_i}}{\sqrt{\frac{\rho_a A}{I_{e\eta}}}R^2}}{\frac{\pi^2}{\sqrt{\frac{\rho_a A}{I_{e\eta}}}\ell_r^2}} = \frac{\bar{\vartheta}^2\sqrt{\lambda_i}}{\pi^2}\,. \tag{56}$$

This is the connection between the natural frequencies of curved and straight beams with the same length ($\ell_b = R\bar{\vartheta}$), cross-section and material. In Figure 2, this ratio is plotted against the central angle $\bar{\vartheta}$ of the curved beam. Four different values of the parameter m were picked: $(1, 3.4, 12, 100) \cdot 10^3$. Observe that the ratio of the even natural frequencies are independent of m, while the odd ones are not. It is also important to mention that the frequency spectrum changes as $\bar{\vartheta}$ increases – e.g.: the first eigenfrequency becomes the second one in terms of its magnitude if $\bar{\vartheta}$ is sufficiently great.

When dealing with the free longitudinal vibrations of fixed-fixed rods, the natural frequencies assume the form [11]

$$\hat{\alpha}_i = \frac{K_{i\,\mathrm{char}}}{\ell_r}\sqrt{\frac{E}{\rho_a}}\pi\,, \tag{57}$$

where the constant $K_{i\,\mathrm{char}} = i$; $(i = 1, 2, 3, \ldots)$ and ℓ_r is the length of the rod. If we recall Eq.(14)$_2$, we can compare the eigenfrequencies of curved beams (given that $|\varepsilon_{o\xi}| = |\varepsilon_{o\xi\,\mathrm{crit}} \cdot 10^{-6}| \simeq 0$ when calculating λ_i) to those of rods by

$$K_{i\,\mathrm{char}}\frac{\alpha_i}{\hat{\alpha}_i} = \frac{1}{\sqrt{m}}\frac{\bar{\vartheta}}{\pi}\sqrt{\lambda_i}\,. \tag{58}$$

These quotients for $i = 1, 2$ are plotted in Figure 3. We found that the ratios do not depend on the parameter m and these are equal to 1 and 2 respectively if the central angle is sufficiently small. We remark that these tendencies are the same with a good accuracy for pinned-pinned and for fixed-fixed curved beams as well.

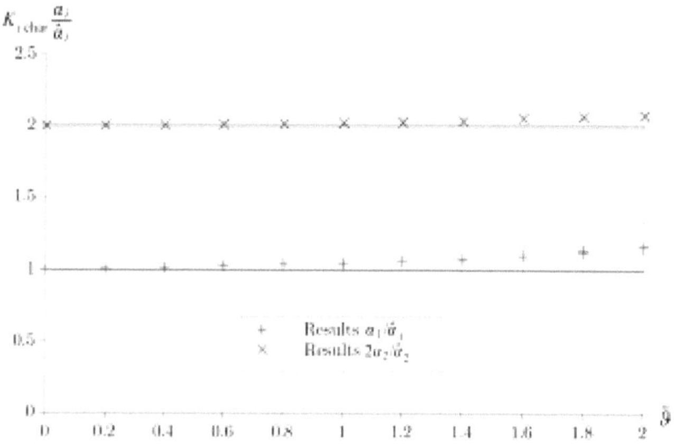

Figure 3. Results for pinned-fixed beams when $\varepsilon_{o\xi} \simeq 0$

In the forthcoming, the effect of the concentrated load to the length of the centerline is taken into account. In what follows, regarding our notations, α_i is i-th eigenfrequency of the loaded curved beam, while the eigenfrequencies that belong to the free vibrations (the beam is unloaded) are denoted by $\alpha_{i\,\mathrm{free}}$.

Figure 4 represents the quotient $\alpha_1^2/\alpha_{1\,\mathrm{free}}^2$ against the quotient $|\varepsilon_{o\xi}/\varepsilon_{o\xi\,\mathrm{crit}}|$ both for a negative and a positive P_ζ. We remark that this time the subscript 1 always refers to the lowest frequencies (which do not coincide with the first one every time – see Figure 2). The frequencies under [compression] <tension> [decrease] <increase> almost linearly and independently of m and ϑ, given that $m >\sim 10\,000$ and $\bar{\vartheta} >\sim 1$. These relationships can be approximated with a very good accuracy by

$$\frac{\alpha_1^2}{\alpha_{1\,\mathrm{free}}^2} = 1.000\,848\,535 - 0.983\,386\,732\frac{|\varepsilon_{o\xi}|}{\varepsilon_{o\xi\ crit}} - 0.174\,018\,254\left(\frac{\varepsilon_{o\xi}}{\varepsilon_{o\xi\ crit}}\right)^2, \quad \text{if} \quad \varepsilon_{o\xi} < 0\,, \tag{59}$$

$$\frac{\alpha_1^2}{\alpha_{1\,\mathrm{free}}^2} = 1.000\,198\,503 + 0.986\,131\,634\frac{|\varepsilon_{o\xi}|}{\varepsilon_{o\xi\ crit}} - 0.008\,370\,551\left(\frac{\varepsilon_{o\xi}}{\varepsilon_{o\xi\ crit}}\right)^2, \quad \text{if} \quad \varepsilon_{o\xi} > 0. \tag{60}$$

We again remark that these results are almost the same, when the curved beam is pinned-pinned or fixed-fixed.

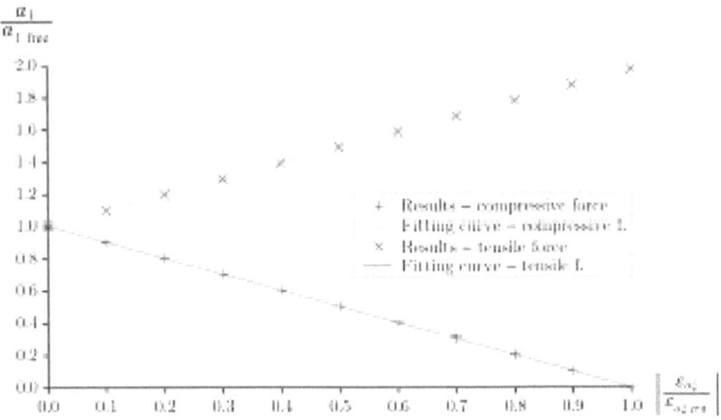

Figure 4. Results for the two loading cases of pinned-fixed beams

8. Concluding Remarks

In accordance with our aims, we have investigated the vibrations of curved beams with cross-sectional heterogeneity under a centra load (a vertical force) exerted at the crown point. We have derived the governing equations of the boundary value problems, which make it possible to clarify how the radial load affects the natural frequencies.

For pinned-fixed beams, we have determined the Green function matrices assuming that the beams are prestressed by a radial load. When computing the corresponding matrices, we had to take into account that the system of differential equations that govern the problem are degenerated. Making use of the Green function matrices, we have reduced the self-adjoint eigenvalue problems set up for the eigenfrequencies to eigenvalue problems governed by homogeneous systems of Fredholm integral equations. These integral equations can be used for those loads, which results in a constant axial strain on the E-weighted centerline.

Numerical solutions were provided graphically. For the loaded beams considered, the quotient $\alpha_1^2/\alpha_{1\,\mathrm{free}}^2$ depends linearly with a good accuracy on the axial strain $\varepsilon_{o\xi}$. With the knowledge of the relationship $\varepsilon_{o\xi} = \varepsilon_{o\xi}(\mathcal{P})$, we can determine the strain, which belongs to a given load and then the natural frequencies of the loaded structure.

Acknowledgements by the second author: This research was supported by the **European Union** and the **State of Hungary, co-financed by the European Social Fund** in the framework of TÁMOP 4.2.4.A/2-11-1-2012-0001 'National Excellence Program'.

References

[1] A. E. H. Love. *Treatese on the mathematical theory of elasticity*. New York, Dower, 1944.

[2] S. Márkus and T. Nánási. Vibration of curved beams. *The Shock and Vibration Digest*, 13(4):3–14, 1981.

[3] P. A. A. Laura and M. J. Maurizi. Recent research on vibrations of arch-type structures. *The Shock and Vibration Digest*, 19(1):6–9, 1987.

[4] P. Chidamparam and A. W. Leissa. Vibrations of planar curved beams, rings and arches. *Applied Mechanis Review, ASME*, 46(9):467–483, 1993.

[5] G. Szeidl. *Effect of Change in Length on the Natural Frequencies and Stability of Circular Beams*. Ph.D Thesis, Department of Mechanics, University of Miskolc, Hungary, 1975. (in Hungarian).

[6] C. S. Huang, K. Y. Nieh, and M. C. Yang. In plane free vibration and stability of loaded and shear-deformable circular arches. *International Journal of Solids and Structures*, 40:5865–5886, 2003.

[7] Ray Lawther. On the straightness of eigenvalue iterations. *Computational Mechanics*, 37:362–368, 2005.

[8] G. Szeidl and L. Kiss. A Nonlinear Mechanical Model For Heterogeneous Curved Beams. In S. Vlase, editor, *Proceedings of the 4th International Conference on Advanced Composite Materials Enginnering, COMAT*, volume 2, pages 589–596, 18 - 20 October 2012, Braşov, Romania.

[9] L. P. Kiss. Free vibrations of heterogeneous curved beams. *GÉP*, LXIV(5):16–21, 2013. (in Hungarian).

[10] Christopher T. H. Baker. *The Numerical Treatment of Integral Equations – Monographs on Numerical Analysis edited by L. Fox and J. Walsh*. Clarendon Press, Oxford, 1977.

[11] K. Kelemen. Vibrations of circular arches subjected to hydrostatic follower loads – computations by the use of green functions. *Journal of Computational and Applied Mechanics*, 2:167–178, 2000.

THE LIGHTWEIGHT OPTIMIZATION OF A COMPOSITE SHELL – CASE STUDY OF A HELICOPTER CRASHWORTHY SEAT

I. Tesula[1], I. Balcu[2], G. Dima[3]

[1]Nuarb Aerospace SRL, Brasov, ROMANIA, i.tesula@nuarb.ro
[2]Transylvania University, Brasov, ROMANIA, balcu@unitbv.ro
[3]Transylvania University, Brasov, ROMANIA, dumitru.dima@unitbv.ro

Abstract: *The topology optimization started to be implemented in aerospace engineering in the last years. Composites optimization is not a standard procedure in aerospace, being only in current research. Article presents a case study of weight saving using topology optimization for a helicopter seat shell. The optimization process results were tested with static analysis in order to have an accurate assessment. Even weight saving were obtained, a decision for design changing has to consider also manufacturing limitations and costs. Design recommendations with comments and conclusions are given.*

Keywords: *aerospace, composites, lightweight, topology, optimization*

1. INTRODUCTION

Among the aircrafts, helicopters are more threatened to vertical crash due to vertical flight capability. Most of accidents occur at low altitude, in landing or take-off maneuvers [2]. For this reason, the survivability of crew is obtainable. One condition is to have a maximum load over the lumbar spine of 6.7 kN [9].

A crashworthy seat consists in a base structure attached to the floor and a sliding seat shell, dampened by a mechanical, hydraulic or pneumatic system. The crashworthy seat relies on an energy absorption system to dampen the movement between the two parts in a crash case scenario [1].

Paper presents a case study of a helicopter crashworthy seat lightweight optimization. The whole program investigated also struts, fracture elements and floor attachment fittings. Because the optimization techniques are not mature and the engineering teams have not sufficient experience, procedures and work methodologies are still not set up. Present case study is completed by discussions and conclusions to be integrated in a future composites optimization work methodology.

2. STRUCTURAL OPTIMIZATION

Design teams main task is to obtain the structure to withstand the certification loads for a minimum empty mass.

Many authors reported significant operational cost reduction for every pound of weight saving for aerostructures and equipments, too [6], [16]. According to Hertel, each pound of structure saving leads to a maximum take off weight saving of four pounds [5]. The lightweight is not only a factor of efficiency, one aircraft which is heavier than specified having less useful load of fuel onboard (leading to a smaller range). Following this trend, together with environment friendly policies , the technical requirements continue to ask for new methods to further decrease the weight.

In the traditional approach, the lightweight is obtained following design guidelines and previous experience, in approx. three design loops. The parts are shaped according functional requirements (aerodynamic, kinematic, ergonomic, etc) while structure is as usual arranged to be orthogonally and diagonal stiffened.

In the last 20 years, efforts were made to develop optimization algorithms and to pack them into commercial software. After 2000, software producers worked on case studies in mixed team with structure engineers to test the software capabilities. The results were non-intuitive, but the static analysis showed that the new designs were significantly improved. Thus, Bombardier reported weight savings of 10% [3] in one comparison between a classical design and the optimized one for one wing rib and a weight saving of 19% [4] for the Fairchild Dornier 728 passenger door arm. Using topology optimization for a limited amount of components, Airbus A380 is 1,000 kg lighter, while the Boeing 787 wing obtained a weight saving of 25 – 30%, by comparison with Boeing 777 [12]. In 2006 Airbus reported weight savings of 20% to 40% for A350 and A380 planes using topological optimization software [10]. Lufthansa succeeded to reduces the seats' weight by nearly 30%, making them 4.3 kg lighter than its predecessor, having a total weight of 11kg [11].

Although the topological optimization was successfully applied in the automotive industry for a considerable time, it hasn't become on the spot of the aerospace industries because of the more complex load cases, boundary conditions and stability problems like buckling [7], or aero-elastic problems which are a big concern in the aerospace [8].

From another perspective, the topology optimization can find the optimal structure using the minimum material by following the optimal load path and material distribution [12].

This paper presents a composite parts optimization method, with a case study on the seat shell of a helicopter crashworthy seat.

2.1. Stress compliance requirements

The main regulations considered in this article for the load case selection are FAR/JAR 27 for small rotorcraft, FAR/JAR 29 for transport rotorcraft and MIL STD 1290A for military seats [13], [14], [15]. A combination of them, combined with manufacturer and client requirements, led to the selection of the load cases.

According the above mentioned regulations, the selected main load cases are listed below (g = 9.81 m/s^2):

- LC1 – Down Crash, Nz = -30g
- LC2 – Up Crash, Nz = 2g
- LC3 – Side Crash, Ny = 8g
- LC4 – Forward Crash, Nx = -18.4g
- LC5 – Rear Crash, Nz = 1.5g

3. THE ANALYSED COMPONENT

The composite shell is mounted on two aluminium seat struts, being attached with four studs, two on each side (Fig. 1). The shell is 6,0 *mm* thick, being made of carbon fiber reinforced plastics with rigid foam filler as follows:

Table 1: Materials characteristics

Material	E_1 [GPa]	E_2 [GPa]	ν	F_{1Tu} [MPa]	F_{2Tu} [MPa]	ε [μ]	ρ [kg/m^3]
Standard CF UD	70	70	0.10	600	600	0.85	1600
Standard CF Fabric	135	10	0.30	1500	50	1.05	1600
Rigid Foam	0.14	-	0.69	1.15	-	-	110

All 20 plies have the thickness of 0.25 *mm,* with a stacking sequence of (45°/ -45°/ 0°/ 45°/ 0°/ -45°/ 90°/ 45°/ 0°/ -45°)$_S$. The 0° rosette orientation corresponds to Y axis of seat. The stacking sequence was designed according best practices and design methods used in composite aerostructures. The 0° plies are designated to major bending loads in the symmetry plane of the seat. The plies oriented at 90° are designated to stiffen of the seat base and back because they may behave not only as beams but as plates (membranes), too. The 90° plies increase the transverse rigidity of the 0° plies. All 0° and 90° plies are from unidirectional standard carbon fiber (UD) in order to provide the maximum rigidity to the shell. The 0° plies are placed closer to the external surface of the shell to improve bending behaviour. The +/- 45° plies insure a smooth transition between the main loaded plies and a better loads transfer. The whole plies package is balanced (same number of +45° and -45° plies) and symmetric to avoid the under load and thermal shell un-desirable deformations. The placement of plies with same orientation was avoided as much as possible considering composite curing recommendation (the best bonding is obtained for different orientation of plies). The external pair of +/- 45° plies tighten the shell, securing the whole package against delaminating of impact damage.

Loads were distributed using a mannequin. Using the free body diagram, there were calculated the loads on seat shell for each load case (Fig. 2). Loads are applied directly on shell surface or via seat belts slots or bolts.

The FE model consists of 2D shell elements (mesh size of 4 mm) and 3D hexa elements for foam volume. The boundary conditions in the studs attaching the shell to the struts have restraint the displacements along X, Y and Z axis (Fig. 3). For preprocessing and analysis, Altair Hypermesh 10.0 and Radioss were used.

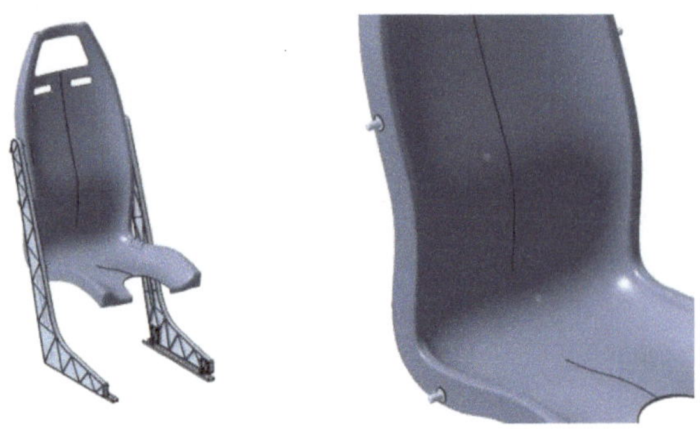

Figure 1: The crashworthy seat, with details of attachment studs

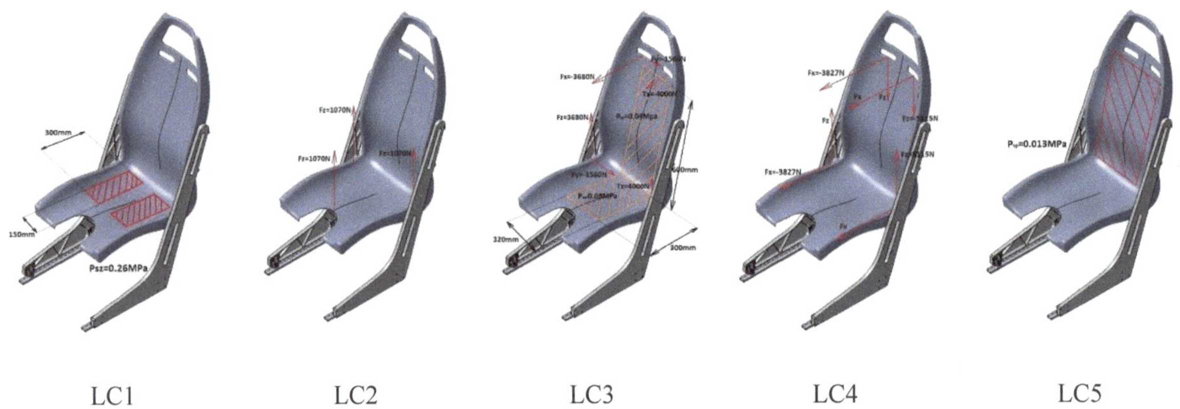

LC1 LC2 LC3 LC4 LC5

Figure 2: The forces on the seat shell corresponding to LC01 ÷ LC05 Load Cases

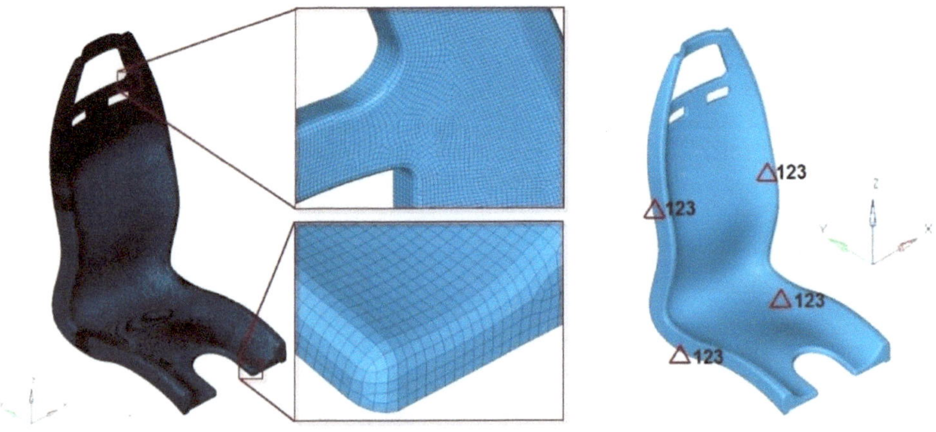

Figure 3: The FE Model and the boundary conditions

4. CASE STUDY – THE INITIAL DESIGN

The initial design (M01) consists of a constant thickness shell, in order to obtain the smallest manufacturing cost. This design was analyzed in order to identify the areas presenting stress concentrators and to have a reserve factor bigger than 2.0. There are many approaches in composites calculations: stress criteria, failure index and strain criteria. According to existing aerospace methodologies, for this application, the strain criteria (ε) was considered. The first design iteration (Fig. 4 a) indicated a hot spot stress in the joining area between seat web and the side foam longerons. For the second design iteration (M02), a web was attached to spread out the stresses concentrated in the first iteration (Fig. 4 b). The stresses were migrating, with a growth of the reserve factor values (Table 01). Adding a flange to the control stick cutout (M03 - Fig. 5), one may see the smallest value of R_F remained below the values of 2.0, even all the other load cases presented compliant values.

In order to further increase the reserve factor, a new design iteration (M04) was created, by adding six plies of UD and fabric oriented at 0°, 45° and 90°. The (M04) stacking sequence is (45°/ -45°/ **0°$_2$**/ 45°/ 0°/ -45°/ 90°/ 45°/ 0°/ -45°/ **90°/ 45°**)$_S$. The new reserve factors are bigger than 2.2, therefore, the design was considered compliant with certification requirements. The (M04) design iteration was considered the reference design. Based on it, the optimisation iterations will be further developed.

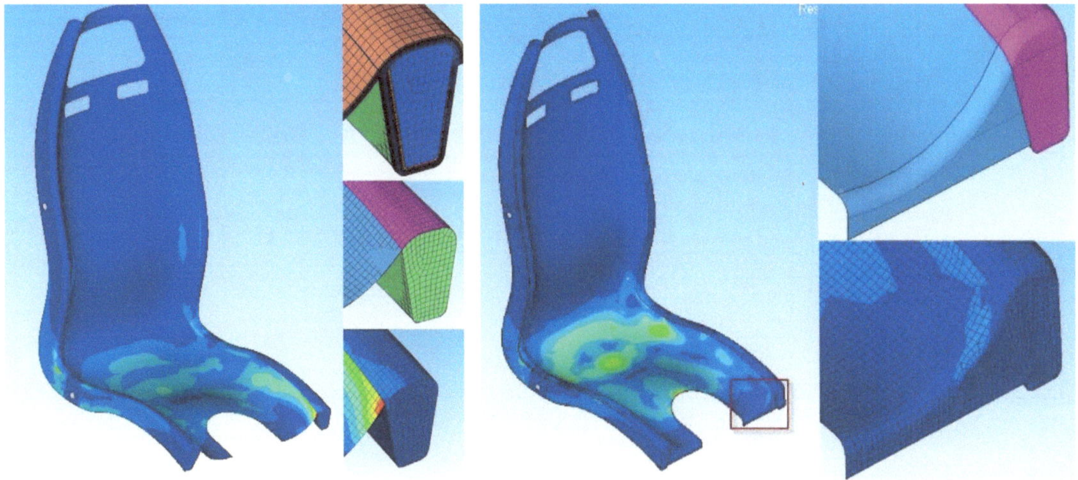

Figure 4: a) First design iteration (M01); b) The second design iteration (M02)

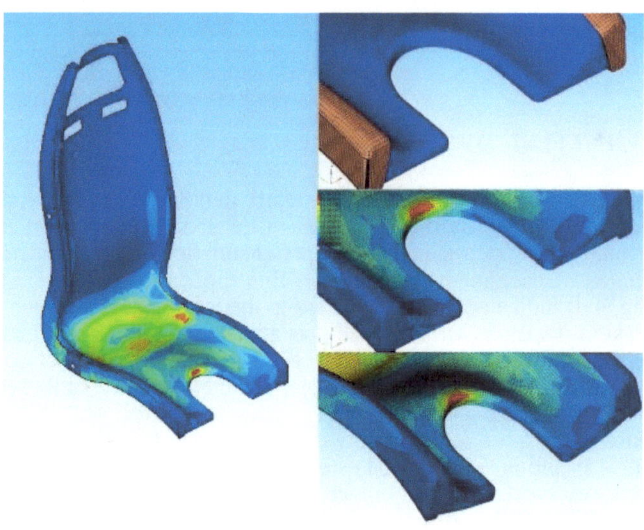

Figure 5: The third design iteration (M03)

Table 2: Strain values and reserve factors for (M01) ÷ (M04) design iterations

Iteration	LC01 Down Crash		LC02 Up Crash		LC03 Side Crash		LC04 Front Crash		LC05 Rear Crash		mass [Kg]
	ε [μm]	R_F	ε [μm]	R_F	ε [μm]	R_F	ε [μm]	R_F	ε [μm]	R_F	
M01	3084	1,2	3049	1,2	1998	1,9	1097	3,4	145	25,5	8,0
M02	1818	2,0	2538	1,5	1510	2,5	1096	3,4	145	25,5	8,2
M03	2505	1,5	574	6,4	1365	2,7	1407	2,6	264	14	8,3
M04	1668	2,2	421	8,8	987	3,7	1057	3,5	176	21,0	10,8

5. SHELL OPTIMIZATION

The designer possibility to generate a well dimensioned part is very limited because of complexity of shape (surfaces with double curvature, cutouts, etc), different points for loads introduction (seat back and base surfaces, safety belts bolts and cutouts) and of the five load cases.

Starting from the initial design, the optimization steps will allow the identification of number of plies for different areas, thickness and orientation of plies. For optimization, Altair Optistruct 12 software was used.

5.1. Free Size Optimization (FS)

The free-size optimization erase or introduce new plies depending on the response, constraints and objectives. The response is the mass, having the constraint of value being less than 6 kg. The objectives were mass and cost minimization, for stiffness maximization.

After identification of the most loaded areas, the software calculated a new configuration of plies. The number of carbon plies depends of the stress intensity on each area. The plies thickness is not controlled, being selected in a range of 0.01 to 0.40 mm.

After the first free size iteration (FS1), the areas with different number of plies follows the stress gradient, having not a regulated contour (Fig. 6). The transition from thicker to thinner areas is not smooth. With the information from this first iteration, the contour of the areas were redesigned by hand to comply also with manufacturing constraints (FS2). One may see that all the areas with thicker walls have a technological shape (square or rectangle), with a smooth transition form thicker to thinner areas (Fig. 7). The (FS2) mass increased relative to (FS1), while reserve factors increased for the majority of load cases (Table 3).

5.2. Size Optimization (Sz)

If previous Free Size Optimization step provided a quantitative solution, Size optimization allows the dimensioning of the plies. This optimization step calculates the optimum thickness of plies. In the first iteration, the proposed solution has 32 plies with thickness in a wide range of 0.070 to 0.43 mm (Fig. 8). In the second iteration, the plies thicknesses were set to values available on the market, like 0.127, 0.254 and 0.380 mm. The new solution keeps the number of plies and orientation, but with the new material thickness (Fig. 9).

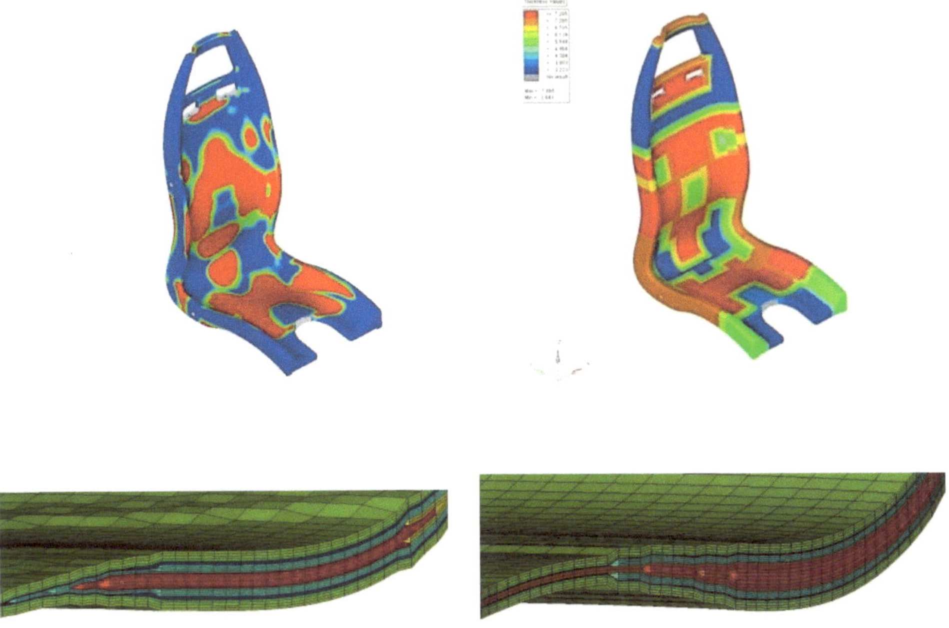

Figure 6: Free size – first iteration (FS1) **Figure 7:** Free size – second iteration (FS2)

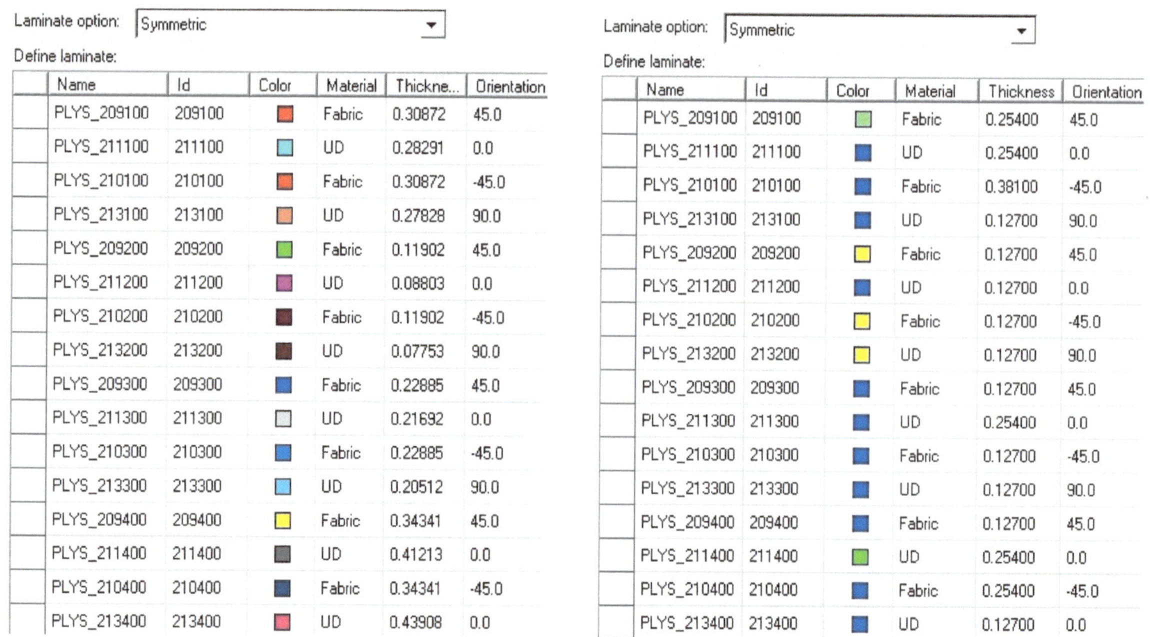

Laminate option: Symmetric

Define laminate:

Name	Id	Color	Material	Thickne...	Orientation
PLYS_209100	209100		Fabric	0.30872	45.0
PLYS_211100	211100		UD	0.28291	0.0
PLYS_210100	210100		Fabric	0.30872	-45.0
PLYS_213100	213100		UD	0.27828	90.0
PLYS_209200	209200		Fabric	0.11902	45.0
PLYS_211200	211200		UD	0.08803	0.0
PLYS_210200	210200		Fabric	0.11902	-45.0
PLYS_213200	213200		UD	0.07753	90.0
PLYS_209300	209300		Fabric	0.22885	45.0
PLYS_211300	211300		UD	0.21692	0.0
PLYS_210300	210300		Fabric	0.22885	-45.0
PLYS_213300	213300		UD	0.20512	90.0
PLYS_209400	209400		Fabric	0.34341	45.0
PLYS_211400	211400		UD	0.41213	0.0
PLYS_210400	210400		Fabric	0.34341	-45.0
PLYS_213400	213400		UD	0.43908	0.0

Figure 8: Size optimization – first iteration

Laminate option: Symmetric

Define laminate:

Name	Id	Color	Material	Thickness	Orientation
PLYS_209100	209100		Fabric	0.25400	45.0
PLYS_211100	211100		UD	0.25400	0.0
PLYS_210100	210100		Fabric	0.38100	-45.0
PLYS_213100	213100		UD	0.12700	90.0
PLYS_209200	209200		Fabric	0.12700	45.0
PLYS_211200	211200		UD	0.12700	0.0
PLYS_210200	210200		Fabric	0.12700	-45.0
PLYS_213200	213200		UD	0.12700	90.0
PLYS_209300	209300		Fabric	0.12700	45.0
PLYS_211300	211300		UD	0.25400	0.0
PLYS_210300	210300		Fabric	0.12700	-45.0
PLYS_213300	213300		UD	0.12700	90.0
PLYS_209400	209400		Fabric	0.12700	45.0
PLYS_211400	211400		UD	0.25400	0.0
PLYS_210400	210400		Fabric	0.25400	-45.0
PLYS_213400	213400		UD	0.12700	0.0

Figure 9: Size optimization – second iteration

5.3. *Shuffle Optimization (Sh)*

The last step of optimization calculates the optimal orientation of plies. Figure 10 presents plies stack-up from Size and Shuffle steps. One may see that, after Shuffle Optimization step, the first two outer layers are +45° and -45° as in the initial design. Layers of 90° were pushed next to the symmetry plane, being kept only four. This indicates the prevailing loads path are due to bending (along the shell), while membrane loads are only secondary (transverse to the shell).

Within this iteration step, because the reserve factors for all load cases have values superior to 2.0, plies of 0.38 *mm* thickness were replaced by 0.254 *mm* with the same material and orientation. Final reserve factors are presented in Table 3.

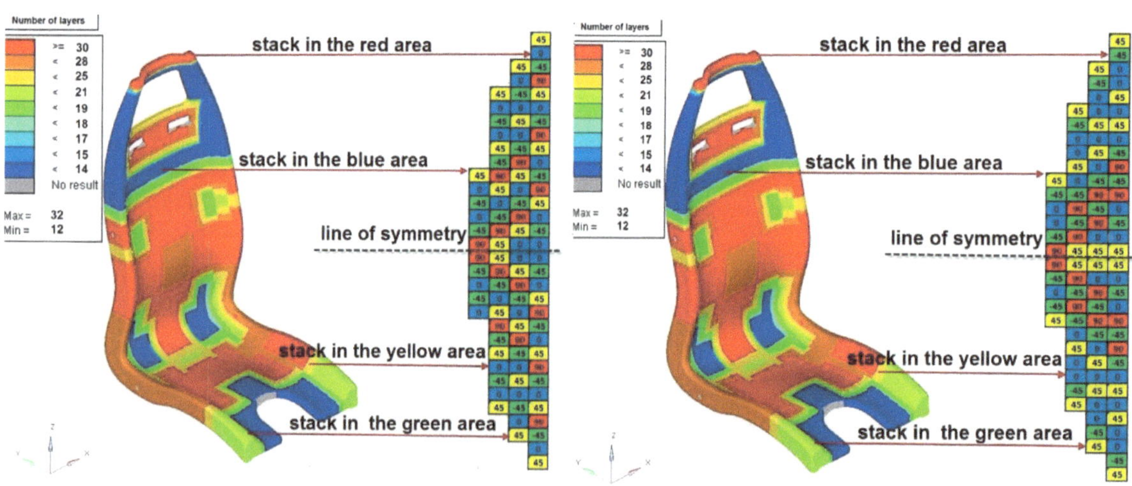

Figure 10: Plies orientation after Size and Shuffle optimization steps

Table 3: Strain values and reserve factors for optimizations steps

Iteration	LC01 Down Crash		LC02 Up Crash		LC03 Side Crash		LC04 Front Crash		LC05 Rear Crash		mass [Kg]
	ε [μm]	R_F	ε [μm]	R_F	ε [μm]	R_F	ε [μm]	R_F	ε [μm]	R_F	
M04	1668	2,2	421	8,8	987	3,7	1057	3,5	176	21,0	9.9
FS1	1849	2,0	1112	3,3	2221	1,7	2729	1,4	2946	1,3	7,8
FS2	1376	2,7	1734	2,1	1729	2,1	1682	2,2	180	21	7,9
Sz	1598	2,3	1381	2,7	1509	2,5	1691	2,2	211	18	7,7
Sh	1580	2,3	1474	2,5	1185	3,1	1671	2,2	235	16	7,6

6. DISCUSSIONS

The evolution of reserve factors along the optimization process is presented in Figure 11. One may see the reserve factors of initial design were too high, with an average of 8.2 which is far form the target of 2.0. After the first step of optimization (iteration FS1), the reserve factors were below 2.0 in three load cases, therefore, solution is considered non-acceptable, even the weight decreased from 9.9 *kg* (initial design) to 7.8 *kg* (Fig. 12). After the second free size iteration (FS2), only for a 0.1 *kg* added weight, the reserve factors are bigger than 2.0 for all load cases. From the mass point of view, FS2 iteration is 20% lighter, with an average reserve factor of 5.9. Therefore only after free size optimization the progress is significant. The results are still inconsistent because of random thickness of plies.

Continuing with size optimization, the average reserve factor decreases to 5.4, for a mass of 7.7 *kg*. The last optimization step gives a final improvement of 5.2 average reserve factor for a mass of 7.6 *kg*. Thus, versus initial

design (M04), the final step optimized design (Sh) offers a 5.2 average reserve factor versus 8.2, for 23% weight saving.

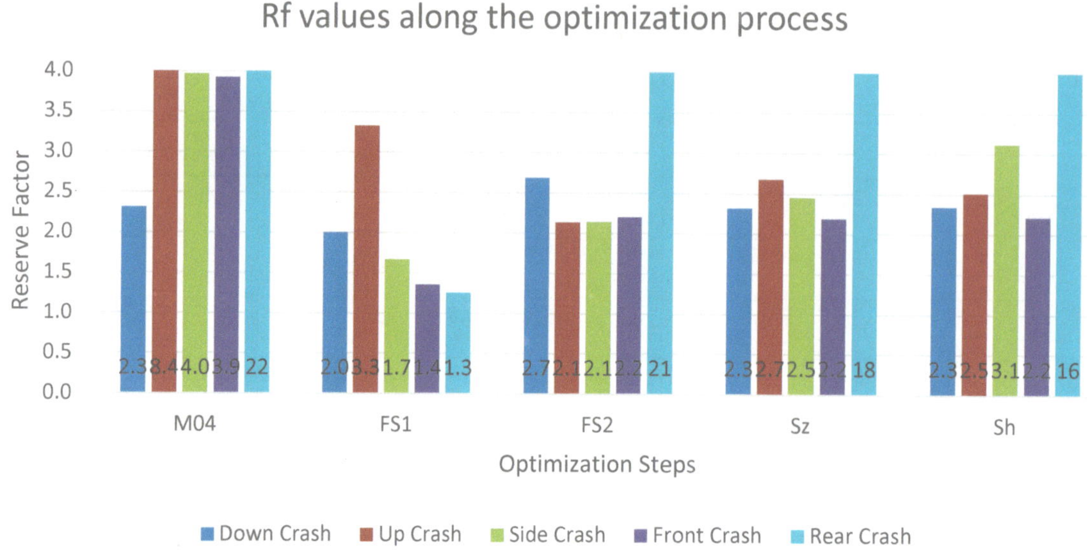

Figure 11: The reserve factors along the optimization process

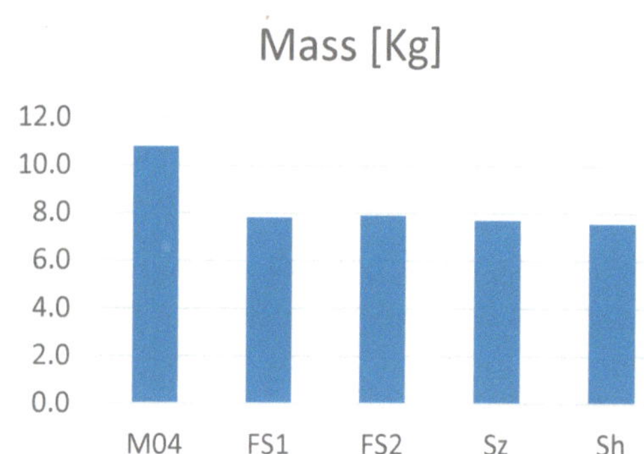

Figure 12: Structural mass along the optimization process

The reserve factor average in all optimization steps present average values far from 2.0 because the stresses are small in load case LC05 (Fig. 13). For a better characterization of the average R_F, the average for LC01 ÷ LC04 (Av* R_F) is given. The values for Av* are 4.6 for initial design (oversized) , while for last optimization steps it is within the range of 2.3 ÷ 2.5, indicating a good sizing of the structural layout.

The reserve factor is one parameter characterizing the compliance with certification requirements. Structural mass is a characteristic of the economic efficiency of the operational flight. An ideal structure will have the minimal mass presenting R_F = 2.0 for all load cases. Because the complexity and diversity of load cases, this model is far from being obtainable. In order to have a global characterization of optimization steps, a structural efficiency index is defined as follows:

$$\xi = [\, m \, x \, (Av \, R_F \, - 2.0 \,) \,]^{-1}$$

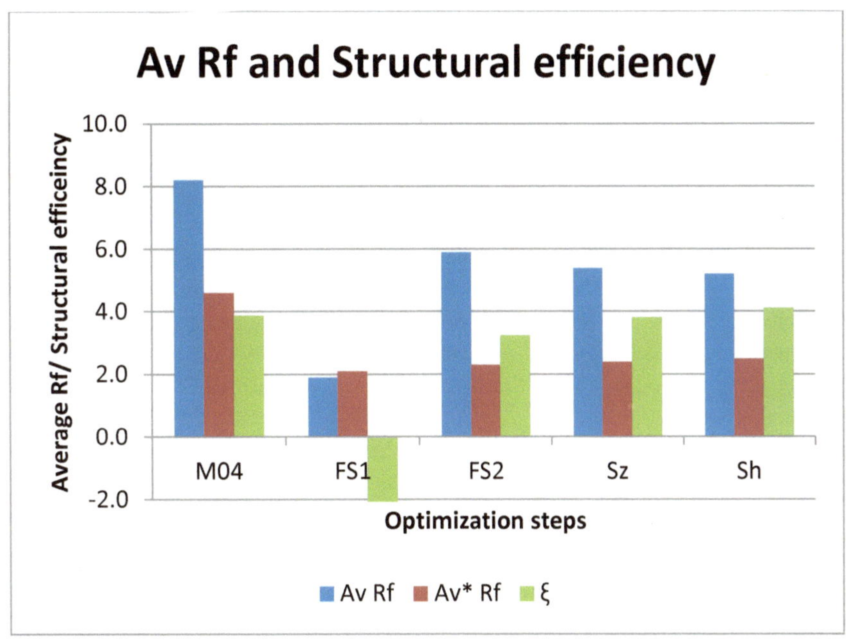

Figure 13: The average reserve factor and structural efficiency index along optimization steps

Figure 13 presents the values of structural efficiency index (ξ). Starting with 3.9 form initial design (M04), it has negative values for (FS1) (because of Average R_F = 1.9). For last optimization steps, one may see ξ increased up to 4.1.

The above results were obtained after much more design and optimization steps, therefore the process was not a smooth one. For additional weight saving, a slight improvement may be possible by decreasing in (Sh) step the reserve factor from 2.2 to 2.0 (LC04 front crash). Another possible improvement may be obtained by redesign of plies areas in (FS2) step.

For further improvements, replacing the high loaded areas with sandwich with foam core may be an option.

7. CONCLUSIONS

Paper presented a case study of weight optimization of a composite shell, starting from an initial design compliant with the certification requirements. The materials and design rules used for stake-up definition are according the aerostructures best practices, taking into consideration stress, fatigue, damage tolerance, thermal deformation and manufacturing constraints and recommendations.

Using the relative displacement (ε) criteria, three optimization steps were made, for every step the reserve factors and mass being registered.

First optimization step (Free Size) gave the most loaded areas, by grouping the plies according loads resulting the biggest weight saving. The second optimization step (Size) offered to the designer a real solution, all plies having standardised thicknesses. The third optimization step (Shuffle) may be considered optional. Even the weight saving is theoretically zero, the reserve factors became more balanced, indicating a better using of material. In the end, an improved design was obtained, with reserve factors in the range of 2.2 ÷ 2.5 for three load cases, with 23% weight saving.

Because along the Free Size step the designer has to manually adjust the plies contour, this leads to a dependency of the engineer's skills, being also a time consuming process.

One major factor to be considered is the manufacturing costs. If the initial design has only all around plies, the final optimization step presents a complicated map of areas with different thickness and slopes, leading to the need of an accurate configuration and production tooling management for proper plies orientation and placement. These costs may be partially compensated by the economy of material (at least 20%) and the grouping of the plies flat pattern in order to minimise the raw material lose.

Acknowledgements

Author Gabriel Dima. This paper is supported by the Sectoral Operational Programme Human Resources Development (SOP HRD), ID134378 financed from the European Social Fund and by the Romanian Government.

REFERENCES

[1] Anghileri M, et al., Multi-Objective genetic optimization of helicopter seats under crashworthiness requirements, Italy, 2004.

[2] Bolukbasi A, et al. Full Spectrum Crashworthiness Criteria for Rotorcraft, RDECOM TR 12-D-12, US Army Research, Development & Engineering Command, Fort Eustis, 2011.

[3] Buchanan S. Development of a Wing Box Rib for a Passenger Jet Aircraft using Design Optimization and Constrained to Traditional Design and Manufacture Requirements", CAE Technology Conference/ Altair Engineering, 2007.

[4] Cervellera P. Optimising Driving Design Process: Practical Experience on Structural Components. Proc. 14th Convegno Nazionale ADM/AIAS, Bari, 2004.

[5] Hertel H. Leichtbau., Bauelemente, Bemessungen und Konstructionen von Flugzeugen und anderen Leichtbauwerken, Springer, New York, 1980.

[6] Kaufmann M. Cost/ Weight Optimization of Aircraft Structures, KTH School of Engineering Sciences, Stockholm, 2008.

[7] Krog L, Tucker A, Rollema G. Application of Topology, Sizing and Shape Optimization Methods to Optimal design of Aircraft Components, Airbus UK Ltd, 2002,Bristol.

[8] Maute K, Allen M. Conceptual design of aero-elastic structures by topology optimization, Structural and Multidisciplinary Optimization, vol.27, no.1-2, pp. 27 – 42, 2004.

[9] H Nauman, Abdul L. Analysis and Optimization of a Crashworthy Helicopter Seat, The University of Texas at Arlington, May 2008.

[10] Schuhmacher G. Optimizing Aircraft Structures, Concept to Reality, Altair Engineering, 2006.

[11] Gubisch M. New Lufthansa seat saves nearly 30% in weight, Flightglobal, 2010, flightglobal.com

[12] Wang Q, Liu ZZ, Gea HC. New topology optimization method for wing leading-edge rib, Journal of Aircraft,vol.48, no. 5, 2011

[13] * * *, JAR/ FAR Part 27 – Airworthiness Standard: Normal category rotorcraft, Federal Aviation Administration, 2013

[14] * * *, JAR/ FAR Part 29 – Airworthiness Standard: Transport category rotorcraft, Federal Aviation Administration, 2013

[15] * * *, MIL-STD-1290A, Military Standard: Light Fixed & Rotary Wing Aircraft Crash Resistance, 2006

[16] * * *, What's a few pounds here and there?, International Conference Innovative Aircraft Seating Proceedings, Hamburg, 2011

THE FINITE ELEMENTS ANALYSIS VS. EXPERIMENTAL BUCKLING BEHAVIOR OF THE THINWALLED PLANE GUSSETS

G. Dima[1], H. Teodorescu- Drăghicescu[2]

[1]Transylvania University, Brasov, ROMANIA, dumitru.dima@unitbv.ro
[2]Transylvania University, Brasov, ROMANIA, draghicescu.teodorescu@unitbv.ro

Abstract: *The lightweight tubular latticed beams are used in aerospace structures as well as in civil engineering applications. Depending of manufacturer, gussets may be used to increase the stiffness and rigidity of the beams. As a continuation of previous researches to obtain lighter structures by using gussets, this article presents the results of a research regarding the plane gussets buckling. Using analytical, finite elements calculation and experimental testing, the buckling behavior of tubular T joints reinforced with gussets was studied. The experimental testing was made for gussets subjected to both compression and tensile loads. The gusset buckling was analyzed together with the joint members failure, conclusions and design recommendation being made.*

Keywords: *aerospace, buckling, welded structures, gusset, finite element analysis*

1. INTRODUCTION

Aircraft structures has to comply with two major contradictory constraints: to withstand to in flight and landing loads, for a minimum weight. Being used from the beginning of the aviation industry, latticed beams was one of the most extensive employed structure on aircraft because its functionality, reliability and low manufacturing costs. Many materials, joining techniques and layouts were used up to 1935 – 1940, when monocoque structures finally demonstrated their superiority, for affordable manufacturing costs.

Loads carrying structure whose failure leads to a catastrophic event is called primary structure. Latticed beams were employed as primary structure for wing and fuselage, having circular section tubes as members. Under the constraint of the minimum weight, all structures are thinwalled, one of the main failure mode being the lose of stability.

In order to decrease the structural weight, previous researches were done to replace the beams with bracing (Pratt and Warren) with beams free of bracing (Vierendeel) [6]. In order to decrease the stress level in joints area and to increase rigidity, the Vierendeel beam were reinforced with gussets. Even from static, dynamic and fatigue point of view, the effect of adding gussets was studied, the local buckling of gussets remained a concern.

This article presents the results of a research regarding the tubular latticed beam gussets. Theoretical and experimental researches were done on joints reinforced with gussets. Both theoretical and experimental studies conducted to conclusion that the gusset buckling occurs only after the failure of beam members.

2. BEAMS REINFORCED WITH GUSSETS

In the design of aircraft welded structures, the main task is to obtain a structure presenting details with the lowest stress concentration factor (SCF). This target allows longer fatigue life and lighter structures.

The tubular welded structures are met in automotive and civil engineering as well. In aerospace, the verification of the structures consists in weld seams static assessment and members buckling. For civil engineering, due to the dimension of the structures and members, the stress concentration factors are calculated with more accuracy, considering also stress distribution in joint area.

In literature there are presented a lot of solutions to decrease the stress concentration factors (SCF) as inner diaphragm, external collar [1], chord doublers [2], simple or double gussets [3], [4]. Other solutions consist in tubular structures with formed end tubes for brace, longitudinally or transversal placed [2], [5] and [8].

There is no unified theory of gussets shape and placement available, a lot of solutions being met on different aircrafts and manufacturers, gussets being external attached or inserted in joint members [9], [10].

In civil engineering (crane booms, offshore platform, masts) gussets are used in *T* or *K* joints to decrease the stress concentration factors. Reference [11] presents significant stress reduction in *K* joints using gussets inserted in the plane of joint, for axial load, in plane and out of plane bending. In reference [12] there are presented the effects of using gussets vs. the reinforcement of joint with chord doublers and brace collar, finding that the gussets provide a lower stress level in joint.

Recent researched were done to asses different shapes and placement of double and simple gussets [7]. Figure 1 presents few types of the analyzed gussets. In figure 2, one may see gusset *D*, which is very usual, has the lowest critical buckling load. Therefore, if gusset *D* does not lose stability under beam loading, therefore all other gusset shapes will be buckling proof.

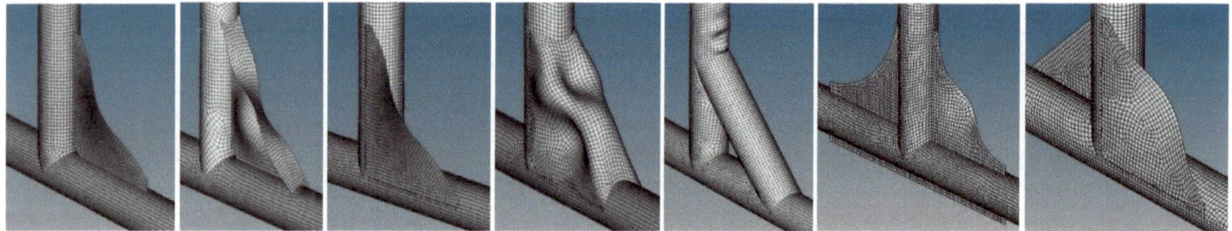

Figure 1: Different shapes of buckling analyzed gussets (*A/ B Fl/ D/ E/ G/ O/ S*) [7]

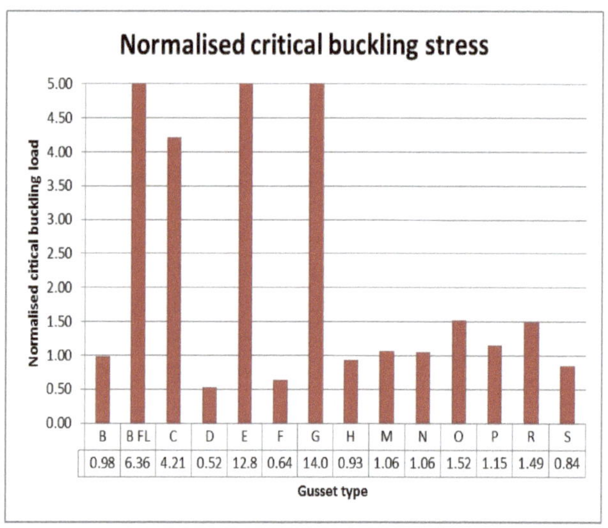

Figure 2: Different gussets buckling assessment [7]

3. THE FEA VS. ANALITICAL RESULTS OF TRIANGULAR PLATE SUBJECTED TO COMPRESSION

According to experimental studies from [13], the elastic buckling stress for a triangular gusset (Fig. 3. a) is given by:

$$f_b = K \cdot E \cdot (t/w)^2,$$

with *K* factor related to boundary conditions of edges as follows (*w/v = 1*):
K = 3.50 for fixed edges (embedded) and

$K = 0.52$ for simple supported edge (pinned).

In order to asses numerical finite elements (FE) solutions vs. analytical calculations, a FE model was generated. Plate dimensions are in a range of 30 x 30 ÷ 110 x 110 *mm*, with the wall thickness of 0.5 ÷ 2.5 *mm*. The material is low alloyed steel, with ultimate tensile strength of 980 – 1080 *MPa*, Poisson ratio of 0.3 and Yield Modulus E = 2.1E5 *MPa*. All boundary conditions are corresponding to the fixed edge condition (Fig. 3).

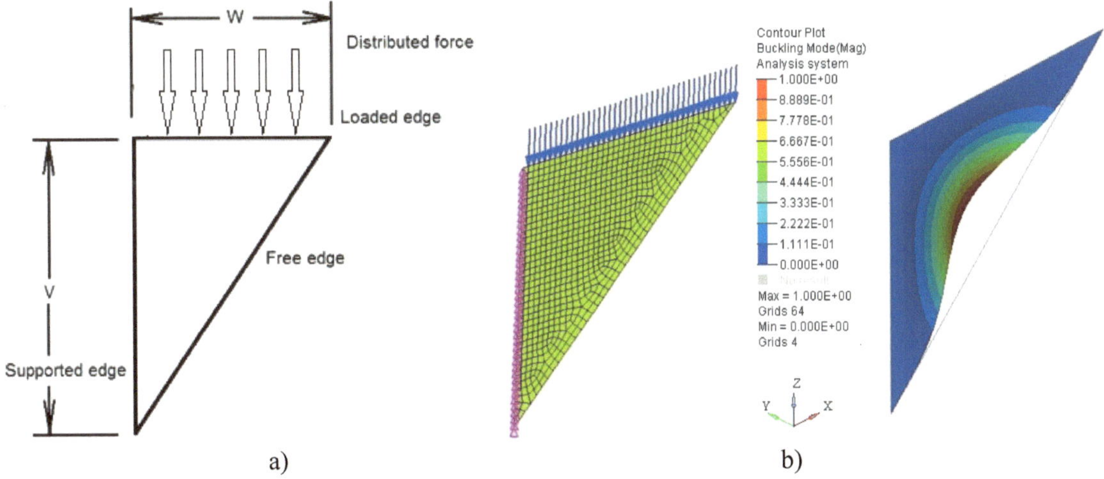

Figure 3: a) Triangular plate planar loaded [13]; b) The FE model and the deformed plate [7]

The analytic calculations used the fixed edge hypothesis, corresponding to the welded attachment of the gusset to the vertical member of joint. A commercial software solution was used (Altair Hypermesh/ Radios 10). The FE model consists of a 2D shell mesh, with the mesh size of 2.0 *mm*. Weld geometry was not taken into consideration. The FE model has all degree of freedom restrained on vertical edge and all degree of freedom unconstrained for the horizontal edge. Vertical load was distributed on horizontal edge's nodes of the model.

With results obtain from FE analyses and analytical calculation according [13], graphs for geometrical parameters over triangular plate stability behavior were raised. For different wall thickness, the differences are in the range of 12 ÷ 14% (Fig. 4. a). For different plate dimensions, the gaps are in a range of 10 ÷ 12% (Fig. 4. b).

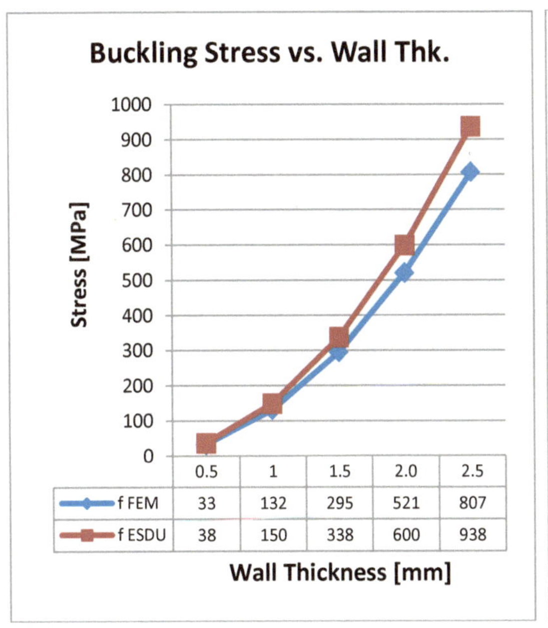

	0.5	1	1.5	2.0	2.5
f FEM	33	132	295	521	807
f ESDU	38	150	338	600	938

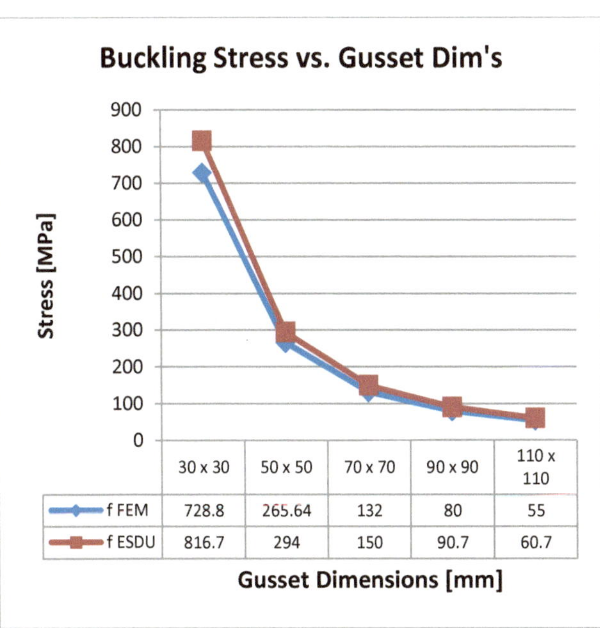

	30 x 30	50 x 50	70 x 70	90 x 90	110 x 110
f FEM	728.8	265.64	132	80	55
f ESDU	816.7	294	150	90.7	60.7

Figure 4: Analytical vs. FE calculated buckling stress for different: a) Plate wall thickness; b) Plate dimensions [7]

In figure 5 one may see the FE vs. analytical critical buckling force for different gusset dimensions ratio. For ratio of 1.0, the first buckling mode occurs. For ratio values different of 1.0, gussets buckling correspond to one of superior buckling mode [12]. Thus, for 0.4, 0.5 and 1.2 ratios, the second buckling mode was considered. For ratio of 2.0 and 2.5. the third and the fifth buckling modes were considered respectively. For up to 1.5 ratio value, the differences are below 12%, while for other ratio values, the differences are growing up to 22%.

Considering all above, one may formulate the conclusion that FE and analytical calculation of critical buckling stress presents good similarities.

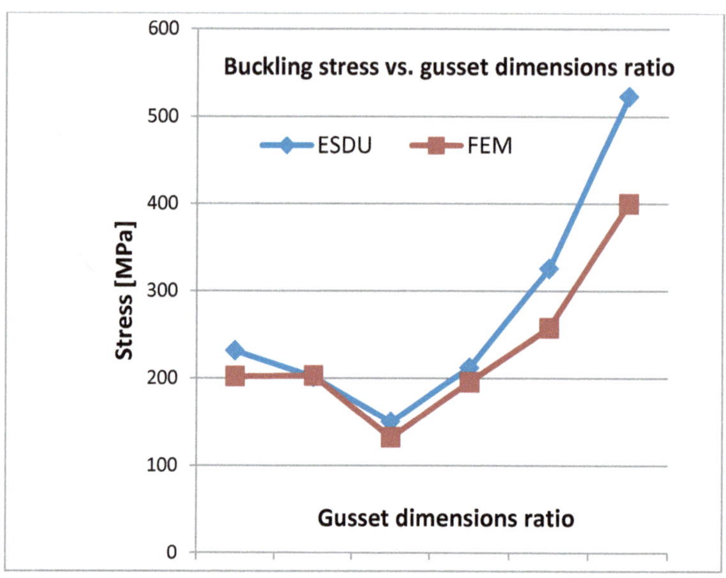

Figure 5: a) Analytical vs. FE calculated buckling stress for different gusset dimensions ratio

4. THE FEA VS. ANALITICAL RESULTS OF A JOINT GUSSET SAMPLE SUBJECTED TO COMPRESSION

For experimental results, a sample made of steel tubes with 22 *mm* diameter and 2 *mm* wall thickness was considered. Dimensions of gusset are 80 x 80 x 2.0 *mm*, gusset being tangent placed to the joint members (Fig. 6).

According to graphs shown in figure 4 b, for a 80 x 80 x 1.0 *mm* gusset (made from OL37 STAS 530/1) the critical buckling stress is 120 *MPa*. In figure 4 a, for a 2.0 *mm* gusset, the critical buckling load is four times bigger (also the inertia momentum is four time bigger), therefore, the critical buckling stress for the 80 x 80 x 2.0 *mm* gusset will be 120 x 4 = 480 *MPa.* Considering that the allowable stress values for OL37 steel are σ_{tu} = 370 *MPa* / σ_{ty} = 240 *MPa*, the conclusion is the joint members will fail before the gusset buckling.

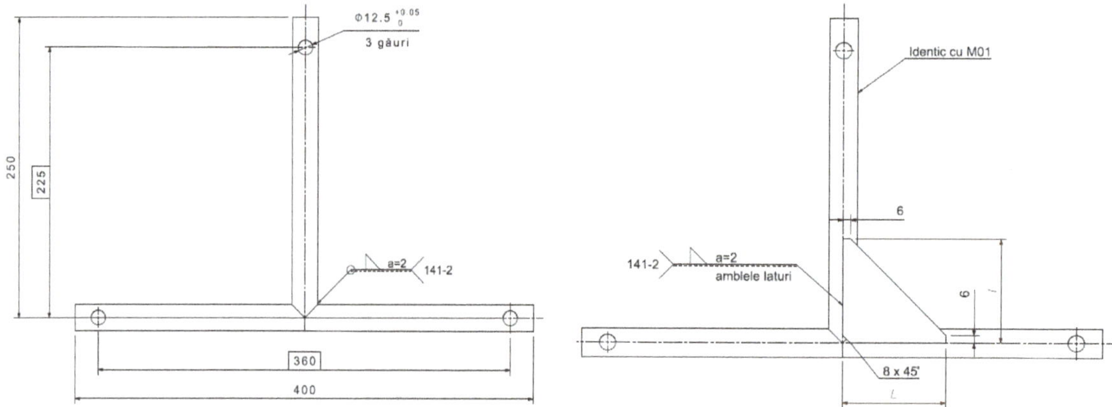

Figure 6: The dimensions of the joint members and the gusset sample

The FE analysis revealed a 0.97 *kN* bending load for a value of 240 *MPa* for the Von Misses stress in joint members (Fig. 7 a). This load, generates a 220 *MPa* von Misses stress in gusset, which is far smaller than the previous calculated 480 *MPa* critical buckling stress. Therefore, the joint's members plastic failure will occur before the gusset buckling. The FE analysis loads are 7.5 *kN* for the first buckling mode, and 9.8 *kN* for the second buckling mode respectively (Fig. 7 b & c). The conclusion is analytical calculations and FE analysis lead to the same conclusion related to static vs. buckling behavior of the joint gusset sample.

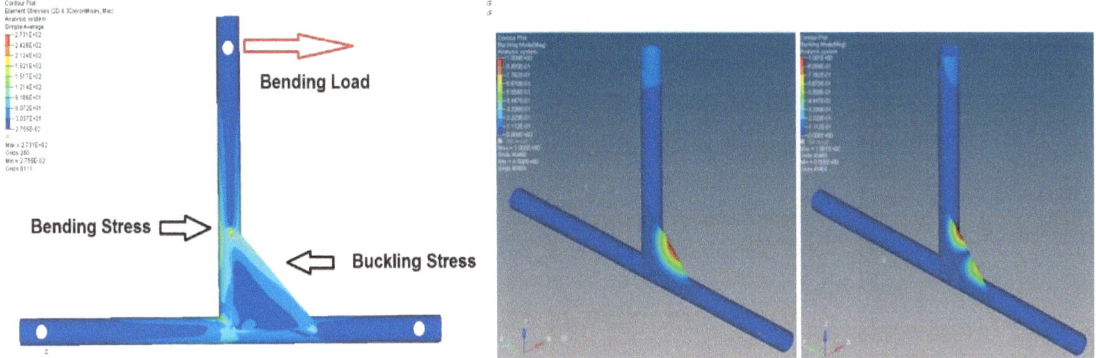

Figure 7: FE analysis results: a) static stress distribution; b) first, and c) the second buckling mode

5. THE EXPERIMENTAL TESTING

For experimental testing four samples were considered (Fig. 8 a). In order to assess the compression vs. tensile behavior of the joint gusset, two samples were subjected to tensile and two to compression loading. For mounting in an universal testing machine, an interface was used (Fig. 8 b, c). For compression/ tension loading of sample, same interface was used with different mounts of sample (gusset upward and downward, respectively). All tests were conducted up to the total failure of the samples. All the deformations phases and failure were registered.

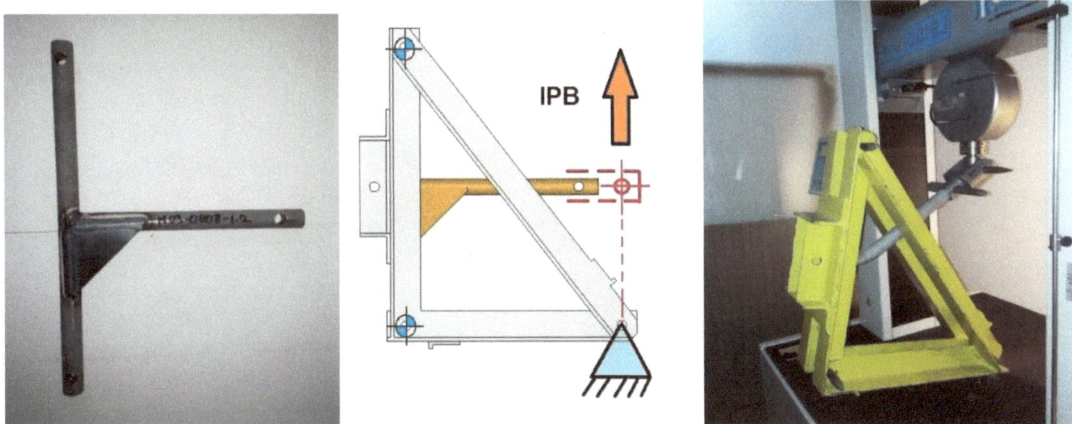

Figure 8: a) Test sample; b) Loading diagram c) Machine interface (yellow)

The tensile loaded gusset samples presented the initial deformation of the vertical member in the vicinity of gusset. Only after the total deformation of the tube section, the gusset free edge started to stretch. The second fail was in the upper margin of the gusset weld seam. After that, came the complete failure of the weld seam, the joint vertical and horizontal members coming to close contact (Fig. 9).

The compressed gusset sample presented the initial deformation of the vertical member in the same area as the stretched gusset. After vertical member complete deformation, the gusset started to lose stability until the vertical and horizontal tubes of the joint come to close contact (Fig. 10).

Therefore, both compression and tensile samples failed in the same manner, the tubes being less strength than the gusset. it is important to emphasize that the compression gusset welds remained intact even after total failure of joints, while for tension gusset the weld seams presented a total failure (Fig. 11 – note one tensile sample were not conducted to total failure for the better identification of the fracture initiation).

Figure 9: Joint gusset sample tensile testing

Figure 10: Joint gusset sample compression testing

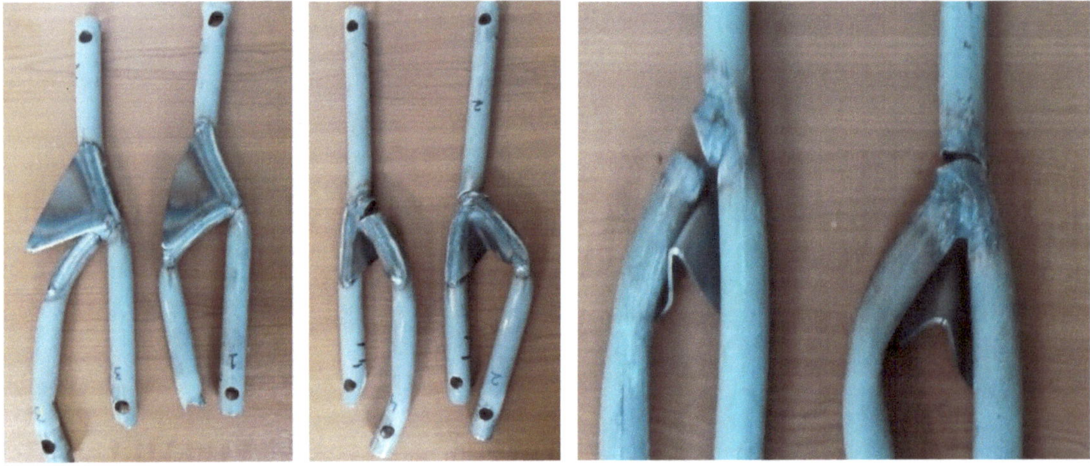

Figure 11: Tensile and compression subjected joint gusset samples

6. EXPERIMENTAL RESULTS & DISCUSSIONS

All Load/ Deformation graphs were similar, with a long plastic range, corresponding to vertical tube rotation (Fig. 12 a). Curves are very similar for all four samples (Fig. 12 b) presenting the yield stress of 1.0 *kN* for three samples and 0.8 *kN* for one sample. The bending moment will be:

$$M_b = F_b \, x \, b = 1.0 \, kN \times 145 \, mm = 1.45 \, E^5 \, Nmm$$

The inertia modulus of the vertical tube is:

$$W_y = (\pi/32D) \times (D^4 - d^4) = 576 \, mm^3$$

Therefore, the bending tensile at the vertical tube contact with the gusset will be:

$$\sigma_b = M_b / W_y = 252 \, MPa$$

From figure 12, the buckling load is within the range of 7.5 ÷ 8.0 *kN*.

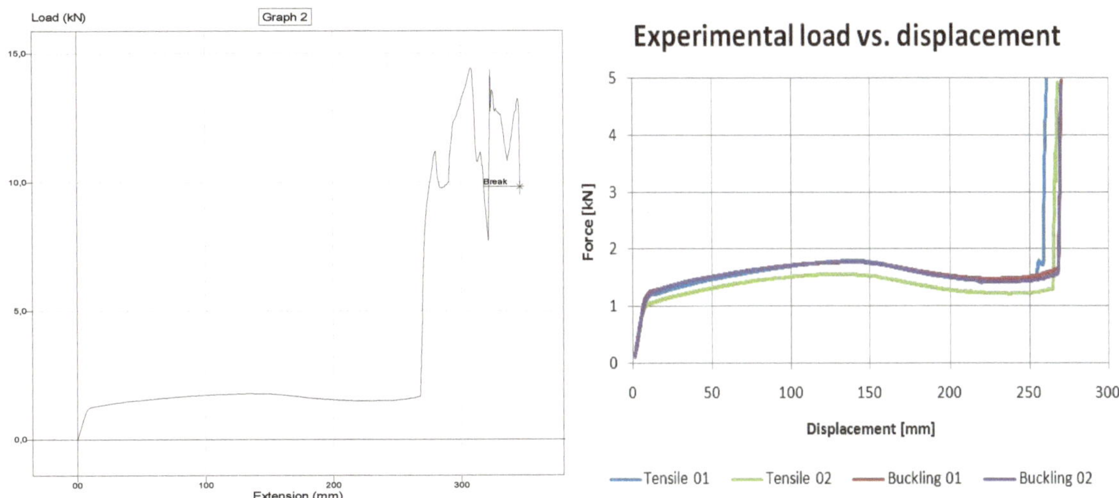

Figure 12: Load/ deformation graphs for all samples

7. CONCLUSIONS

Paper presented a numerical and experimental assessment of planar gussets used to increase the stiffness of the lightweight tubular latticed beams. The main conclusion of this research are:

- The finite elements analysis results comply with the ESDU theory for the triangular plate buckling.
- The experimental, FE analysis and analytic results are very similar.
- For a gusset with the same wall thickness as tubes, buckling occurs only after joint's members failure. A research worth to be continued is to consider thinner gussets, in order to determine the thickness where the gusset will fail very close to the failure moment of the joint tube. This gusset thickness will be limited by manufacturing considerations (welding assembly is reccomended between parts with thickness ratio of maximum 2:1).
- The buckled gusset welds are more stiff than those of the tensile subjected gusset. This conclusion is important for structures calculated to preserve a certain structural integrity after failure (for instance those who need to preserve an minimum inner volume).
- For structural limit integrity, compression gussets are more recommended than tension gussets. Even this is not an intuitive behavior, this conclusion was demonstrated by experimental testing
- The gussets may increase stiffness of latticed beams with minimal added weight, without the risk of buckling. Therefore, for lightweight reasons, the Pratt beam could be replaced by the Vierendeel beam just adding corner gussets.

Actual and future research are focused to the dimensions, shape and placement of corner gussets to decrease the stress concentration factor in joints.

Acknowledgements

Author Gabriel Dima. This paper is supported by the Sectoral Operational Programme Human Resources Development (SOP HRD), ID134378 financed from the European Social Fund and by the Romanian Government.

REFERENCES

O. W. Blodgett, *Design of Welded Structures*, The James F. Lincoln Arc Welding Foundation, 1976.

O. W. Blodgett, *Using gussets and other stiffeners correctly*, weldingdesign.com, 2005.

E. F. Bruhn, *Analysis and design of flight vehicle structures*, Tri-State Offset Company, 1973.

J. J. Cao, et al., *Design Guidelines for Longitudinal Plate to HSS Connections*, Journal of Structural Engineering, 1998, pp. 764-791.

C.C. Chou, P. J. Chen, *Compressive behaviour of central gusset plate connections for a buckling restrained braced frame*, Journal of Constructional Steel Research No. 65, 2009, pp. 1138-1148.

G. Dima, *Elastic Buckling Behaviour of Aerospace CHS Gusseted "T" Connections,* Transactions of FAMENA, 2014, XXXVIII-2, pp. 67-76

G. Dima, I. C. Rosca, I. Balcu, *The Influence of Corner Gussets over the Lightweight Tubular Latticed Beams,* Interdisciplinarity in Engineering INTER ENG 2014 Proceedings, Tg Mures

S. K. Duggal, *Design of Steel Structures*, Tata McGraw Hill, New Delhi, 2009.

Eurocode 3, Part 1.8. *Design of Joints*, CEN, 2002.

I. Grosu, *Calculul si constructia avionului*, Editura didactica si pedagogica, Bucuresti, 1965.

Y. Kurobane, et al., *Design guide for structural hollow section column connections* – CIDECT/ TUV Verlag, 2004.

L. H. Martin, J. A. Purkiss, *Structural Design of Steelwork*, Butterworth-Heinemann, Oxford, 2008.

* * *, *Avoidance of buckling of some engineering elements (struts, plates and gussets)*, Engineering Science Data Unit, ESDU 88034, 2000.

STEADY PRECESSION OF A ROLLING DISK OR RING

G.Deliu[1], D.Botezatu[2]

[1]Transilvania University, Braşov, ROMANIA, deliumec70@gmail.com

[2]Transilvania University, Braşov, ROMANIA, dan.botezatu@gmail.com

Abstract): *The paper presents the deduction of motion equations for any circular body moving in a turn, with application to the particular case of a monocycle vehicle. In this sens, the authors start from the simplest case of steady turning rolling disk. Inspite of the impression of over simplified model, the results of this paper may be largely applied to the study of the motion of an actual monocicle vehicle[1]. The most important conclusion is that such a vehicle is more stable in the turn than some might think.*

Keywords: *steady 1, motion 2, fixed point 3, precession 4, stability 5*

1. INTRODUCTION

The present paper deals with the equations describing a particular case of motion, namely the steady precession of a free rolling body. By steady precession [2], we mean a rotation about a fixed point, having the following particularities: the nutation angle θ = ct., the magnitude of the precession angular velocity $|\dot{\psi}| = |\omega_1|$ = ct., the magnitude of the spin angular velocity $|\dot{\varphi}| = |\omega_0|$ = ct.(Fig.1).

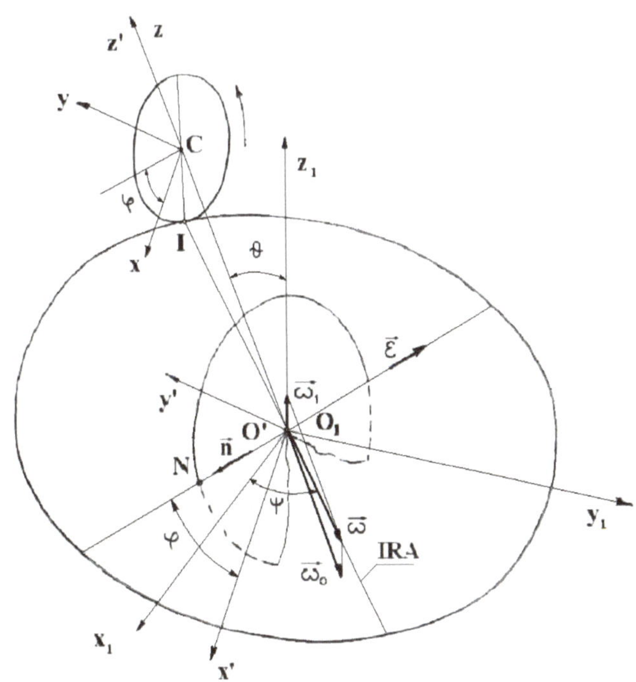

Figure 1: Steady Precession of a Disk

2. REFERENCE FRAMES AND NOTATIONS

To describe the motion of the circular body (disk or ring), we have to use three **reference frames** (Fig.1):

R_1 - *the space frame* $O_1x_1y_1z_1$, fixed to the ground, and having the O_1z_1 – axis vertical upward;

R - *the body frame* $C\,x\,y\,z$, bound to the moving body, having the origin C in the centre of mass of the body and the $C\,z$ – axis passing by the fixed point O_1;

R' – *the reflected frame* $O'x'y'z'$, parallel to the body frame **R**, but having the origin O' over to O_1.

Notations:

I – contact point between disk and ground; IRA – instantaneous rotation axis; φ - own rotation angle (spin angle); ψ - precession angle; θ - nutation angle; $\vec{\omega}$ - instantaneous (resultant) angular velocity vector; $\vec{\omega}_0$ - spin velocity vector; $\vec{\omega}_1$ - precession velocity vector; $\vec{\varepsilon}$ - angular acceleration vector; $\vec{n}$ - unit vector of the line of nodes.

3. KINEMATIC RELATIONS

3.1. Velocities

As it is well known, a motion about a fixed point may be interpreted as a rolling without slippage of two surfaces one over another – the loci of the instantaneous axis of rotation- that is the *Poinssot's cones:* the body cone rolls on the space cone (Fig.2).

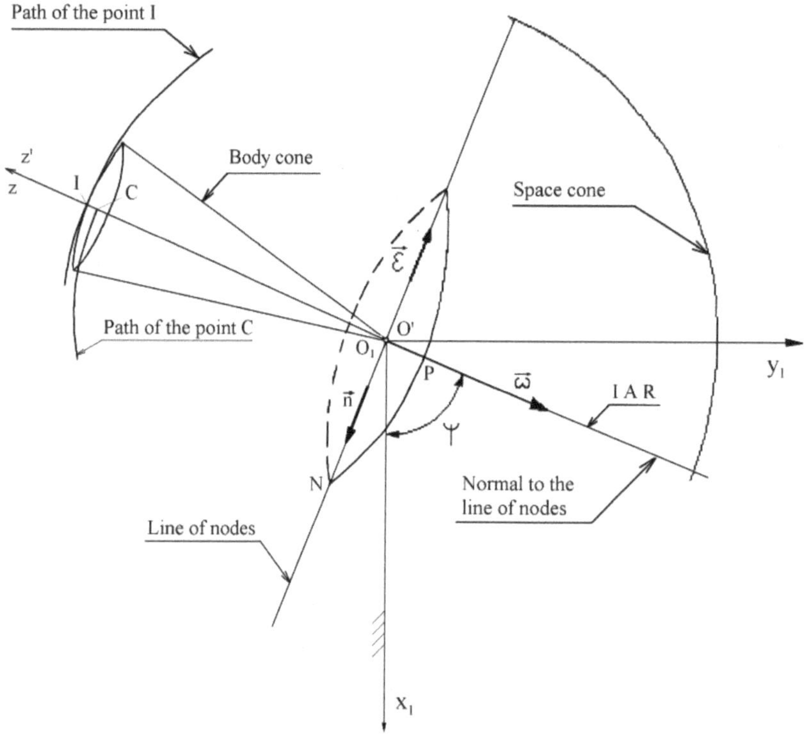

Figure 2. Poinssot's cones

This rotation has the angular velocity (Fig.3)

$$\vec{\omega}=\vec{\omega}_1+\vec{\omega}_0 . \quad (1)$$

The velocity of the point *C* may be written now as

$$\vec{V}_C = \vec{\omega}\times \vec{h} = (\vec{\omega}_0 + \vec{\omega}_1)\times \vec{h} = \vec{\omega}_1 \times \vec{h} . \quad (2)$$

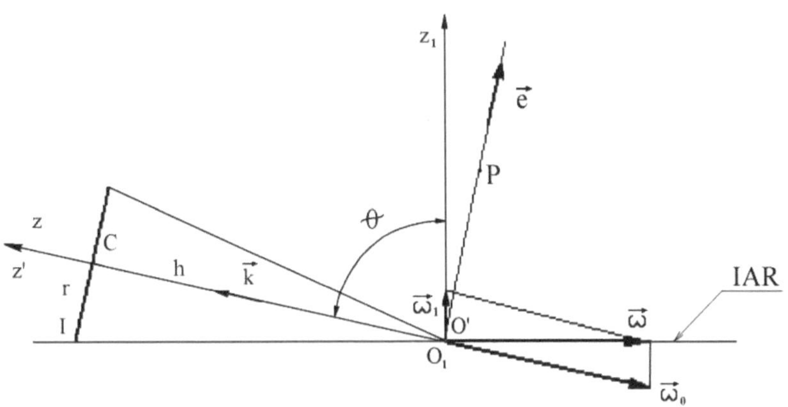

Figure 3. View along the line of nodes

In order to simplify the expressions of the implied vectors, here is useful to introduce another reference frame, namely the *frame of three-orthogonal unit vectors* $\vec{h},\vec{e},\vec{k}$ - frame denoted by **R”** (Fig.2 and Fig.3).
In this frame, the velocity of the centre of mass C will have a simpler expression:

$$\left[\vec{V}_C\right]_{R''} = h\omega_1 \sin\theta \cdot \vec{h}, \quad (3)$$

or, in terms of body frame ***R*** components:

$$\left[\vec{V}_C\right]_R = h\omega_1 \sin\theta \cdot \begin{bmatrix} \cos\varphi \\ \sin\varphi \\ 0 \end{bmatrix} . \quad (3')$$

Now, we can see (Figure 3), the relations between the magnitudes of angular velocities:

$$|\omega_1| = |\omega_0| \cos\theta , \quad (4)$$

$$|\omega| = |\omega_0| \sin\theta . \quad (5)$$

3.2. Accelerations

a) Angular acceleration

From relations above, we can observe that the magnitude of the angular velocity $\vec{\omega}$ is constant. But, because it direction is variable in time, results that the time derivative of the vector $\vec{\omega}$ will be

$$\dot{\vec{\omega}} = \vec{\omega}_1 \times \vec{\omega} = \vec{\omega}_1 \times \vec{\omega}_0 . \quad (6)$$

Then, we shall obtain:

$$\vec{\varepsilon} = \omega_1 \vec{k}_1 \times \omega_0 \left(-\vec{k}\right) = -\omega_1 \cdot \omega_0 \cdot \sin\theta \cdot \vec{h}, \quad (7)$$

meaning a vector parallel to the line of nodes and having the magnitude $|\vec{\varepsilon}| = |\omega_1| \cdot |\omega_0| \cdot \sin\theta$.

b) Linear acceleration of centre of mass

The acceleration of the centre of mass C can be written starting from the fact that the center of mass has a circular path of radius $R_C = h \cdot sin\theta$, and that its speed is constant in time. Then, we get

$$\vec{a}_C = \vec{\omega}_1 \times \vec{v}_C, \tag{8}$$

or, in terms of the ***R"*** frame components:

$$\vec{a}_C = \begin{vmatrix} \vec{h} & \vec{e} & \vec{k} \\ 0 & \omega_1 \sin\theta & \omega_1 \cos\theta \\ h\omega_1 \sin\theta & 0 & 0 \end{vmatrix}, \quad \Rightarrow [\vec{a}_C]_{R''} = h\omega_1^2 \sin\theta \begin{bmatrix} 0 \\ \cos\theta \\ -\sin\theta \end{bmatrix}. \tag{9}$$

Finally, the magnitude of the acceleration of the centre of mass will be simply

$$|\vec{a}_C| = h\omega_1^2 \sin\theta\,. \tag{10}$$

4. DYNAMICS OF THE STEADY PRECESSION OF A DISK

4.1. Momentum and Angular Momentum about the Centre of Mass

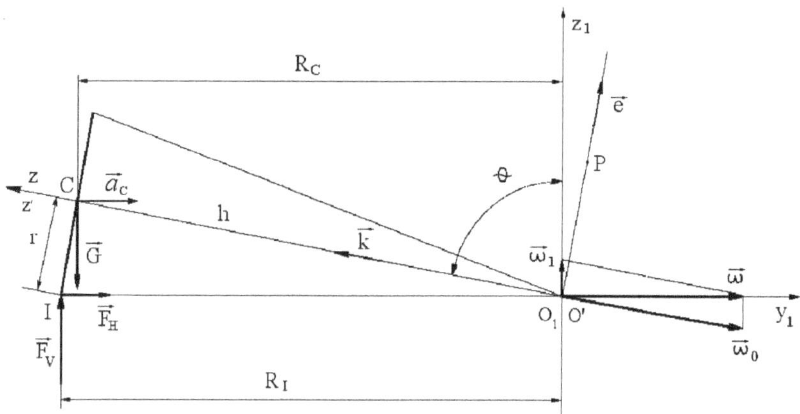

In order to find the differential equations of motion, we have to write first the expressions of :

a) *Resultant Momentum*,

$$\vec{p} = m \cdot \vec{v}_C \text{, or according to equation (3),}$$

$$\vec{p} = m \cdot h \cdot \omega_1 \sin\theta \cdot \vec{h}. \tag{11}$$

b) *Angular Momentum about C* :

$$\left[\vec{K}_C\right]_R = \left[\hat{J}_C\right]_R \cdot [\vec{\omega}]_R, \tag{12}$$

where the **inertia tensor** about *C* in the frame *R* is

$$\left[\hat{J}_C\right]_R = \begin{bmatrix} J_x & 0 & 0 \\ 0 & J_y & 0 \\ 0 & 0 & J_z \end{bmatrix}. \tag{13}$$

Then, we shall have

$$\left[\vec{K}_C\right]_R = \begin{bmatrix} J_x & 0 & 0 \\ 0 & J_y & 0 \\ 0 & 0 & J_z \end{bmatrix} \cdot \begin{bmatrix} \omega_x \\ \omega_y \\ \omega_z \end{bmatrix} = \begin{bmatrix} J_x\omega_x \\ J_y\omega_y \\ J_z\omega_z \end{bmatrix}, \text{ meaning } \left[\vec{K}_C\right]_R = \begin{bmatrix} J_x\omega_1 \sin\theta \sin\varphi \\ J_y\omega_1 \sin\theta \cos\varphi \\ J_z\left(\omega_0 + \omega_1 \cos\theta\right) \end{bmatrix}. \qquad (14)$$

4.2. Time Derivatives

In order to apply the **theorem of momentum variation**, now we must take the time derivatives of the momentum, and respectively of the angular momentum. These will be:

a) $\dot{\vec{p}} = m \cdot \vec{a}_C = m\left(\vec{\omega}_1 \times \vec{V}_C\right).$ (15)

In accordance to equation (9) and Figure 4, we shall obtain in the frame ***R*** the vector

$$\left[\dot{\vec{p}}\right]_R = mR_C\omega_1^2 \begin{bmatrix} -\cos\theta \sin\varphi \\ \cos\theta \cos\varphi \\ -\sin\theta \end{bmatrix}. \qquad (16)$$

b) $\dot{\vec{K}}_C = \dfrac{\partial \vec{K}_C}{\partial t} + \vec{\omega} \times \vec{K}_C,$ (17)

where $\left[\dfrac{\partial \vec{K}_C}{\partial t}\right] = \left[\hat{J}_C\right] \cdot \left[\vec{\varepsilon}\right],$ (18)

with

$$\left[\vec{\varepsilon}\right]_R = -\omega_0\omega_1 \sin\theta \begin{bmatrix} \cos\varphi \\ \sin\varphi \\ 0 \end{bmatrix}. \qquad (19)$$

Now, the relation (18) becomes

$$\left[\frac{\partial \vec{K}_C}{\partial t}\right] = \begin{bmatrix} J_x & 0 & 0 \\ 0 & J_y & 0 \\ 0 & 0 & J_z \end{bmatrix} \cdot \begin{bmatrix} -\cos\varphi \\ -\sin\varphi \\ 0 \end{bmatrix} \cdot \omega_0\omega_1 \sin\theta = \begin{bmatrix} J_x\varepsilon_x \\ J_y\varepsilon_y \\ 0 \end{bmatrix}. \qquad (20)$$

Here, the inertia tensor for a disk has the expression $\left[\hat{J}_C\right]_R = \dfrac{mr^2}{4} \begin{bmatrix} 1 & 0 & 0 \\ 0 & 1 & 0 \\ 0 & 0 & 2 \end{bmatrix}.$ (21)

Then, relation (20) takes the form

$$\left[\frac{\partial \vec{K}_C}{\partial t}\right]_R = \frac{mr^2}{4} \omega_0\omega_1 \sin\theta \cdot \begin{bmatrix} -\cos\varphi \\ -\sin\varphi \\ 0 \end{bmatrix}. \qquad (22)$$

Now, returning to the equation (17), the last term in the right member can be written as

$$\vec{\omega} \times \vec{K}_C = \begin{bmatrix} 0 & -\omega_z & \omega_y \\ \omega_z & 0 & -\omega_x \\ -\omega_y & \omega_x & 0 \end{bmatrix} \cdot \begin{bmatrix} J_x\omega_x \\ J_y\omega_y \\ J_z\omega_z \end{bmatrix} = \begin{bmatrix} \omega_y\omega_z\left(J_z - J_y\right) \\ \omega_z\omega_x\left(J_x - J_z\right) \\ \omega_x\omega_y\left(J_y - J_x\right) \end{bmatrix}. \qquad (23)$$

Finally, introducing (20) and (23) in (17), we get:

$$\left[\dot{\vec{K}}_C\right]_R = \begin{bmatrix} J_x\varepsilon_x + \omega_y\omega_z\left(J_z - J_y\right) \\ J_y\varepsilon_y + \omega_z\omega_x\left(J_x - J_z\right) \\ 0 \end{bmatrix}. \qquad (24)$$

4.3. External Forces

As we can see from Fig.4, on the disk are acting the following external forces: the weight $G=mg$, the vertical component of the reaction of the ground F_V and the horizontal component of the reaction of the ground F_H.

Note. Rolling resistance is assumed to be negligible in this stage of the study, as well as the tangential to the disk component of the ground reaction.

4.4. Motion Equations

According to the *Theorem of Momentum Variation*, we can write:

$$\dot{\vec{P}}=\sum F_i \text{, that is } m\vec{a}_C=\vec{G}+\vec{F}_V+\vec{F}_H. \quad (25)$$

Now, as the acceleration of the centre of mass has no component on the vertical axis, we have the obvious relation

$$\vec{G}+\vec{F}_V=\vec{0}, \quad (26)$$

which gives further

$$m\vec{a}_C=\vec{F}_H. \quad (27)$$

Now, in terms of *the angular momentum variation,* we can write: $\dot{\vec{K}}_C=\vec{M}_C$, that is to say

$$\begin{bmatrix} J_x\varepsilon_x-\omega_y\omega_z(J_y-J_z) \\ J_y\varepsilon_y-\omega_z\omega_x(J_z-J_x) \\ 0 \end{bmatrix}=\begin{bmatrix} M_x \\ M_y \\ 0 \end{bmatrix}, \quad (28)$$

where:

$$M_x=\vec{M}_C\cdot\vec{i}=mr(a_c\sin\theta-g\cos\theta)\cos\varphi, \quad (29)$$

$$M_y=\vec{M}_C\cdot\vec{j}=mr(a_c\sin\theta-g\cos\theta)\sin\varphi. \quad (30)$$

Now, introducing in (30) the relations (3') and (19), we get:

$$-J_x\omega_0\omega_1\sin\theta\cos\varphi+\omega_1\sin\theta\cos\varphi(\omega_0+\omega_1\cos\theta)(J_z-J_y)=M_C\cos\varphi, \quad (31)$$

$$-J_y\omega_0\omega_1\sin\theta\sin\varphi+\omega_1\sin\theta\sin\varphi(\omega_0+\omega_1\cos\theta)(J_z-J_x)=M_C\sin\varphi. \quad (32)$$

The most important result of these relations is: <u>the resultant moment of the external forces is constant in magnitude and is pointing along a parallel to the line of nodes.</u> As we can easily see after some simple calculi, the magnitude of this moment is for a rolling disk:

$$M_C=J_x\omega_1^2\sin\theta\cos\theta=\frac{mr^2}{4}\omega_1^2\sin\theta\cos\theta. \quad (33)$$

On the other hand, from the relations (29) and (30), we can write the same moment as

$$M_C=mr(a_C\sin\theta-g\cos\theta). \quad (34)$$

Equalizing these values, <u>in accordance with the *angular momentum variation theorem*,</u> we get the equilibrium relationship equation between the precession speed ω_1 and the nutation angle θ:

$$tg^2\theta=1+\frac{g}{r\omega_1^2}. \quad (35)$$

3. CONCLUSION

1. Under the conditions of steady precession motion, the lateral inclination of the rolling body (disk or ring) will be constant thanks to inertial (gyroscopic) properties of the motion.

2. The above result is valid also for a monocycle in steady turning, under good adherence conditions, and with accurate keeping of the centre of mass in the symmetry plane of the vehicle,after the turn was started.

REFERENCES

[1] Deliu G.,Deliu M., Rolling Disk Dynamics, *RECENT,* Vol.10,nr.3(27), Braşov, November, 2009

[2] Deliu G., Mecanica, Editura Albastră, Cluj-Napoca, 2003

WAVE VELOCITY ESTIMATION BY APPLICATION OF THE INTRINSIC TRANSFER MATRIX

Nicolae Creţu[1], Ioan Călin Roşca[2]

[1] IEFA/Physics Department, Transilvania University, Brasov, Romania, cretu.c@unitbv.ro

[2] Department of Mechanical Engineering, Transilvania University, Brasov, Romania, icrosca@unitbv.ro

Abstract

The work presents experimental data of wave velocity measurements in wood samples, obtained by using intrinsic transfer method. The method is based on the properties of the behavior of the eigenvalues of the transfer matrix in resonance cases, this means the method is a particular modal approach of a resonance method. To find the wave velocity in a sample, respective sample is built-in an embedded system containing gauge material and the sample under tests. A numerical analysis applied on the analithycal expression of the eigenvalue permits the wave velocity estimation.

Keywords: *transfer matrix; eigenmodes; elastic wave velocity in wood*

1.INTRODUCTION

If we consider a simple solid homogeneous elastic rod with the length l, much larger than its diameter, and characteristic impedance $Z = \rho c$, placed between two semiinfinite media with characteristic impedances Z_{in} and Z_{out}, which propagates a longitudinal plan wave, TM which connects the amplitudes of the Fourier components of the incident and reflected displacements in the sample $A(\omega)$ and $B(\omega)$ has the expression:

$$\begin{pmatrix} A_{out}(\omega,l) \\ B_{out}(\omega,l) \end{pmatrix} =$$

$$\frac{1}{4}\begin{pmatrix} 1+\dfrac{Z}{Z_{out}} & 1-\dfrac{Z}{Z_{out}} \\ 1-\dfrac{Z}{Z_{out}} & 1+\dfrac{Z}{Z_{out}} \end{pmatrix} \begin{pmatrix} e^{-\beta l} e^{i\frac{\omega}{c}l} & 0 \\ 0 & e^{\beta l} e^{-i\frac{\omega}{c}l} \end{pmatrix} \begin{pmatrix} 1+\dfrac{Z_{in}}{Z} & 1-\dfrac{Z_{in}}{Z} \\ 1-\dfrac{Z_{in}}{Z} & 1+\dfrac{Z_{in}}{Z} \end{pmatrix} \begin{pmatrix} A_{in}(\omega,0) \\ B_{in}(\omega,0) \end{pmatrix} \quad (1)$$

with β attenuation factor related to amplitude, ω angular frequency and c the speed of the wave. In the stationary case, when the stationary wave is confined inside the sample, only the intrinsic part of the TM is involved, i.e.

$$TM(\beta,\omega) = \begin{pmatrix} e^{-\beta l} e^{i\frac{\omega}{c}l} & 0 \\ 0 & e^{\beta l} e^{-i\frac{\omega}{c}l} \end{pmatrix} \quad (2)$$

Intrinsic part of the TM has the eigenvalues:

$$\lambda_{1,2} = \left[cos\left(\frac{\omega l}{c}\right) \cdot cosh(\beta l) - i \cdot sin\left(\frac{\omega l}{c}\right) \cdot sinh(\beta l) \right] \pm \sqrt{\left[cos\left(\frac{\omega l}{c}\right) \cdot cosh(\beta l) - i \cdot sin\left(\frac{\omega l}{c}\right) \cdot sinh(\beta l) \right]^2 - 1} \quad `(3)$$

The eigenmodes of the rod, which correspond to the real values of the eigenvalues, are given by: $sin\left(\frac{\omega l}{c}\right) = 0,$ and

therefore $l = \frac{n\lambda_w}{2}, \quad n = 1,2,...$

2. TRANSFER MATRIX IN THE CASE OF A TERNARY SYSTEM

Consider a ternary system which contain three layers having the thicknesses l_1, l_2, l_3 and characteristic impedances Z_1, Z_2, Z_3 which propagates longitudinally waves with wavenumbers k_1, k_2, k_3, placed between two semiinfine elastic media, with characteristic impedances Z_{in} and Z_{out}. The spectral amplitudes of the waves at input and output connected by a TM are given below[1][2]:

$$\begin{pmatrix} A_{out}(\omega) \\ B_{out}(\omega) \end{pmatrix} = \frac{1}{16} \begin{pmatrix} 1+\frac{Z}{Z_{out}} & 1-\frac{Z}{Z_{out}} \\ 1-\frac{Z}{Z_{out}} & 1+\frac{Z}{Z_{out}} \end{pmatrix} \begin{pmatrix} e^{ik_3 l_3} & 0 \\ 0 & e^{-ik_3 l_3} \end{pmatrix} \cdot \begin{pmatrix} 1+\frac{Z_2}{Z_3} & 1-\frac{Z_2}{Z_3} \\ 1-\frac{Z_2}{Z_3} & 1+\frac{Z_2}{Z_3} \end{pmatrix} \cdot \begin{pmatrix} e^{ik_2 l_2} & 0 \\ 0 & e^{-ik_2 l_2} \end{pmatrix} \cdot$$
$$\cdot \begin{pmatrix} 1+\frac{Z_1}{Z_2} & 1-\frac{Z_1}{Z_2} \\ 1-\frac{Z_1}{Z_2} & 1+\frac{Z_1}{Z_2} \end{pmatrix} \cdot \begin{pmatrix} e^{ik_1 l_1} & 0 \\ 0 & e^{-ik_1 l_1} \end{pmatrix} \begin{pmatrix} 1+\frac{Z_{in}}{Z} & 1-\frac{Z_{in}}{Z} \\ 1-\frac{Z_{in}}{Z} & 1+\frac{Z_{in}}{Z} \end{pmatrix} \begin{pmatrix} A_{in}(\omega) \\ B_{in}(\omega) \end{pmatrix} \quad (4)$$

The intrinsic part of the TM, taking into consideration attenuation is:

$$TM(\omega) = \frac{1}{4} \cdot \begin{pmatrix} e^{i\frac{\omega}{c_3} l_3 - \beta_3 l_3} & 0 \\ 0 & e^{-i\frac{\omega}{c_3} l_3 + \beta_3 l_3} \end{pmatrix} \cdot \begin{pmatrix} 1+\frac{Z_2}{Z_3} & 1-\frac{Z_2}{Z_3} \\ 1-\frac{Z_2}{Z_3} & 1+\frac{Z_2}{Z_3} \end{pmatrix} \cdot \begin{pmatrix} e^{i\frac{\omega}{c_2} l_2 - \beta_2 l_2} & 0 \\ 0 & e^{-i\frac{\omega}{c_2} l_2 + \beta_2 l_2} \end{pmatrix} \cdot$$
$$\cdot \begin{pmatrix} 1+\frac{Z_1}{Z_2} & 1-\frac{Z_1}{Z_2} \\ 1-\frac{Z_1}{Z_2} & 1+\frac{Z_1}{Z_2} \end{pmatrix} \cdot \begin{pmatrix} e^{i\frac{\omega}{c_1} l_1 - \beta_1 l_1} & 0 \\ 0 & e^{-i\frac{\omega}{c_1} l_1 + \beta_1 l_1} \end{pmatrix} \quad (5)$$

The eigenvalues of $TM(\omega)$ with identical materials of the layers 1 and 3 are:

$$\lambda_{1,2}(\omega)=\left[\left(\frac{Z_1+Z_2}{2\sqrt{Z_1Z_2}}\right)^2\left[\begin{array}{l}\cos\left(\frac{\omega}{c_1}l_1+\frac{\omega}{c_2}l_2+\frac{\omega}{c_1}l_3\right)\cdot\cosh(\beta_1l_1+\beta_2l_2+\beta_1l_3)-\\ i\cdot\sin\left(\frac{\omega}{c_1}l_1+\frac{\omega}{c_2}l_2+\frac{\omega}{c_1}l_3\right)\cdot\sinh(\beta_1l_1+\beta_2l_2+\beta_1l_3)\end{array}\right]\right]-$$
$$\left[\left(\frac{Z_1-Z_2}{2\sqrt{Z_1Z_2}}\right)^2\left[\begin{array}{l}\cos\left(\frac{\omega}{c_1}l_1-\frac{\omega}{c_2}l_2+\frac{\omega}{c_1}l_3\right)\cdot\cosh(\beta_1l_1-\beta_2l_2+\beta_1l_3)-\\ i\cdot\sin\left(\frac{\omega}{c_1}l_1-\frac{\omega}{c_2}l_2+\frac{\omega}{c_3}l_3\right)\cdot\sinh(\beta_1l_1-\beta_2l_2+\beta_1l_3)\end{array}\right]\right]\pm\sqrt{F(\omega)} \tag{6}$$

where:

$$F(\omega)=\left\{\begin{array}{l}\left[\left(\frac{Z_1+Z_2}{2\sqrt{Z_1Z_2}}\right)^2\left[\begin{array}{l}\cos\left(\frac{\omega}{c_1}l_1+\frac{\omega}{c_2}l_2+\frac{\omega}{c_1}l_3\right)\cdot\cosh(\beta_1l_1+\beta_2l_2+\beta_1l_3)-\\ i\cdot\sin\left(\frac{\omega}{c_1}l_1+\frac{\omega}{c_2}l_2+\frac{\omega}{c_1}l_3\right)\cdot\sinh(\beta_1l_1+\beta_2l_2+\beta_1l_3)\end{array}\right]\right]-\\ \left[\left(\frac{Z_1-Z_2}{2\sqrt{Z_1Z_2}}\right)^2\left[\begin{array}{l}\cos\left(\frac{\omega}{c_1}l_1-\frac{\omega}{c_2}l_2+\frac{\omega}{c_1}l_3\right)\cdot\cosh(\beta_1l_1-\beta_2l_2+\beta_1l_3)-\\ i\cdot\sin\left(\frac{\omega}{c_1}l_1-\frac{\omega}{c_2}l_2+\frac{\omega}{c_3}l_3\right)\cdot\sinh(\beta_1l_1-\beta_2l_2+\beta_1l_3)\end{array}\right]\right]\end{array}\right\}^2-1 \tag{7}$$

A modal analysis combined with a numerical method able to study the behavior of the eigenvalues can be proposed to find elastic constants of the solid materials. Such a method can be applied to study simple embedded systems containig gauge materials and the sample under test. Such simple systems are characterized in the case of a longitudinal plan wave propagation by a simple distribution of the eigenmodes and also by a simple and convenient analytical expression TM. Examples of such simple systems are binary or ternary built-in systems, which contain gauge materials and materials for investigation[3].

3. APPLICATION OF THE METHOD FOR SOME SPECIAL MATERIALS

A good application of the method is to characterize the elastic properties of the wood samples[4] Because the elastic and mechanical properties of wood (elastic module, mass density, and Poisson's ratios) are random variables that vary significantly for the same wood species [5], the intrinsic transfer method offers a fast and convenient method to characterize such materials. Instead of huge poles with the ends connected to emitters and receivers [6], the samples used in the transfer matrix method are much smaller. The wood samples consisting of small cylinders were built-in ternary systems brass-wood-brass, taking the brass as gauge material. Moreover, the small cylinders can be cut so as to comply the cylindrical geometry used in characterization of the orthotropic behavior of wood. Tables 1 and 2 express the configuration of the experimental setup, samples sizes and the obtained values of sound velocity along the fiber c_l and perpendicular to it c_r. ν represents the value of the frequency of the eigenmode taken into consideration for numerical analysis.

Table 1. Elastic wave velocity along the fiber estimated by intrinsic transfer method

No.	Species of wood	l_2 (mm)	Diameter (mm)	ρ (Kg/m3)	l_1 (brass) (mm)	l_3 (brass) (mm)	ν (Hz)	c_l along (m/s)
1	Fir tree	29.19	10	473.14	160.2	93.57	3300.78	4563
2	Oak	29.98	9.95	724.85	199	129.16	3164	4161
3	Beech	29.90	9.98	703.77	123.4	123.4	4355	4905
4	Spruce	29.45	9.47	450.73	160	93.39	3808	5437
5	Ash	29.5	9.52	801.11	198.4	129.15	3398.44	4253

Table 2. Elastic wave velocity transversal to the fiber estimated by intrinsic transfer method

No	Species of wood	l_2 (mm)	Diameter (mm)	ρ (Kg/m3)	l_1 (mm)	l_3 (mm)	ν (Hz)	c_r Radial (m/s}
1	Fir tree	31.37	10.02	473.14	160.2	93.57	1269	1512
2	Oak	31.07	10.0	724.85	199	129.16	1562.5	1707
3	Beech	31.15	10.1	703.77	123.4	123.4	1738.28	1692
4	Spruce	16.03	9.8	450.73	159.64	92.72	1171.88	1019
5	Ash	22.14	9.93	801.11	198.07	129.42	1972.66	1850

4. CONCLUSIONS

The work proposes a resonance method based on the properties of the eigenvalues of the wave transfer matrix combined with a numerical method, in order to find the velocity of elastic waves in solid elastic samples. The study also considers the attenuation and shows that attenuation affects the frequency of eigenmodes for an embedded system. The ternary system is preferred because a such system preserves much better the longitudinally plan wave, special case for which the transfer matrix has a simple mathematical form. Although the theory is valid for plane longitudinal waves, we consider that the experimental model consisting of three solid rods connected in line by adhesion, with extreme rods made from identical materials and the sample of interest placed between these gauge materials, approaches quite well the theoretical assumption. All experiments were done using noncontact methods based on Doppler interferomety

REFERENCES

[1] Song, B. H., Bolton, J. S., A transfer matrix approach for estimation the characteristic impedance and wave number of limp and rigid porous materials, Journal of the Acoustical Society of America, Volume 107, Issues 3, Pages 1131-1152, 2000

[2] Nayfeh, A. H., The general problem of elastic wave propagation in multilayered anisotropic media, Journal of the Acostical Society of America, Volume 89, Issue4, pages 1521-1531,1997

[3] Cretu, N., Nita, G., A simplified modal analysis based on the propeties of the transfer matrix, Mechanics of Materials, Volume 60, pages 121-128, 2013

[4]Green, D.W., Winandy, J.E., Kretschmann, D.E. Mechanical properties of wood Wood Handbook. Wood As An Engineering Material. General Technical Report FPL GTR 113, pp. 1-45, 1999.

[5] Tallavo, F., Cascante, G., Pandey, M. D., Estimation of the Probability Distribution of Wave Velocity in Wood Poles, Journal of Materials in Civil Engineering, Volume 23, Issue 9, Pages 1272-1280, 2011

[6] Bucur , V., Feeney, F., Attenuation of ultrasound in solid wood, Ultrasonics, Volume 30, Issue 2, Pages 76-81,1992

RESEARCH ON THE CONSTRUCTION OF GREENHOUSES LOCATED ON THE ROOFS OF BUILDINGS

E. C. Badiu[1], Gh. Brătucu[2]
[1]Transilvania University of Braşov, Braşov, ROMANIA, eb@cebb.net
[2]Transilvania University of Braşov, Brasov, ROMANIA, gh.bratucu@unitbv.ro

Abstract: *The paper refers to the benefits of building of green roofs and greenhouses located on the buildings' roofs in the context of increased greenhouse effect and continued growth pollution on our entire planet. In this sense, a trend that is continue growing is the creation of green roofs or buildings which can be used not only for pitch, but also for flowers or vegetable, creating very real organic greenhouses. There are also presented some of the research conducted around the world in order to improve the infrastructure of this kind of constructions.*
Keyword: *green city, greenhouse roof, legislation, pollution*

1. INTRODUCTION

The concept of *green city* is increasingly used in the context of rapidly increased of the greenhouse effect and the continuous increase of the level of pollution throughout our planet.

In a green city can be found the features that make urban living healthier, more pleasant and friendlier with surrounding ecosystems.

Green cities widely use renewable energy, host many companies which use clean technologies, promote an ecological way of life and had adopted rules for the protection of the environment, but also innovative strategies to promote new environmental concerns. Among the concerns which raise an increasingly interest is the creation of green roofs or buildings on which can be grown not only pitch but also flowers or vegetable, prosing really organic greenhouses.

By growing flowers or vegetables on rooftops we actually render to the nature a part of the area which is occupied by the construction of such buildings. Until recently, the idea of setting up greenhouses on the roofs of apartment buildings, businesses, and educational institutions was considered an utopia. Nowadays, this idea gets more adepts, the arrangement being not only an oasis of tranquillity, but also a way to reduce pollution, noise, and the amount of dust and carbon dioxide in the atmosphere.

2. THE CONCEPT OF GREEN CITY AND ECOLOGICAL ROOFS

Due to the emergence of a multitude of smart materials it is possible the directly crops' establishment on the soil laid on the roof surface.

But must be taken into account the structure of the building, the possibility of draining of the excess precipitation, and the compensation of their absence, as appropriate. Also, must be selected plants whose dynamics could match such a project, which can be viable for a long time and can be easily replaced with another similar.

An example of fitting the grows of small plants on the roof of buildings is shown in Figure 1. In this figure can be see that over concrete slab sits an antiroot membrane, a decking drainage geotextile and then a layer of soil can be put [1].

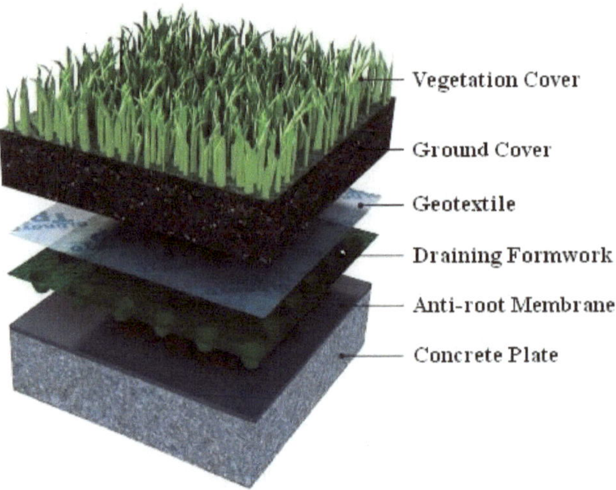

Figure 1: Materials used for small plants' growing on roofs []

Toronto is the first city in the world which has passed the legislation that require for new buildings to have '*eco*' roofs. *Eco-roofs* program is launched by the city's authorities who were receptive to alarming signals over global warming and have decided to take this measure. The distribution of green roofs in Toronto is shown in Figure 2.

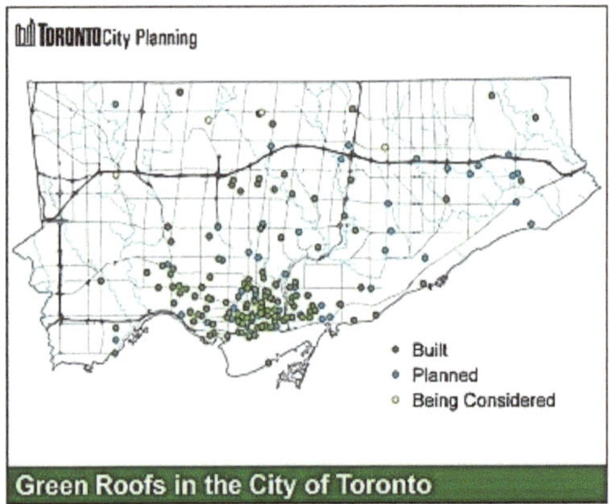

Figure 2: The repartition plan of green roofs in Toronto [2]

In Tokyo, the largest metropolis in the world, green spaces are few and the massive use of air conditioning apparatus worsens the hot weather during the summer time. For several years, all new skyscrapers in Tokyo should have mandatory a garden on their roof. This requirement is difficult to meet because of earthquake prevention regulations which limit the maximum authorized load for roofs and walls.

Japanese company Suntory has introduced an artificial type of soil, more porous, stronger and lighter than the ground soil, which can be applied on the roofs and walls of buildings to plant vegetation on and thus to refresh the temperature in major cities.

According to the Japanese company, a sample of 450 grams of artificial soil called 'Pafcal' (a new urethane foam material based on urethane, specially developed by Suntory), can absorb water as much as a kilo of ground soil. Applied on the roof of a house, it allows the grow of the grass and deciduous plants, which significantly reduces the temperature inside the building.

The concerning for the development of green roofs appeared also in Romania thus, for several years lectures deal with this topic. The purpose of these courses is to assist architects, manufacturers and solution providers to understand the design principles of green roofs, the options available and the needed steps.

According to studies conducted by the manufacturer of additives and building materials, green roofs can reduce the cost of heating or air conditioning in homes by up to 26%, providing excellent thermal insulation during all the year.

Currently, the level of promotion and implementation of '*green*' roofs is growing worldwide. At the present time, in Germany, 10% of the roofs are green and in Switzerland the legislative regulations require that any newly constructed covering with an area greater than 500 square meters to be produced using such a system.

Throughout the world there have imagined many futuristic projects, a significant example being shown in Figure 3.

Figure 3: The 'eco' construction project

One of the greatest manufacturers of greenhouses is the North American company Nexus Corporation. Among other things, this company produces greenhouses designed to be located on the roofs of the building (Figure 4) for over 10 years [4].

Figure 4: Types of greenhouses produced by Nexus Corporation USA

One such project, shown in Figure 5, was designed for Florida State University.
Following the calls, greenhouses were also placed on other structures, such as Arkansas State University, University of California, Centralia Community College etc.

Figure 5: Greenhouses placed on the roof of the Florida State University [4]

The material used for the structure is extruded aluminum and acrylic glass was used for coating. The plants' growing in greenhouses is done mainly in hydroponic system. In Figure 6 is shown a crop developed on a rooftop greenhouse at Gotham Greens, Greenpoint, New York, USA.

Figure 6: Greenhouse developed on a rooftop at Gotham Greens-Greenpoint, New York, SUA [8].

Another example of a greenhouse placed on the roof is shown in Fig. 7. This is located on the roof of a warehouse in Montreal, Canada and has a total area of 3,000 m2. It is used for the growing of tomatoes, eggplants, carrots, arugula, and other herbs and vegetables in hydroponic system [5].

In the context of the importance of growing plants in greenhouses located on roofs, significant is that, at least for now, the paid price for the rent of the roof is modest.

Figure 7: Greenhouse in Montreal, Canada (Lufa Farms) [5]

A greenhouse made of polycarbonate lightweight structure, located on the roof of a parking (garage) in Tucson, USA, can be seen in Figure 8.

Figure 8: Greenhouse in Tucson, Arizona

A greenhouse made of polycarbonate on lightweight structure, located on the roof of a parking (garage) in Tucson, USA can be seen in Figure 8. UrbanFarmers AG company from Switzerland launched in 2013 a pilot greenhouse (UF001 LokDepot) with an area of 250 m^2, located on the roof of a warehouse in Basel (Figure 9) [6].
This was done for commercially purposes, can produce up to 5 tons of vegetables a year and has been tested by the Swiss company together with a team from the University of Applied Sciences (ZHAW), in order to verify the functionality, robustness and quality of production.

Figure 9: UF001 LokDepot farm from Basel, Switzerland [6]

A futuristic micro model, intended to be located on the roof (Globe / Hedron), and designed by Italian architect Antonio Scarponi along with UrbanFarmers will be made on the structure of bamboo and will be commercialized in the near future.

Figure 10: Globe/ Hedron farm [7]

The greenhouse shown in Figure 10 may provide the necessary of herbs and vegetables for 4 families of 4 people throughout the year [7].

3. CONCLUSION

By the analysis performed on the importance of green roofs and roof-mounted greenhouses, we can draw the following conclusions:

- *green cities* widely use renewable energy, host many companies which use clean technologies, promote an ecological way of life and had adopted rules for the protection of the environment, but also innovative strategies to promote new environmental concerns;

- among the concerns which raise an increasingly interest is the creation of green roofs or buildings on which can be grown not only pitch but also flowers or vegetable, prosing really organic greenhouses;
- due to the emergence of a multitude of smart materials it is possible the directly crops' establishment on the soil laid on the roof surface;
- green roofs can reduce the cost of heating or air conditioning in homes by up to 26%, providing excellent thermal insulation during all the year;
- the idea of developing greenhouses on the roofs of apartment buildings, businesses and educational institutions, get more importance, this arrangement being not only an oasis of tranquility, but also a way to reduce pollution, the degree of noise, the quantity of dust and carbon dioxide from the atmosphere;

REFERENCES

[1] http://ideipentrubucuresti.blogspot.ro/
[2] http://www1.toronto.ca/
[3] http://4.bp.blogspot.com/
[4] www.nexuscorp.com
[5] http://lufa.com/fr/
[6] http://urbanfarmers.com/
[7] http://www.conceptualdevices.com/

THEORETICAL RESEARCH REGARDING HEAT TRANSFER BETWEEN GREENHOUSES AND ENVIRONMENT

Bodolan Ciprian [1], Gh. Brătucu [2]

[1] Transilvania University of Brasov, Braşov, ROMANIA, bodolan.ciprian@gmail.com

[2] Transilvania University of Brasov, Braşov, ROMANIA, gh.bratucu@unitbv.ro

Abstract: *Because the greenhouse heating costs represents approximately 40% of the total cost production, heat loss must be kept under control and even lower, which leads us first to understand the way in which heat transfer occurs between the greenhouses and the environment. Heat loss from a greenhouse usually occurs by all three modes of heat transfer: conduction, convection and radiation. Usually many types of heat exchange occur simultaneously. The heat demand for a greenhouse is normally calculated by combining all three losses as a coefficient in a heat loss equation. In the paper the author's presents from the theoretical point of view the impact and the calculation of heat loss in greenhouses.*

Keywords: *greenhouse, energy, heat, losses*

1. INTRODUCTON

In greenhouses complex thermal phenomenon occur due to the exchange of hot or cold air, it also signaled the ventilation airflow through frequent ventilation of farmed environment. Solar energy which penetrates the greenhouse suffers significant changes that occur largely by the passage of radiation through the material that covers construction. Inside the greenhouse, solar radiation contributes to the overall heat exchange, to establish the balance of heat and radiation, complex phenomena involving temperature and humidity, evaporation and condensation, plants, soil, the design, heating ducts. Losses and heat accumulations occur, frequent exchanges of air that strongly influences the greenhouse microclimate.

The transfer and the heat exchange are made through three distinct mechanisms:

Thermal conductivity, mainly characterized by the coefficient of thermal conductivity λ, expressed in Kcal/m^2 • grd. sau W/m^2 °C;

Convection, the heat is transported once with the conductive environment in motion and can be:

a) spontaneous convection (natural and free) due to natural causes such as temperature difference that causes difference of air density, giving rise to it;
b) forced convection due to imposed external causes;

Radiation, when the heat transport takes place by means of electromagnetic waves between the areas having different temperatures; the two surfaces can be separated, such as greenhouses, by an environment more or less "leaky" to thermal radiation.

2. HEAT TRANSFER BETWEEN GREENHOUSE AND ENVIRONMENT

Heat transfer and heat exchange of the greenhouse is illustrated by the overall heat transfer coefficient or heat transfer (K), which characterizes very well convection between greenhouse air and the inner surface of the roof (α_i), conduction between the two surfaces of the roof (δ / λ) and convection between the outer surface of the roof of the greenhouse and adjacent air layer (α_e).

The coefficient K is determined with the following expression:

$$K = \frac{1}{\frac{1}{\alpha_i}+\frac{1}{\alpha_e}+\frac{\delta}{\lambda}} \qquad (1)$$

where: α_i - inner convection coefficient in kcal / m2 * h * deg;

α_e - external convection coefficient in kcal / m2 * h * deg;

δ - thickness of the transparent cover, in mm;

λ - coefficient of conductivity, in kcal / m2 * h * deg;

Gac (1967) shows that the ratio δ / λ ranges from 0.02 to 0.001 kcal / m2 * h * degrees, and the ratio 1 / α_i is equal to 0.25 kcal / m2 * h * degrees, wich demonstrate that in greenhouse the heat transfer by convection is frequently and totaly limited by conduction. It depends on the movement of the inside air flow, by the outside wind speed, on the nature of constructive elements, by the radiation coefficient of the cover surface and the difference between adjacent air temperature and surface temperature of the glass or plastic cover of the greenhouse.
In greenhouse, the convection coefficient α is a function of the wind speed and frequency, the ratio between these two factors are directly proportional.
Coefficient K grows by increasing wind velocity through the coefficient α, determination worthy of taking into consideration for greenhouses (Figure 1)

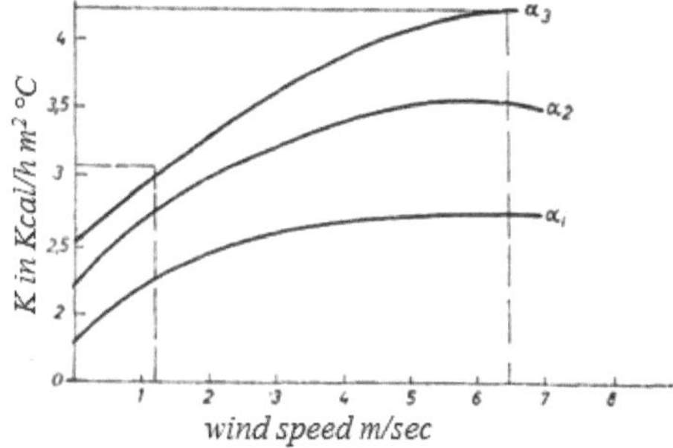

Figure 1 : Variation of the overall heat transfer coefficient in a glass greenhouse, depending on the wind speed, for three convection coefficient

The coefficient K has different values depending on the materials used in the construction of the greenhouse and some of them are given in table 1.

Table 1: The heat transfer coefficient (K)

Name of building and coverage elements	K (Kcal/m2.h.grd)
A. Infrastructure	
Concrete base with thickness of:	
20 cm	2,7
25 cm	2,45
30 cm	2.2
B. Over-glass structure	
Simple glass with wooden structure	5,0
Simple glass with steel structure	5,5
Double glass with wooden structure	2,8
Double glass with steel structure	3,0
C. Over-plastic structure	
A polyethylene foil layer	6,2
Two layers of polyethylene foil	3,0
Rigid PVC plate	4,9
Polyethylene plate	4,7
Polyester plate doubled with polyethylene layer	2,8

The direction of heat exchange in greenhouses is variable in that a certain phenomenon or energy transfer mechanism, which contributes to the accumulation of heat during the day can become a loss of heat at night and vice versa. The amount of heat introduced in a greenhouse should be at least equivalent to the amount of heat that is lost to the outside (the greenhouse effect). By raising the temperature will increase the necessary of heated to be added by conventional way , which will inevitably leads to the increase of losses, that otherwise is explained by the increased temperature gradient. In the same way once the outside air temperature will be lower the heat losses will be higher. These losses increase by 5-15% during the year period in which the cold and strong winds are blowing.
Cultivated plants actively participate in the sensible and latent energy exchange with the environment, through the mechanisms of heat transfer. Temperature of plant is one of the essential elements of the heat transfer mechanisms in greenhouses. This can be determined by knowing the ambient temperature. However, there is a difference between the plant temperature, which may be higher during the day and lower at night, than the ambient temperature. Usually, the calculations are taking into account the temperature of inner air, since it can be measured more easily than the temperature of the plant.
Heat losses occur through the transparent cover of the greenhouse due to irradiation, through over structure and greenhouse soil by conduction and through construction joints leak, by convection (Figure 2).

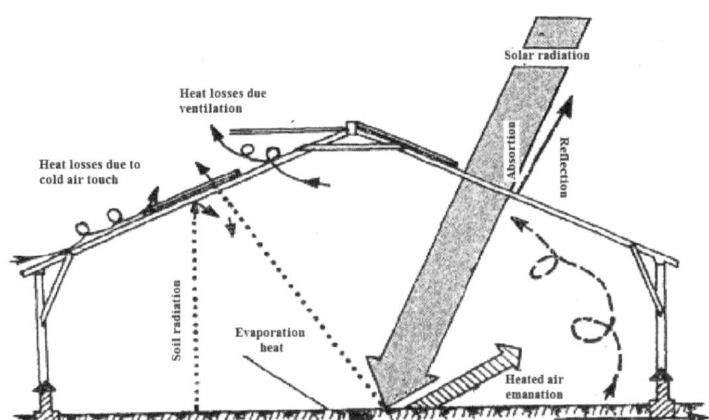

Figure 2 : The main scheme of heat loss in the greenhouse (Mad Seeman)

Heat losses by radiation

Among the mechanism of heat transfer an important role plays infrared radiation, which influences temperature distribution in the greenhouse. Heat exchange caused by this phenomenon can be written as a radiative balance. There is always an exchange of radiation (longer wavelengths) between plants, soil in greenhouses, superstructure and exterior space.

During the night the conditions are of particular interest, since during this period the heat deficit is maximum. Over the night the outside atmosphere is cooling thanks to that the greenhouse constantly emits radiant heat flux. Radiation losses can be calculated according to Walker's relationship as follows:

$$Q_{rad} = 4{,}4 * 10^{-8} * S * P(T_i^4 - T_e^4) \qquad (2)$$

where: Q_{rad} - the quantity of heat lost by irradiation Kcal / h;

S - area under the cover of the greenhouse m^2;

P - Transparency empirical factor (values: Polyethylene = 0.8; glass = 0.04);

T_i - Indoor temperature in ° C;

T_e - the outdoor temperature in ° C;

Irradiation losses are less than the losses by convection, but considerably higher than the losses due to conduction through infrastructure and the soil of greenhouse.

Heat losses by convection

A substantial contribution to heat loss in greenhouses brings the convective exchange between the internal and external atmosphere. These losses have the largest share in the global balance equation, for which almost all sizing calculations using formulas losses that takes into account only the losses by convection.

For example according to Romanian standard the heat loss is given by:

$$Q = K * S(T_i - T_e) \qquad (3)$$

where: Q - amount of heat lost (convection) in Kcal / h;

K - overall heat transfer coefficient in Kcal / m^2;

S - surface construction elements in m^2.

Besides the heat lost through structural elements and warm air infiltration through atmosphere due to imperfect tight of joints, cracks, defective joints of constructive elements, broken windows, broken plastic, a certain amount of the heat (Q_{inf}.) is lost to heat the cold air that enters in greenhouse from the surrounding atmosphere. In this case,

$$Q_{inf} = 0{,}31 * V * (t_i - t_0) \qquad (4)$$

The heat consumed for heating the infiltrated air is conditioned by the wind speed, the length of leaks, the quality factor of glass or plastic and the pressure difference between inner and outer air. For greenhouses, Nafrady (1953) proposes the following mathematical expression:

$$V = L * a * (p_k - p_e)^{\frac{3}{2}} \qquad (5)$$

where: Q_{inf} - the heat consumed for heating the infiltrated air in greenhouse, in kcal / h;

0.31 - specific heat of air Kcal / m^3 degrees;

V - the volume of infiltrated air m^3 / h;

L - length of cracks in m;

a - quality factor of the windows (3 ... 10);

$(P_k - P_e)$ - pressure difference between inner and outer air in mm mercury column.

Heat losses through greenhouse soil

Represent a weigh higher or lower depending on the convective heat transfer, by the regime of water evaporation from the ground and physico-thermal characteristics of cultivated land. Heat loss through the soil is conditioned by the fact that the land is heated artificially or by solar energy. The loss of the heat in the soil may be in its depth, on vertical, or perimeter, horizontally. From observations can be noted that the nature of the soil, ground water level, irrigation system and land moisture directly influence the heat losses, uniform temperature distribution in the soil. The convective heat transfer increases with the speed of air movement at soil surface which leads to an heat loss increased by 5-20%, depending on the coating system of the greenhouse.

Also, the reduced values of the air temperature above the ground and elevated temperature on the ground surface, increases the amount of heat loss due to convection $Q_{conv.s}$. In this case,

$$Q_{conv.s.} = \alpha'(t_{ss} - t_{gp}) \quad (6)$$

Where (t_{ss} -t_{gp}) represents the difference between temperatures at the soil surface and ground proximity and the coefficient α 'characterizes the conditions of the exchange of heat between the soil surface and the layer of air adjacent to a wind speed up to 5 m / s.

Heat loss through evaporation of the soil is determined using the equation proposed by Korolkov (1955);

$$Q_{ev} = 0{,}6 * G \quad (7)$$

where 0.6 is a characteristic number and G results from Dalton's equation:

$$G = \eta(21{,}9 + 17{,}8v) * (p_{ss} - p_{gp}) \quad (8)$$

where: Q_{EV} - the amount of heat lost through the evaporation of ground water in Kcal / h;

G - the amount of water evaporated in g / m^2h;

v - wind speed in m / s;

η - coefficient equal to 0.3 showing reduced evaporation from the soil surface compared to slick surface;

p_{ss} - water vapor pressure at soil surface in mm Hg;

p_{gp} - pressure of water vapor in the vicinity of the ground in mm Hg; 22.9, 17.8 - Characteristic numbers.

However, the heat lost from the soil through evaporation does not greatly affect the heat balance of the greenhouse because it is found in moist air closed by roof. Some of this heat is released to transparent cover through condensation, and when the inside air reaches saturation and settle on glass or plastic roof in the form of a continuous layer or droplets of water.

3. CONCLUSION

1. Heat loss from a greenhouse usually occurs by all three modes of heat transfer: conduction, convection and radiation.
2. Heat transfer and heat exchange of the greenhouse is illustrated by the overall heat transfer coefficient or heat transfer (K).
3. The direction of heat exchange in greenhouses is variable in that a certain phenomenon or energy transfer mechanism, which contributes to the accumulation of heat during the day can become a loss of heat at night and vice versa.
4. Heat losses occur through the transparent cover of the greenhouse due to irradiation, through over structure and greenhouse soil by conduction and through construction joints leak, by convection.

4. REFERENCES

[1] BAKKER, J.C., S.R. ADAMS, T. BOULARD, and J.I. MONTERO, “Innovative technologies for an efficient use of energy”, Acta Horticulturae 801: 49–62, 2008.

[2] ELBATAWI, I.E. “Heating inside a greenhouse at night using solar energy”. ASAE Annual Meeting, Paper number 034040. 2003.

[3] MANESCU, B. “Greenhouses microclimate”, CERES publisher, Bucharest, 1977.

This work was partially supported by the strategic grant POSDRU/159/1.5/S/137070 (2014) of the Ministry of National Education, Romania, co-financed by the European Social Fund – Investing in People, within the Sectorial Operational Programme Human Resources Development 2007-2013.

RESEARCH REGARDING THE GRAIN COMPLIANCE IMPORTANCE ON THE QUANTITY OF FLOUR OBTAINED BY GRINDING

Gh. Brătucu[1], D. D. Păunescu[1]

[1]Transilvania University of Brasov, Brasov, ROMANIA, gh.bratucu@unitbv.ro

Abstract: *The paper presents the influence of technological equipment quality wheat conditioning prior to milling by removing foreign bodies, debarking and wetting the percentage of flour made from wheat ground. The version I use a line conditioner to older technology, and the version II equipped with modern technical equipment. It is noted that the percentage of flour obtained, so that the overall and for the most common categories of the flour (F450, F650 and F1350) is greater than the version II compared to the version I, as debarking is better, and the humidification is very accurate [3]. Also, ash content and specific energy consumption are more favorable version II.*

Keywords: *wheat conditioning by humidification, percentage of meal energy consumption.*

1. INTRODUCTION

The technology of milling wheat must ensure as high a percentage to obtain a high-quality flour, as well as the optimum specific energy consumption[6]. Of quality indicators of wheat flour in this paper were taken into account the ash content and the amount of flour extracted from the raw material and economically determined the specific consumption of electricity [3]. The quantity of flour obtained, the amount of ash and specific energy consumption is required, or stop the optimization calculation, because the increase in the percentage of flour needs to increase the number of passages of grinding and screening, which leads to an increase in the specific consumption of energy per tonne of wheat ground [7]. By upgrading technology to remove foreign bodies, peel and humidification can improve these indicators simultaneously, provided that the equipment used for grinding and sieving continue to be upgraded [1],[5].

2. MATERIAL ŞI METODĂ

The object of experimental research was the common wheat variety **Apulum** grain reddish or yellowish, oval, with long beards and visible. From the point of view of the degradation of this kind fall into class A, Group I, weight per storage volume greater than 75, impurities content less than 6% and less than 15% moisture. Samples were taken from the same point of the technological process. To highlight the influence of technical equipment used in the conditioning of wheat on technical and economic results obtained after milling were used two production lines, which will be called *variants*. By variant I were used: a volumetric dispenser; table type densimeter MD; debarking type DD 714; tarar Magheru a type; BT selector battery type 8; simple screw conveyor type humidifier bunk; a cascade of aspiration. In the variant II embodiment were used: SDT type dispenser weight; SRD double rotational type separator; intensive debarking type SPO; TCR recirculation channel type; stone separator type SPT; Agromatic intensive type humidifier [2]. Every technological variant were taken every 6 samples, which were determined ash content was applied peeling I was determined again ash content; humidification I was performed after the wheat was allowed to rest for homogenisation moisture. Next step was done peeling II was determined ash content of the six samples was checked and corrected moisture, then enter the grinding wheat. In addition to the amount of ash of the two types of samples were measured amounts of flour and all kinds F 480, F 650 and F 1350, a percentage of the amount of wheat powder, obtained by applying the two types of technology conditioning of wheat before milling. Also, there was energy use and specific energy consumption was calculated in the two technological variants [4].

3 RESULTS AND DISCUSSION

Particular attention during the measurements was given moisture content of the grain mass.
In Table 1 and Figures 1 and 2 is the moisture content of the wheat meal, by gross wheat and by wheat neto wheat.

Table 1: The moisture content of the wheat meal gross

Sample	Wheat moisture bruto,%	Wheat moisture neto,%	Sample	Wheat moisture bruto,%	Wheat moisture neto,%
A1	11.77	14.84	B1	11.74	15.10
A2	12.19	15.04	B2	12.60	15.08
A3	11.17	14.92	B3	13.50	15.10
A4	11.78	14.99	B4	12.40	15.10
A5	11.20	14.50	B5	13.40	15.10
A6	11.55	14.70	B6	12.92	15.09

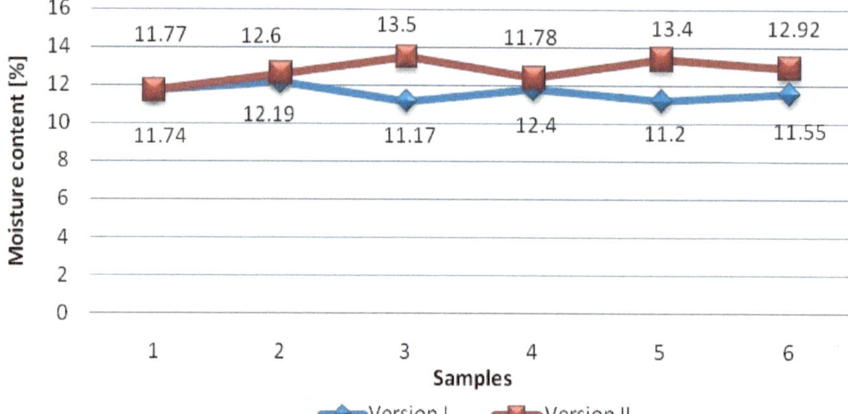

Figure 1: Wheat moisture bruto variation

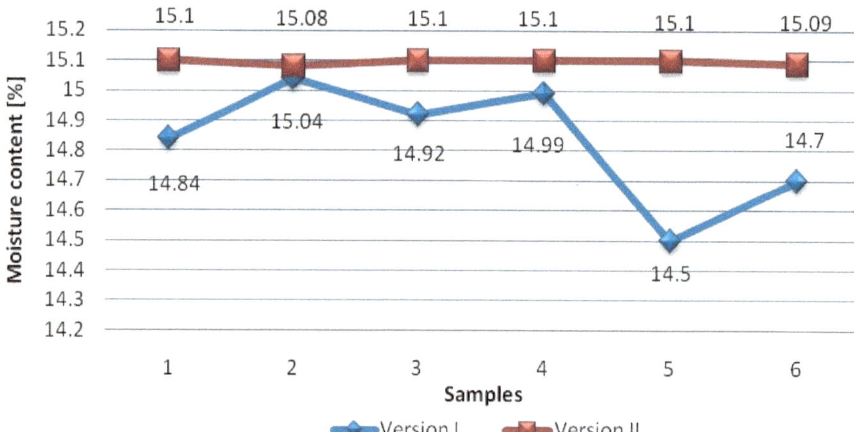

Figure 2: Wheat moisture neto variation

Following the measurements and data processing resulted in the following:

- desired humidity in both versions was 15.1%;
- moisture content in the version I prior to the introduction went between 11.17-12.19%;
- moisture content version II ranged before placing the wetting between 11.74-13.50%;
- moisture wheat content version I ranged between 14.50-15.04% by neto wheat;

- moisture wheat content version II ranged between 15.08-15.10% by neto wheat;

- average moisture content was 14.83% in the version I;
- average moisture content was 15.095% in the version II;
- version II can find a constant humidification resulting values around 15.1%.

After entering the grinding economic indicators were:

- the quantity of total flour;
- the quantity of flour in three types: F 480, F 650 and F 1350;
- the total ash content;
- the ash content of each type of flour.

Table 2 shows the values of the quantities of ashes and the quantity of flour obtained in version I and in Table 3 the values obtained in the version II. Figure 3 presents the quantity of flour total and is shown in Figure 4 compared to the average value of the amount of flour obtained, as a percentage of milled wheat.

Table 2: Variation in ash content and the amount of flour (version I)

Sample	Ash wheat neto, %	Flour total		Flour 480		Flour 650		Flour 1350	
		%	Ash, %	%	Ash %	%	Ash, %	%	Cenuşa %
A1	1.90	75.68	0.71	0.00	0.00	69.00	0.65	6.68	1.35
A2	1.88	73.03	0.66	11.42	0.48	58.00	0.65	3.61	1.35
A3	1.89	74.78	0.69	7.65	0.48	61.38	0.65	5.75	1.35
A4	1.92	74.84	0.71	10.00	0.48	55.98	0.65	8.86	1.35
A5	1.99	74.77	0.72	10.01	0.48	54.62	0.65	10.14	1.35
A6	1.94	74.77	0.70	11.89	0.48	54.21	0.65	8.64	1.35
Media aritmetică		74.645	0.698	8.495	0.480	58.865	0.650	7.280	1.350

Table 3: Variation in ash content and the amount of flour (version II)

Sample	Ash wheat neto, %	Flour total		Flour 480		Flour 650		Flour 1350	
		%	Ash, %	%	Ash %			%	Ash, %
B1	1.80	76.5	0.68	0	0.00	72	0.64	4.5	1.30
B2	1.79	76.0	0.67	10	0.46	60	0.64	6.0	1.28
B3	1.75	77.0	0.66	12	0.46	58	0.62	7.0	1.35
B4	1.75	77.0	0.65	10	0.46	62	0.63	5.0	1.30
B5	1.74	77.4	0.65	9	0.47	63	0.62	5.4	1.30
B6	1.68	76.5	0.62	10	0.47	63	0.60	3.5	1.31
Media aritmetică		76.733	0.652	8.5	0.464	63	0.625	5.233	1.307

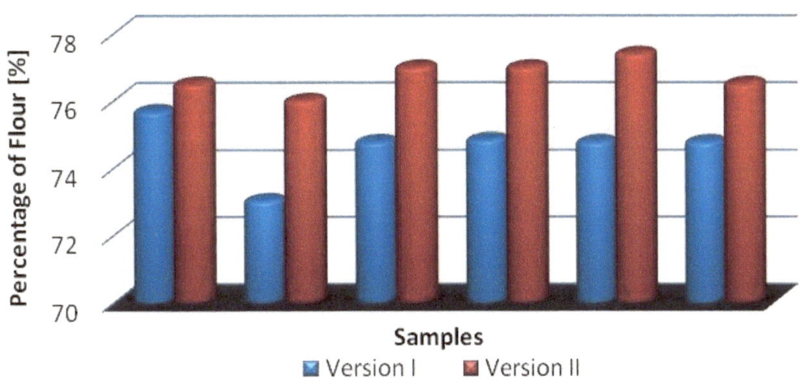

Figure 3: The amount of flour total in the two versions

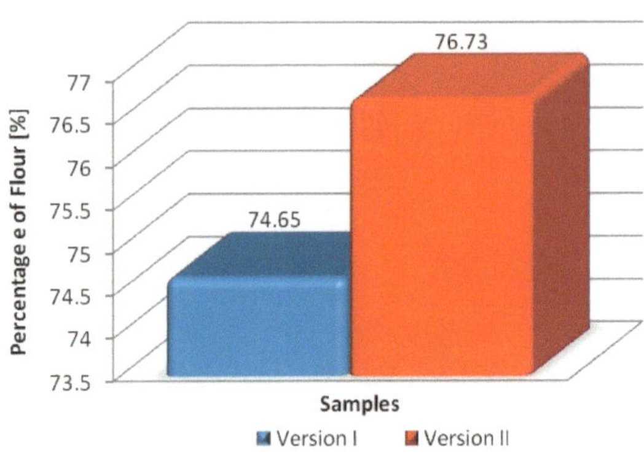

Figure 4: The average amount of flour total in the two versions

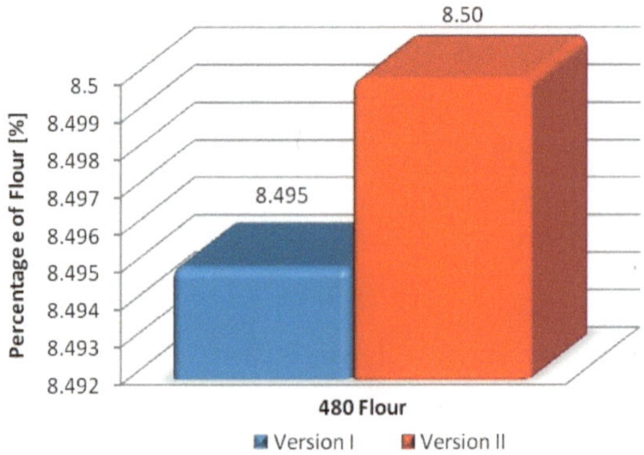

Figure 5: The average amount of flour type 480 in the two versions

According to the values listed in Tables 2 and 3, as well as the graphical representations in Figures 3 and 4 results in the following:

- the quantity of flour total in the version I ranged from 73.03-75.68% of the total amount came to the milling of wheat;
- the quantity of flour total average of the 6 samples taken into account in the version I is 74-65.5% of the total quantity came to the milling of wheat;
- the quantity of flour total of the version II embodiment varied from 76-77.40% by weight of the total grain milling entered;
    - the average cantity of flour total of the 6 samples taken into account in the version II is 76.733% of the total quantity came to the milling of wheat;
    - the total quantity of flour obtained from the milling was higher in each of the samples studied for use in the machine of the Version II;
    - the maximum quantity of flour obtained when using the version I is less than the lowest quantity of flour obtained when using the version II.

Increasing the quantity of flour extracted is found in the three types of flour made and analyzed.

Figure 5 shows the average value of the quantity of flour type 480 obtained for the two technological variants conditioning, and in Figures 6 and 7 are the average values of the quantities of flour type 650 and type 1350 obtained for the two variants.

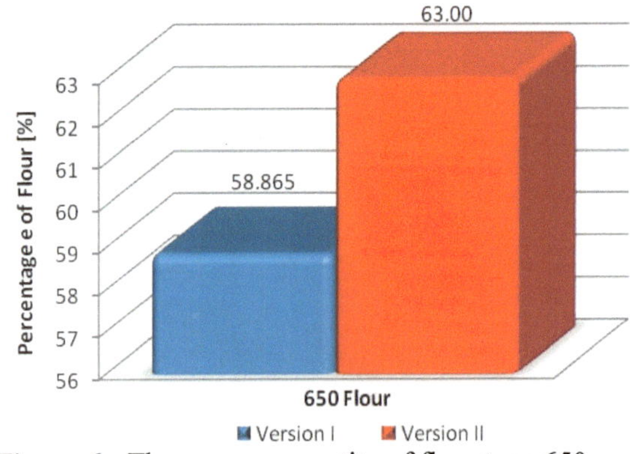

Figure 6: The average quantity of flour type 650 in the two versions

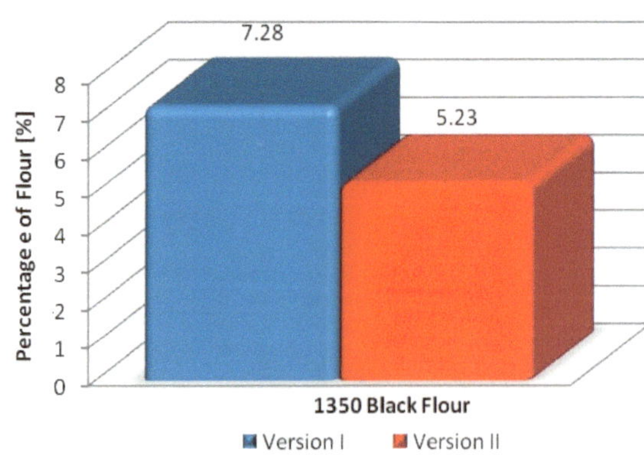

Figure 7: The average quantity of flour type 1350 in the two versions

The analysis results presented in Tables 1 and 2 and as graphical representations in Figures 5, 6 and 7 shows the following:

- the quantity of flour type 480 obtained in version I varied between 7.65-11.89% of the total quantity came to the milling of wheat;

- the quantity of flour type 480 obtained in version II varied between 9.00-12.00% of the total quantity came to the milling of wheat;

- the quantity of flour type 650 obtained in version I varied between 54.21-69.00% of the total quantity came to the milling of wheat;
- the quantity of flour type 650 obtained in version II varied between 58.00-72.00% of the total quantity came to the milling of wheat;
- the quantity of flour type 1350 obtained in version I varied between 3.61-10.14% of the total quantity came to the milling of wheat;
- the quantity of flour type 1350 obtained in version II varied between 3.50-7.00% of the total quantity came to the milling of wheat;

- increasing the average quantity of flour type 480 of 8.495% into version I at 8.50% in version II;
- increasing the average quantity of flour type 650 of 58.865% into version I at 63,00% in version II;
- decreasing the average quantity of flour type 1350 of 7.28% into version I at 5.23% in version II;

Another indicator was watching all sorts ash content of flour obtained. Based on the data presented in Tables 1 and 2, Figure 8 shows the average ash content of the flour obtained from the total, and Figures 9, 10 and 11 is shown ash content for each type of the flour produced.

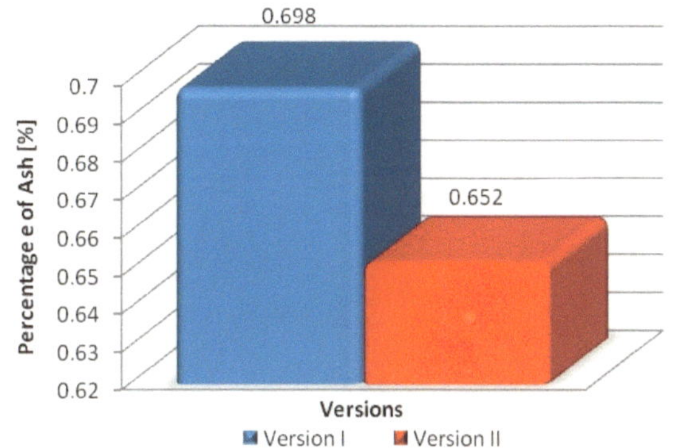

Figure 8: The average content of ash in the flour total in the two versions

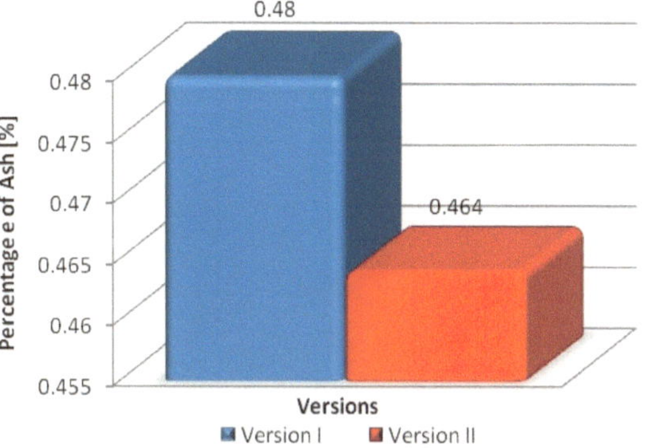

Figure 9: The average ash content of the flour type 480 in the two versions

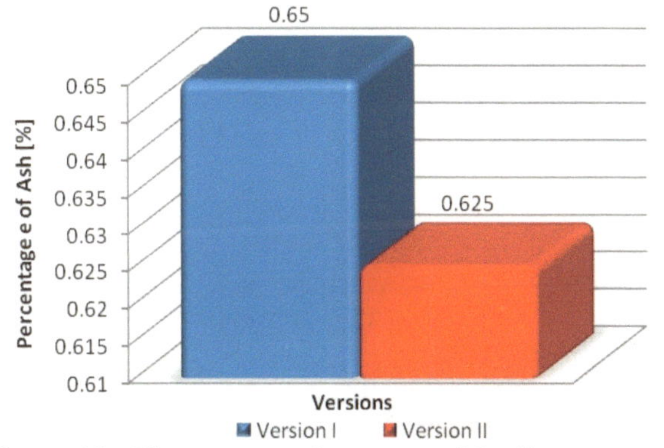

Figure 10: The average ash content of the flour type 650 in the two versions

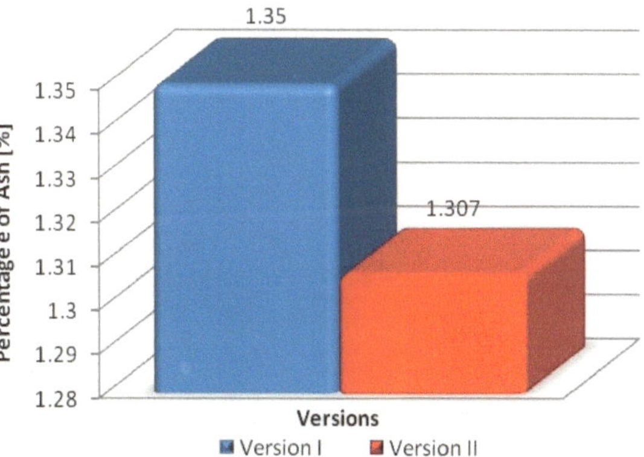

Figure 11: The average ash content of the flour type 1350 in the two versions

Analyzing the results presented in Tables 5 and 6 according to graphic representations in Figures 14, 15, 16 and 17 shows the following:
- the ash content of flour total in version I ranged between 0.66-0.72%;
- the ash content of flour total in version II ranged between 0.62-0.68%;
- the ash content of flour type 480 in version I was constantly, respectively 0.48%;
- the ash content of flour type 480 in version II ranged between 0.46%-0.47%;
- the ash content of flour type 650 in version I was constantly, respectively 0.65%;

- the ash content of flour type 650 in version II ranged between 0.60%-0.64%;
- the ash content of flour type 1350 in version I was constantly, respectively .35%;
- the ash content of flour type 1350 in version II ranged between 1.28%-1.35%;

Specific energy consumption for milling were determined for the samples analyzed are shown in Table 4 and Figures 12 and 13.

Table 4: Specific consumption of electric energy

Version I	Specific consumption, kW/t	Quantity flour total, %	Version II	Specific consumption, kW/t	Quantity flour total, %
A1	78.23	75.68	B1	79.20	76.50
A2	82.76	73.03	B2	80.50	76.00
A3	81.60	74.78	B3	78.12	77.00
A4	75.50	74.84	B4	77.85	77.00
A5	78.80	74.77	B5	77.45	77.40
A6	77.8	74.77	B6	78.25	76.50
Average	**79.38**	**74.65**	**Average**	**78.56**	**76.73**

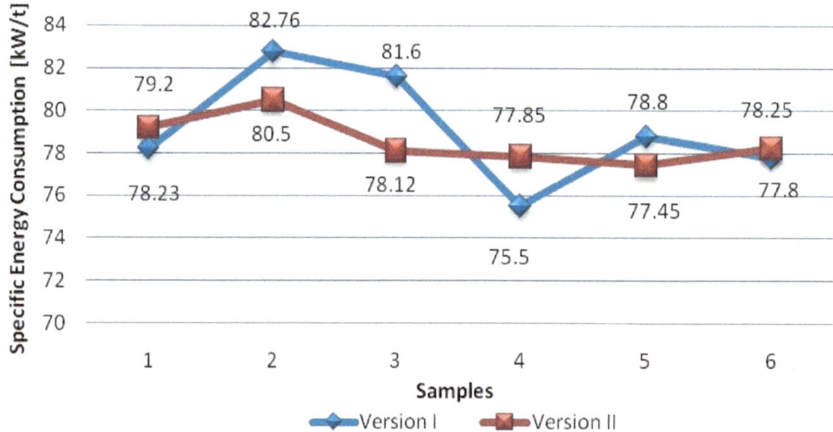

Figure 12: The variation of specific energy consumption in the the two versions, for the 6 samples

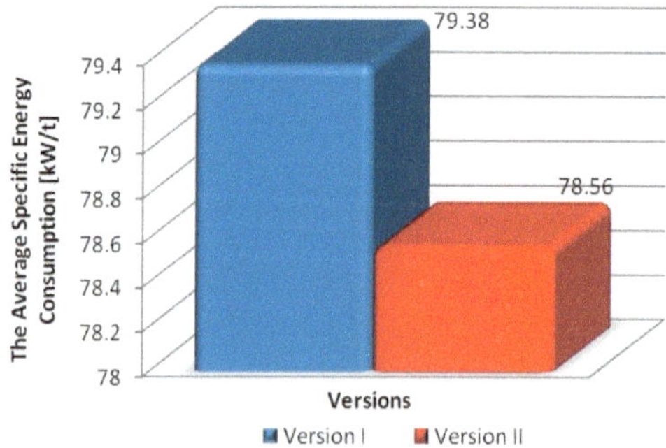

Figure 13: Average specific of electric energy consumption in the two versions

3. CONCLUSION

1. Use of a wetting system of the version II has resulted in a much greater uniformity in the percentage of moisture; humidity in version I ranged between 14.50-15.04% by neto wheat and version II of humidity ranged between 15.08-15.10%.

2. Due to ensure an almost constant humidity on one hand, and on the other hand, to improve the conditioning process by removing the grain mass of a much larger fraction of foreign bodies, as well as a considerable reduction of the ash content of the two steps of debarking, and good results have been marked in the milling, such as:

- the average quantity of flour total increased from 74.655% by version I to 76.733% in the version II;
- the average quantity of flour type 480 increased from 8.495% by version I to 8.50% in the version II;
- the average quantity of flour type 650 increased from 58.865% by version I to 63.0% in the version II;
- the average quantity of flour type 1350 decreased from 7.28% by version I to 5.23% in the version II;
- the average quantity of ash of flour type 480 decreased from 0.48% by version I to 0.464% in the version II;
- the average quantity of ash of flour type 650 decreased from 0.65% by version I to 0.625% in the version II;
- the average quantity of ash of flour type 1350 decreased from 1.35% by version I to 1.307% in the version II;
- the average quantity of ash of flour total decreased from 0.7% by version I to 0.652% in the version II;

3. Elimination of a bigger quantity of foreign bodies of wheat mass and ensuring a nearly constant moisture to wheat neto led to a process of grinding that the specific of electric energy consumption fell from 79.38 kW/t in the version I, from 78.56 kW/t of the version II.

REFERENCES

[1] Arsene C., Moraru C., Theoretical Basis of Hydrothermal Treatment Process, University of Galati, 1998.
[2] Banu C. et. al., Handbook of Food Engineer. Vol. I and Vol. II, Technical Publishing House, Bucharest, 1999.
[3] Brătucu Gh. et. al., High and Transport Machinery in Food and Agriculture, University of Braşov, 2011.
[4] Brătucu Gh., Lupea, I.D., Istrate A.M., Automating the Process of Moistening of Wheat for Flour Production in the Journal Agricultural Mechanization, No. 2/2004, p. 28-31, ISSN 1011-7296, Bucharest, 2004.
[5] Costin I., Influence of Technologicaly Processes on Food Products Quality, Technical Publishing House, Bucharest, 1979.
[6] Leonte M., Technologies and Tools in the Milling Industry, Millennium Publishing, Piatra Neamt, 2001.
[7] Rus F., Bases of the food industry operations, University of Braşov, 2001.

SOIL DEFORMATION PROCESS ANALYSIS UNDER THE AGRICULTURAL TOOLS TYRES

Hodirnau Marius[1], Dinu Liviu[2]

[1] Universitatea Transilvania din Braşov, Braşov, ROMÂNIA, mariushod@yahoo.com
[2] Universitatea Transilvania din Braşov, Braşov, ROMANIA, dinuliviu2002@yahoo.com

Abstract. *The paper present the contact surfaces between tyre and soil, of the pressure on the soil in the contact surface, and of the deformation depth of the soil, depending on wheel load, tyre constructive elements, air pressure in the tyres and traffic intensity..*
Keywords : *diameter, width, wheels, track*

1. INTRODUCTION

Soil compaction can be defined as the reduction of the degree of soil aeration under the influence of external factors. By changing the aeration and moisture regime, compaction strongly influence the activity of soil microorganisms especially those nitrifying where oxygen is one of the limitations of their work. The direct effect of soil compaction on porosity is manifested by reduced ability to retain water and air sustainably.

In general soil compaction due to agricultural traffic, results in damage to the soil surface and deep layers, changing the pore distribution by size, pore continuity and thus the permeability for water and air. These can have negative effects on soil biological activity on soil physical and chemical equilibrium and finally the production.

2. MATERIAL AND METHODS

In the vertical soil application of a rigid body in the form of a plate (Figure 1). Stress distribution in the soil is more complex than for soil application in the horizontal direction (eg tensile or running wheel). Figure 1 shows the situation in which the board is down on the ground and in Figure 1, b is shown the penetration into the soil outside of a plate pressed with a small force N, the soil is subjected to shear stress, the whose value is less than the amount which results in the occurrence and development of breakage. After product overload plaque rupture occurs by shear (Fig. 1, c), and by (slice) located beneath the deformed structure continues to fall and the soil deforms.

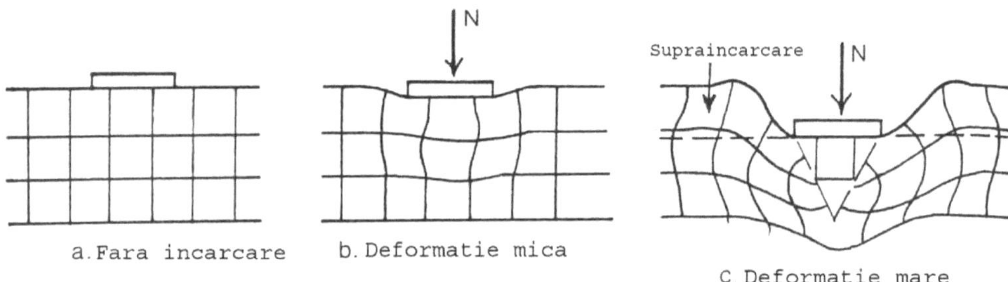

Fig. 1 The ground pressure of a plate:
a- without pressing force; b - small load that produces ground breaking;
c - the local press that produces ground breaking shear.

Natural soils are characterized by a gradual change of the load required for the transition from (b) and (c) of Figure 1, which leads to a gradual change (smooth) the immersion penetration by force.

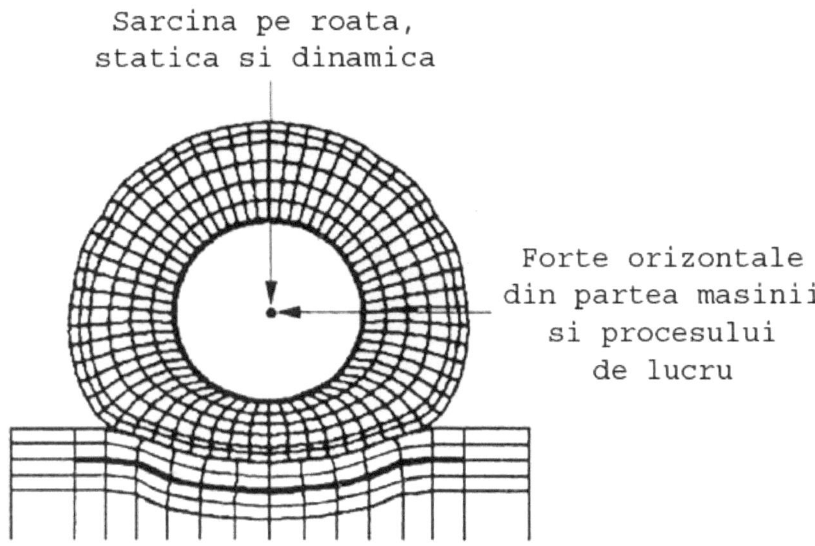

Fig.2. Scheme tire and soil deformation in the contact area between tire and deformable ground

In the process of moving the wheel to the road surface external forces acting on the machine frame on wheels produce elastic deformation of the tire contact area between the wheel and the road (Fig. 2) [3]. When the rolling wheel, the outer surface of the tire produces a strain deformed, and the soil. As the wheel passes over the contact area there is a spring-back to its original tire surface, whose elasticity is determined by the specific construction of the carcass of the tire and the air pressure in the tire. The variation of air pressure in tire size will affect not only the surface, but also a partial modification of the tire stiffness and thus the ability of the tire to take the peak pressure in the forming the contact surface.

By running (passing) wheel tractors and agricultural machinery deformable soils occurring movements of soil particles (Figure 3). Both in the interface between the wheel and the ground and in deep soil layers both horizontally and vertically [4]. As a result of various experimental researches carried out by the researchers it was found that the removal of soil particles horizontal deflection decreases with increasing depth of the vertical movement of the particles is influenced by air pressure in the tire, the geometrical characteristics of the tire, the load of the pneumatic tire load speed travel and spinning tire surface contact area (where the drive wheels).

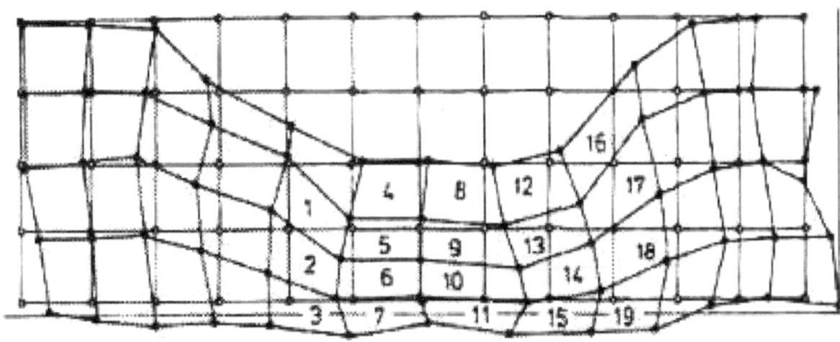

Fig. 3 Scheme Explanatory ground deformations induced by wheel traffic.

After the wheel on the surface of the soil occurs deformable soil residual deformation due to the plastic deformation of the soil, due to the irreversible movement (flow) into the soil particles. Deformations elastic (reversible) soil only appear when moving the wheel on the ground and can only be quantified at this time, by using an appropriate method of measurement

To assess the correlation between the geometry of the tire and wheel geometry trace of ground deformations due to residual soil in Figure 4 is shown the geometric model of the tire-ground interaction. This scheme allows highlighting quantities (parameters) characterizing the elastic deformation of the tire (radial deformation) z, the contour surface soil all at runtime b and residual soil deformation br. Also, the scheme is envisaged and vertical components of strains and wheel geometrical parameters.

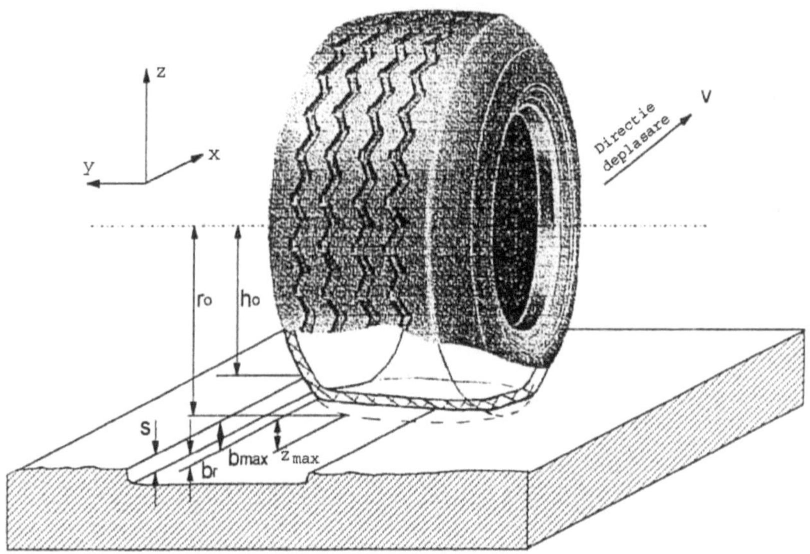

Fig. 4 The geometry of the tire-ground in the running wheel.

The relationship between the maximum sizes that appear in the plan xOz wheel-soil interaction, geometrical model shown in Figure 4 is described by the following relationship:

$$r_0 = h_0 + b_{max} + z_{max},, \tag{1}$$

which shows the relationship:

$$b_{max} = r_0 - h_0 - z_{max} \tag{2}$$

in which:

r0 is the radius of the tire undeformed (free range);

h0 - height from the ground to the wheel axis;

bmax - maximum deformation of the soil;

zmax - the maximum deformation of the tire casing;

s - all in soil depth;

br - residual deformation of the soil.

The maximum deformation of the soil can be determined from the relationship:

$$b_{max} = s + b_r, \tag{3}$$

For accurate assessment of soil deformation shift (run) wheel is to be determined (measurements) of the maximum vertical deformation of the soil bmax. Because the residual strain of the soil br not be measured directly, the value of the maximum deflection b max of the land can not be determined from equation (3), thus to determine the maximum deformation of the ground zmax is necessary to measure the components of equation (2), i.e., the height h0 wheel axis to the ground and undeformed tire radius r0 (free range). On the basis of the tds used in the method, the elastic deformation of the tire is determined experimentally by methods based on laser technique, and the values of r0 and h0 can be determined through manual measurements. As shown in figure 4.4 the height from the ground to the wheel axis h0 is the only parameter that does not depend on the transverse coordinate y, because the wheel is charged only in the vertical direction and moves only in the horizontal direction.

CONCLUSIONS

• Reduce the area of contact between the wheels and the ground is generally achieved by increasing the contact surface, a process that can be achieved by the following methods: the use of low tire pressure, tire fitting wheel tractors with large width or twin wheels, the use of special low-pressure tires.
• Maximum pressure of the tire and the ground contact surface occurs. The effect of the load weighing on the wheel is the layer of soil under the ground tire pressure application to the three-dimensional spread in vertical direction (producing a compaction process, the compaction of soil) in the longitudinal direction (the shearing process of the soil) and laterally (soil discharge process).
• Assessment of how to apply soil compaction process is generally to the shape of distribution lines equal ground pressure, called isostatic curves, depth distribution in soil layers isostatic curves wheel load depends on the size, size tire (special year balonajului tire width), internal pressure of the tire size wheel spin (an event wheels) ..
• Depth propagation curves isostatic soil layer under wheel increases with increasing duration. Therefore, by increasing the running speed of the wheels on the ground is required shorter, and his request that depth is reduced.

REFERENCES

1. MCK YES, E. Soil cutting and tillage. Elsevir, Amsterdam, 1985.
2. MCK YES, E. Agricultural Engineering Soil Mechanics Elserier Sciens Publishers, Amsterdam – Oxford - New York - Tokyo, 1989.
3. SCHWIEGER, H. Untersuchung neuartiger Laufwerke und lasergestützte Erfassungen Reifen-/Bodenverdichtung. Dissertation. MEG 289, Kiel, 1996.
4. MOLNAR, I. Cercetări privind influenţa maşinilor agricole şi tractoarelor asupra compactării solului. Teză de doctorat, Universitatea Tehnică din Cluj Napoca, 2008

ACKNOWLEDGEMENT: This paper is supported by the Sectoral Operational Programme Human Resources Development (SOP HRD), financed from the European Social Fund andby the Romanian Government under the project number POSDRU/159/1.5/S/134378.

COMPARATIVE RESEARCH ON THE TECHNOLOGY AND EQUIPMENT USED FOR DECONTAMINATION BERRIES

D. D. Păunescu[1], Gh. Brătucu[1]
[1]Transilvania University of Braşov, Braşov, ROMANIA, paunescu.dan@unitbv.ro

Abstract: *No matter how careful berries would be harvested, they will not have the physical-biological purity required for processing and for direct transmission to consumers. Therefore, various types of technology are applied in order to remove the foreign or degraded plant material, the mineral or organic impurities and even for the destruction of some microorganisms or germs, which shouldn't come in finished products.*
Keywords: *alteration, berries, decontamination*

1. INTRODUCTION

The specific conditioning technology for each plant should include operations that simultaneously satisfy the requirements of purity, keeping valuable items undegraded and minimal costs, so that the final product has the best characteristics at affordable prices. Important roles in this ensemble have the equipment used which must be continuously improved and adapted to the specific requirements of each plant [1].

2. TECHNOLOGY AND EQUIPMENT USED FOR DECONTAMINATION BERRIES

After harvesting, the berries are directed to a particular technological process according to the purpose of use, i.e. immediate outlet for food consumption, storage and keeping fresh for staggered distribution in human food consumption, for seed or further processing. Each of the destinations of use requires a certain way of previous preparation, summary or more complex, known as conditioning.
After breaking the link with the parent plant, berries remain living organisms maintaining their natural immunity and continuing to carry out life slowly. To prolong this state appropriate measures are required to be taken to ensure the equilibrium between berries internal factors and the external factors of the environment. However, even in these circumstances, the berries can maintain their original quality only a certain time, because a number of physicochemical, chemical, biochemical and microbiological changes occur gradually [2].

2.1. Decontamination of berries by washing

Soft textured fruits that have a low degree of mechanical impurities and microbiological contamination are cleaned with washing machines by splashing. This type of machine can operate independently or be part of a complex technological line of industrialization, making the final washing operation. Washing is done only by sprinkling.
Equipment shown in Figure 1, produced by CESARE TAVALAZZI SRL, ITALY, is entirely made of stainless steel except for the elevator belt, is used for effectively washing different types of little fruit.
The product is washed by means of the forced turbulent water flow and by the air that is sent from a compressor on the bottom of the tank.

Figure 1: Washing tank with elevator manufactured by CESARE TAVALAZZI, Italy [3]

In Figure 2 is shown a washer equipment produced by the company NIKO, Slovenia. The equipment is designed for washing and cleaning fruit before processing.

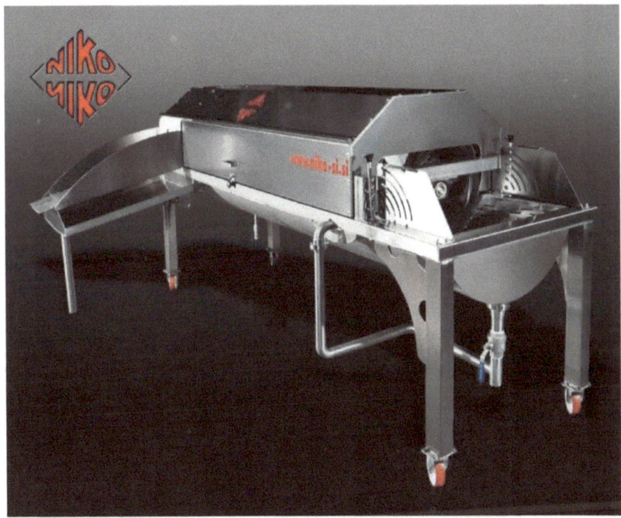

Figure 2: Washer equipment produced by the company NIKO, Slovenia [4]

Fruit in the washer is washed with water and soft brushes which leave it intact.

2.2. Decontamination of berries through chemical methods

Chlorine has been used for a long time as a disinfectant in the food industry. Several studies have evaluated its effectiveness for the decontamination of fresh produce and the concentrations used are generally about 5 to 20 ppm, for 1 to 2 minutes of decontamination.
Some of the problems that may occur at the disinfection of surface products using clorine depend on the products' topography. The presence of fissures, cracks or other natural openings may cause the formation of hypochlorous acid and decreases the efficiency of accessing microbial cells.
Treatments carried out with *hydrogen peroxide* (H_2O_2) to the fruit and vegetables such as carrots, broccoli, cauliflowers, strawberries, and raspberries were found to be ineffective, and even more, may lead to changes in the organoleptic properties of the product [5]

Using *water electrolysis* (AcEW) in the decontamination of fruit is a relatively new phenomenon. This method involves the electrolysis of an aqueous solution of sodium chloride (NaCl) between an anode and a cathode. AcEW generally has a pH below 2.7, a high oxidation power and can provide between 30 to 50 ppm of free chlorine [6].

2.3. Decontamination of fruits using ultrasound

By travelling through water with a frequency of 30 to 50 kHz, *ultrasonic waves* produce millions of microscopic air cavities loaded with vacuum energy. These air cavities, when getting in contact with surfaces create a vacuum effect. The result is the quick separation of greases, carbons, chemicals, blood, particles from these surfaces. In the mean time, these surfaces are also cleaned from any *bacteria* or living organism [7].

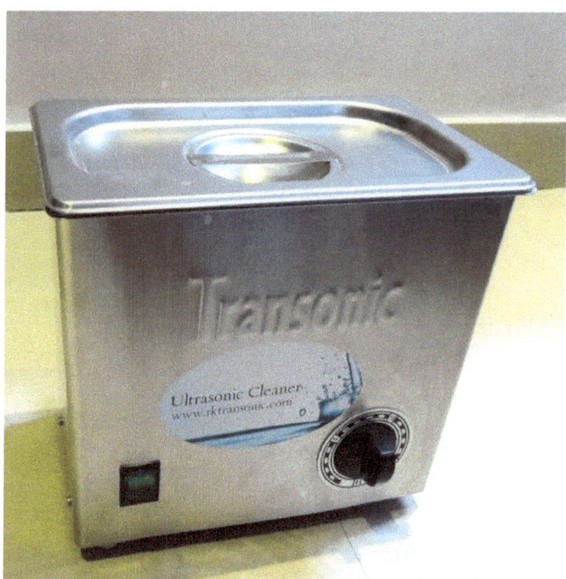

Figure 3: Ultrasonic equipment for decontamination RKT 2000

An equipment that uses ultrasound to decontaminate fruit (Figure 3) is manufactured by RKTransonic Engineers Pvt. Limited, India. It has a load capacity of 25 kg of fruit or vegetables and an average power of 4000 Watts.

2.4. Decontamination of berries using pulsed electric fields (PEF)

The cell membranes of microorganisms, plant or animal tissue can be made permeable by using Elea PEF technology. This process of electroporation is suitable for use in a broad range of food and bio-processes using low levels of energy and for this means the equipment shown in Figure 4 is used. [8].

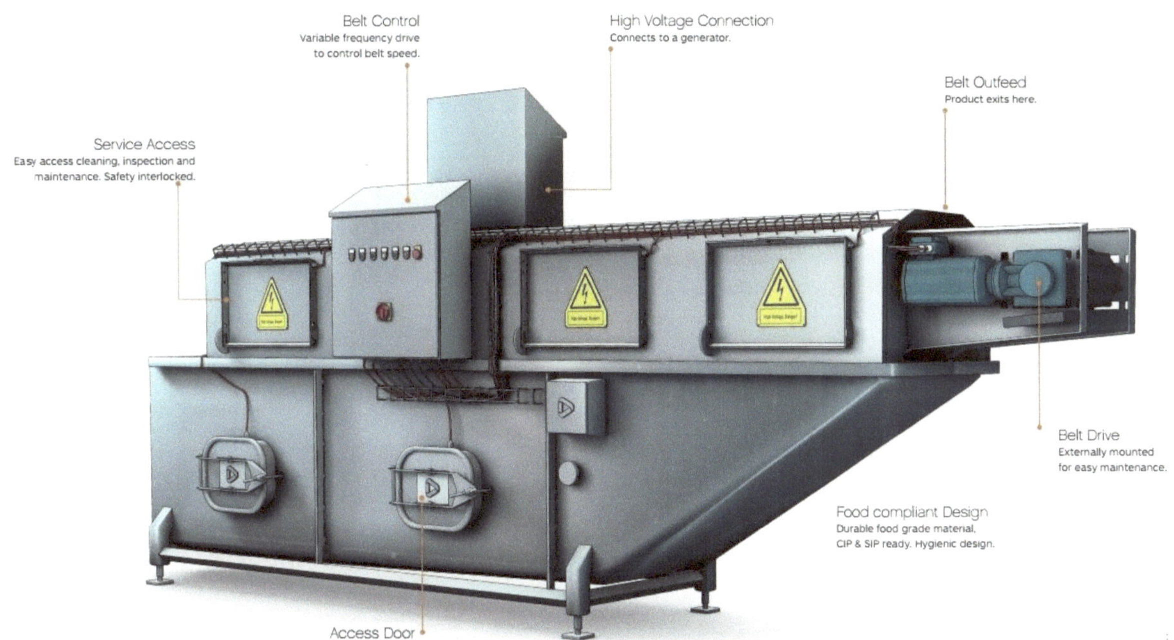

Figure 4: Technical equipment PEF used for microbiological decontaminating [8]

2.5. Decontamination of fruit using ultraviolet radiation UV-C

Worldwide there is an increasing concern related to the use of UV-C radiation for decontamination of food (cheese, meat, bread) and various fruits, companies involved perfecting technologies every year and always pulling new equipment on the market.

Generally, for the food decontamination, as a source of UV-C, low pressure UV-tubes are used, which have a maximum monochromatic emission at the wavelength of circa 254 nm, or average pressure UV-tubes, which produce a polychromatic light on a wider frequency spectrum. Their power ranges from 10 to 20 W to 25 kW.

Through the production and distribution of UV-C decontamination tunnel (Figure 5) North American company DDK SCIENTIFIC CORPORATION has allowed a wide range of food producers from the bakery industry, the fruit processors to enjoy the benefits of these new types of equipment [9].

Figure. 5: The decontamination tunnel produced by DDK SCIENTIFIC CORPORATION [9]

The company DARO UV SYSTEMS LTD from the United Kingdom produces equipment primarily used for the meat and bread surface decontamination, and makes tunnels on request, which can be used by the customer specifications for any food product (Figure 6).

Figure 6: The UV-C decontamination tunnel produced by DARO UV SYSTEMS LTD [10]

Also, UV TECHNOLOGY LIMITED company from the United Kingdom has made equipment using UV-C technology with applications in fruit processing (Figure 7).

Figure 7: Decontamination tunnel produced by UV TECHNOLOGY LIMITED [11]

An equipment based on bactericidal effect of UV-C has been manufactured by the company REYCO SYSTEMS U.S.A. (Figure 8). It consists basically of a rotating stainless steel drum, where the products are rotated and exposed to the radiation over their entire surface. Thus viruses, bacteria, molds, fungi and yeasts are distroyed, by disrupting the structure of DNA or RNA of the cell.

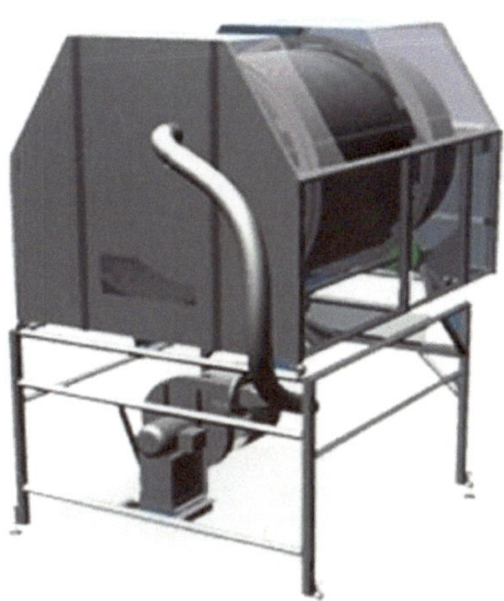

Figure 8: The UV-C Trumbling Drum decontamination equipment produced by REYCO SYSTEMS [12]

Germicidal lamps and some special manufactured LEDs used for food decontamination emit UV-C radiation. This type of radiation induces the dimerization of nitrogen base pairs of DNA (for example, dimerization of thymine (Figure 9), welding the DNA strands in that place. As a result, gene transcription (copying of information from DNA to RNA) is blocked, resulting in stopping the cell division and ultimately the cell death (bactericidal).

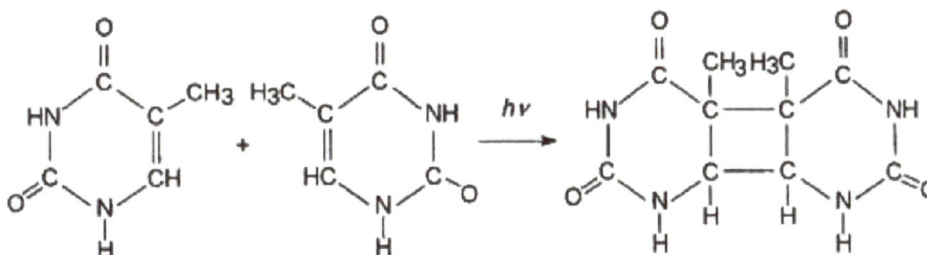

Figure 9: Dimerization of thymine

The amount of energy can be controlled to get the desired effects in terms of conservation, while maintaining quality, safety and nutritional caracteristiques of the fruit.

3. CONCLUSION

- No matter how careful berries would be harvested, they will not have the physical-biological purity required for processing and for direct transmission to consumers;
- The specific conditioning technology for each plant should include operations that simultaneously satisfy the requirements of purity, keeping valuable items undegraded and minimal costs, so that the final product has the best characteristics at affordable prices;
- After harvesting, the berries are directed to a particular technological process according to the purpose of use, i.e. immediate outlet for food consumption, storage and keeping fresh for staggered distribution in human food consumption, for seed or further processing.;
- Soft textured fruits that have a low degree of mechanical impurities and microbiological contamination are cleaned with washing machines by splashing;
- Other methods for decontamination berries are: chemical methods, ultrasound, pulsed electric fields (PEF) and non-ionizing radiation (UVC).

REFERENCES

1. C.C. Florea, Contributions to the improvement of technology and equipment used for the conditioning of herbs and berries before processing, PhD thesis, Transilvania University of Brasov, BRASOV, 2013
2. Gheorghe, A., et al .: The biochemistry and physiology of fruit and vegetables, Academy Publishing House, Romania, Bucharest, 1983.
3. *** http://www.tavalazzi.com
4. *** http://niko-si.si
5. *** www.herbs2000.com
6. *** www.science.gov
7. *** http://rktransonic.wordpress.com
8. *** http://www.elea-technology.eu
9. *** www.ddkscientific.com
10. *** http://www.surfacedisinfection.co.uk
11. *** http://www.uvtechnology.co
12. *** www.reycosystems.com

ACKNOWLEDGEMENT: This paper is supported by the Sectoral Operational Programme Human Resources Development (SOP HRD), financed from the European Social Fund and by the Romanian Government under the project number POSDRU/159/1.5/S/134378.

UTILIZATION OF TANGENTIAL FILTERS FOR INCREASING ECONOMIC AND QUALITATIVE PERFORMANCE OF WASTEWATER TREATMENT PROCESSES

Zârnoianu Daniela, PhD Student,[1] Prof. Simion Popescu, PhD,[2] Marin Radu, PhD[3]
[1] Transilvania University of Brasov, Romania, e-mail: daniela.zarnoianu@yahoo.com
[2] Transilvania University of Brasov, Romania, e-mail: daniela.zarnoianu@yahoo.com
[3.] CCMMM Bucharest, Romania, e-mail: office@ccmmm.ro

Abstract: *The paper discusses basic aspects of the filtering process and the construction of tangential filters used in wastewater treatment. Further presented are the schematic, flow and construction of an experimental (pilot) test rig of the filtration system efficiency, consisting of 4 filtering modules equipped with stainless steel sieves of 475 μm, 100μm, 80μm and 20 μm mesh size, respectively, included in the system subsequently to pre-filtering through a 1μm mesh size filter. In order to determine filtering efficiency for a given feed flow rate, liquid pressure and the concentration of suspended particles in clear w ater (in mg/l) were measured at the filter inlets and exits after certain functioning times (within an interval of 60 minutes). Eventually the variation graphs of the determined parameters versus time are presented, followed by conclusions concerning the efficiency of the analysed filtration system.*
Keywords: *wastewater, tangential flow filtration, filtering efficiency, experimental rig, fluid pressures, suspended particle concentrations*

1. INTRODUCTION

Filtering is a separation process of solid particles (impurities) from fluids when the suspension flows through porous permeable media called filtering media of filters. The flow of the suspension at a rate Q through the filtering medium is caused by a pressure drop Δp between pressure $p_1$preceding and pressure p_2 succeeding the filtering medium (fig. 1) [2; 5; 6].

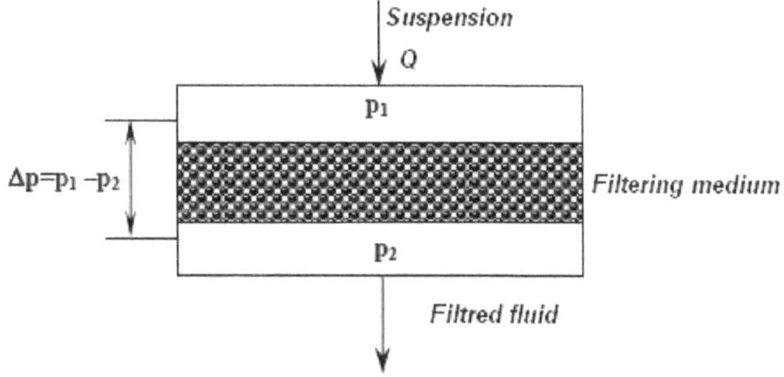

Figure. 1 . Principle schematic of the filtering process

A series of physical, physical-chemical and biological phenomena occur in the filtering processes of fluids (considering the general case of wastewater), the complexity of which is influenced by a number of factors: type and characteristics of the filtering medium, type and characteristics of the suspension in the filtered fluid (size, shape and type of the particles, density and concentration), as well as by the filtering conditions (filtering flow rate, pressure, temperature, filtering procedure) [1; 2;5;6].
The filtering media most used for wastewater treatment are granule beds, sieves and membranes. The filtration mechanism can be explained by two approaches: retention of the solid particles at the surface of the filtering medium pores and retention of the particles within the filtering medium. Depending on the direction of flow and the particle size (d_i) – to – pore size (D_p) ratio, fluid filtration can be conducted in three distinctive ways (fig. 2): surface, depth and, respectively, tangential filtration. The graphs of figure 2 present the variation versus time of the liquid flow rate Q, the pressure drop Δp and the thickness of the particle layer built up in the filter, wherefrom

follows that tangential filtration (bottom figure) is more efficient than surface filtration and depth filtration (top figures) [5; 6].

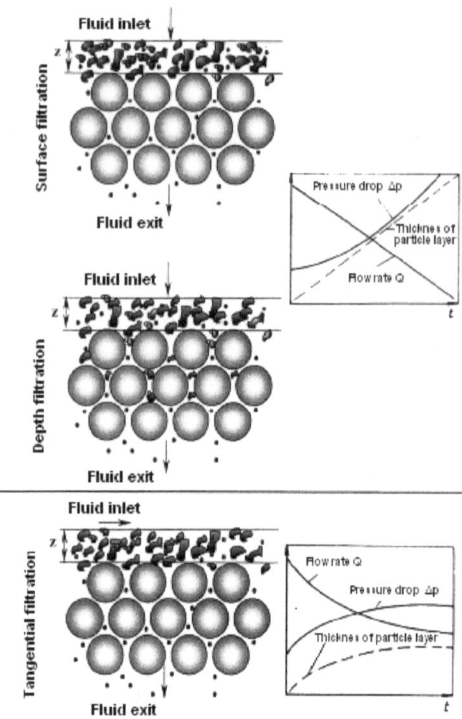

Figure 2. Filtration methods through permeable media (surface, depth and tangential filtration) and variation versus time of the basic parameters of these methods

Surface filtration occurs for $d_i > D_p$, when practically all particles are retained at the inlet surface of the permeable porous medium in the direction of flow.

In *depth filtration* the finer particles can penetrate the thickness of the permeable porous layer, while being gradually removed from the fluid consequently to their impact with a large number of obstacles within the filtering medium.

As in *tangential filtration* the fluid's direction of flow is parallel to the permeable porous medium, the fluid „flow-sweeps" the filter surface, thus diminishing clogging and preventing the building up of a particle layer at the filter surface. The structure of the permeable medium can be regenerated by counter-current flushing, thus rendering tangential more efficient than surface and depth filtration. As in the latter two the flow rate falls abruptly to zero, the pressure drop increases exponentially consequently to the growing thickness of the particle layer. Tangential filtration allows closed circuit recirculation and simultaneous recovery of both filtrate and separated concentrate, a further benefit under the aspect of specific energy consumption.

The developments in high performance synthetic membrane manufacturing technology led to increased utilisation of *membrane filtration of liquids* in various configurations [3; 5; 6]. A membrane can be seen as a barrier separating two compartments, being a selective filtration medium that allows the preferential transfer of a particle, molecule, phase or substance under the action of a motor transporting force. The filtering membranes are manufactured from various materials: organic (polymers), inorganic (metallic, ceramic, glass, active coal), mixed (hybrid polymers, composites, etc.).

Wastewater treatment utilises the following types of filtering membranes: large pore *microfiltration* membranes *(MF)* that retain large solid particles and various microorganisms, smaller pore *ultrafiltration* membranes *(UF)*that in addition to large particles and microorganisms can also retain bacteria and soluble macromolecules; *reverse osmosis* membranes *(OI)* that retain particles as well as numerous species of small molecular weight, like salt ions, organic compounds, etc.) and *nanofiltration* membranes *(NF)* with pore sizes of 10 Å or smaller, that have an operational behaviour between that of *OI* and *UF* membranes.

The relative flow of the liquid in relation to the membrane surface can be parallel (tangential flow) or perpendicular (frontal or piston type flow). In the case of *tangential flow filtration (TFF)* most of the filtered flows in parallel to the membrane surface, and a much smaller part flows through the membrane. The *„sweeping"* and cleaning effect of such flow prevents premature colmatation and differences in concentration. TFF is increasingly used in industrial processes, as it allows the utilisation of significantly greater flows than perpendicular (normal) flow.

In wastewater filtration technology multiple tangential flow filters are used (fig. 3), which includes several parallel-connected filtering elements. Such a filter consists of a cylindrical casing 3 enveloping a system of pipes with tangential flow filters (*TFF*) elements. These pipes are welded to two caps size perforated pipes that provide tangential flow filtration fluid subject. The cylindrical housing is closed with two lids that separate three chambers: chamber *A* for filtered liquid collection (the liquid that has passed through the membrane), chamber *B* for feeding of the liquid to be filtered and distribution to the filtering elements, and chamber *C* for evacuation of the unfiltered liquid that has not passed through the membrane (or the concentrated liquid in relation to the substance the concentration of which is to be increased). After lengthy operation the membrane pores can clog (colmatation), thus affecting process efficiency. This inconvenient is eliminated (decolmatation) the construction of the filter allows reverse filtering, by means of a special circuit of the liquid to be filtered, that under a greater pressure passes through the membrane in the opposite direction. Thus the liquid is fed to chamber *A* and evacuated through chamber *C*. In order to determine when reverse filtration has to be induced, the installation was equipped with fittings 5 connected to the pressure gauges that measure the pressure in chambers *A* and *B*. In practice membrane cleaning by reverse filtration is recommended when the gauge indicates a pressure drop between inlet and exit of filter that exceeds 2 bar.

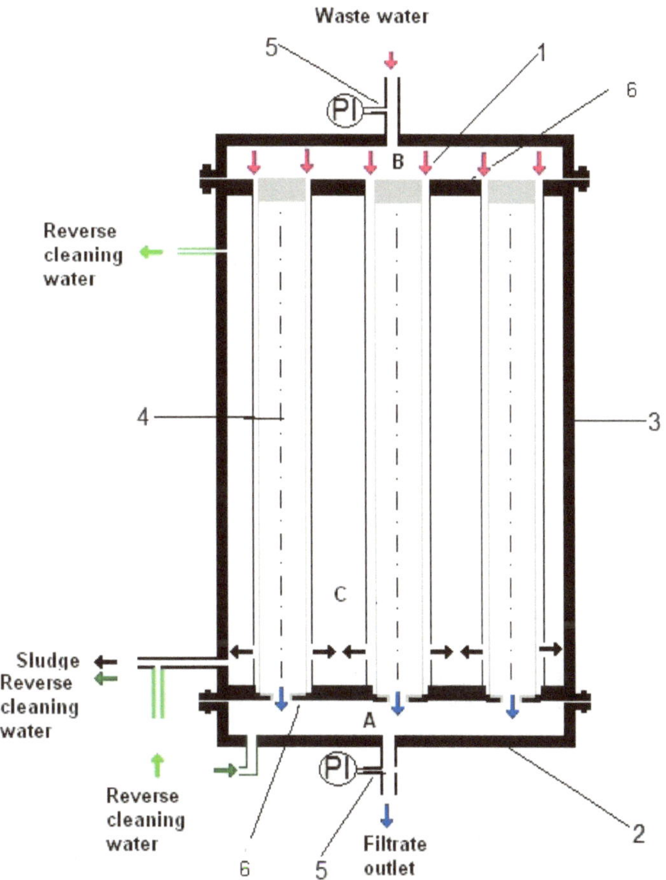

Figure 3. Constructive and operational schematic of a multifunctional tangential flow filter (TFF): 1- upper lid; 2- lower lid; 3- filer body; 4- filtering element; 5- fittings to the pressure gauge; 6- holding plate of filtering elements.

3. EXPERIMENTAL RESEARCH

The experimental research of the operational behaviour of tangential flow filtration systems of industrial wastewater was conducted on a specially built multi-level filtration pilot rig that ensures the retention of suspended particles up to 400 mg/l. The installation consists of a modular tubular filtration system operating under conditions specific for coarse filtration and microfiltration. The filtration system (fig. 4), consists of 4 filtering modules *F1, F2, F3* and *F4*

*(*of TFF type) endowed with stainless steel sieve filters [4] of 475 µm, 100 µm, 80 µm and 20 µm mesh size, respectively, included in the system after pre-filter *M1* of 1µm mesh size. The installation also includes a wastewater collection/neutralizing vessel *VS* where also the initial separation of coarse suspended particles is conducted; pump *P* and an heat recovery system, for collection of the wastewater resulted from the first filtration level, consisting of filters *F1* and *F2* (endowed with series-connected 475 µm and 100 µm mesh size filtering elements). Filters *F3* and *F4* form the second filtration level. Filters *F1, F2, F3, F4*, are endowed each with 7 filtering elements like the one in figure 3. The installation was equipped with the pressure gauge *PR1, PR2, PR3, PR4* and *PR5* that measure the pressures at the inlets and the exits of the filters *F1, F2, F3, F4.*
Figure 5 shows a view of the pilot rig and figure 6 presents the schematic of the flow of operations for the experiments carried out on the tested installation.

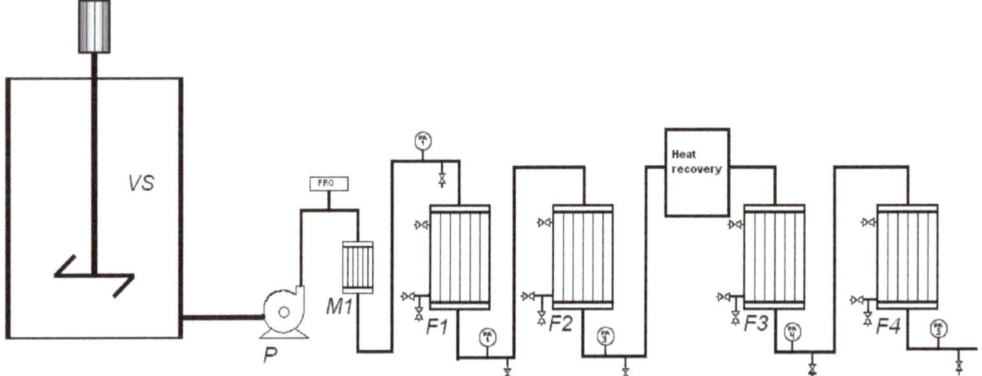

Figure 4. Flow diagram of the pilot rig used in experimental research

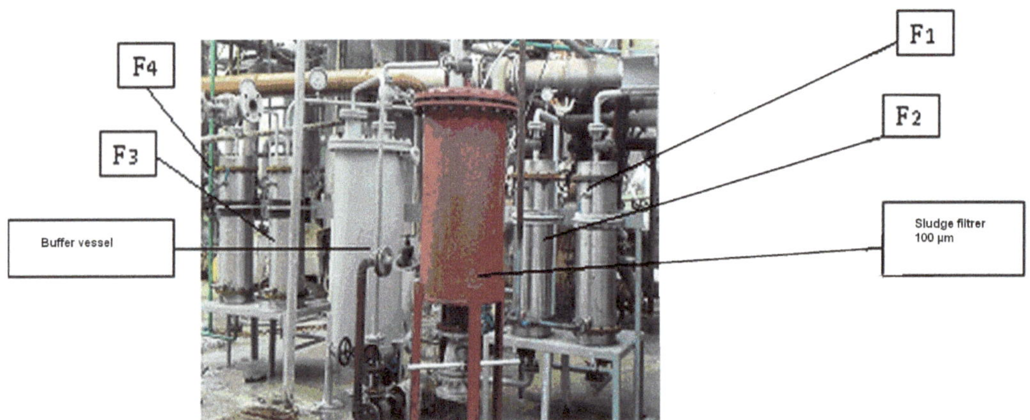

Figure 5. General view of the pilot filtration rig used in experimental research

Table 1: Variations versus time of pressures and suspension (suspended particle) concentrations in clear water at the inlets and exits of the filters.

Time	Pressure at filters					Concentrationof suspension (suspended particle) concentration in clear water				
	Before F1	After F1	After F2	After F3	After F4	Before F1	After F1	After F2	After F3	After F4
min	bar	bar	bar	bar	bar	mg/l	mg/l	mg/l	mg/l	mg/l
0	3	2.7	2.5	1.2	0.4	15878	5264	4568	3129	587
20	3.2	2.9	2.6	1.4	0.4	15878	3171	2825	2042	361
40	3.6	3.4	3	1.8	0.4	15878	2800	2350	1850	336
60	4	3.9	3.4	1.9	0.3	15878	1224	640	570	269

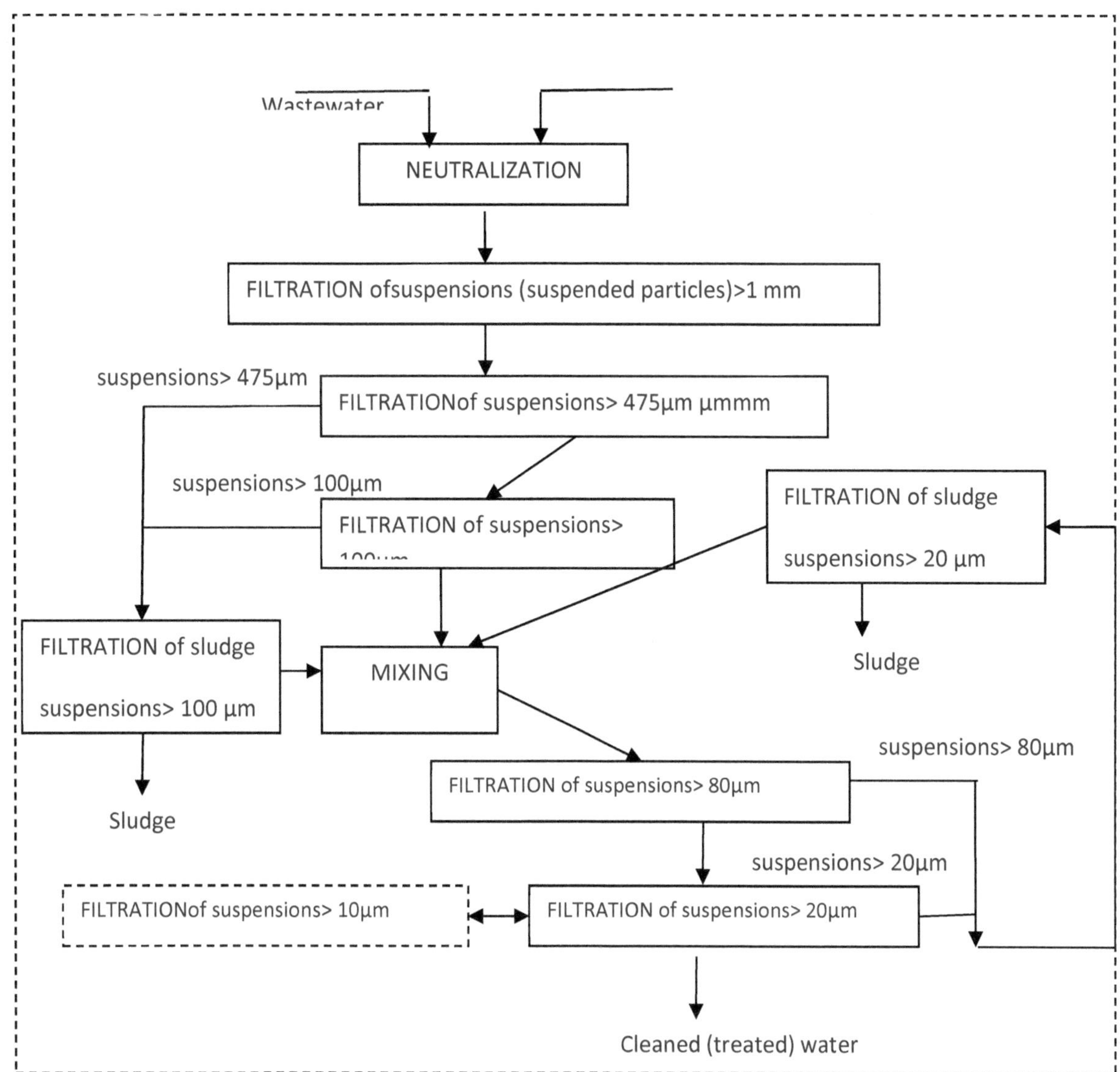

Figure 6. Flow of operations of the filtering process in the pilot rig

3. ANALYSIS OF RESULTS

In order to determine the *efficiency* of the tested filtration system, wastewater containing suspended particles of 16,000 mg/litre initial concentration (concentration in clear water) was fed into vessel *VS* and circulated by means of pump *P*. After certain operation times, within an interval from 0 to 60 minutes, the pressures were measured at the inlet of filter *F1* and the exits of filters *F1, F2, F3* and *F4*; also samples of suspension were collected in order to determine the concentrations in clear water (mg/l). The pressures were measured by the gauges included in the circuit (s. fig.4), and the concentration of the suspension was determined of the collected samples. Table 1 features the obtained experimental data.
Based on the experimental data of table 1 the graphs of variation versus time of pressures (fig. 7) and suspension (suspended particle) concentrations in clear water (fig. 8) were plotted at the inlets and exits of the filtration rig, allowing the analysis of the filtration process achieved by the studied pilot rig.

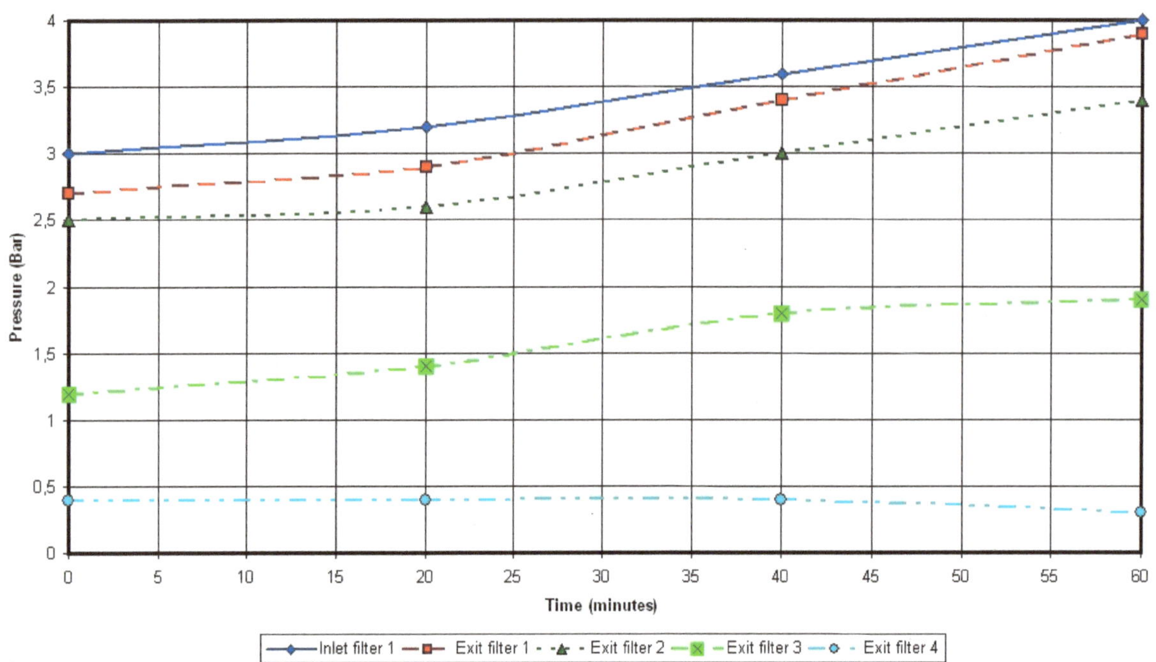

Figure 7. Evolution versus time of pressures (in bar) at the inlet of the filter *F1* and the exits of filters *F1, F2, F3* and *F4*, over a 60 minute interval

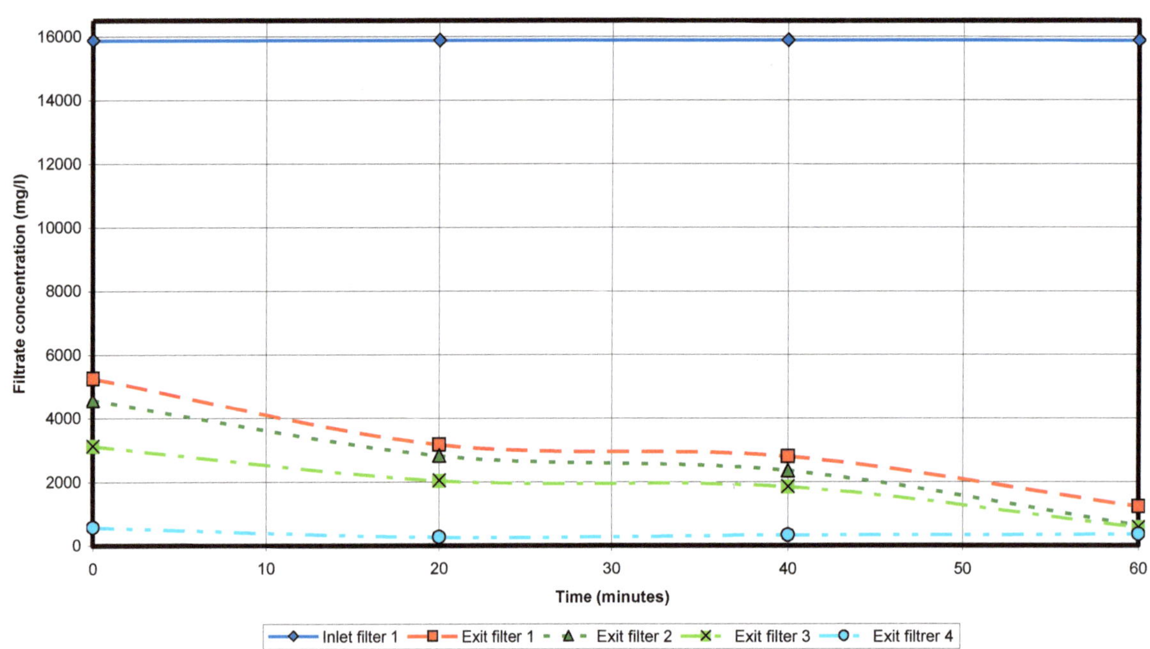

Figure 8: Evolution versus time of suspension concentration in clear water (in mg/l) at the inlet of the filter *F1* and the exits of filters *F1, F2, F3* and *F4*, over a 60 minute interval

4. CONCLUSIONS

An analysis of the graphs of figure 7 reveals that system pressures increase at the inlet of filter *F1* and the exits of filters *F1, F2* and *F3* and remain approximately constant at the exit of final filter *F4*.
An analysis of the graphs of figure 8 shows that the studied filtration system is operational, does not clog (colmatation is avoided) and has the following characteristics:

- after filtering module *F1* (with sieves of mesh size 475 µm) retention was of 80% and no particles larger than 475 µmwere found in the filtered water); the content of suspended particles decreased from initial approx. 16000 mg/l to 5264 mg/l;
- after filtering module *F2* (with sieves of mesh size 100 µm) retention was of 82,5% and in the filtered water 24,82% particles larger than 100 µm were found (because of sieve defects); the content of suspended particles decreased from initial approx. 2800 mg/l to 1224 mg/l;
- after filtering module *F3* (with sieves of mesh size 80 µm) retention was of 95,3% and in the filtered water 6,68% particles larger than 80 µm were found (because of sieve defects); the content of suspended particles in wastewater decreased from initial approx. 3100 mg/l to 570 mg/l;
- after filtering module *F4* (with sieves of mesh size 20 µm) retention was of 98,3%.the content of suspended particles in wastewater decreased from initial approx. 570 mg/l to 269 mg/l;

Overall it could be established that after the first 3 filtering modules (*F1, F2, F3*) about 95% of the initial total of suspended particles in wastewater are retained. Thus, in the completed experiment the initial concentration of the filtrate of about 16000 mg/l was reduced to 570 mg/l.

REFERENCES

[1]. Bratu, E.A., Operaţii unitare în ingineria chimică, vol. I, vol. II Editura Tehnică, Bucuresti, 1984, 1985.
[2]. Robescu, Diana et al. Tehnici de epurare a apelor uzate, Editura Tehnica, Bucuresti, 2011
[3]. El-Bourawi M. S. A framework for better understanding membrane distillation separation process, Journal of Membrane Science, vol. 285, nr. 1-2, noi. 2006, pp. 4-29.
[4]. Euro Inox. Performance of stainless steels in waste water installations, Material and Application Series, vol. 13, 2010.
[5]. Moise G. Contribuţii privind procesele de separare-concentrare în industria alimentară utilizând tehnica membranelor, Teză de doctorat, Universitatea „Lucian Blaga" Sibiu, 2008.
[6]. Rus F. Operaţii de separare în industria alimentară. Editura Universităţii Transilvania, Braşov, 2001 [7]. Techobanoglous, G., Burton, F.L., and Stensel, H.D. Wastewater Engineering. (4th Edition ed.). McGraw-Hill Book Company, 2003.
ACKNOWLEDGEMENT: This paper is supported by the Sectoral Operational Programme Human Resources Development (SOP HRD), ID134378, financed from the European Social Fund and by the Romanian Government

KINETICS OF ZINC IONS ADSORSOPTION FROM WATER AND WASTEWATER BY ION EXCHANGE RESINS

C. Modrogan [1], A. R. Miron[1], O. D. Orbulet[1], D. E. Pascu[1], C. Costache[1]

[1] *Departament of Analytical Chemistry and Environmental Engineering, Faculty of Applied Chemistry and Materials Science University "Politehnica" of Bucharest, 1-7 Polizu Street, Romania, c_modrogan@yahoo.com*

Abstract:*Innovative processes for treating industrial wastewater containing heavy metals often involve technologies for reduction of toxicity in order to meet technology-based treatment standards.*
The capacity of ion exchange resins, Purolite MN 500 and Purolite C 100 H, for removal of Zn^{2+} *from aqueous solution has been investigated under different conditions namely initial solution pH, initial metal-ion concentration, and contact time. The equilibrium data obtained in this study have been found to fit both the Langmuir and Freundlich adsorption isotherms. The adsorption of* Zn^{2+} *on these resins follows first-order reversible kinetics.*
Keywords: *ion exchange, zinc ions, wastewater, adsorbtion*

1. INTRODUCTION

Heavy metal pollution has become one of the most serious environmental problems today. The treatment of heavy metals is of special concern due to their recalcitrance and persistence in the environment. In recent years, various methods for heavy metal removal from wastewater have been extensively studied [1].

Pollutant load of industrial waste waters is the most massive and harmful type of pollution. After the Romanian adjuration to the European Union, heavy metal retention from mine waters will become a mandatory operation from at least two points of view. First, because heavy metals ions are toxic and are presently being dumped in water sources, polluting surface water as well, rendering it not healthy for drinking. Secondly, aquatic flora and fauna are affected by this pollution. Using those waters as irrigation sources is not recommended because the heavy metals accumulate in the soil and in some plants. Animals that consume these plants carry the accumulated heavy metals further in milk and meat products that we humans consume. Retaining the heavy metals is also useful from an economic point of view. Retaining those rare and expensive to produce metals could lead to a decrease in purification costs.

Various processes (i.e. adsorption, chemical precipitation/ coagulation, ultrafiltration, electrodialysis, etc.) have been practiced to remove the dissolved metal ions from aqueous solution [2]. Among these processes, adsorption is known as one of the most applicable treatment techniques, and it offers flexibility in operation accompanied with a satisfaction of lower levels of metal ion concentration in effluent [3,4]. Mostly, surface characteristics that provide the higher adsorption capacity including ion exchange ability are important to select materials as adsorbents [5–10].

Industrial effluents from electroplating industries contain high amounts of heavy metal ions, such as chromium, nickel, copper, cadmium and zinc. These heavy metal bearing wastewaters are of considerable concern because they are non-biodegradable, highly toxic and probably carcinogen. Only 30-40% of all metals used in plating processes are effectively utilized i.e. plated on the articles. The rest contaminates the rinse waters during the plating process when the plated objects are rinsed upon removal from the plating bath. Electroplating rinse waters may contain up to 1000 mg/L toxic heavy metals which, according to environmental regulations worldwide must be controlled to an acceptable level before being discharged to the environment. Several treatment processes have been suggested for the removal of heavy metals from aqueous waste streams: adsorption, biosorption, ion exchange, chemical precipitation and electrochemical methods: electro winning, electro deionization, membrane-less electrostatic shielding electro dialysis/electro deionization and electro coagulation [6]. Chemical hydroxide precipitation is the most economic and the most commonly utilized procedure for the treatment of heavy metal-bearing industrial effluents but after this treatment the wastewater stream can still contain up to 5ppm heavy metals, which is an unacceptable concentration for discharge to the environment. In order to remove heavy metals down to the ppb concentration level, the wastewater stream must be further treated using a second sulfide precipitation as a polishing step or a series of ion exchange columns.

The large amounts of precipitated sludge which contains concentrated heavy metal hydroxides or sulfides are an extremely hazardous waste and must be disposed of using special facilities at great expense to industry. From the environmental protection and resource saving point of view, effective recycling and reusing of the heavy metal wastewater is strongly expected. Closed-recycle system or so-called effluent-free technology should be developed [7].

Chemical coagulation is a quite effective method for treating heavy metal bearing wastewaters but may induce secondary pollution by adding coagulants, such as aluminum or iron salts or organic poly-electrolytes to remove colloidal matter as gelatinous hydroxides. Also this wastewater treatment process produces large amounts of sludge.

The use of physicochemical treatment generally enables the legislation concerning liquid industrial effluent to be respected but this conventional treatment does not completely remove pollution. However, as it has to cope with an increasingly strict framework, the industrial sector continues to look into new treatment methods to decrease the levels of pollution still present in the effluent, the aim being to tend towards zero pollution out flow. [8]

Zinc is one of the most important pollutants for surface and ground water. Because of its acute toxicity and non-biodegradability, zinc-containing liquid and solid wastes are considered as hazardous wastes [3,4]. Among the materials used in ion exchange processes, synthetic resins are commonly preferred as they are effective and inexpensive [9]. Cation exchange resins generally contain sulphonic acid groups.

These groups can also be carboxylic, phosphonic or phosphinic. Certain general rules for cation exchange are: (i) the exchanger prefers ions of high charge, (ii) ions of small hydrated volume are preferred and (iii) ions, which interacts strongly with the functional groups of the exchangers are preferred [10,11].

In literature, there have been various investigations about removal of heavy metals by ion exchange resins. Halle et al. observed that macro porous carboxylic cation exchanger Wofatit CA-20 in the sodium form exhibites high removal efficieny for treatment of Ni(II) ions from washings formed during the nickel plating [12]. Cu (II) ions are effectively removed from sea and river water by Amberlyst A-27 and Diaion PA-318, strongly basic anion exchangers and batocuproinodisulphonate as a chelating agent [13]. Also, Dowex HCR S/S and Dowex Marathon C resins provided adsorption capacities of 26.27mg/g and 46.55 mg/g for copper removal [14]. Macroporous strongly basic anion exchanger-Lewatit MP-500A is characterized by high selectivity for Cr(VI) ions [15].

2. MATERIALS AND METHODS

2.1 Chemicals

All chemicals used during the experiments, were purchased from Merck (Germany). All the experiments were carried out in duplicate sets. All measurements were performed in parallels in each set. The removal efficiency reported are the average of the parallel measurements of the duplicate sets and the parallel measurements.

Analytical grade reagents were used in experimental studies. Nitrate salts of test metals ($Zn(NO_3)_2 \cdot 6H_2O$ were used for preparing certain concentrations of synthetic solutions. pH adjustments were carried out by using 0.1N HCl.

Ion exchange studies were done by contacting synthetic aqueous solutions, containing 100 mg/L Zn^{2+}, with an ion exchange resin, in a batch system. In these studies, a weakly basic anion exchanger, Purolite MN 500 and Purolite C 100 H type was used.

MN 500 is a hyper crosslinked strong acid resin and C 100 H is a premium gel, polystyrenic, strong acid cation exchange resins.

2.2. Apparatus

Cintra 5 spectrophotometer was used for the determination of remaining metal concentrations in solutions. Used Zincon Method (Method no: 8009) for zinc analyses. Batch experiments were carried out in *NUVE* shaker. Testo pH-meter was used for pH measurements.

3. RESULTS AND DISSCUDION

3.1. Equilibrium studies

In order to study the ion exchange equilibrium, the ion exchange isotherms were determined. For this purpose the $Zn(NO_3)_2$ solutions (7 samples) of different concentrations (between 5 and 100 mg/L) were prepared.

Each experiment consists of the contact between the 50 mL $Zn(NO_3)_2$ solution of known concentration and 1 g resin. The resin-solution system was stirred in the thermostated vessel during 24 hours, enough to reach the equilibrium. Each sample was spectrofotometrically analyzed, and the resin loading was calculated using equation (2).

3.2. Kinetic study

1g ion exchanger was put in contact with 50 mL $Zn(NO_3)_2$ (concentration 100 mg/L) in an Erlenmayer jar. The samples were stirred at 300 rpm for 5 minutes, 10 minutes, 20 minutes, 30 minutes, 60 minutes and 90 minutes using a Heidolph Unimax shaker. The experiments were carried out at 20°C. After mixing, samples were filtrated using blue ribbon filter paper and brought to a 50 mL flask. 10 mL samples from the filtrate were taken and spectrofotometrically analysed. The coloured samples were blue. The calibration curve was achieved first.

Equilibrium studies

The equilibrium adsorption isotherm is important in the adsorption system design. The distribution of Zn^{2+} between the liquid and the solid phase in equilibrium is expressed by the Freundlich and Langmuir models. These equations are widely used, the former being empirical while the second assumes that the maximum adsorption occurs when the surface is covered by the functional groups.
Langmuir developed a theoretical equation, considering that a monomolecular adsorption layer occurs on an energetically homogeneous surface, and that there is no interaction between the adsorbed molecules. The Langmuir adsorption isotherm plot (q_e vs. C_e). C_e (Figures 1 and 2) indicates the applicability of Langmuir adsorption isotherm. The values of q_m and b were calculated from the slope and the intercept of the linear plots C_e/q_e vs. C_e (Table 1). To predict the adsorption efficiency of the adsorption process, the dimensionless equilibrium parameter was determined by using the following equation:

$$q_e = \frac{q_m \cdot b \cdot C_e}{1 + b \cdot C_e} \quad (3)$$

where: q_e - the adsorbent equilibrium concentration (mg/g); q_m – the adsorbent capacity for a monolayer adsorption (mg/g); b – equilibrium constant; C_e – the equilibrium concentration in liquid phase (mg/L).
The Langmuir adsorption isotherm model may be rearranged as follows:

$$\frac{C_e}{q_e} = \frac{1}{q_m \cdot b} + \frac{C_e}{q_m} \quad (4)$$

Table 1 shows the thermodynamic parameters calculated using the Langmuir isotherm at 20°C.
The Freundlich model is an indicative of the sorbent surface behaviour. The Freundlich equation, which is an empirical one, is as follows:

$$q_e = K_f \cdot C_e^{(1/n)} \quad (5)$$

where: K and n are constants which were experimentally determined.

$$\log q_e = \log K_f + \frac{1}{n} \cdot \log C_e \quad (6)$$

where: K_f and 1/n are Freundlich constants in respect of the adsorption capacity and adsorption intensity, respectively.
The K_f and 1/n are Freundlich constants related to the adsorption capacity and adsorption intensity, respectively. The resin behavior showed a better fit to the Langmuir isotherm than the Freundlich one.
Table 1 shows the empirical Freundlich equation coefficients (4) as obtained by logarithmic transformation.

Table 1. Langmuir equilibrium parameters at 20°C

Ion exchanger	Langmuir, 20°C			Freundlich, 20°C		
	q_m	b	R^2	R^2	k	n
MN 500	38.5	0.046	0.90	0.91	0.046	0.98
C 100 H	40.5	0.049	0.84	0.80	0.049	0.97

The isotherms obained by fitting the experimental data using the two isotherms reveal that Langmuir isotherm are more in line with the experimental results.
Comparing isotherms 1-2 the high exchange capacity of the Purolite MN 500 for Zn^{2+} ions versus the Purolite C 100 H could be observed.
The experimental data are presented in Fig. 1-2 in q_e –C_e coordinates, where q_e is the ion exchange loading at equilibrium, mg/g, and C_e – $Zn(NO_3)_2$ concentration at equilibrium, mg/L.

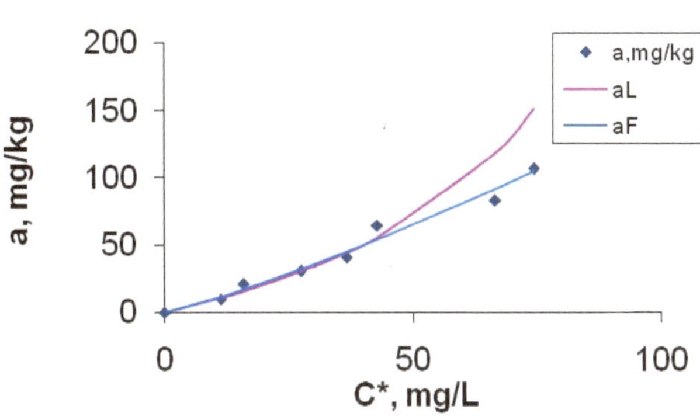

Figure 1. Equilibrium isotherm for loading Zn^{2+} onto Purolite MN 500 at 20°C

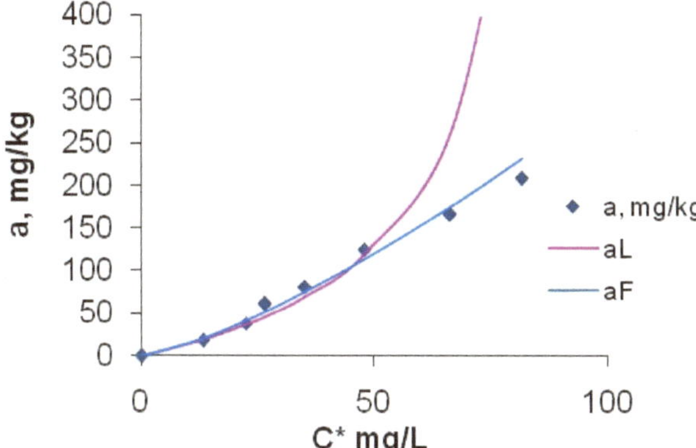

Figure 2. Equilibrium isotherm for loading Zn^{2+} onto Purolite C 100 H at 20°C

As expected, the adsorption capacity, q_m, in monomolecular layer, is higher for C 100 H cationite. This fact was verified using the linearized Langmuir equation. The resin showed a better fit for the Langmuir isotherm than the Freundlich isotherm. The results are virtually identical to those obtained by fitting the experimental results presented in Figure 1 and 2.

The aspect of the ion exchange isotherms shown in Figure 1 and 2 indicates that both cation exchangers in H^+ form have a good capacity of Zn^{2+} retention by ion exchange, but the values are strongly influenced by the exchanger type. The cation exchanger C 100H retains a larger amount of Zn^{2+} than the cation exchanger MN 500. The removal efficiency increases to an optimum dosage beyond which the removal efficiency is negligible. The Zn^{2+} concentration at equilibrium in solution increases with the increasing initial concentration of Zn^{2+} in initial solution.

The ion exchange isotherms shown in figures 1 and 2 have the same form up to 30 mg Zn^{2+}/L concentration and show similar values for the capacity of Zn^{2+} retention from water under static conditions, although total capacities and morphologies are different. On the other hand the amount of Zn^{2+} removed from water is almost identical in the case of both ion exchangers.

3.3. The kinetic study of ion exchange process

In order to characterize the process from the kinetical point of view, curves were obtained using the experimental apparatus described above.

The process efficiency is controlled by the kinetics of adsorption and hence the several kinetic models are available to predict the mechanism involved in the sorption process. Among these models, there are pseudo first order, pseudo second order and interparticle diffusion, rate equation. [15]

Pseudo-first-order:

$$\log(q_e - q_t) = \log q_e - \frac{k_1}{2,303} t \qquad (7)$$

Pseudo-second-order:

$$\frac{t}{q_t} = \frac{t}{q_e} + \frac{1}{k_2 q_e^2} \qquad (8)$$

The interparticle diffusion model was applied to different adsorption systems and was adequate to the adsorption systems [15]

$$q_t = k_p t^{0,5} \qquad (9)$$

Samples were taken after 5, 10, 20, 30, 60 and 90 minutes, the results of the experiments being presented in Fig. 3 and 4.

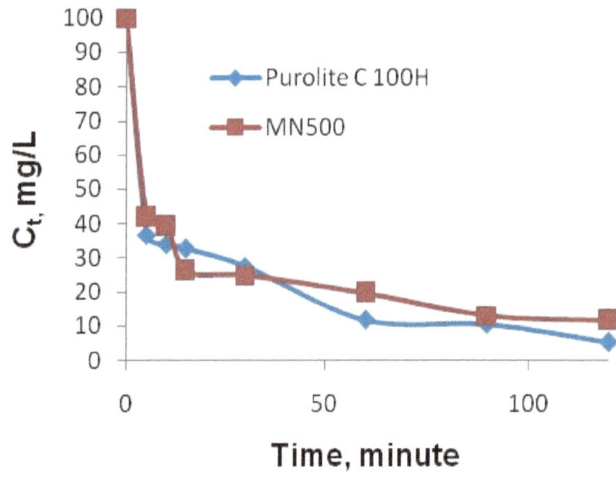

Figure 3. Variation of the nitrate concentration in the solution

The curves from figures 3 reveal that the ion exchange rate is higher in the case of C 100 H, having a higher exchange capacity. The experimental data on ion exchange kinetics for cation exchanger samples with Purolite MN 500 and C 100H are shown in figures 3, as a function of Zn^{2+} concentration in solution versus time. The kinetic curves have a similar shape and reveal a high ion exchange rate at the beginning of the process. The decrease of the ion exchange rate is correlated to the decrease of Zn^{2+} concentration in solution and with the progress of the system towards the equilibrium state.

Several kinetic models including pseudo first order, pseudo second order and pore diffusion were used for the simulation of the experimental data. The kinetic parameters calculated from equations 7-9 for the adsorption of Zn^{2+} at concentrations on Purolite MN 500 and C 100 H are given in Table 2. The pseudo second order successfully fit the adsorption kinetic.

Table 2. The Kinetic model parameters for the adsorption of nitrates on MN 500 and C100 H

Kinetic model parameters	Temperature, 20°C		Kinetic model parameters	Temperature, 20°C	
MN 500			C 100 H		
Pseudo first order	k_1	0,238	Pseudo first order	k_1	0,023
	r^2	0,876		r^2	0,913
Pseudo second order	k_2	0,033	Pseudo second order	k_2	0,041
	r^2	0,989		r^2	0,998
Pore diffusion	k_p	5,849	Pore diffusion	k_p	6,136
	r^2	0,891		r^2	0,889

4. CONCLUSIONS

The results showed that the MN 500 and C 100 H cation-exchange resins can be used as an adsorbent for the effective removal of Zn^{2+} from aqueous solution. Quantitative removal of these metals from synthetic water confirmed the validity of the results obtained in these batch mode studies. Equilibrium data of Zn^{2+} was well fitted by both the Langmuir and Freundlich models, and the order of affinity was followed as C 100 H > MN 500. The kinetic data would be useful for the fabrication and designing of wastewater treatment plants. Application of this ion-exchanges resin to wastewater treatment is expected to be economical and efficient. Further studies on equilibrium and kinetics of multi-metal systems at low concentrations are under progress.

Kinetically, adsorption of Zn^{2+} was predicted by using the pseudo-second order model with higher correlation coefficients ($r^2 > 0.98$ from C 100 H and > 0.95 from MN 500 respectively). Moreover, initial adsorption rate, h (mg/g min), linearly increased. As a result obtained from large-scale reactor in series, the removal efficiency of Zn^{2+} was approximately 78–98% from C 100 H and 70-95 % from MN 500, respectively, and the proposed process containing C 100 H was applicable to treat a real plating wastewater.

REFERENCES:

1. Nganje T.N, Adamu C.I, Ukpong E.E, Heavy metal concentrations in soils and plants in the vicinity of Arufu lead-zinc mine, Middle Benue Trough, Nigeria, Department of Geology, University of Calabar, Nigeria, 2010
2. Fenglian Fu, Qi Wang, Removal of heavy metal ions from wastewaters: A review, Journal of Environmental Management Volume 92, Issue 3, 407–418, 2011
3. Çoruh, S., Turan, G., Akdemir, A., ve Ergun, O.N., The influence of chemical conditioning on the removal of copper ions from aqueous solutions by using clinoptilolite, Environmental Progress & Sustainable Energy, 28, 2, 202-211, 2009
4. Tabata M., Ghaffar A., Eto Y., Nishimoto J., Yamamoto K., Distribution of heavy metals in interstitial waters and sediments at different sites in Ariake Bay, Japan. E-Water,1-24, 2007
5. Charles J., Bradu C., Morin-Crini N., Sancey B., Winterton P., Torri G., Badot P. M., Crini G., Pollutant removal from industrial discharge water using individual and combined effects of adsorbtion and ion-exchange processes: Chemical abatement, Journal of Saudi Chemical Society, 2013
6. Deleanu C., Simionescu C. M., Constantinescu I., Adsorption Behavior of Cu2+ ions from aqueous solutin on chitosan, Rev. Chim., nr. 6, Bucureşti, 2008
7. Macoveanu M., Bâlbă D., Bâlbă N., Gavrilescu M., Şoreanu G., Procese de schimb ionic în protecţia mediului, Editura Matrix ROM, Bucureşti, 2002Mămăligă I., Tehnici moderne de separare a sistemelor omogene, 33 - 40, 2011
8. Magdi Selim H.,. AmacherM. C, Sorption and Release of Heavy Metals in Soils: Nonlinear Kinetics, In: Heavy Metals Release in Soils, H. Magdi Selim, D. L. Sparks, (Eds.), CRC Publishers, Boca Raton, 1 - 29, 2001
9. Fawell J. K., Lund U., Mintz B., 2nd ed. Vol.2, Health criteria and other supporting information, World Health Organization, Geneva, 1996
10. Kanawade S. M., Gaikwad R. W., Removal of zinc ions from industrial effluent by using cork powder as adsorbent, International Journal of Chemical Engineering and Applications, Vol. 2, No. 3, 199-201, 2011
11. Bradl H.B., Adsorption of heavy metal ions on soils and soils constituents,Journal of Colloid and Interface Science, 277, 1 - 18, 2004
12. Bejan M., Rusu T., Avram S., Metode performante de recuperare a metalelor grele din apele de mină, Buletinul Agir, nr. 1/2007
13. Ghomri F., Lahsini A., Laajeb A., Addaou A., The removal of heavy metal ions (cooper, zinc and cobalt) by natural bentonite, Larhyss Journal, 12, 37- 54, 2013
14. Gakwisiri C., Raut N., Al-Saadi A., Al-Aisri S., Al-Ajmi A., A Critical Review of Removal of Zinc from Wastewater, Proceedings of the World Congress on Engineering, Vol I, July 4 - 6, 2012

15. Modrogan C., Apostol D. G., Butucea O. D., Miron A. R., Costache C., Redha K., Kinetic study of hexavalent chromium removal from wastewaters by ion exchange, 12, 5, 929-936, 2013

THE EFFECT OF SALT WATER TREATMENT ON BENDING TEST OF JUTE TISSUE/EPOXY COMPOSITES

F. R. Coterlici[1], V. Geaman[1]

[1]Transilvania University, Brasov, Romania, e-mail: coterliciradufrancisc@yahoo.com

[1]Transilvania University, Brasov, Romania, e-mail: geaman.v@unitbv.ro

Abstract: *Nowadays composites can be made by replacing synthetic fibres with different types of natural fibres. Main factor in the composite is to have a good strength of adhesion between polymer matrix and fiber. More unidirectional samples with different number of jute tissue layers was immersed in salt water with different concentrations for 90 days (2.160 hours) at room temperature. This paper present the bending test before and after treatment in salt water for some samples. The bending test evaluation was performed with Universal Testing Machine – WDW-150S type, from Jinan TE Corporation.*

Keywords: *jute fibre, natural composites, polymer-matrix composites, salt water, flexural strength.*

1. INTRODUCTION

During the last years, the natural fiber based polymer composites have become an great interests among scientists in order to developing derivate biodegradable ecologic materials and partly reducing the dependency of synthetic fibers [1].

In the past, the use of synthetic materials dominated the industry of confectioning composite materials; nevertheless, the use of natural fibers has become an important point in their development for obtaining composites for various applications.

The interest in the use of the polymeric composites, reinforced with natural fibers, has increased rapidly due to their mechanical properties, to their significant processing advantages, to their density and reduced costs [2]. The use of natural fibers in obtaining composites is very benefic as their resistance and hardness are higher than those of the plastic materials, which are not reinforced [3].

It is good to know the fact that the composites performance depends in a big extent on the individual properties of each material and their interface regarding the compatibility between them. A significant impediment in obtaining fiber reinforced composite materials is the weak connection between the fiber and the matrix, which is due to the hydrophilic characteristic of the cellulose and the hydrophobic nature of the matrix material, which leads to a weak adhesion on the interface between the materials [4].

2. EXPERIMENTAL DETAILS

For this research, hemp canvas and epoxy resin type materials were purchased (ROPOXID 501). The manual formation technique, presented in Fig. 1 (a), known as the oldest method, was used to obtain composite materials. Plates (with 1-3-5-7 layers) were obtained, where, following polymerization, they were left for drying several days in order to obtain the optimal mechanical properties.

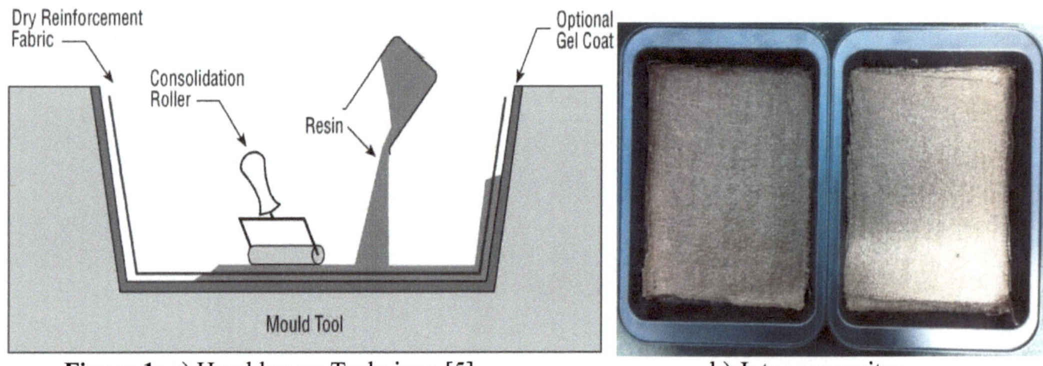

Figure 1: a) Hand lay-up Technique [5] b) Jute composites

Specimens were obtained and subject to treatments with saline water of different concentrations (3, 4, 5 g NaCl / 100 ml H_2O), on a 3-month period (2160 hours). For the saline treatment with a concentration of 3,4,5 g NaCl / 100 ml H_2O, 5 specimens were prepared for each composite with 1-3-5-7 layers, as presented in Fig. 2.

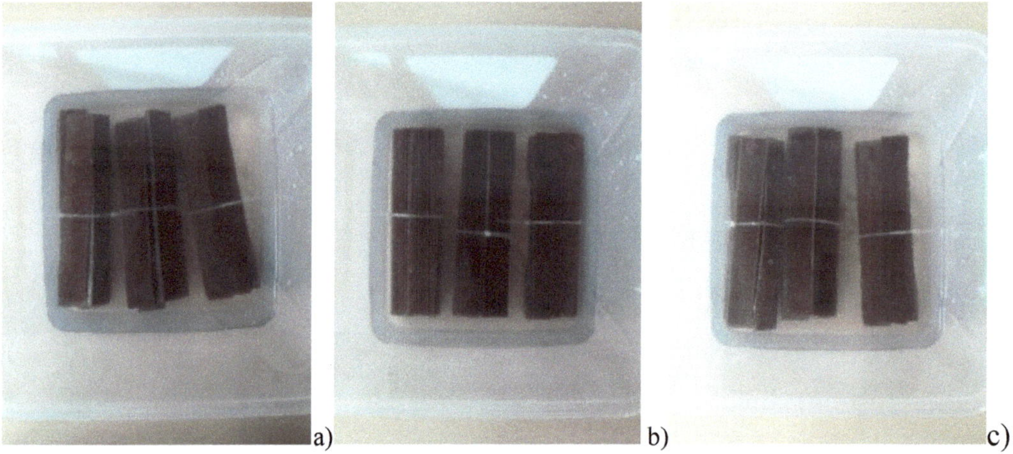

a) b) c)

Figure 2: Composites with 1-3-5-7 layers under treatment with different concentration:
a) 3g NaCl /100ml H_2O,4g NaCl / 100ml H_2O;b) 4g NaCl /100ml H_2O,4g NaCl / 100ml H_2O;
c) 5g NaCl /100ml H_2O,4g NaCl / 100ml H_2O

In order to measure the water absorption, the specimens were took out and periodically weighed with a precise electronic balance, for 2 weeks (336 hours), every 24 hours, for monitoring the variation of the sample mass during the ageing process. The absorption process was expressed with the relation (1) [3]:

$$Water\ uptake\ (\%) = \left(\frac{P_w - P_o}{P_o}\right) * 100 \quad (1)$$

Where, P_w is the wet weight, P_o is the dry weight of the specimen.

3. RESULTS AND DISCUSSION

3.1. Water uptake

In most cases, the humidity absorption process may be described by Frick's law, where the absorbed water mass grows linearly with time's square root and then decreases gradually up to a constant equilibrium. The diffusion coefficient (D) is the most important parameter, where it shows the easiness with which the water molecules penetrate inside the composite, which is calculated by the relation (2) [7].

$$\frac{M_t}{M_s} = \frac{4}{h} * \left(\frac{D}{\pi}\right)^{1/2} t^{1/2} \quad (2)$$

Where, h is the specimen's thickness, M_t and M_s represent the absorbed and desorbed solution mass.

The absorption curves for the specimens obtained from reinforced hemp tissue in epoxy resin with 1-3-5-7 layers are presented in Figure 4. Each diagram represents the average water absorption for each type of layer in a different concentration.

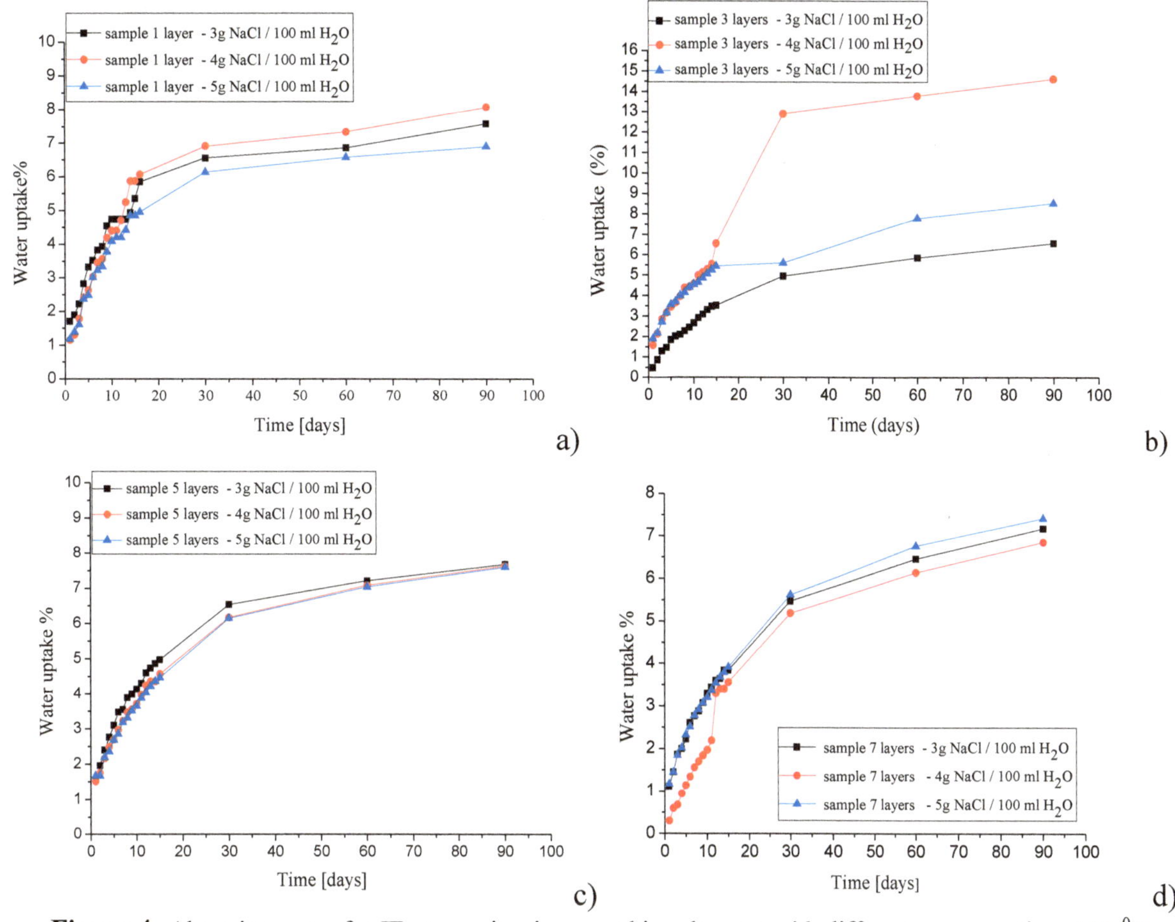

Figure 4: Absortion curve for JE composites immersed in salt water with different concentration at 22 ^{0}C: a) 1 layer in 3g NaCl /100ml H_2O, 4g NaCl / 100ml H_2O, 5g NaCl/100ml H_2O; b) 3 layer in 3g NaCl /100ml H_2O,4g NaCl / 100ml H_2O, 5g NaCl/100ml H_2O; c) 5 layers in 3g NaCl /100ml H_2O,4g NaCl / 100ml H_2O, 5g NaCl/100ml H_2O; d) 7 layers in 3g NaCl /100ml H_2O, 4g NaCl / 100ml H_2O, 5g NaCl/100ml H_2O

The absorption curves for the specimens obtained from reinforced hemp tissue in epoxy resin with 1-3-5-7 layers are presented in Figure 4. Each diagram represents the average water absorption for each type of layer in a different concentration. From the absorption curves in Fig. 4 a),b),c),d) it results that, for each concentration in the first two weeks, the water content absorbed by the specimens increased with the increase of the immersion time. We can notice that, in Fig. 4 a),b),c),d) regardless of the number of curves, the water absorption curves tend to equilibrium, which results upon the appearance of the saturation moment. The significant increase of the water absorption curve in Fig. 4 b) with a concentration of 4g NaCl / 100 ml H_2O highlights the presence of certain defects occurred during obtaining the composite material. The significant justification of the percentage of absorbed water by all types of specimens obtained by the manual method may be drawn from the lack of a lamination process.

3.2. Flexural strength

In order to determine the bending stress in three points, specimens were obtained according to SR EN ISO 7438:2005 and tested with the Universal Testing Machine – WDW-150S, from Jinan TE, presented in Fig. 5 [6].

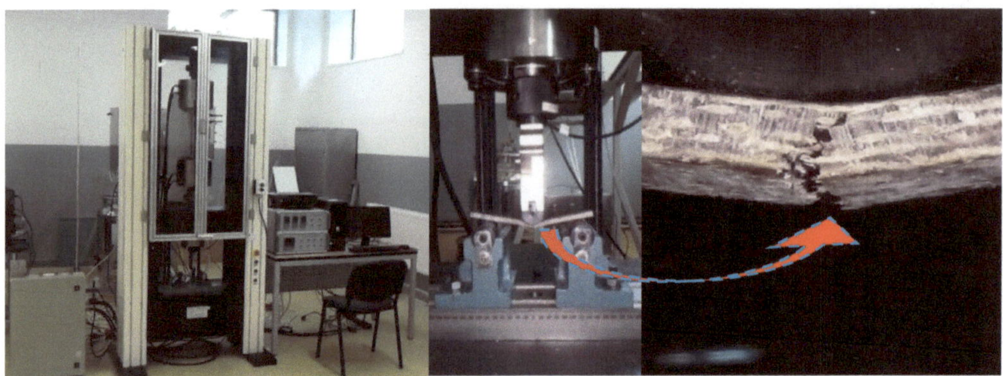

Figure 5: Universal Testing Machine – WDW-150S

The results obtained represent the average of tests made for 5 specimens of each type of composite formed with 1-3-5-7 layers for each type of treatment.

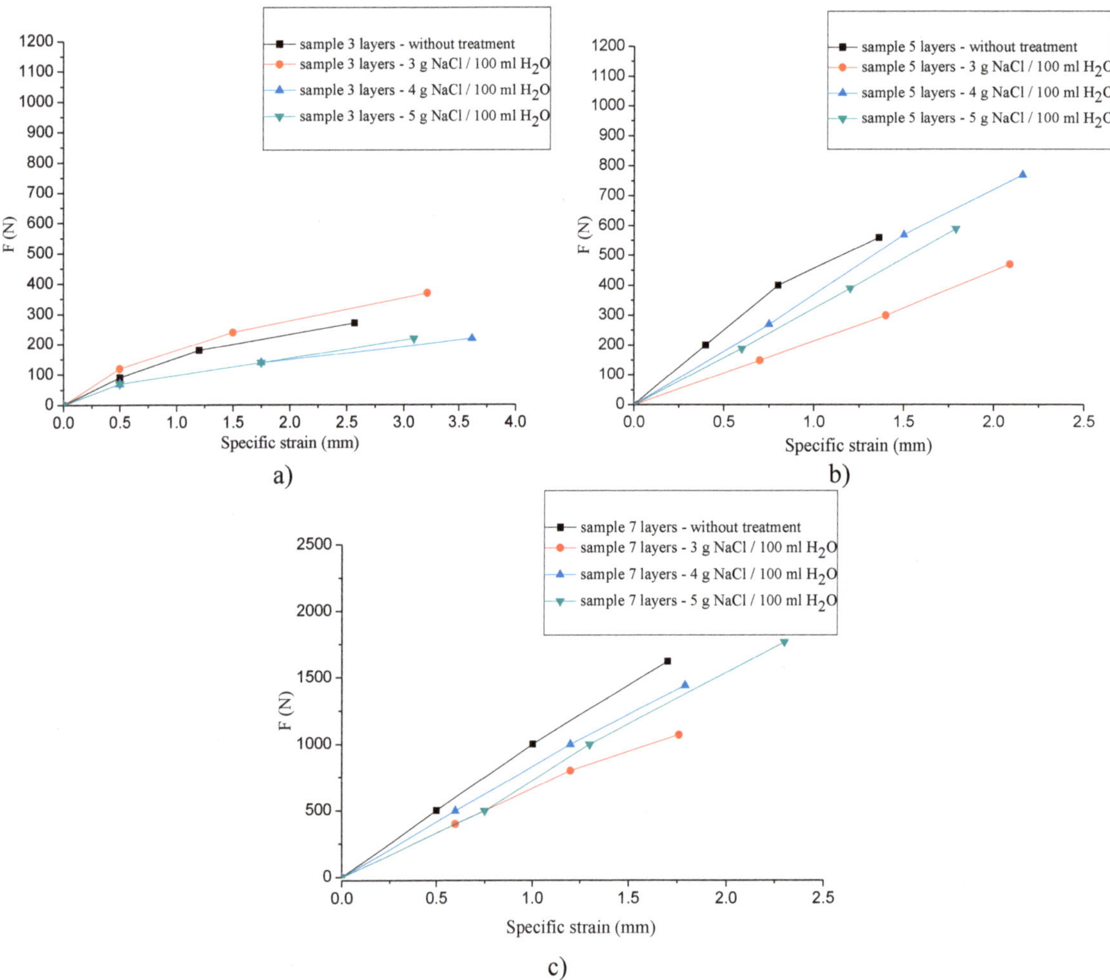

Figure 5: Diagrams regarding the average bending strength for specimens without treatment and with saline treatment: a) with 3 layers for 3,4,5 g NaCl / 100 ml H_2O; b) with 5 layers for 3,4,5 g NaCl / 100 ml H_2O; c) with 7 layers for 3,4,5 g NaCl / 100 ml H_2O

Figure 5 shows the results of the bending strength for each type of specimen without treatment and with saline treatment with a concentration of 3,4,5 g NaCl / 100 ml H_2O. Determination of flexural strength for single-layer specimens were removed due to their high elasticity. We could notice that, for all specimens, the bearing strength increased with the increase of the number of layers, the deformation degree being smaller for the dry ones and higher for the immersed ones. With regards to the bending strength, we can notice that the values for the immersed specimens are smaller as compared to the dry ones, leading to the deduction that saline treatments of different concentrations do not help in the increase of the resistance, but they highlight a significant increase of the elasticity degree.

4. CONCLUSION

The effect of the saline water absorption over the mechanical properties of the composites formed of hemp tissue and epoxy matrix was studied. It was revealed that the water absorption increased with the immersion time on the entire treatment period, according to Fick's diffusion process.

With the increase of the number of layers, it was also noticed an increase of the properties of composites without treatment, but also the effect of all treatments applied to composites led to an increase of elasticity.

ACKNOWLEDGEMENT

This paper is supported by the Sectoral Operational Programme Human Resources Development (SOP HRD), ID137516 financed from the European Social Fund and by the Romanian Government and by the Romanian Government and the structural founds project PRO-DD (POS-CCE, O.2.2.1., ID 123, SMIS 2637, ctr. No 11 / 2009) for providing the infrastructure used in this work.

REFERENCES

[1] M.R. Ishak, Z. Leman, S.M. Sapuan, M.Y. Salleh and S. Misri, The effect of sea water treatment on the impact and flexural strength of sugar palm fibre reinforced epoxy composites, IJMME, vol.4, nr.3, pag. 316-320, 2009.

[2] Abdul Khalil H.P.S., Bhat I.U.H., Jawaid M., Zaidon A., Hermawan D., Hadi Y.S., Bamboo fibre reinforced biocomposites : A review, Materials Science, vol. 42, pag. 353-368, 2012.

[3] Seki Y., Sever K., Sarıkanat M., Şen İ. , Aral A., Jute/Polyester Composites: The effect of water aging on the interlaminar shear strength (ILSS), IATS'11, pag. 368, 2011.

[4] Franco H., Gonzalez P.J.V., A fibre-matrix adhesion in natural fibre composites, In Natural Fibres, Biopolymers, and Biocomposites, pag..177 - 230, 2005.

[5] Cripps D., Searle T.J, Summerscales J., Open Mold Tehniques for Thermosets Composites, Comprehensive composite materials encyclopædia 2, pag. 3, 2000.

[6] Pop M.A., Cercetări asupra tehnologiilor şi materialelor moderne pentru confecţionarea garniturilor de model, pag. 30, 2009.

[7] Pradeep K.K., Rakesh K., Studies on Water Absorption of Bamboo-Polyester Composites: Effect of Silane Treatment of Mercerized Bamboo, Polymer-Plastics Technology and Engineering, pag. 45–52, 2010.

TRANSOM STRENGTHENING OF A RIGID INFLATABLE BOAT (RIB) TO INCREASE PROPULSION POWER

Ionel Chirică, Professor, NA, PhD.[1], Dumitru Lupașcu, Naval Architect [2]

[1]University "Dunărea de Jos" - Galați;
[2]Romanian Naval Authorithy

Abstract: *When in use, boats could be modified provided that their safety of construction is not affected. For this purpose, verification calculations are carried out for the elements affected by changes and when ascertained that safety of construction is not ensured, adequate strengthening is made. This article describe that kind of strengthening to the transom of a rigid inflatable boat (RIB), following the replacement of the original engine of 90 HP by one of 130 HP. Strength calculations for transom made of sandwich type composite material, were made using COSMOS/M software and laminated plate type elements SHELL4L.*
Keywords: *composite material, small craft, scantling, transom*

1. INTRODUCTION

During the use of boats made of composite materials, these could be subject to modifications such as increase of length, breadth or depth, increase of displacement, modification of superstructure, rise of power, etc., which could impact upon hull strength and for this reason the verification of the resistance of structural elements affected by modifications is necessary, and if not suitable, these elements shall be strengthen.
Such kind of modification was carried out to a RIB (Rigid Inflatable Boat), to which the propulsion outboard engine type EVINRUDE E90DSL of 90 HP was replaced by a new engine type EVINRUDE E130DSL of 130 HP (but keeping same propeller pitch thus, at equal rate of revolutions, the boat speed is not modified) in order to increase boat acceleration from stop for decreasing the transitory regime to reach the aquaplanning speed of towed water skiers.
The particulars of this boat are given in Table 1.

Table 1: Boat particulars

Craft Type	RIB
Design Category	C
Material	GRP single-skin laminates and GRP sandwich composites
Displacement, m_{LDC}	1200.00kg
Length of Hull, L_H	4.52m
Hull Beam, B_H	2.20m
Depth of Bulkhead, D_b	1.16m
Maximum speed, V	40.00kts

The above-mentioned modification could be accepted if we prove that the transom strength is sufficient to stand the new efforts introduced by the 130 HP engine and the boat stability and maneuverability are ensured.
Transom strength was verified hereinafter through direct strength calculations based on finite element method and taking into account the international standard ISO 12215-5: 2008 – Small kraft. Hull construction and scantlings. Part 5: Design pressures for monohull, design stresses and scantlings determination.

2. STRENGTH ASSESSMENT OF NON-MODIFIED TRANSOM

The transom of initial boat, having the shape and dimensions according to Figure 1, is built of sandwich type material as shown in Figure 2, with 5 alternate layers of GRP and teak wood (MULTI-LAYER 1) whose dimensions are given in Table 2.

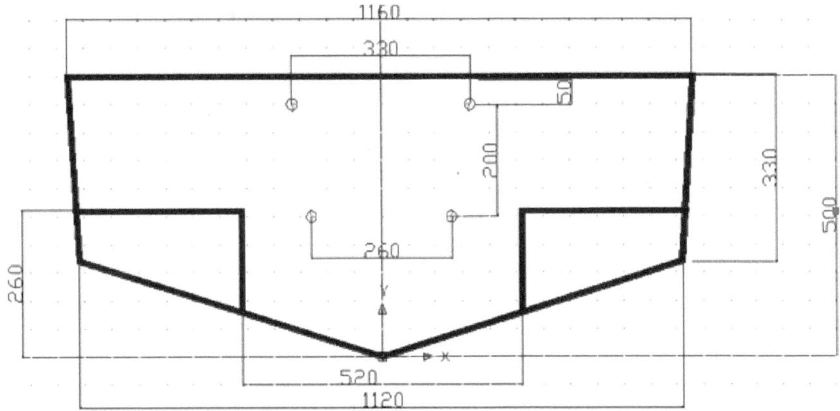

Figure 1: Initial transom

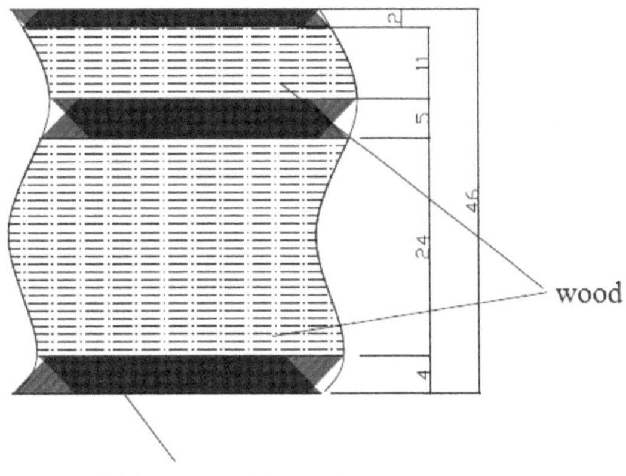

Figure 2: Structure of transom material

Table 2: Mechanical and physical characteristics of materials: teak wood, GRP and steel

Characteristic	m.u.	Teak wood	GRP ISO-12215-5	Steel
E1	MPa	10500 - 15600	45000	210000
E2	MPa	3000	12000	210000
E3	MPa	1000	10000	210000
μ12		0.32	0.32	0.32
μ23		0.08	0.32	0.32
μ31		0.06	0.32	0.32
G	MPa	5000	4800	80000
Ultimate tensile strength	MPa	95 - 155	95	4000
Ultimate compression strength	MPa	48 - 91	95	4000
Ultimate flexural strength	MPa	86 - 170	140	4500
Density	kg/m3	400	1900	7800

The calculation of forces acting on transom is shown thereinafter:

- Engine weight, G =1840 N – is uniform distributed on 4 fixing bolts;
- Steady point thrust on propeller was determined according to BV Rules, Pt. E, Ch. 14, Sect.2: P=20000 N

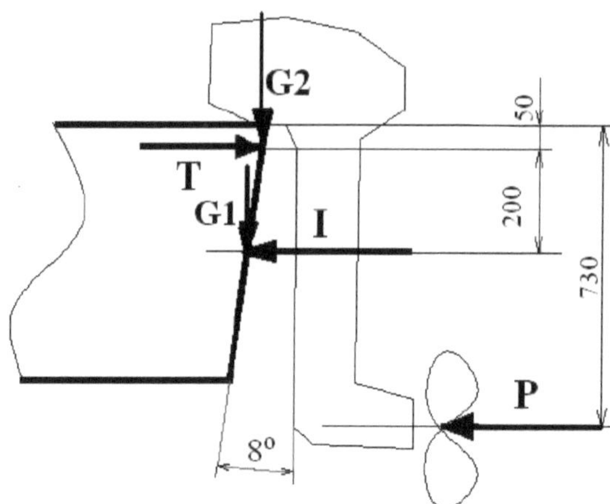

Figure 3: Diagram of efforts acting on transom

- Efforts T and I are distributed on each of 4 fixing points on transom as follows:
  on each of 2 bolts on upper line:
  T/2=23935.35 N
  G2/2=460 N
  on each of 2 bolts on lower line:
  I/2=33935.35 N
  G1/2=460 N

Strength calculation was performed with COSMOS/M software.

As finite element mesh (FEM), SHELL4L type of laminated plates were used.
Number of elements: 14920.
Number of nodes: 16110.

Bearing constraints:

- Simply supported in the longitudinal and vertical direction of the boat:
  - on the boundary of transom: bottom and sides
  - on the boundary of longitudinal girders
- Constraint of symmetry in the symmetry axis (angle of rotation around the vertical symmetry axis is zero).

Results of calculations are shown in Table 3 and in the diagram of Figure 4.

Table 3: Maximum equivalent stresses on layers

Layer	Thickness [mm]	Material	Node	Node position	Equivalent stress, σ_{vM} [MPa]	Permissible stress [MPa]
1	4	GRP	11684	Corresponding to the bolt on the lower row	68.36	47.0
2	24	Wood	11684	Corresponding to the bolt on the lower row	42.78	47.5
3	5	GRP	11684	Corresponding to the bolt on the lower row	49.77	47.0
4	11	Wood	11684	Corresponding to the bolt on the lower row	44.91	47.5
5	2	GRP	11684	Corresponding to the bolt on the lower row	77.01	47.0

Maximum transom deflection is 4.54 mm and is located in center line, where transom intersects deck.

Following assessment of these results, we find out that actual stresses in transom are greater than safe working stress in points where efforts from engine are transmitted. And, in consequence, strengthening of the transom was necessary.

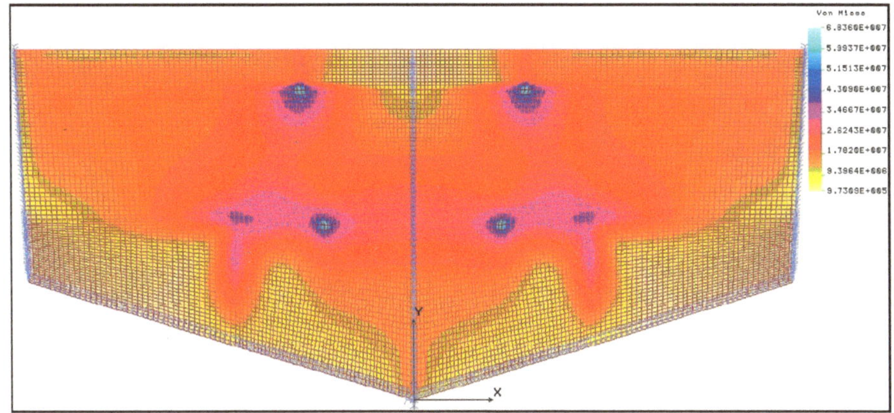

Figure 4: Diagram of von Mises equivalent stresses [Pa] in 1st layer – in case of non-modified transom

3. STRENGTH ASSESSMENT FOR STIFFENED TRANSOM

In order to have satisfactory strength in new working conditions with propulsion engine of 130 HP, the boat transom was stiffened as shown in Figure 5, as following:

- one stiffener made by two layers of wood and stainless steel in the central area, outside the boat,
- two stiffeners of stainless steel, inside the boat.

These stiffeners have been glued together and on the transom by a polyurethane glue for naval use.

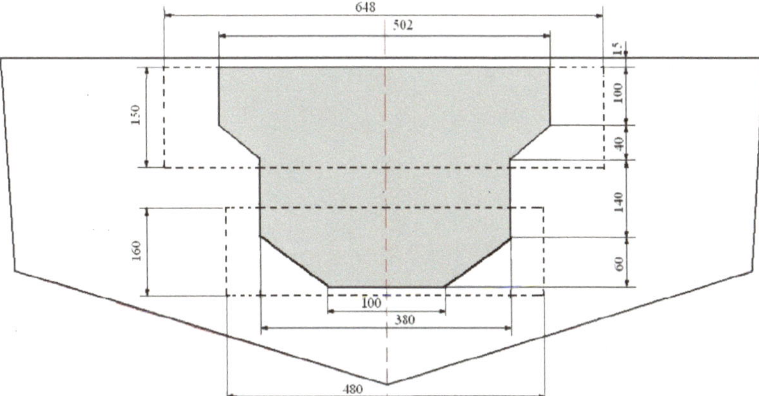

Figure 5: Transom stiffened area, by doubling with wood and stainless steel (central area in dark color – on the outside) and stainless steel stiffeners (doted areas – inside)

All structural elements of the transom, materialize in 4 laminates, as shown in Figure 6 and Table 4, situated in different places, dependent on overlapping of stiffeners with initial transom, see Figure 7.

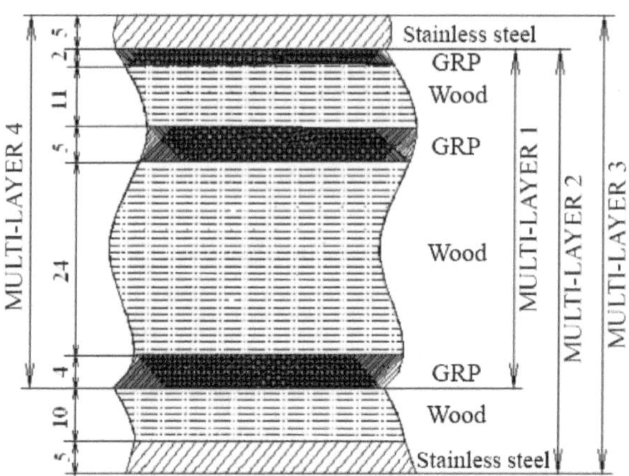

Figure 6: Structural layers of the transom

Table 4: Type of lamination for surfaces

Laminate type	Multi-layer 1	Multi-layer 2	Multi-layer 3	Multi-layer 4
Surfaces	4,5,6,7,9,11,13,49,50, 51,52,16,17,18,27,28, 29,30,32,36,53,54,55, 56,57,58,40,41	19,42	2,3,26,27,1,24,20,43	8,10,21,14,23,15, 38,46,37,44,31,33

Strength calculation was performed with COSMOS/M software.

As FEM, SHELL4L type of laminated plates were used.
Number of elements: 19026.
Number of nodes: 11043.
External efforts were applied on outside stainless steel plate.

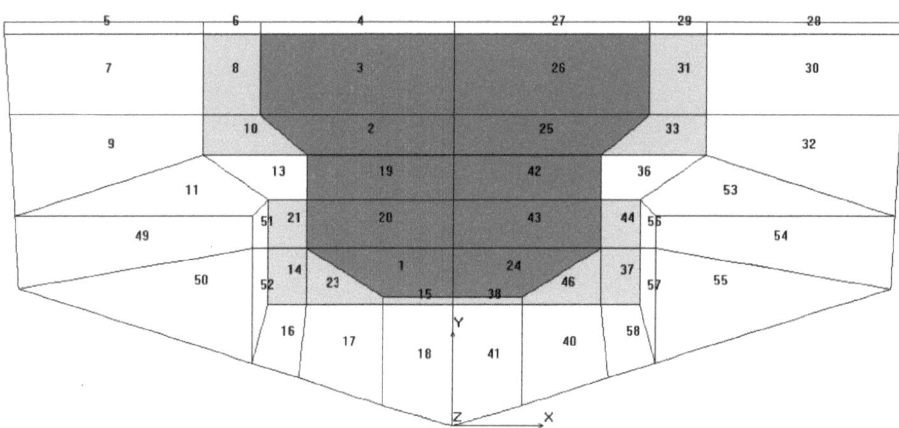

Figure 7: Transom meshing surfaces

Bearing constraints:

- Simply supported in the longitudinal and vertical direction of the boat on the exterior perimeter of transom, bottom and sides;
- Simply supported in the longitudinal direction of the boat on the boundary of longitudinal girders;
- Constraint of symmetry in the symmetry axis (angle of rotation around the vertical symmetry axis is zero).

Results of calculations are shown in Table 5 and in the diagrams of Figures 8 and 9.

Table 5: - Maximum equivalent stresses on layers

Laye	Thickness [mm]	Material	Hot spot position	Equivalent stress σ_{vM} [MPa]	Permissible stress [MPa]
1	5	Steel	Corresponding to the bolt on the lower row	68.87	200
2	2	GRP	Corresponding to the bolt on the lower row	7.48	47
3	11	Wood	Corresponding to the bolt on the lower row	10.80	47.5
4	5	GRP	Corresponding to the bolt on the lower row	8.22	47
5	24	Wood	Corresponding to the bolt on the lower row	21.79	47.5
6	4	GRP	Corresponding to the bolt on the lower row	17.51	47
7	10	Wood	Corresponding to the bolt on the lower row	47.35	47.5
8	5	Steel	Corresponding to the bolt on the lower row	40.48	200

Maximum transom deflection is 0.19 mm is located in center line, in way of the bolts on the lower row.

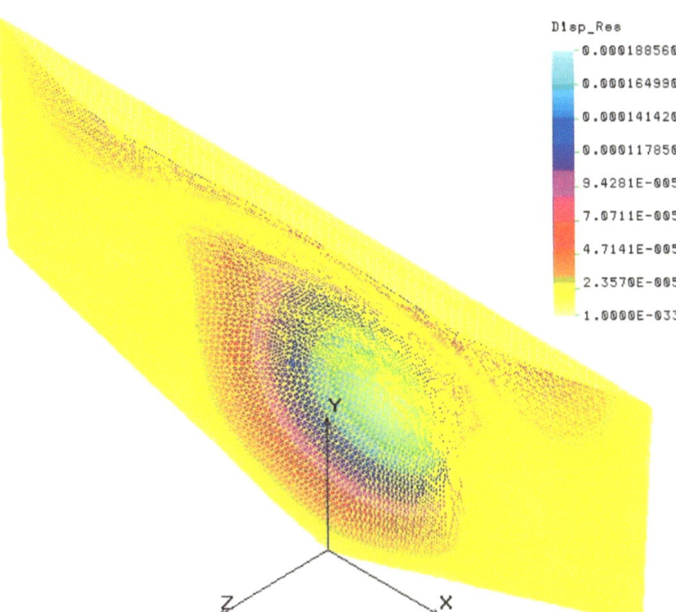

Figure 8:- 3D deflection of the transom in case 1

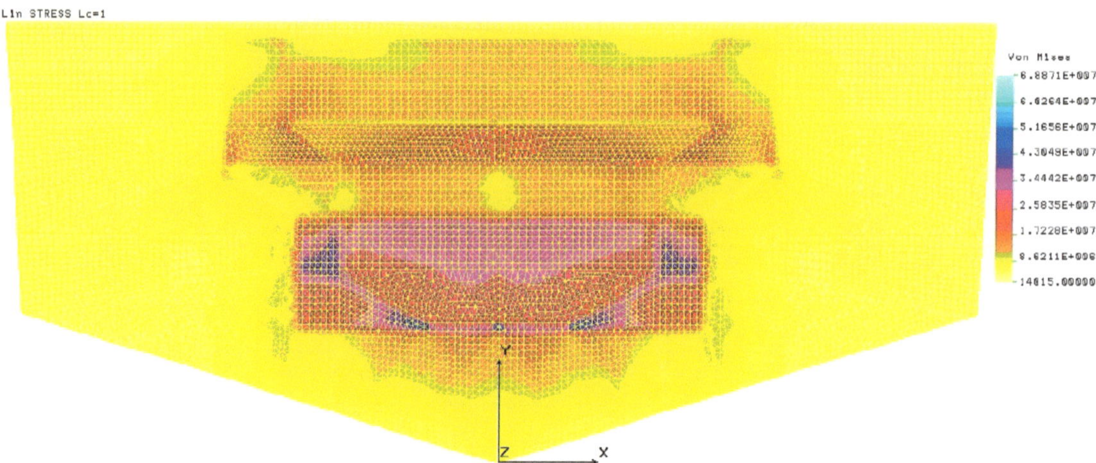

Figure 9:- Diagram of von Mises equivalent stresses [Pa] in 1st layer in case 1

By analyzing these results, we can conclude that actual stresses in transom are not greater than safe working stress and consequently it could be supposed that strengthen solution was effective.

Furthermore, after 2 years of use, owner of the boat is satisfied of boat behavior after modifications made, the stiffened structure of the transom having not permanent deflection, fractures or other mechanical damages due to greater forces transmitted by outboard engine.

REFERENCES

1. Chirica I., Beznea E. F., Elasticity of anizotrop materials, Ed. Fund. Univ. Dunărea de Jos, Galaţi, 2004;
2. Chirica I., Beznea E. F., Boazu, D., Gavrilescu, I., Ship hull Repairing by using composite materials, Proceeding s of the Seventh International Conference on Marine Science and Tehnology BLACK SEA, Varna, 7 - 9 oct. 20 04;
3. Det Norske Veritas, Classification notes no. 30.11 - Steel sandwich panel construction, 2012;
4. International Organization for Standardization, ISO 12205 – Part5. Small craft. Hull construction and scantlings. Design pressures for monohulls, design stresses, scantlings determination, 2008;
5. Lloyd's Register, Rules and Regulations for Classification of ships, 2014.

ANALYSIS OF PHASE TRANSFORMATIONS IN EUTECTOID ZN-AL ALLOYS

M. Agapie[1], B. Varga[2]

[1]Transilvania University, Brasov, Romania, e-mail: agapiemirela@yahoo.co.uk
[2]Transilvania University, Brasov, Romania, e-mail: varga.b@unitbv.ro

Abstract: The paper concentrates on the phase transformations of $ZnAl_{22}$ and $ZnAl_{22}$Cu2TiB eutectoid alloys obtained under various processing and cooling conditions. The results obtained by dilatometric (DIL) and differential calorimetric (DSC) analyses at various heating and cooling rates are presented. The presence of eutectic transformation in theses compositions indicates the intensity of the segregation processes during solidification. The analyses also aim at identifying the phase transformations specific for the two variants of thermal equilibrium diagrams for the analysed system.

Keywords: Zn-Al alloys, dilatometric analysis, differential calorimetric analysis, phase transformation.

1. INTRODUCTION

Vast amounts of literature are dedicated to the analysis of the structure of Zn-Al alloys [1-7]. This interest is explained by the complexity of the structural transformations in these alloys, as well as by the current tendency of promoting Zn-Al alloys with high aluminium contents (10 - 30 %). The complexity of structural transformations is amplified by the strong segregation tendency of these compositions. It needs be pointed out that because of the mentioned phenomena, scientific approaches to structural transformation analysis utilise two variants of thermal equilibrium diagrams, Fig. 1.

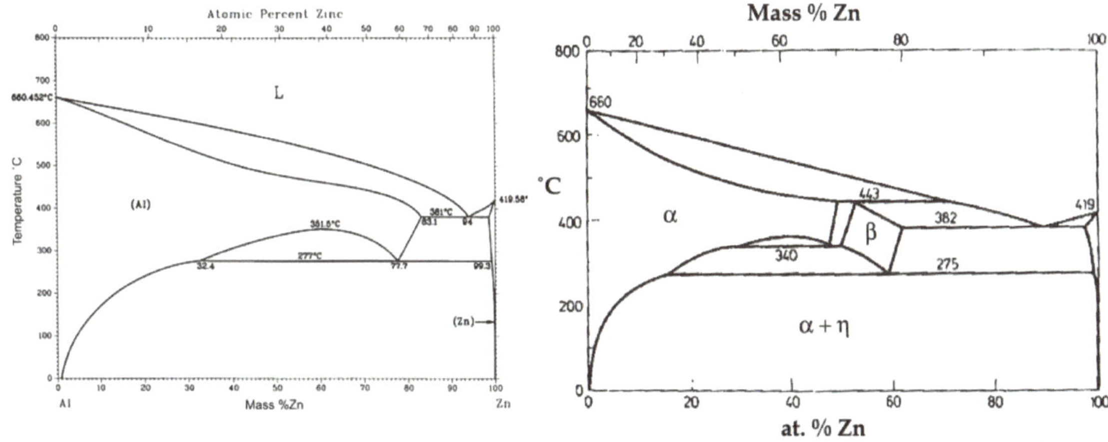

Figure 1: Variants of thermal equilibrium for the Zn-Al system [8,9]

The main difference between the two diagrams is the existence/absence of the peritectic transformation at 443 °C and the presence/absence of phase β, stable in the temperature interval of 275 - 443 °C and at Al concentrations of 17.2 - 29 %:

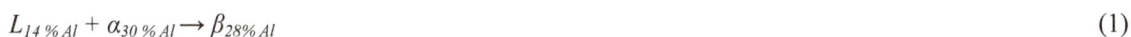

$$L_{14\,\%\,Al} + \alpha_{30\,\%\,Al} \rightarrow \beta_{28\%\,Al} \qquad (1)$$

The paper analyses the structure of a series of eutectoid Zn-Al alloys, binary ($ZnAl_{22}$) and alloyed/modified with Cu – Ti + B ($ZnAl_{22}Cu_2TiB$).

A LINSEIS L75PT/1400 °C dilatometer and a DSC 200 F3 – Maia - Netzsch differential scanning calorimeter were used for the analysis of microstructure modifications determined by phase transformations as well as by the diffusion processes brought about by the thermodynamic tendency of restoring the state of equilibrium. The structural analyses were conducted by means of a Nikon – Omnimet - Buehler microscope.

2. EXPERIMENTAL DETERMINATIONS

The following primary metals were used for the two compositions: 99.95% pure Zn, 99.995% pure Al, Al-Cu_{33} pre-alloy and $AlTi_5B_1$ for finishing the structure. The weight of the charges was of 2000 g. Melting was achieved in an electric furnace with silit bar heaters (electrical resistances) in a graphite crucible.

The alloy melted at 650° C was cast in metal chills, such as to obtain ingots of 14x160x80 mm size. Cylindrical samples of 40 mm diameter and 50 mm height were cast in refractory brick moulds, Fig. 2. Temperature was recorded by means of K – TPN - 101 coaxial thermocouples of 0.6 mm diameter protected by means of a refractory paste.

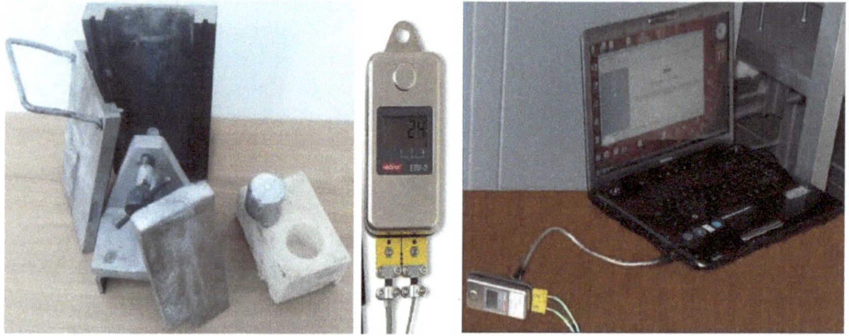

Figure 2: Casting moulds and assembly for cooling curve recording

Figs. 3(a) and 3(b) show the solidification curves obtained for the two compositions cast in metal moulds (OL).

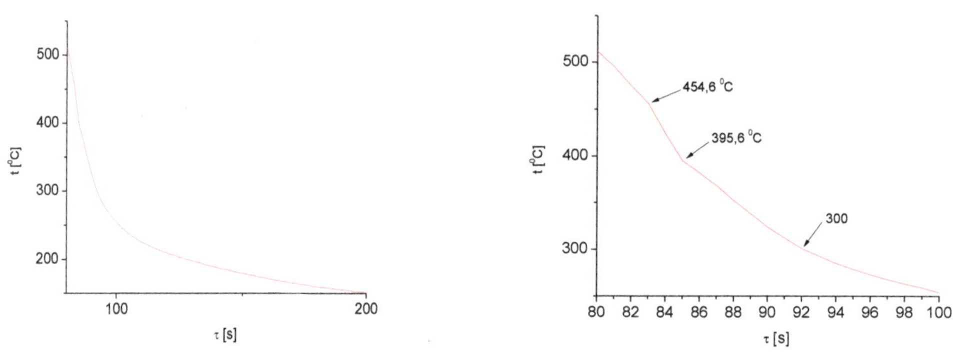

Figure 3: (a) Cooling curves for Zn-Al 22 alloy cast in a metal mould

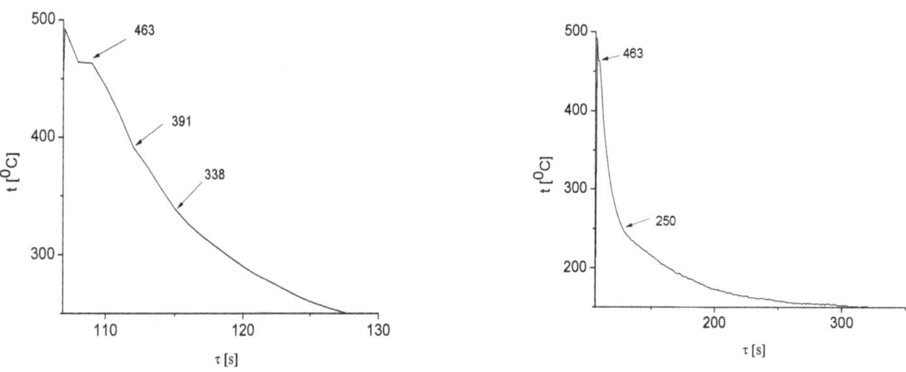

: (b).

Figure 3 (b) Cooling curves for Zn-Al 22-Cu2-TiB 22 alloy cast in a metal mould

The points of inflexion on the cooling curves were considered as corresponding to phase transformations. Table 1 shows the cooling rates for the two compositions cast in metal moulds. The cooling rates were computed prior to each phase transformation and the transformation temperatures indicated on the cooling curves of Fig. 3 (a) and Fig. 3 (b) respectively, are given in brackets.
Cooling rates and transformation temperatures corresponding to the points of inflexion of the cooling curves of Fig. 3.

Table 1: Cooling rates for $ZnAl_{22}$, $ZnAl_{22}Cu_2TiB$ respectively

Alloy	Cooling rate (°C/s) and transformation temperatures (°C)			
	Beginning of solidification	Eutectic transformation	Unidentified transformation	Eutectoid transformation
$ZnAl_{22}$	18.45 (454,6)	30.34 (395.5)	10 (300)	-
$ZnAl_{22}Cu_2TiB$	29.24 (463)	26 (391)	18.22 (338)	4.3 (250)

The samples taken from the cast ingots were subjected to DIL, DSC and structural analyses. H_2SO_4+HF +H_2O reactive was used for highlighting the structure.
Table 2 presents the parameters of DIL, DSC and structural analyses and the corresponding codes of the samples.

Table 2: Parameters of the DIL and DSC analyses

Sample state	$ZnAl_{22}$			$ZnAl_{22}Cu_2TiB$		
	Parameters		Sample code	Parameters		Sample code
	DIL	DSC		DIL	DSC	
Cast	-	-	a - 0	-	-	b - 0
	V_h = 20 V_c = 10 T = 420 τ = 0.1h	V_h = 10 V_c = 0,3 T = 385 τ = 2h	a - 1 - 1	V_h = 1 V_c = 1 T = 430	-	b - 1 - 1
One heating	-	V_h = 10 V_c = 1 T = 510 τ = 0.1h	a - 1 - 2	V_h = 20 V_c = 5 T = 430 Softening (beginning of melting)	-	b - 1 - 2
Two heatings	V_h = 1 V_c = 5 T = 430 τ = 0.1h Partially melted	-	a - 2	V_h = 1 V_c= 1 T = 430	V_h = 10 V_c = 10 T = 510 τ = 0.1h	b - 2

V_h = heating rate [°C/min]; V_c = cooling rate; T = heating temperature [°C] and τ = maintaining time

The dilatometric analyses reveals significant dimensional modifications upon repeated heating of the $ZnAl_{22}$ samples cast in steel moulds, Fig. 4. Upon the first heating beyond the eutectoid temperature the sample contracts, what does not occur anymore upon the second heating. It can be noticed that the reverse eutectoid transformation in both heatings occurs at the same temperature but with different intensities, what indicates a modified share of eutectoid in the structure of the alloy. Both heatings were conducted at a rate of 1 °C/min.

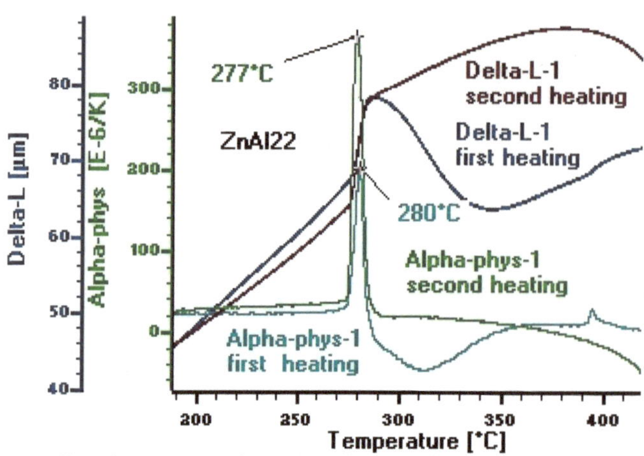

Figure 4: Dilatation curves of $ZnAl_{22}$ alloy cast in steel moulds, second sample

Figure 5 presents the dilatation curve for $ZnAl_{22}Cu_2TiB$ alloy versus time as well as versus temperature, in order to highlight the dimensional modifications determined by the diffusion phenomena generated by both (eutectoid) phase transformations and the structure's tendency of achieving equilibrium.

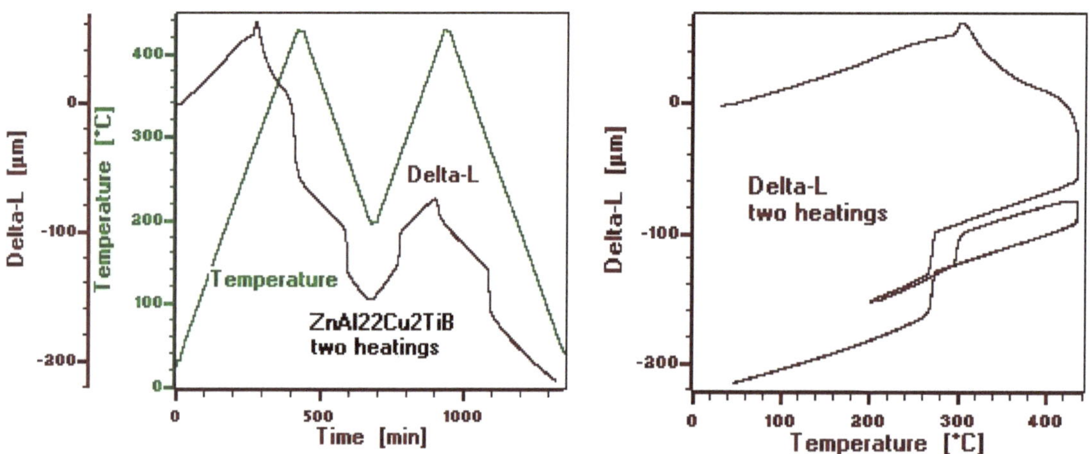

Figure 5: Dilatation curve versus time and temperature for heating and cooling rates of 1 °C/min, sample b - 2

Figure 6 presents the processed dilatation curves with indication of the transformation temperatures on their derivatives, actually representing the variation of the physical dilatation coefficient versus temperature. The heating rates applied while conducting the dilatometric analyses are written on the curves, namely 1 and 20 °C/min, respectively.

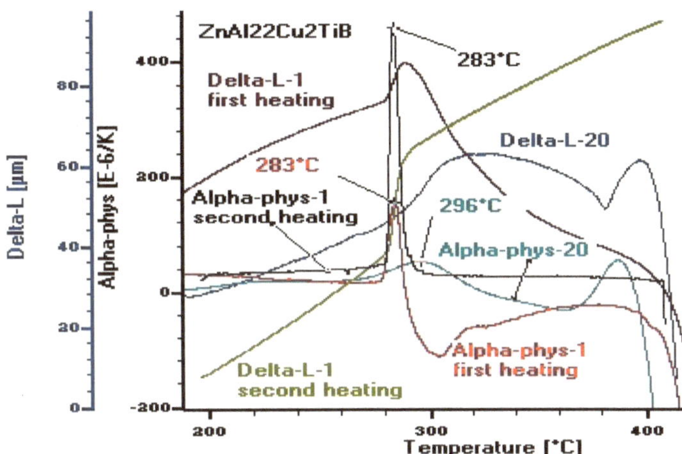

Figure 6: Dilatation curves of $ZnAl_{22}Cu_2TiB$ alloy cast in steel moulds, samples b - 1 - 2 and b - 2

The dilatometric measurements were further completed by DSC analyses. These determinations allowed tracing of the phase transformations until the melted phase is reached.

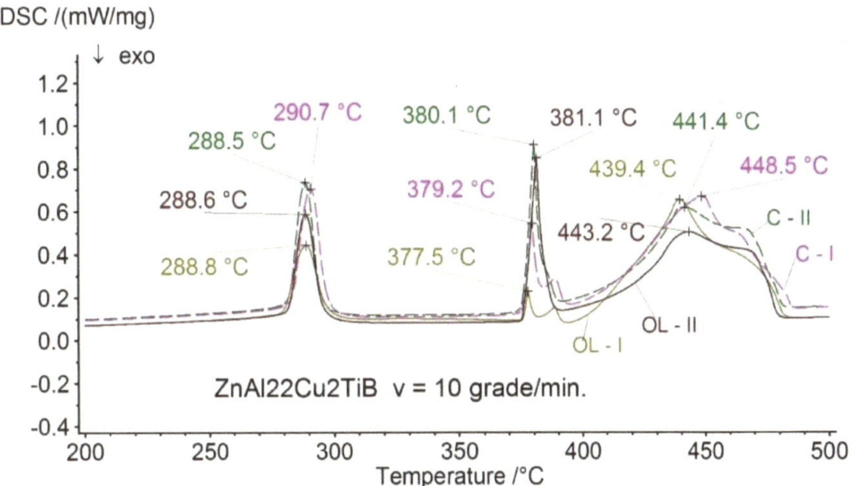

Figure 7: DSC curves for the heating of the samples cast in steel moulds (OL) - sample b - 2, brick moulds (C), heated once (I) and heated twice (II)

At heating rates of 10°C applied in DSC analysis the peaks corresponding to the eutectoid, eutectic transformations and to melting by liquid solution appear on the heating curves, Fig. 7. The peak corresponding to the melting interval has the characteristic shape of melting with liquid solution [10]. It needs be pointed out that the 3 transformations occur at different intensities for the two heatings, what explains the thermodynamic tendency of restoring the state of equilibrium. It can be noticed that in the first heating the peak corresponding to the eutectic transformation is less obvious and is followed by a peak that suggests the existence of yet another solid state phase transformation prior to the beginning of melting through solid solution. For the second heating the first peak of the eutectic temperature becomes significantly more obvious, while the immediately subsequent peak disappears.

This observation is in agreement with the results of the dilatometric analyses that reveal the closeness of the real structure to the equilibrium one by disappearance of the peak corresponding to the eutectic transformation.

This observation is supported also by the DSC curve recorded for both heating and cooling of a $ZnAl_{22}$ sample cast in a metal mould, heated to a temperature slightly greater than that of the eutectic transformation, Fig. 8 (a). These curves obtained for sample a-1-1 are comparable to similar curves recorded for sample a-1-2, Fig. 8 (b) and 8(c).

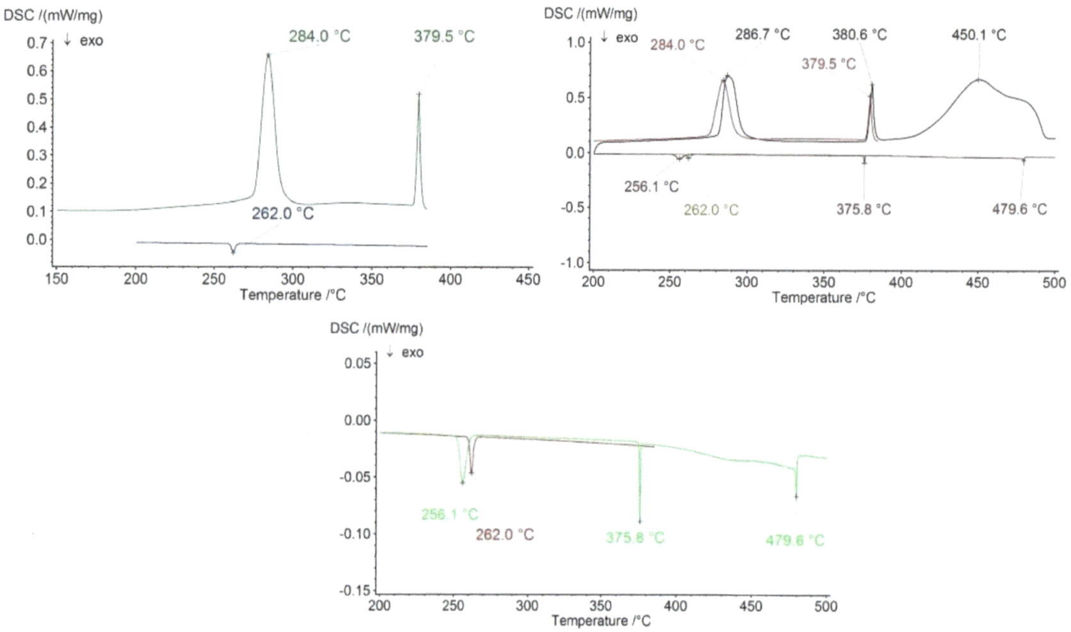

Figure 8: DSC curves for metal mould cast $ZnAl_{22}$ alloy (a) sample a - 1: heating + cooling; (b) samples a - 1 - 1 and a - 1 - 2 heating + cooling, overlapped and (c) samples a - 1 - 1 and a - 1 - 2 cooling, overlapped

The results of the DIL and DSC analyses are confirmed by the structural analyses conducted on the cast and heat treated samples (we considered that both analyses, when melting does not intervene, can be assimilated to homogenisation heat treatment). Figure 9 presents the structures observed in a binary alloy with 22 % Al.

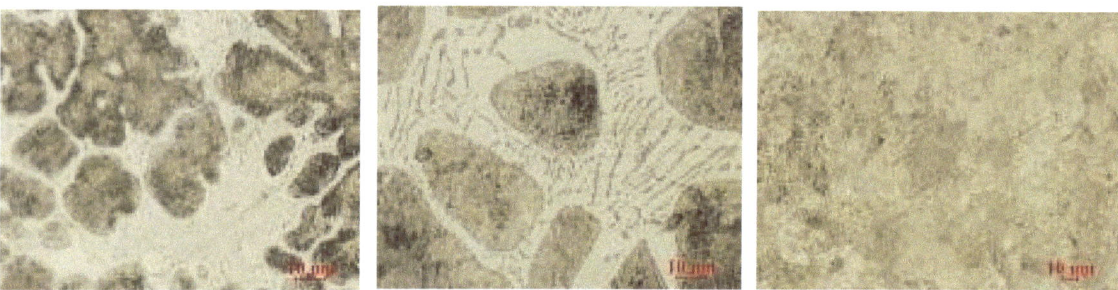

Figure 9: Structure of steel mould cast $ZnAl_{22}$ alloy: (a) sample a - 0 - in cast state; (b.) sample a - 2 - DIL – partial melting, and (c) sample a - 1 - 1-DIL, no melting

3. CONCLUSION

The results of the DIL analysis reveal the intensity of the segregation processes during solidification.The results of the dilatometric analyses underline the importance of heat treatment in view of restoring the equilibrium structure such as to avoid dimensional modifications of components during operation.

Because of the high cooling rates during the liquid-solid phase transformation the point corresponding to the maximum solubility of aluminium in zinc at eutectic temperature is significantly displaced to the left, towards higher aluminium concentrations, what explains the presence of the eutectic structure (transformation) in alloys with 22% Al, in which, according to the thermal equilibrium diagram eutectic is not supposed to appear.

The cooling rate of 1 °C/min does not ensure achievement of the equilibrium structure according to the thermal equilibrium diagram.

ACKNOWLEDGEMENT: *This paper is supported by the Sectoral Operational Programme Human Resources Development (SOP HRD), ID137516 financed from the European Social Fund and by the Romanian Government and the structural founds project PRO-DD (POS-CCE, O.2.2.1., ID 123, SMIS 2637, ctr. No 11 / 2009) for providing the infrastructure used in this work.*

REFERENCES

[1] Savas M., Altintas S., The microstructural control of cast and mechanical properties of zinc-aluminium alloys, J. Materials Science 28 (1993) 1775-1780.

[2] Zhu Y. H, Phase transformations of eutectoid Zn-Al alloys, J. Materials Science, 36 (2001) 3973-3980.

[3] Zhu Y. H., Chan K. C., G. Pang K. H., Yue T. M.,. Lee W. B, Structural Changes of α Phase in Furnace Cooled Eutectoid Zn-Al Based Alloy, J. Mater. Sci. Technol., 23/3 (2007) 347-352.

[4] Dorantes-Rosales H. J., Lopez-Hirata V. M., Moreno-Palmerin J., Cayetano-Castro N.,. Saucedo-Munoz M. L, Torres Castillo A. A., β' phase decomposition in Zn-22 mass % Al and Zn-22 mass % Al-2 mass % Cu alloys at room temperature, Materials Transactions, 48/10 (2007) 2791-2794.

[5] Varga B., Peter I., Structural transformations in Zn-Al22 alloy solidified by ultra-rapid cooling, in Metalurgia International, XVII Special Issue 6 (2013) 9-12.

[6] Krajewski W. K., Zak P. L., Orava J.,. Greer A. L, Krajewski P.K., Structural stability of the high-aluminium zinc alloys modified with Ti addition, Archives of Foundry Engineering 12/1 (2012) 61-66.

[7] Dorantes-Rosales H. J.,. Lopez-Hirata V. M, Mendez-Velazquez J. L.,. Saucedo-Munoz M. L, Hernandez-Silva D., Microstructure characterization of phase transformations in a Zn-22wt%Al-2wt%Cu alloy by XRD, SEM, TEM and FIM, Journal of Alloys and Compounds, 313 (2000) 154-160.

[8] Torres-Villasenor G., Martinez-Flores E., Hybrid materials based on Zn-Al, Metal, Ceramic and Polymeric Composites for various uses, Published online 20 July 2011 http://www.intechopen. com/books/metal-ceramic-and-polymeric-composites-forvarious-uses/hibrid-materials-based-on zn-al-alloys.

[9] Wang Y., Zeng J., Eutectoid transformation in Zn alloy with high Al contant, in Advanced Materials Resarch, 652-654 (2013) 1111-1114.

[10] Dumas J. P., Gibout St., Cézac P., Franquet E., Haillot D., Model for the DSC thermograms of the melting of ideal binary solutions, Thermochimica Acta 571 (2013) 64– 76.

CAPACITY OF BALL BEARINGS ANGULAR CONTACT FULL CERAMIC MATRIX COMPOSITE BALL BEARINGS

Ioan Enescu
Transilvania University Brasov, Faculty of Mechanical Engineering, Mechanical Engineering Departament
50027 Brasov, Romania, E/mail enescu@unitbv.ro

Abstract *The ball bearings are universal used . He assume the radial force but and the axial force and compose force (radial and axial). The calculus of the angle of contact and respectively the zone of contact is very importante.*
Full ceramic-matrix composite angular bearings made enterly of ceramic composite /material and are superior to common stell angular contact bearings in many way.
Keywords: *contact ,angular, ball, bearings, ceramic, composite*

1.INTRODUCTION

The normal tensions whoappears in the interaction between the balls and the rolling way in the case of radial load of the bearing are distributed into an ellipsoidal area.
In fig.1 is presented the ellipsoidal contact area between the ball and the rolling way of inter inner.The major axis is symmetry in the width of rolling way.
In the case of combined load (radial-axial) or pure axial, the ball moves to the board of the rolling way, at a constant angle α , the contact ellipse may be imperfect , respectively the contact tension is maxim to the board of rolling way.

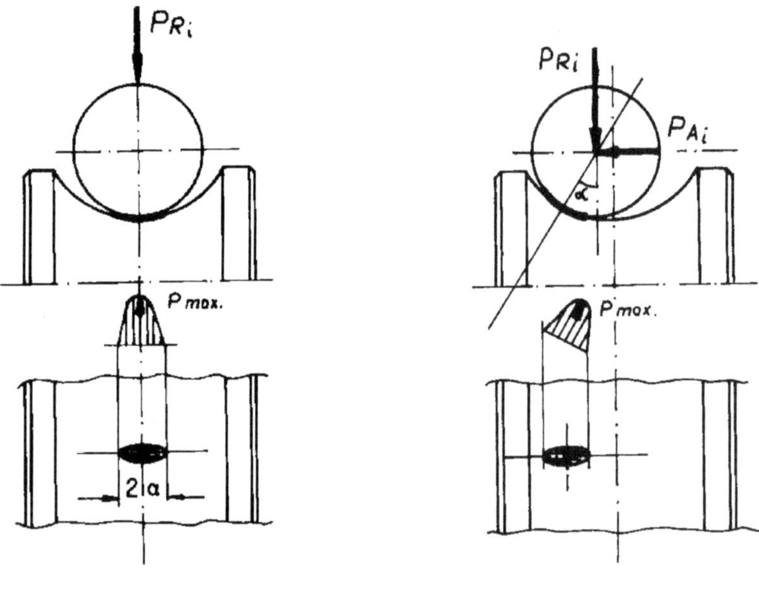

Figure.1 **Figure.2**

The tension in the imperfect contact ellipse are major that in the case of the tension in the contact ellipse at radial load of bearing.

For avoid the appearance of same excessive tensions in the board zone of rolling way is major necessary to calculation the angle of contact α than the major axis of ellipse is perfectly.

2.THE ADMISIBLE ANGLE OF CONTACT CALCULATION

In the case of the problems is:- what major is the α angle than the pressure ellipse are perfect (the major axe rest in the rolling way).

In fig.1 is presented the situation that the radial bearing are loaded with radial load P

In fig.2 the load is P_1+P_2 for the most loaded ball.

For the eccentric contact ellipse do not pass the boarder of the rolling way to be satisfied by the inequality (fig.3)

$$\alpha + \alpha_e \langle \alpha_s \quad (1)$$

$$\cos\alpha_s = \frac{r_b - h}{r_b}$$

$$\sin\alpha_e = \frac{a}{d_b}$$

Respectively the inequality (1) are verify with

$$\alpha + \arcsin\alpha_e \leq ar\cos\alpha_s$$

If not, the contact ellipse pass the board of rolling way and implicit the tension in marginal zone are over-fulfilled.

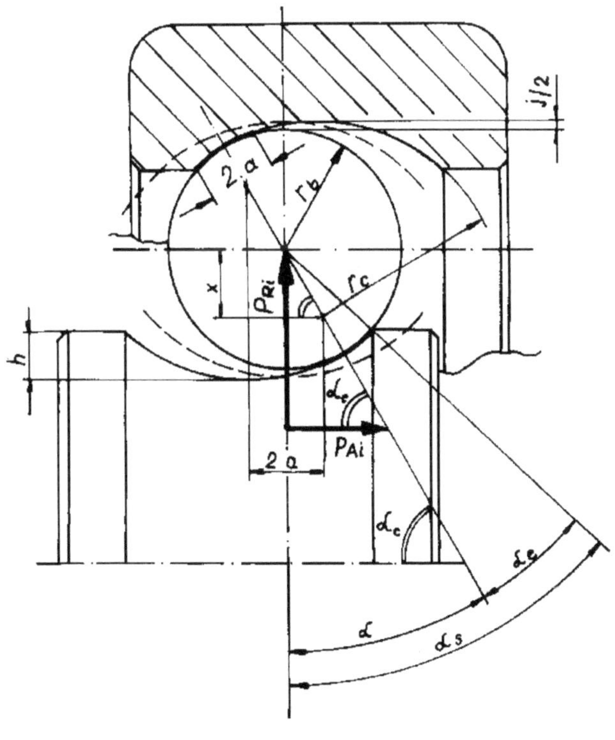

Figure.3

Notation fig.3

i = number of ball
P_R = total radial load
P_A = axial load/ball
$r_b = d_b/2$-ray ball
r_c = ray of rolling way
a = major axis of contact ellipse
j = diametrical working
h = depth of rolling way just at inner shoulder
α_c = angle between the center of contact ellipse on exterior way
α = angle between the center of contact ellipse with symmetry axis of bearing
α_s = angle between the symmetry axis of inner and the board of rolling way
α_e = angle determined by the center of contact ellipse with the board of rolling way

3. THE CALCULATION OF LIMIT AXIAL LOAD

The relation Hertz-Beliaevpermit the calculation of the axis of the ellipse of contact, elastic deformation and the maximum unitary effort

$$a = 23.6\,10^{-3} a^* \sqrt{\frac{P_i}{\sum \rho}} \text{ [mm]} \qquad (2)$$

$$P_i = \sqrt{P_{R_i}^2 + P_{A_i}^2} \text{ [N]}$$

and

$$\sum \rho = (\rho_1 + \rho_2)_I + (\rho_1 + \rho_2)_{II} \ [1/mm]$$

- ρ_1 and ρ_2 are the curvatures of the ball in the planes I and II , respective the minim invers value of curvatures rays
- a^* parameter of function F(ρ) –the parameter a^* is done by diagram (fig.5)

$$tg\alpha_e = \frac{P_{R_i}}{P_{A_i}} = \frac{x}{a} \qquad (2)$$

$$x^2 + a^2 = (r_c - r_a)^2$$

$$P_{A_i} = P_{R_i} \frac{a}{x} = \frac{P_R}{i} \frac{a}{x}$$

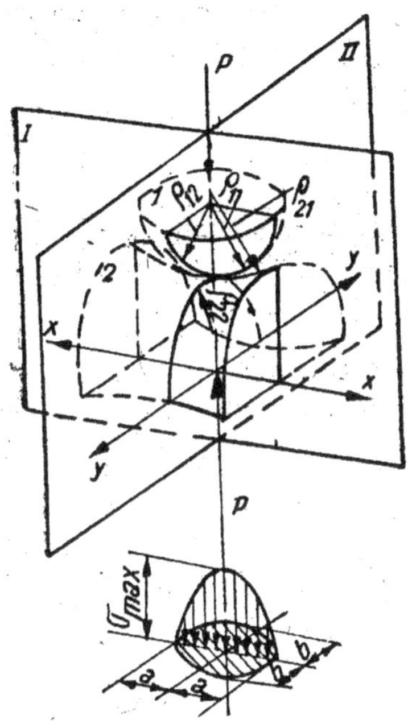

Figure.4

For the most loaded ball we have:

$$P_A = \frac{P_{A_i}}{i} \; and \; P_R = \frac{5P_{R_i}}{i}$$

-and from (2) result

$$x = \sqrt{(r_c - r_b)^2 - a^2}$$

$$P_{A_{lm}} = P_{A_i} i = \frac{5P_R a}{\sqrt{(r_c - r_a)^2 - a^2}}$$

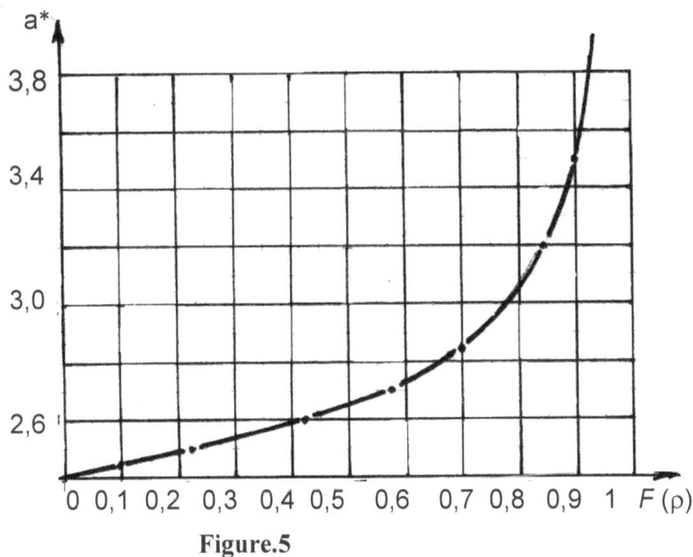

Figure.5

In the case of ball bearings (one road), the limit axial load depend by measure of dynamic equivalent radial load P_R, the measure of major axis of contact ellipse (a) , geometry
Of ball and rolling way, materials, etc.

4. ANGULAR CONTACT FULL CERAMIC MATRIX COMPOSITES (CMCS) BEARINGS

Full Ceramic Angular Contact Ball Bearings are made entirely of ceramic material and are superior to common Steel Angular Contact Bearings in many ways. Ceramic is the perfect material for any application seeking to achieve higher RPM's, reduce overall weight or for extremely harsh environments where high temperatures and corrosive substances are present. Applications such as cryopumps, medical devices, semiconductors, machine tools, turbine flow meters, food processing equipment, robotics and optics. Ceramic materials commonly used for angular contact bearings are Silicon Nitride (Si3N4), Zirconia Oxide (ZrO2), Alumina Oxide (Al2O3) or Silicon Carbide (SiC.)

Because ceramic is a glass like surface it has an extremely low coefficient of friction and is ideal for applications seeking to reduce friction. Ceramic balls require less lubricant and have a greater hardness than steel balls which will contribute to increased bearing life. Thermal properties are better than steel balls resulting in less heat generation at high speeds. Full Ceramic bearings can have a retainer or full complement of balls, retainer materials used are PEEk and PTFE.

Full Ceramic Angular Contact Bearings can continue to operate under extremely high temperatures and are capable of operating up to 1800 Deg. F. Ceramic is much lighter than steel and many bearings are 1/3 the weight of a comparable steel bearing. Full ceramic bearings are highly corrosion resistant and will stand up to most common acids, they will not corrode in exposure to water or salt water. And finally full ceramic bearings are non-conductive.

Full Ceramic Angular Contact Bearings are designed such that there is an angle between the races and the balls when the bearing is in operation. An axial load passes in a straight line through the bearing, whereas a radial load takes an oblique path that tends to want to separate the races axially. So the angle of contact on the inner race is the same as that on the outer race. Full Ceramic Angular Contact Bearings are typically assembled with a thrust load or preload. The preload creates a contact angle between the inner race, the ball and the outer race. The preload can be done while manufacturing the bearing or it can be done when the bearing is inserted into an application.

The contact angle is measured relative to a line running perpendicular to the bearing axis. Full Ceramic Angular Contact Bearings are capable of withstanding heavy thrust loads and moderate radial loads. The larger the contact angle (typically in the range 10 to 45 degrees), the higher the axial load supported, but the lower the radial load. In high speed applications, such as turbines, jet engines, dentistry equipment, the centrifugal forces generated by the balls will change the contact angle at the inner and outer race.

Ceramix-matrix composites (CMCS) comprise a ceramic matrix reinforced by a refrectory fiber, such as silicon carbide (SIC) fiber. CMCS offer low density, high hardness and superior thermal and chemical resistance,

REFERENCES

[1] I.Enescu, Gh. Ceptureanu, D.Enescu RulmentiEdituraUniversitatiiTransilvania 2005
[2] K.L.Johnson, Contact mechanics Cambridge University Press 1985
[3] Boca Bearings Miniature Bearings For Robotics , Industry Recreations'

EXPERIMENTAL INVESTIGATION ON WIND TURBINES BLADES USING VIDEO IMAGE CORRELATION (VIC)

Raluca Dora Ionescu[1], Botond-Pál Gálfi, Péter Dani, Mircea Ivănoiu, Ildikó Renata Munteanu, Ioan Száva

[1] University Transilvania of Brasov, Brasov, ROMANIA, idora@unitbv.ro, janoska@clicknet.ro

1. INTRODUCTION. MAIN ASPECTS REGARDING VERTICAL AXIS WIND TURBINES

The wind turbines are systems that are transforming the kinetic energy from wind in mechanical energy – through the blades. The mechanical energy is then transformed in electrical energy by means of an electric generator.

The wind turbines can be divided into two main categories according to the axis orientation: vertical axis (VAWT) and horizontal axis (HAWT). The most popular ones are the horizontal axis. This design is more appropriate for open areas with more constant wind direction and behavior. When we talk about urban environment, the wind turbines need a different positioning, and the rotor design should be made carefully. Among the small power wind turbines, implementable in urban environment, the vertical axis ones are proving to be more appropriate due to their ability to capture the wind from any direction, to work in turbulent wind conditions, and to facilitate the access to the control system.

There are two main types of VAWTs, according to the functioning principle: Savonius and Darrieus.

The Savoius is a drag type machine, where the wind is "pushing" the blades, which leads to limitations in rotational speed of the turbine, and thus in the power it produces. Between the cup-shaped blades there is a small gap which brings a small component of lift. The classical configuration is with two semi-cylinder blades and two circular plates at the ends. The blades can be positioned with different overlapping ratios. The most common example of this solution is the anemometer. These turbines have been thoroughly studied and we can find in literature several studies that are aiming to improve the power coefficient, and hence a better performance. Thus, the best result has been given by the 2 and 3 bladed solutions. The best power coefficient has been given by the 2 bladed rotor: 0.25, and the 3 bladed one gives only 0.16. Results that this design can convert in mechanical power – at the rotor level - maximum 11% of the undisturbed air flow.

The Darrieus design is a lift-type device. The blades have an aerodynamic profile which gives the possibility to reach rotational velocities higher than the wind speed. In [1] Mertens shows that, for a lift-type rotor, the maximum power coefficient is$C_{P,max} \approx 0.59$.

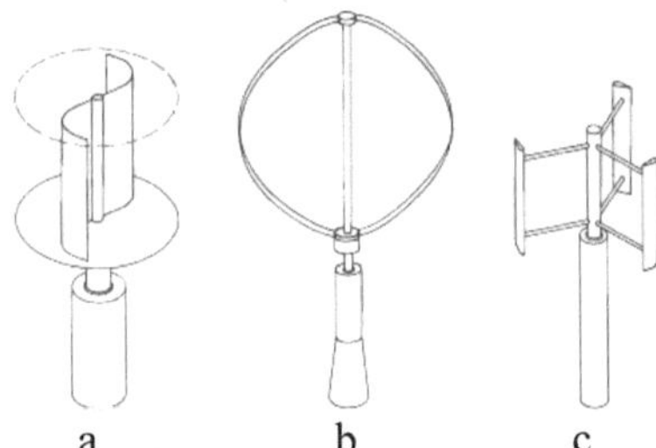

Figure 1: The two VAWT types: a. Savonius, b,c. Darrieus [2]

From the two VAWT types, the Savonius has the advantage of easily self-starting at low wind speeds, but provides low power coefficient. On the other hand, Darrieus has self-starting problems but reaches higher power coefficient. Thus, it is preferably to use Darrieus wind turbines and find solutions for improving the self starting capacity.

2. REQUIREMENTS IMPOSED FOR DARRIEUS TURBINE BLADES

Based on the state of art thorough analysis for the Darrieus VAWT type blades have been formulated the following basic requirements:

- To have optimized aerodynamic profile, in order to convert as efficiently as possible the wind energy in electrical energy
- To be as light as possible, in order to start easily in low wind speeds and have higher efficiency due to lower inertia.
- To be flexible, such as to adapt as good as possible to the environment conditions (i.e. to a wind with a certain intensity and frequency)
- Do not have natural frequencies in the functioning domain, conditioned also by the wind and the pole.
- To have higher fatigue resistance, because during functioning the blades are subjected to periodical loadings; let us just think about the effect of centrifugal force in deformations and stress.

In this regard, from the last requirement point of view could be mentioned the fact that, together with rotational speed change, takes place also a change in relative position of the blade ends, which leads to a bending of a different intensity.

Obviously it should be taken into consideration also the effect of centrifugal forces and aerodynamic forces, as the blades are subjected to extremely complex compound stresses, which are significantly varying with the change in working regime.

Considering even only these minimal requirements, the authors have investigated in this first stage, the degree of 3D deformation of the blades with the rotational speed. Based on the results, have decided to formulate more strict criteria regarding the choice of blades material (which usually is a composite one).

3. ORIGINAL TESTIG STAND

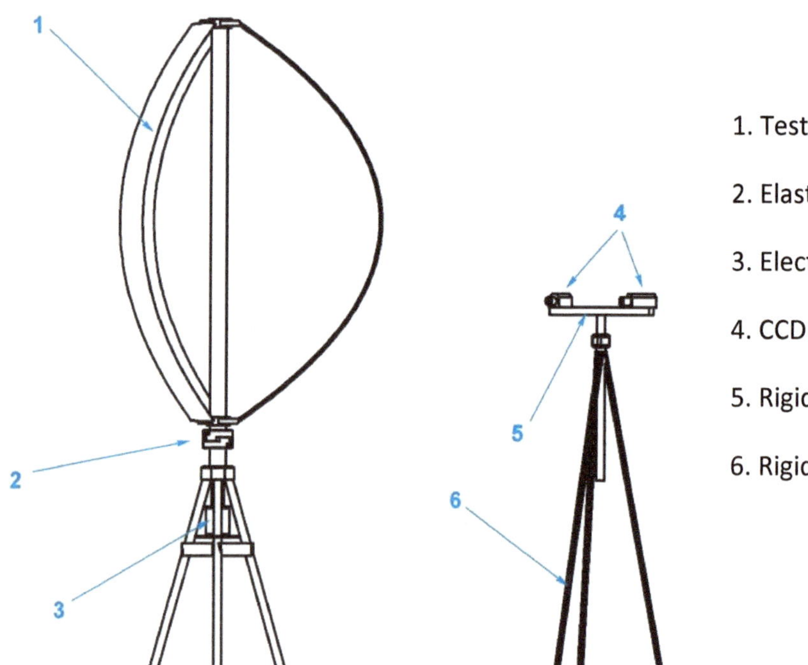

1. Tested turbine
2. Elastic coupling
3. Electrical stepper motor
4. CCD camera
5. Rigid support
6. Rigid tripod

Figure 2. Testing stand for monitoring the centrifugal forces effect by rotating the turbine shaft

3. VIC METHOD

In order to perform a high-accuracy 3D evaluation of the displacement field, during the rotation of the blades, the authors have chosen the Video Image Correlation (VIC) method.
Its strobo-module (the *3D-Vibro-correlation module*) assures these requirements.
The VIC method is a full-field contact-less method and its 3D version practically eliminates all disadvantages or limitations of most used experimental methods.
Mainly, the system consists of two high-resolution video cameras, mounted on a tripod by means of a high-precision connecting rod (see Figure 3).

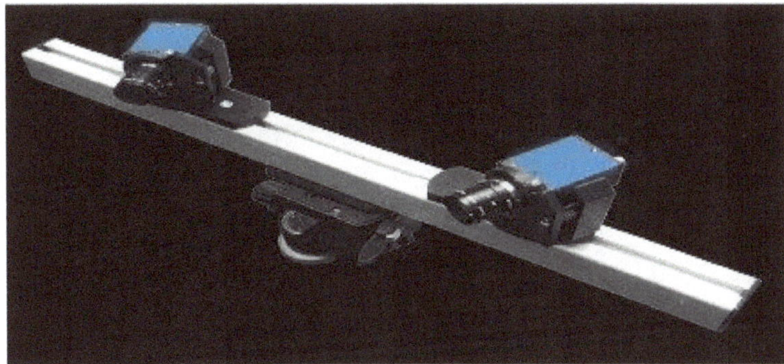

Figure 3. The VIC-3D setup [3, 4]

One can be applied in normal working conditions (not only in laboratories), because its software allows *eliminating the rigid body movements from the displacements field*, which represents one of its main advantages!
In advance, the tested object is sprayed with a water-soluble paint, in order to obtain a non-uniform dotted surface; the sizes of dotes depend on the surface sizes. In this way one can assure different grey-intensity of each pixel from the analyzed surface.
Before beginning of the tests, one has to perform a calibration, using some special targets/plates provided with a number of some high-accuracy set of dots (Figures 4 and 5), disposed adequately (in the plane corresponding to the predictable median plane of the tested object's surface). The target is rotated in horizontal and vertical plane in order to allow to the program a high-accuracy recognizing the 3D displacements of the significant pixels of the captured images (see explanations at Figure 6).
After the calibration, the cameras will perform the image acquisition in an $[n*m]$ matrix of pixels, firstly for the unloaded tested specimen (where one has to define *the area of interest*) and after then: for the loaded one.

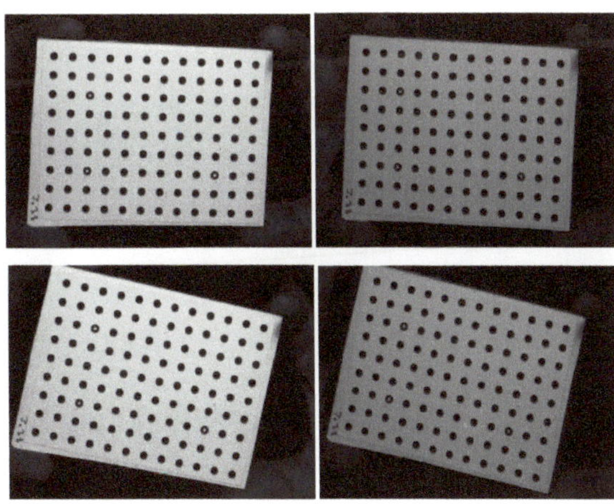

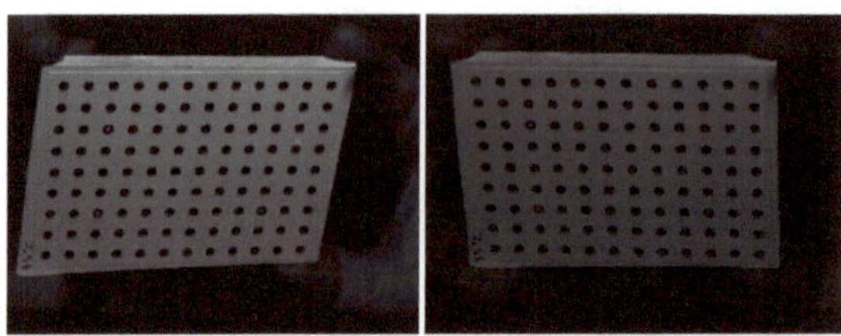

Figure 4. Different stages of the calibration process of each camera [6]

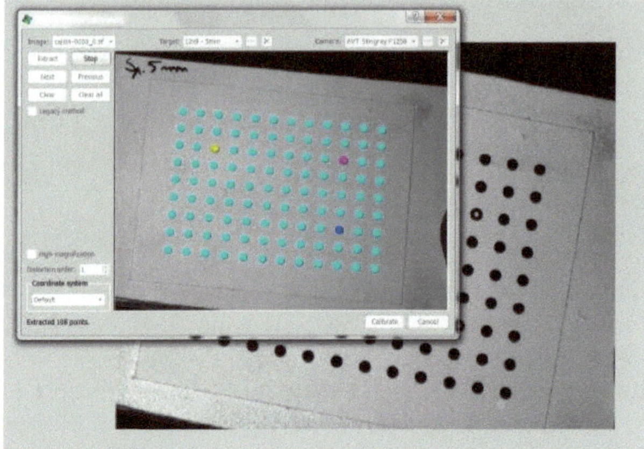

Figure 5. The extracting of the calibration points for the stereo calibration [6, 5]

Each captured image (by these two cameras), corresponding to the initial state of the object (more exactly: only the predefined area of interest), will be analyzed step-by-step (based on the principle schema from Figure 4).
So, the program allows the pre-selecting of a *Subset* (primarily cell) sizes (here: 5.5=25 *pixels*), respectively the step-magnitude (step size) for moving/translating of the Subset in horizontal and vertical plane).
For this Subset the program will establish /determine a unique grey-code, correlated to its median pixel high-accuracy 3D positioning.

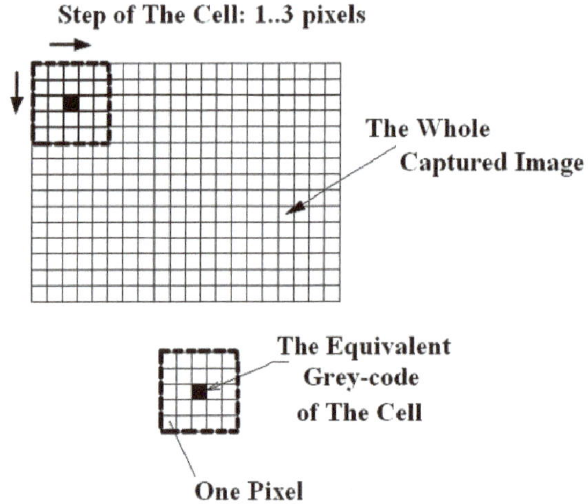

Figure 6. The measuring principle based on the scanning procedure [4, 5]

By analyzing of the whole image (by crossing over it with a pre-selected step: a number of pixels), each Subset cells will obtain a nominated (unique) high-accuracy spatial positioning and also a unique grey-code, too.

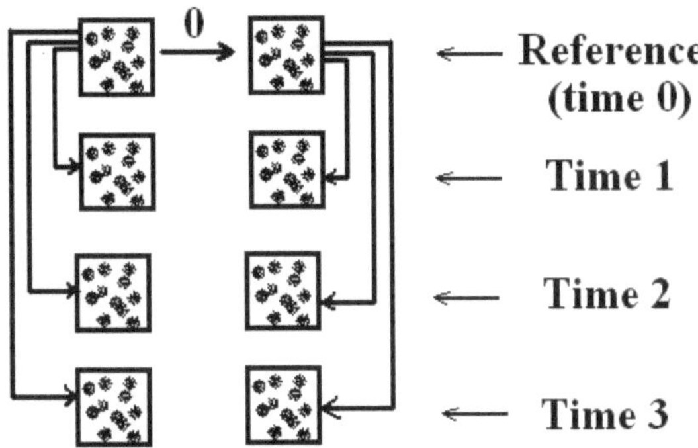

Figure 7. The stereo images analysis [4, 6]

After loading of the tested specimen, for all captured images (only in the area of interest, of course!), the program will identify the new positions of these Subsets, by performing an adequate comparison: *only once are compared the left and right images* (at time 0), after then, the succeeding left images are compared to the left reference, and succeeding right images are compared to the right reference (see Figure 7).
In order to perform an adequate evaluation/analysis of the captured images, the software requests, on the reference states, one single point's (meaning: one Subset's) identification on the left and right captured images; based on this single identification, the software will perform the identification of all Subsets in all captured images-pairs.
The same procedure will be applied in the strobo-mode, too, where the first image-pair will be captured in the static state of the object.
After that, by means of the stroboscopic image capturing, the 3D displacement capturing and analysis will be performed similarly.

5. A FEW PRELIMINARY RESULTS

Based on the previously described VIC methodology, have been prepared the blades. The monitoring system software allows the capturing of the images with the rotational speed of the blades. Hence, we will have the possibility to follow the 3D deformations field (and to save the data) at the level of a blade.
Also, the software allows the immediate establishment of specific linear and angular deformations for the same field of vision. These obtained data through experiment will serve as validation parameters for the numerical modeling.
In this case, the illustration of VIC method efficiency has been made by monitoring the deformations of the blades due to centrifugal forces at angular constant speeds (respectively at the corresponding frequencies of 1.0559, 1.47, 1.6447, 1.84, 2.0357 Hz).
A numerical analysis in Ansys has been performed together with the monitoring of scaled model deformations. The comparative results have been illustrated at the level of point with coordinates: R=333mm, H=397mm, with the coordinate system origin placed at the upper end of the blade.
For the numerical analysis have been introduced as input data the mechanical characteristics of the composite material from which has been made the scaled model.
As a result of the forced rotation of the turbine, the effect of centrifugal forces will lead, to the deformation of the blades.
In this simplified version of the stand (when there are not considered the aerodynamic forces by a wind tunnel), it is possible only the deformations evaluation of the blades.
In case is needed the experimental evaluation of the oscillating moment effect on the system, then we eliminate the elastic coupling from the system.

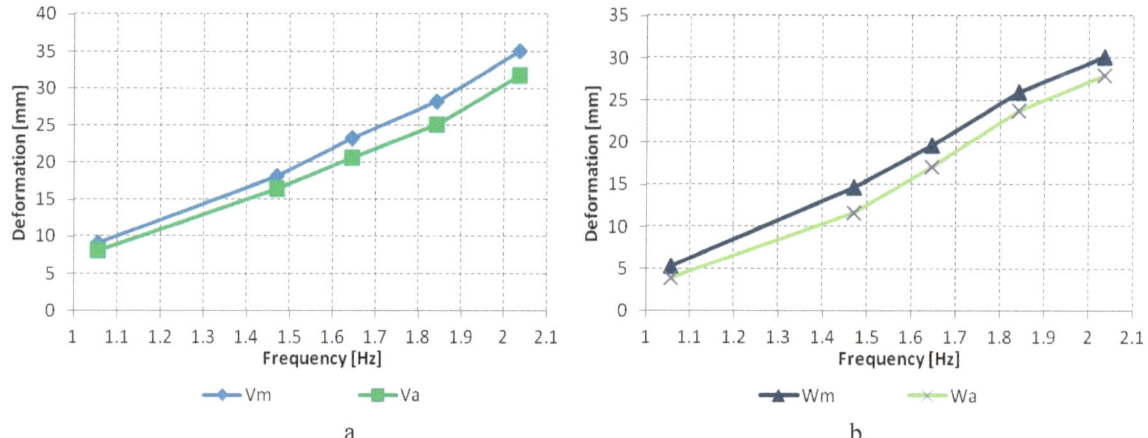

Figure 8. a. Variation of the point deformation on axial direction: Vm – experimental values, Va – numerical calculus values; b. Variation of the point deformation on radial direction: Wm – experimental values, Wa – numerical calculus values

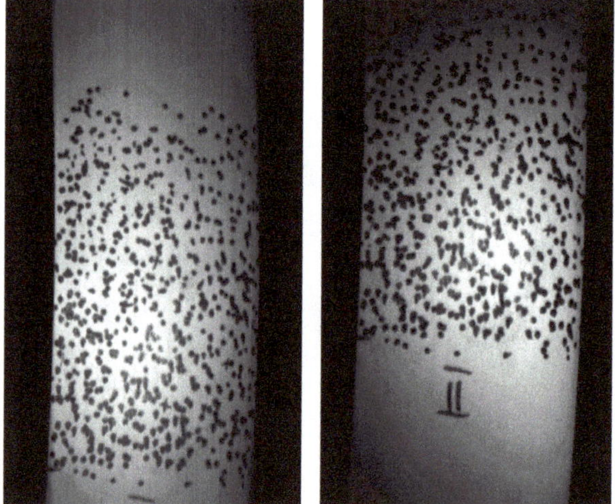

Figure 10. Blades deformations recorded by a camera: initial position and at maximum rotational velocity

3. CONCLUSIONS AND PERSPECTIVES

The original method presented, and illustrated by measurements, is, in authors' opinion of a great efficiency. Also, as we know, this methodology has not been previously used in analysis of wind turbines, so, it is a premiere.

The previously described methodology it maintains its remarkable qualities not only when applied to scaled models but also when it is desired to investigate the behavior during functioning for a real wind turbine scale model. Also, the method can be used in case of wind tunnel testing, for scaled models.

Must be acknowledged the special qualities of the VIC system, by its capacity to monitor the total deformation field (3D), as well as the wide range of the deformations (from a few microns to a few cm or more).

The authors are hoping that, based on these preliminary results, there will be a national and international collaboration with companies or universities.

REFERENCES

6. MERTENS, S., Wind energy in the built environment. Concentrator effect of buildings, Multi-Science, 2006, ISBN 0906522-35-8.
7. HAU, E., Wind turbines. Fundamentals, Technologies, Application, Economics, 2nd edition, Springer, 2006, ISBN-13 978-3-540-24240-6
8. *** - VIC-3D 2010, Testing Guide, Correlated Solutions, ISI-Sys GmbH, USA, Germany, 2010.
9. *** - VIC-3D 2010, Reference Manual, Correlated Solutions, ISI-Sys GmbH, USA, Germany, 2010.
10. IONESCU, D.R., PhD Thesis, "Transilvania" University of Brasov, Romania, 2014.
11. SUTTON, A. M., ORTEU, J. J., SCHREIER, W., H., Image Correlation for Shape, Motion and Deformation Measurements, Springer Verlag, 2010
12. KHAN, A., WANG, X., Strain Measurements and Stress Analysis, Prentice-Hall, Inc. Upper Saddle River, New Jersey, 2001
13. KOBAYASHI, A.S., Handbook of Experimental Mechanics, Prentice Hall, Upper Saddle River, New Jersey, 1987
14. JONES, R., WYKES, C., Holographic and Speckle Interferometry, A Discussion of the Theory, Practice and Application of the Techniques, Cambridge University Press, Cambridge, 1983.

ACKNOWLEDGEMENTS

This paper is supported by the Sectoral Operational Programme Human Resources Development (SOP HRD), ID134378 financed from the European Social Fund and by the Romanian Government.
This paper is supported by the Sectoral Operational Programme Human Resources Development (SOP HRD), ID137070 financed from the European Social Fund and by the Romanian Government

A CHEAP AND PORTABLE MOTION ANALISYS SYSTEM

M. Mihălcică[1], V. Guiman[2], V. Munteanu[3]
University „Transilvania" of Brasov, ROMANIA, Department of Mechanics, mircea.mihalcica@unitbv.ro
University „Transilvania" of Brasov, ROMANIA, Department of Mechanics
University „Transilvania" of Brasov, ROMANIA, Department of Mechanics

Abstract: *In this paper we will describe a cheap and portable motion analysis system dedicated to amateur and juniors sports, developed mainly for athletics. We will present the hardware and software components of the system and the way the system works: how the data is collected, stored and analyzed in order to improve the young athletes performances.*
Keywords*: motion analysis, sports, biomechanics*

1. INTRODUCTION

When building a motion analysis system, usually the developers take into consideration domains like sports and medicine. In sports, motion analysis is used in order to improve athletes performance by modifying their motion. That technique is common and it can nowadays should be found in every high-performance sports environment. Unfortunately, the costs of implementing such a system are usually very high for a common sports club which does not train top class athletes: the costs imply buying a professional system (which can reach up to a few hundred thousand Euros) and train and pay dedicated people able to understand the way the system works and able to deliver results based on gathered data. For this reason, many (if not most) sports clubs lack such a system. It is obvious that a cheaper alternative would benefit these clubs a lot, and we are thinking here about juniors and amateurs and any club which does not manage only top performance athletes.

2. MEANS AND METHODS

The system which we designed is a very simple one, having a hardware component and a software component. In order to keep the costs low, the hardware component is composed of only one video camera (any camera able to capture video materials with at least 250 frames per second), fixed support for the camera, long power cables (this is very important, because for many athletic disciplines the recordings should take place on the stadium) and a few markers. The software part, which takes care of analyzing the captured data, will contain one motion capture application and a few MATLAB programs specifically developed for each athletic discipline.
For a better understanding of how things work, let's say we will cover the "long jump" discipline. In brief, the system's capture part works like this:

- An athlete is chosen in order to analyze his performance.
- Markers are attached to the athlete's main joints: knee, ankle, hip, wrists, elbow, shoulder (the markers should be high contrast - if the athlete wears black equipment, we should use light-colored markers).
- The camera is set up perpendicularly to the athlete's running path, so that it captures the athlete's motion from the side.
- As one of the most important points of the discipline is the takeoff point, we will set the camera to primarily focus on that point.

- The athlete starts running, from the starting point. His full motion is captured by the camera. The video is then saved, the performance is measured and noted correspondingly to the athlete's first try (this is not mandatory, but it helps later, making the analysis easier).

In the same session, the athlete jumps again, until we have at least 10 tries (the more, the better).

Now we have 10 videos with the athlete's performances during the session. These videos are then send to analysis. A dedicated software is mandatory here: we strongly recommend Adobe After Effects (mostly because it is a popular application with excellent support), although we also recommend Dartfish or Kinovea as variants. Here, things work like this:

- The motion of each marker is analyzed using the dedicated software.
- A set of frame-by-frame motion coordinates is obtained: let's say the athlete was filmed for 2 seconds at 250 frames per second: now we will have 500 (X,Y) pairs of coordinates for each marker.
- Each coordinate set for each marker is saved in Excel and we will have (if we use 6 markers) 6 Excel tables for each of the athlete's try (one for each marker).
- These tables are analyzed using MATLAB applications developed specifically for analyzing long jump performances (angle of take-off, speed of take-off etc).
- Graphics and tables are offered as results, so that the trainer can see where improvements can be made.

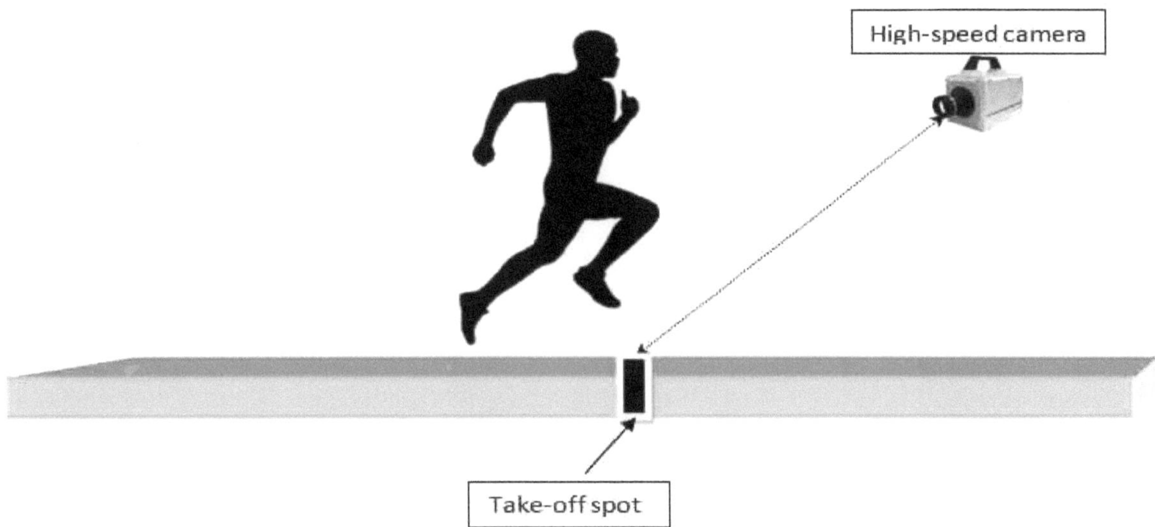

Fig. 1 The way the motion analysis system is set up - the athlete has markers attached on the other side (where the camera captures the motion)

3. BENEFITS

Because of the economical crisis and the huge difference regarding financing between top end sports clubs and juniors and normal clubs, the situation is you either have access to a professional system (which is very rare) or have just the trainer's eye for performance analysis (the most common situation). The most important benefit of using such a system is that you, as a trainer, can have some other data about your athlete's performance besides just the eyes, without having to pay the price of a fully professional system. Also, the system allows the trainer to use whatever camera he has access to (it has to be able to obtain video material at 250 frames per second at least, obviously, but there are no restrictions regarding the model or developer) and the system is portable (most of the professional systems are not, they are a pain to set up for an outside sports discipline)

4. CONCLUSIONS AND FUTURE WORK

There is definitely room on the market for a cheap portable alternative to the professional motion analysis systems used today in sports. In this paper, we presented the way in which such a system works, both hardware and software, and also we made some recommendations regarding the main components which are likely to be used (a 250+ frames per second video camera, Adobe After Effects, MATLAB). For future work, we will begin the analysis of some popular athletic disciplines such as long jump, high jump, hurdles and others using the system which we described above.

ACKNOWLEDGEMENT: This paper is supported by the Sectoral Operational Programme Human Resources Development (SOP HRD), financed from the European Social Fund and by the Romanian Government under the project number POSDRU/159/1.5/S/134378.

REFERENCES

1. Stephen G. Miller, Ancient Greek Athletics. New Haven: Yale University Press (2004)
2. Wendi H. Weimar, Ellen H. Martin, Sarah J. Wall. Kindergarten students' qualitative responses to different instructional strategies during the horizontal jump, Physical Education & Sport Pedagogy 16:3, 213-222 (2011)

THERMAL PROPERTIES OF SOLID WASTE MATERIALS

Liviu Costiuc
Transilvania University of Braşov, Braşov, Romania, lcostiuc@unitbv.ro

Abstract: *This paper presents the results obtained in measurements on heat of combustion of solid waste materials coming from municipal waste to evaluate this as fuel candidate. There are analyzed especially samples of plastic wastes coming from Braşov, Romania. Measurements for the heat of combustion of solid waste were done using the Parr bomb calorimeter. There are presented the sample preparation, the testing procedure, the comparative results on the mixed wastes and usual fuels and also the conclusions of the experiment. That proves the solid waste could be effectively used for energy recovery, by incineration.*
Keywords: *solid waste; heat of combustion; calorimeter.*

1. INTRODUCTION

Since 1999, Romania adopted the Waste Management Evidence and The Waste European Catalogue. At the end of 2005, the 1281 Rule (1281/16.12.2005) of the Environment and Water Management Ministry concerning the waste containers colors was adopted.

Concerning the management of plastics waste, landfilling is the last option, more and more avoided. Other options are: energy recovery by waste incineration, power station and cement kiln, or feedstock recovery by synthesis production and finally mechanical recycling.

A Romanian citizen generates around 5 kilos of household waste per week, half of it is biodegradable, a half of a kilo is glass and another half of a kilo is paper and cardboard. The rest is shared between other types of waste, 250 grams textiles and 200 grams polymers [1].

The reason of this study is to determine the effectiveness of municipal solid waste plastic wastes use as fuel.

2. MATERIALS AND METHODS

The tested plastic wastes come from Urban Enterprise source, Brasov County, Romania. The municipal solid waste was sorted to separate the plastic waste (see Figure 1). Second step was the washing procedure to eliminate organic waste, and after that the waste was dried and cut (see Figure 2). The sample average size after the cutting is about 10-20mm. The average sample size required for heat of combustion measurement with calorimeter is about 0.5-1.0mm. For that reason the plastic waste sample has been frozen in liquid nitrogen and after that was cut in the frozen state with a Retsch ZM200 centrifugal mill.

The resulted average sample size was approximately 0.5 mm. After drying, the samples were weighed using precision analytical balance. Each sample weight was approximately 1 gram with 0.1 mg precision. The resulted samples are a mixture of plastic materials, and there are prepared for measurements about 10 samples of municipal plastic waste to find out an averaged value for the heat of combustion.

The equipment used for the heat of combustion tests were: XRY-1C Oxygen bomb calorimeter, XRY-1C software and Kern & Sohn ABJ 220-4M analytical balance.

The heat of combustion is the energy released as heat when a compound undergoes complete combustion with oxygen in an enclosure of constant volume.

The gross heat of combustion, or calorific value, at constant volume (higher heating value or gross energy or upper heating value or higher calorific value) is the absolute value of the specific energy of combustion, in Joules, for the unit mass of a solid recovered fuel burned in oxygen in a calorimetric bomb under the conditions specified. The products of combustion are assumed to consist of gaseous oxygen and nitrogen coming from the gaseous atmosphere of burning, of carbon dioxide and sulphur dioxide, of liquid water (in equilibrium with its vapor) saturated with carbon dioxide under the conditions of the bomb reaction, and of solid ash, all at the reference temperature [7].

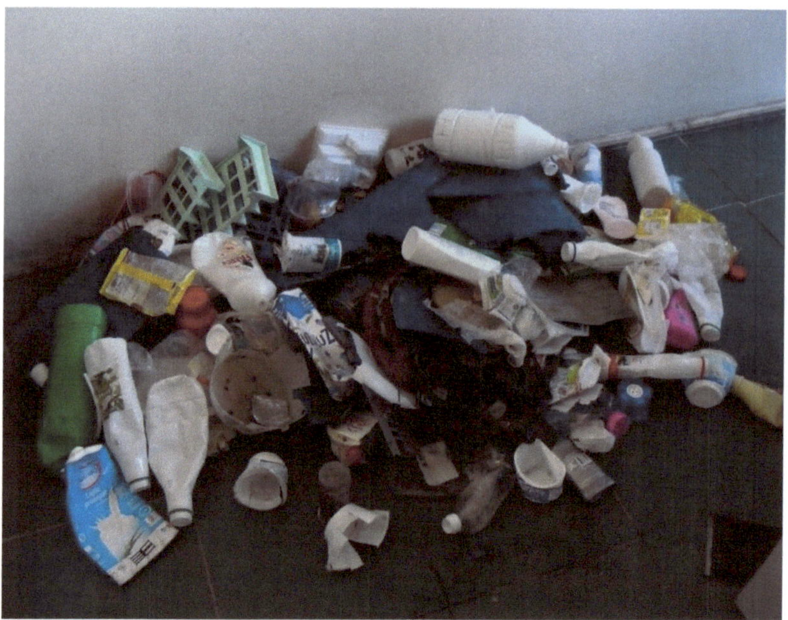

Figure 1. Sorted plastic waste from municipal Urban waste, Romania

Figure 2. Cut and dried prepared plastic waste.

The net heat of combustion, or calorific value, at constant volume (lower calorific value) is the absolute value of the specific energy of combustion, in Joules, for the unit mass of a solid recovered fuel burned in oxygen under conditions of constant volume and such that all the water of the reaction products remains as water vapor in a hypothetical state at 0.1 MPa, and the other products being all at the reference temperature [7]. With the bomb calorimeter is measured the gross calorific value. The testing procedure is presented in [2, 3, 7].

3. RESULTS

Before determinations of the heat of combustion value of samples, it is necessary to do the calibration of oxygen bomb calorimeter. This consists in a reverse procedure of testing. Having the value of the heat of combustion of benzoic acid standardized sample about 26435 J/g, it is determined by the same kind of test the thermal capacitance of the calorimeter, W, burning in crucible the benzoic acid and knowing its mass. The thermal capacitance of the calorimeter determined as average of 3 calibrations was W=12875 J/K.

For each waste plastic sample, in the XRY-1C software panel there were introduced the following input data: the mass of ignition wire in grams; the mass of cotton fuse in grams; the calorific value of wire [J/g]; the calorific value of cotton [J/g] and the mass of test sample in grams. After burning process the software has plotted the output graph containing temperature-time evolution(figure 3) and finally calculates the gross heat of combustion value as output using Regnault-Pfaudler method.

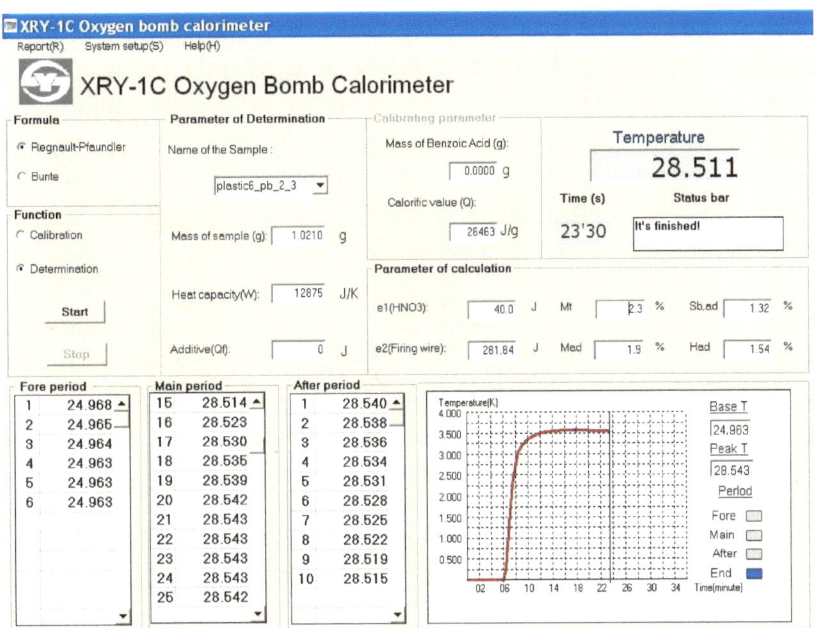

Figure 3. The XRY-1C software result panel of the plastic waste sample

The results obtained for mixed plastic waste are presented in Table 1. The $Q_{gr,ad}$ column is the gross heat of combustion value in adiabatic conditions, and $Q_{net,ad}$ is the net heat of combustion value, adiabatic. In Table 2 are presented the values for the heat of combustion of usual polymers and fuels as reported in literature.

Table 1. Results of calorimetric analysis of plastic waste mixture from Romania

Sample no.	Sample mass [g]	Gross heat of combustion $Q_{gr,ad}$ [MJ/kg]	Net heat of combustion $Q_{net,ad}$ [MJ/kg]
1	1.0654	45.135	44.582
2	0.8243	45.269	44.716
3	0.8204	44.441	43.891
4	0.8232	44.499	43.948
5	0.8350	44.496	43.946
6	0.8778	44.498	43.947
7	0.9211	44.151	43.602
8	0.8720	42.757	42.214
9	0.8418	43.151	42.606
10	0.8563	42.528	41.986

Table 2. Heat of combustion of some polymers and fuels [4]

Material type	Heat of combustion [MJ/kg]
Polyethylene (PP)	46.40
Polypropylene (PE)	46.30
Polystyrene (PS)	41.40
Polyvinyl Chloride (PVC)	18.00
Poly Ethylene Terephtalate (PET)	24.13
Poly carbonate bisphenol A (PC)	31.53
Unsaturated Polyester	26.00
Coal, Anthracite	32.80
Lignite	28.00
Gasoline	46.80
Jet Fuel, JP-4	46.60
Paper, Newsprint	19.70
Wood, Dry, Average	20.00

Comparing the results for heat of combustion for solid waste with the heat of combustion of usual polymers and fuels as reported in literature (Figure 4) can be observed that the all samples have high heat of combustion with an average value of 43.544 MJ/kg which is about 33% higher than anthracite coal, 55% higher than lignite, 7% smaller than gasoline and 6.6% smaller than jet fuel.

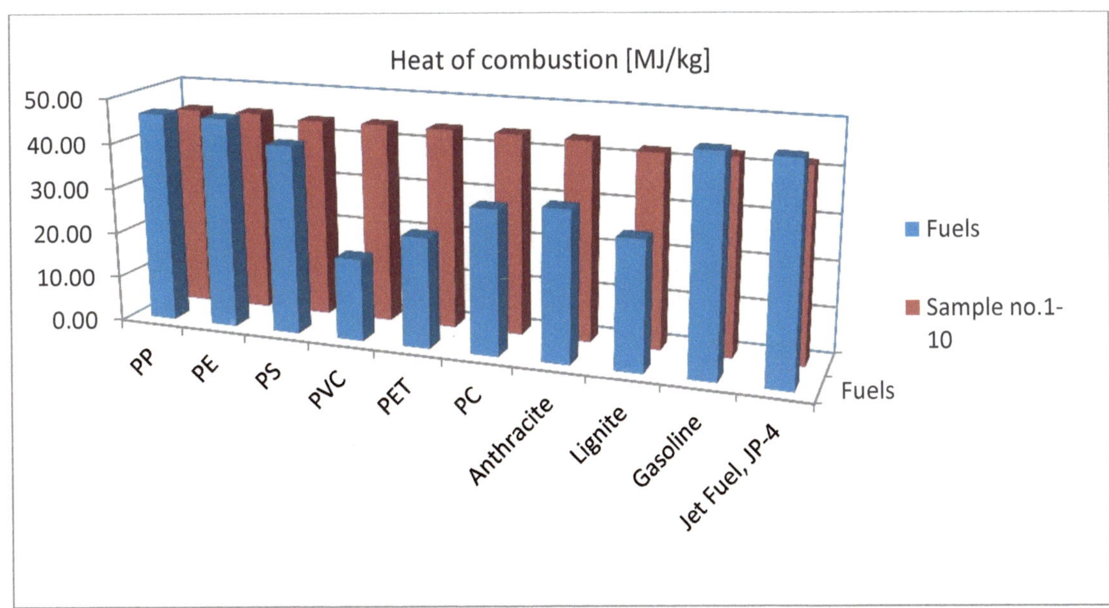

Figure 4. The comparative values for heat of combustion for fuels and solid waste samples.

4. CONCLUSIONS

The calorimetric analysis reveals a high capacity of the mixed plastic wastes to be used as fuel. The heat of combustion of those materials has a high level compared to solid fuels and comparable with petroleum products. Generally the polymers having higher density are characterized by lower calorific power. As the PE and PP were found in plastic wastes as composites, the heat of combustion of these mixtures will be higher, due to the higher polyolefinic calorific power of the PE and PP.

Even after considering the mixture of plastic waste in different concentrations the calorific power of these residues is still greater than those of different sorts of coals and these waste materials could be burned to obtain a useful amount of energy. That proves the solid waste could be effectively used for energy recovery, by incineration.

Acknowledgements: This research was funded by FP7 Grant 212782, „Magnetic Sorting and Ultrasound Sensor Technologies for Production of High Purity Secondary Polyolefins from Waste", acronym W2Plastics.

REFERENCES

1. Baltes, L., Draghici, C., Manea, C., Ceausescu, D., Tierean, M., Trends in Selective Collection of the Household Waste, Environmental Engineering and Management Journal, July/August 2009, Vol.8, No. 4, pag. 985-991.
2. Costiuc, L., Popa, V., Serban, A., Lunguleasa, A., Tierean, M.H., Investigation on heat of combustion of waste materials, Proceedings of the International Conference on Urban Sustainability, Cultural Sustainability, Green Development Green Structures and Clean Cars, Malta, September 15-17, 2010, Published by WSEAS Press, ISSN: 1792-4781, ISBN: 978-960-474-227-1, pag. 165-168.
3. Costiuc, L., Lunguleasa, A., Improving measurement accuracy of biomass heat of combustion using an oxigen bomb calorimeter, Bulletin of the Transilvania University of Brasov, vol.2(51)-2009, ISSN 2065-2119(print), ISBN- 978-973-598-521-9, pag. 467-474.
4. Kittle, P.A., 1993, Alternate daily cover materials and subtitle D-the selection technique, Rusmar Incorporated West Chester, PA, http://www.aquafoam.com/papers/selection.pdf
5. Walters, R.N., Hackett, S.M., Lyon, R.E., Heats of combustion of high temperature polymers, http://large.stanford.edu/publications/coal/references/docs/hoc.pdf.
6. European Commission DG ENV, Plastic Waste in the Environment, Specific contract 07.0307/2009/545281/ETU/G2 under Framework contract ENV.G.4/FRA/2008/0112, Final report, November 2010.
7. European Committee for Standardization, Solid recovered fuels - Methods for the determination of calorific value, DD CEN/TS 15400:2006, 2006.

HYBRID POLYMER BASED COMPOSITE MATERIALS' KINETIC PARAMETERS RETRIEVAL FROM DMA DATA

Dana I. Luca Motoc[1]

[1] Transilvania University of Braşov, Braşov, ROMANIA, danaluca@unitbv.ro

Abstract: *The herein paper aims to present an applied modeling concept directed towards kinetic parameters' estimation during glass transition as retrieved from dynamic mechanical measurements for hybrid particle-fiber polymer based composites. This dynamic process will be described by the aid of Arrhenius law applied to experimental data thus enabling the conversion degree at glass transition retrieval. The concept will be used to model and debate on the temperature-dependent overall complex moduli for the particle-fiber type hybrid polymer composite under the focus.*

Keywords: *thermo-mechanical properties, hybrid, polymer, kinetic, composites*

1. INTRODUCTION

Thermo-mechanical properties of hybrid polymer based composite materials can be ranked as one of the latest interests of both designers and developers since temperature-dependent properties are often misunderstood or ignored. It has been proven that dynamic mechanical analysis (DMA) is one of the effective methods used to study the relaxation phenomena in polymer materials and thereby the behavior of polymer based materials under various loading and temperature conditions.

In terms of theoretical prediction, literature reports few models used to describe the above relation processes.

Next, predicted and retrieved data were used to develop or modify existing empirical or statistical models proposed in literature. Thus, the most referred theoretical model is the one proposed by Mahieux et al. [1]. Their models compromises of Weibull-type functions that were used to describe the complex modulus change over the full range of transition temperatures in case of a polymer material or polymer based composite.

Furthermore, the well-known Arrhenius law was used to predict the kinetic parameters during the transition changes in few polymer based composite materials or solely polymer materials. Experimental data used were retrieved from DSC curves (DSC – Differential Scanning Calorimetry), proven the accuracy of the method. Modified versions of this were the subject of many reference works. Nonetheless, the model can be used in conjunction with thermo-mechanical measurements as can be retrieved from DMA or any other temperature-dependent setups [2].

Overall material properties in case of hybrid polymer composites are rather tricky to be predicted by the aid of theoretical models. The herein author was co-authored several papers regarding the previous issue by approaching the subject of predicting the overall properties, including dynamic ones by the aid of micro-mechanics formalism [3]-[5]. A multi-step procedure was employed and few theoretical models used to predict the overall mechanical properties of particle-fiber or fiber-fiber type reinforced polymer composite tailored architectures. Mathematical formalism associated with these models is difficult to manage due to the employment of complex modulus that requires skills and computer programs to collect the factors.

Besides of above, there were reported the influences of few external loading conditions and changes of theirs upon the kinetic parameters of fiber-reinforced polymer composites, including the water absorbed by this while immersed for a certain time interval [6].

Next, interest on dynamic properties of natural fiber-reinforced composites can be referred to be a recently topic approached by several researchers [7]-[8]. To the best knowledge of the herein author, there are no theoretical models developed to hold in case of these architectures neither in use modified ones.

The herein paper aims to present a framework for kinetic parameters' assessment using experimentally retrieved data retrieved from dynamic mechanical testing. Few influencing factors can be identified to be responsible for values obtained proven the identified sensibilities due to the testing configuration.

2. MATHEMATICAL FORMALISM

Due to experimental related restrictions in DMA data available for the herein hybrid polymer composite architecture (i.e. one heating rate) a modified Coats-Redfern formalism will be deployed to estimate the kinetic parameters during the glass transition (see [1]). Simple mathematical manipulation in the method's expression leads directly to the following:

$$ln\left(-ln\left(1-\alpha_g\right)/T^2\right)=ln\left(A_g R/\beta E_{A,g}\right)\cdot\left(1-2RT/E_{A,g}\right)-\left(E_{A,g}/RT\right) \quad (1)$$

where α_g is the conversion degree of the glass transition, T is the temperature, R is the universal gas constant (8.31J/mol K), A_g is the corresponding pre-exponential factor from Arrhenius law, β is the constant heating rate (set to 1 in this work) and $E_{A,g}$ is the activation energy.

Furthermore, the dynamic mechanical properties (e.g. complex modulus, either storage or loss modulus) can be expressed in function of the conversion degree such as:

$$E_{exp}=E_g\left(1-\alpha_g\right)+E_r\alpha_g\left(1-\alpha_d\right) \quad (2)$$

where E_{exp} is the experimentally retrieved dynamic property and α_d is the conversion degree during decomposition. Since decomposition was not recorded for the herein hybrid polymer based composite material the parameter will be neglected. Thus, expression (2) allows the conversion degree at glass transition estimation proven the E_g and E_r will be taken from the glassy and rubbery plateau in storage/complex modulus variation with temperature.

3. EXPERIMENTAL DATA.DISCUSIONS

3.1 Materials manufacturing and characterization

An open molding method based on wet lay-up technology was used to manufacture the hybrid polymer based composite plates. The particle-fiber type hybrid composite was obtained by intimately mixing alumina particles (10% out of the total reinforcement's volume content) and random E-glass fibers as five layers architecture. The matrix material was selected to be an unsaturated polyester resin, commercially available and employed due to its attractive price that is balancing the overall manufacturing associated costs.

The dynamic mechanical properties were retrieved in accordance with ASTM D5023-07 using a DMA 242 C analyzer from Netzsch GmbH (D) running into a 3-point bending mode at an oscillating frequency of 1 Hz. The temperature range was imposed between - 30°C up to 200°C and applied upon the samples at a scan rate of 3 °C/min under a controlled atmosphere.

3.2 Results and discussions

In Figure 1 were plotted the storage modulus and damping factor variation within the temperature range for the hybrid polymer composite under the focus. As it can be seen, the glass transition occurs mainly between 75° and 150° C. From the experimental data one has to retrieve the dynamic property associated to the glassy and rubbery plateau, respectively. Herein, the storage modulus values were employed and thus retrieved for 25° C and 150° C even mean values of their may be also used.

Based on expression (1), a plot of $ln\left(-ln\left(1-\alpha_g\right)/T^2\right)$ against $1/T$ was provided in Figure 2. Simple linear regression applied to the selected data (correlation factor 0.97) followed by parameter identification from slope value provided enables glass transition associated energy of activation retrieval. The activation energy, $E_{A,g}$ was calculated to be 2.04 kJ/mol as provided in Table 1. Next, this value was substituted in expression (1) allowing the retrieval of pre-exponential factor A_g (see provided values).

With respect to the value predicted for energy of activation, one might debate on the relatively small value retrieved compared with other polymer materials. The latter can be assigned to the polymer used as matrix in this hybrid composite architecture that is an unsaturated polyester resin. Apart for being one of the less costly material often employed by manufacturers it's polymerization process strongly depend on several factors such as: storage environment, manufacturing conditions, content of relaxing agents, etc. The activation energy strongly depends upon the constitutive content within the composite and reveals the effectiveness in stress transfer between the particles, fibers and matrix. Further insights into the issue should be closely approached.

Furthermore, one may further use the kinetic parameters to predict the temperature-dependent complex/storage or loss modulus and deploy an error minimization scheme between theoretical predicted and experimentally retrieved values. Consequently, the above formalism can be applied to the loss modulus experimental values and may be used to model the temperature-dependent viscosity. Since the viscosity in the

leathery and rubbery states appears to be different, the analysis should be individually carried out. It is beyond the purpose of the article to further debate on the issue.

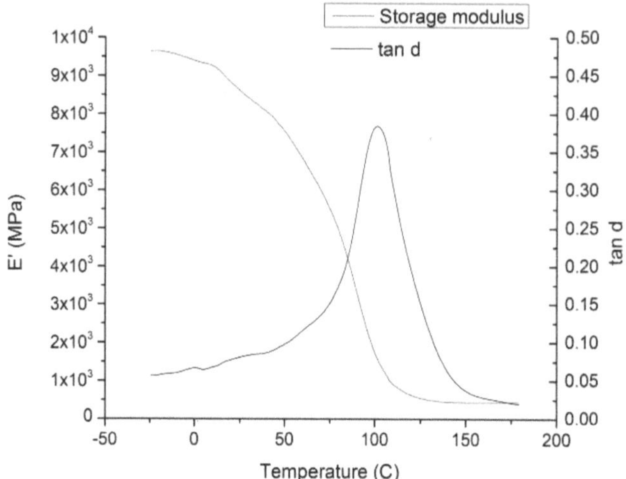

Figure 1: Storage modulus and damping factor variation within the temperature range

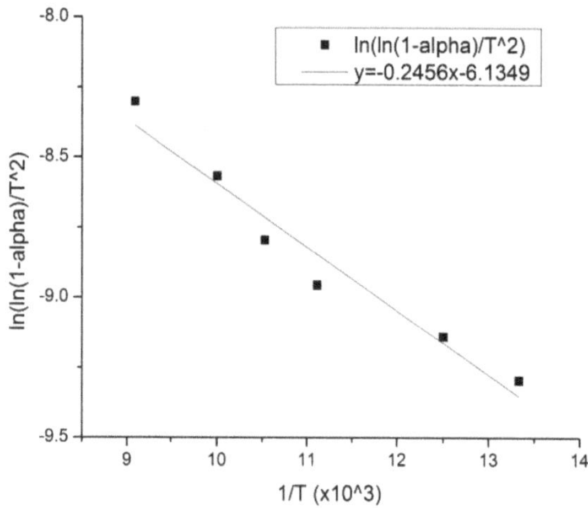

Figure 2: Experimental data related values to estimate $E_{A,g}$ during glass tansition

Table 1: Kinetic parameters associated to glass transition in hybrid particle-fiber composites

T (°C)	α_g (%)	A_g ($x10^{-7}$ min^{-1})	$E_{A,g}$ (kJ/mol)
75	40	37.07	2.04
80	50	40.57	
90	65	43.30	
95	75	48.21	
100	85	57.50	
110	95	68.12	

4. CONCLUSIONS

A mathematical formalism adopted by modifying the well-known Arrhenius law was used to estimate the kinetic parameters of a particular hybrid polymer based composite material architecture. Based on the estimates and their related plots, the following conclusions can be drawn:

- The hybrid composite state changes with temperature increase up to a certain value imposed by settings. Three different temperature-dependent material states can be indentified (glassy, leathery and rubbery) as well as their associated transitions (glass, leathery-to-rubbery).
- Kinetic parameters can be identified using the kinetic theory and Arrhenius equations by considering the experimental values retrieved for one of the dynamic properties (e.g. storage or loss modulus).
- The value of activation energy is relatively small in comparison with other types of polymers and an insight into the reasons relaying beneath this can be identified the relaxations' processes within unsaturated polyester resins.

Further approaches can be led with respect to the kinetic parameters in order to model the temperature-dependent viscosity of the herein hybrid composite architecture and next, directed towards other types of materials. Other temperature-dependent material properties can be identified and subjected to modeling by using the conversion degree of glass transition. Based on author's experience, significant differences are being expected despite the idea of setting the same temperature programs for all composite specimens.

REFERENCES

[1] C. A. Mahieux, K. L. Reifsnider – Property modeling across transition temperatures in polymers: application to thermoplastic systems, J. Mater. Sci., vol. 37, 2002, p. 911-920.

[2] Y. Bai, T. Keller, T. Vallee – Modeling of stiffness of FRP composites under elevated and high temperatures, Comp. Sci. and Technol., vol. 68, 2008, p. 3099-3106.

[3] D. Luca Motoc, I. Curtu. - Dynamic mechanical analysis of multiphase polymeric composite materials, Mater. Plast., vol. 46, 2009, p. 462-466.

[4] D. Luca Motoc - Dynamic mechanical characterization of cf/gf hybrid reinforced polymeric composite structures, Proceedings of the ASME 11th Biennial Conference on Engineering Systems Design and Analysis, 2012, Vol 3, p. 27-32.

[5] D. Luca Motoc – Hybrid particle/fiber polymer based composites analysis based on DMA data vs. material property predictions, Applied Mechanics and Materials, vol. Advanced Concepts in Mechanical Engineering II, 2014, p. 101-106.

[6] E. Faguaga, C. J. Perez, N. Villarreal, E. S. Rodriguez, V. Alvarez – Effect of water absorption on the dynamic mechanical properties of composites used for windmill blades, Mater. Design, vol. 36, 2012, p. 609-616.

[7] H. L.Ornaghi, et al. – Mechanical and dynamic mechanical analysis of hybrid composites molded by resin transfer molding, J. Appl. Polym. Sci., vol. 118, 2010, p. 887-896.

[8] S. Mohanty, S. K. Verma, S. K. Nayak – Dynamic mechanical and thermal properties of MAPE treated jute/HDPE composites, Comp. Sci. Techn., vol. 66, 2006, p. 538-547.

THE ANTIFRICTION LAYER OBTAINED THROUGH FINPLAST TEHNOLOGY - THE COMPOSITE MATERIAL CHARACTERISTIC PROPERTY

Dumitru I. Dascălu[1], Dorin Andrei D. Dascălu[2]

[1]Navy Academy "Mircea cel Batran", Constanta, ROMANIA, dumitru_dascalu2005@yahoo.com
[2]University of Architecture and Urbanism, "Ion Mincu" Bucharest, ROMANIA, dascalu.dorin@yahoo.com

Abstract: The paper presents a new type of composite obtained in the superficial layer of antifriction alloy of the sliding bearings, finished through FINPLAST technology. The paper shows several theoretical and experimental results of the authors, obtained after theoretical and experimental studying of this proceeding. The FINPLAST technology for finishing by cold plastic deformation of antifriction alloy of sliding bearings is an original proposal of the authors.

Key words: antifriction materials, composites, FINPLAST, design.

1. IN-DEPTH MODIFICATION OF CRYSTALLINE STRUCTURE IN THE SUPERFICIAL LAYER AFTER CHIP REMOVING PROCESS

The first aspect is being shown in fig.1, where the modifications of the materials after the chip removing process can be seen [1], [2], [6]. It becomes obvious in the diagram, the fact that after the chip removing process the superficial layer of the processed material has a certain depth of modified structure. According to the experimental tests which are shown in fig.1, the real profile obtained after the chip removing process is being covered by a superficial layer made of crystals pieces that are attached to the basic material with reduced bonds. There are two possibilities to obtain this type of layer, called "absorption layer" marked "*A*'" in the diagram. The first is obtaining it from polar molecules, which are connected to the basic material by Van der Wales connection, and the second possibility is by chemical bonds made between the structure of the profile and the pieces of crystals as a result of the chip removing process.

Immediately under the real profile, an oxide layer can be seen, called reaction layer and marked "*A*" in the diagram. Next, another layer called "BEILBY LAYER" noted marked "B" can be noticed. This is an amorphous layer (without regular crystalline structure) formed on the surfaces of the metal via mechanical working, wearing or mechanical polishing. A BEILBY'S layer is a disorganized molecular surface layer (science Dictionary). Next under the BEILBY layer, a layer marked "*C*" which has a deformed crystalline structure can be noticed.

The last modified layer is marked "*D*", and has normal crystalline structure. The specific characteristic of "D" layer is the fact that the tension in "C" layer determines a strain in it.. This tension makes it possible for some structure modifications in case of a big loading force of contact to the surface to appear. There is a possibility that the C layer will increase and will penetrate the D layer. The dimensional limits of these layers are shown in fig. 1. All these layers determine a specific superficial structure of the antifriction layer. This structure is characterized by a "memory" of generation mode that is obligating us to use the same way of turning of sliding bearings. The Beilby layer presenting a special crystalline structure has been known since 1903, but it was experimentally confirmed after three decades. This is characterized by a very fine structure that contains a high density of dislocations. This is similar to the amorphous structure. By the experimental determinations [1], [4], [6], this layer is very important for a good tribologycal function of bearings. In conclusion, it's very important for this layer to be obtained and maintained after the finishing process [6]. The ensemble of all these layers (A', A, B, C, D) represents an area in the depth of the material, with specific proprieties for each antifriction alloy. Next, this ensemble of layers is called LCRP (Layer-Chip

Removing Process). [1], [2], [6].

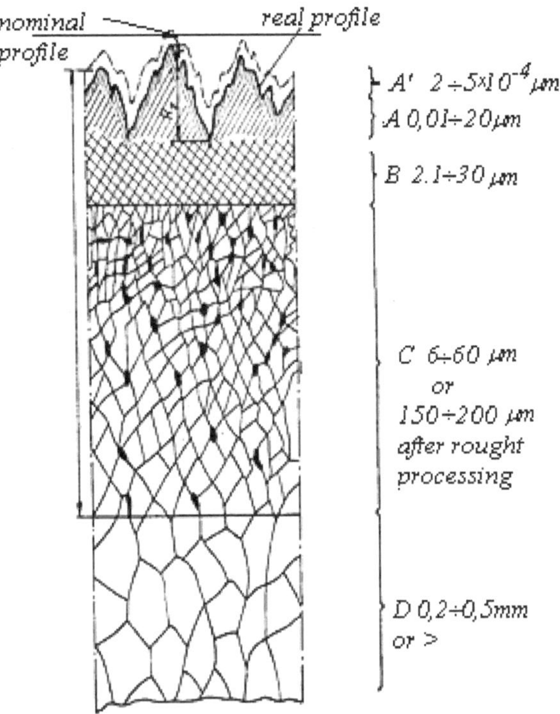

Legend:
A' – Absorption layer;
A –Amorphous structure;
B – Beilby layer (particularly amorphous structure);
C – Deformed crystalline structure layer
D - Intact crystalline structure but tensed, which makes it possible for some changes to occur;

Figure1. The superficial layer modifications of crystalline materials, after the chip removing process

LCRP and the experimental values which can be determined are shown in table 1, in the end of the paper. By analyzing the data from table 1, the influence of the technological process of fabrication and also the final finishing quality of the LCRP thickness can be seen. Due to the specific tenacity and plasticity of the antifriction materials, finishing by chip removing process brings multiple challenges. In order to solve the problem, special technologies like cathode deposit, thermo-chemical treatment, ionic plating, structural modifications with laser, etc. are used. The author has experienced and studied an original technical solution, which consist of finishing the antifriction layer by cold plastic deformation method, called finplast. The procedure is in accordance with the conclusion [5], which shows that tribologycally speaking, actually the surface layers are the ones that influence the durability and reliability of a bearing and not the rest of the material.

Table1. The values of LCRP depending of processing method

Processing method for finishing	Roughness R_a, R_z, [μm]	Beilby layer [μm]	Deformation Layer [μm]	LCRP [μm]
Rough turning, cold stamping, hot rolling, deepening	25-50 100-200	0.1	50-100	130-150
Lathe turning, milling, shaping, drilling, deepening	6.3-12.5 25.-50	0.05	25	50-70 (80)
Finishing turning, finishing milling, shaping, boring, broaching	1.6-3.2 8-12.5	0.01-0.02	5	15-20
Exact milling, rectifying, lapping, boring, fine turning	0.6-1.2 3-6	0.003- 0.008	2-4	4-8
Abrasive fine polishing, shaping, boring	0.4—0.6 2-3	0.003- 0.005	2-4	4-7
Finishing rectifying, diamond fine turning, polishing, honing, lapping	0.2-0.4 1-2	0.003- 0.005	2-3	3-5
Lapping, exact rectifying	0.1-0.2 0.5-1	0.003	1.5	2-2.5
Mechanical super finishing, polishing, lapping	0.025-0.05 0.125-0.25	0.001-0.0015	1.5	1.7
Electrolytic super finishing, polishing	0.012-0.025 0.063-0.125	0.0003	0.0034	0.01-0.2

2. FINPLAST TECHNOLOGY. GENERAL ASPECTS

The originality of the procedure is given by the use of the cold plastic deformation when finishing the contact surface of antifriction material of the sliding tribologycal couplings.

Keeping in mind the fact that the possibilities of finishing the antifriction alloys through cutting are reduced, the improvement of the performances both on the surface plane and in depth plane is solved by the help of other procedures. Besides other procedures known in the literature of specialty, the author considers the proposed procedure very useful. As we can see in figure 1 the procedure is very simple.

This one deals with the sag of the semi-finished material transformed through cutting, under the action of the lay-on roller. Previously, these surfaces were obtained through a cutting procedure. The main technological parameters of the procedure are:

- *F – rolling force;*
- *N – number of passing;*
- *If the contact is or is not lubricated during the processing.*

Besides these parameters, the processing results could be influenced by other measurements too, such as: rotary speed, which is equal to the displacement of the device mass, the asperity of the lay-on roller and its roughness, the hardness of the device and the precision of the relative position between the roller axis and the mobile mass plan, the steadiness of the rotary load and speed, the thickness of the antifriction material, etc.

The roller diameter is necessary to be correlated to the thickness of the antifriction alloy layer first. In order to avoid the adherence of the antifriction material on the roller surface, liquid and solid lubricants can be used.

3. THE DESCRIPTION OF THE DEVICE USED TO OBTAIN STUDY SAMPLES

For the study of the proposed procedure, two rectangular steel (OL 37) plates, one coated through plastic deformation with alloy AlSn10 and the other with alloy CuPb5, deposited through warm sintering were used. Both materials are used in the series production to get sliding bearings.

To obtain the study surfaces and the thickness of the antifriction layer, the frontal facing of the semi-finished materials on a normal lathe has been used. In order to do this, the study plates have been fixed on a rigid plane support, fixed in the lathe universe. The facing of the alloy AlSn10 has been obtained with the following technological parameters: rotation n=60 rot/min, radial advance s=0.14 mm/rot, and the cutting depth t= 0,57mm. For the alloy CuPb5 have been used: n=460 rot/min, s=0.18 mm/rot, t =0.5 mm.

In order to get study samples, a very simple device has been conceived in conformity with the pictures in figure 1.

Figure2. General view of device for finishing test surfaces

4. EXPERIMENTAL RESULTS AND OBSERVATIONS

For observing and analyzing the effect of finishing by FINPLAST technology, the scanned image of the sample study was used. In figure 3, the scanned images of samples for AlSn10, and a few details are represented. By comparison, in the fig. 4 there are represented the scanned images of CuPb5 alloys and also a few details. In order to observe the modifications after the contact with rolled surfaces after finishing by FINPLAST technology, in the detail there are presented and references the standard surfaces. These surfaces result after chip removing process.

As we can see in the picture, the device allows a wide range of modifications of the technological parameters described above.

From the details that are shown, one can observe that the studied surfaces present visible modifications. In the same time, one can also observe differences between details depending on the technological parameters used. This aspect is presented in scanned images.

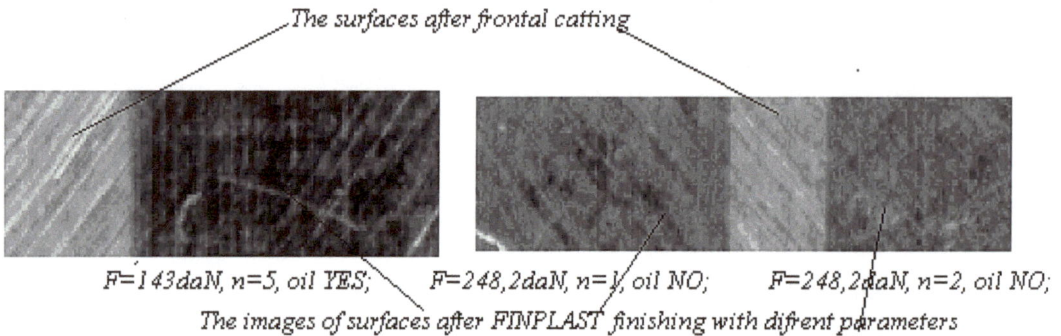

Figure.3 The most representative scanned image of the sample study from AlSn10

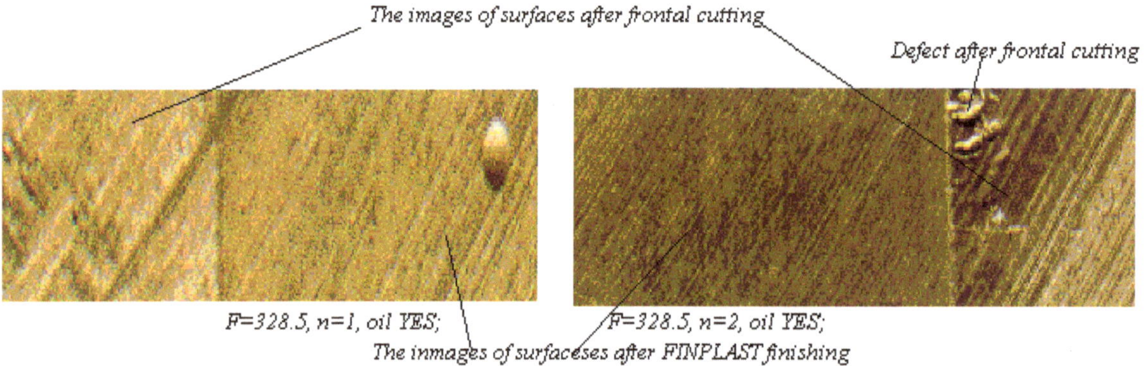

Figure 4. The most representative scanned image of the sample study from CuPb5,

By comparing figure 4 with figure 3, one can observe the different effect of Finplast finishing depending on the antifriction alloy. In the same time, one can also observe the better homogeneity of the finished surfaces in comparison with those obtained through sintering. It can be remarked how the sintering defects shown in fig. 4, detail of defect which is modified after finishing, and surfaces are corrected. The result is a surface which is no different from the surface with no defects.

As we can see in the picture, the device allows a wide range of modifications of the technological parameters described above. With the help of this device more study samples have been obtained under the form of plane bands.

This aspect expresses with no doubt the utility of the proposed procedure and the influence of the technological parameters.

5. CONCLUSION

5.1 General aspects

During chip removing process, we also find the two components of the force: radial and axial forces. After the chip removing process, the material suffers complex transformations in the superficial layer (figure 1 and table 1);

This shows that the obtained material layer will suffer in-depth, a preferential deformation of structure, described in paragraph 1. By creating a plastic deformation to these layers, like the ones presented until now, we have to consider sense of movement of the splinting tool.

Another important aspect of the problem is the LCRP layer. Due to plastic deformations suffered, the couplings between the crystals and the crystal fragments of these layers are becoming stronger. Inside of the layer, the crystals that suffered structural transformations by movements and/or deformations create a tensioned structure with a high number of dislocations. This will give the material's superficial layer superior tribologycal properties compared to the ones of the material before the plastic deformation.

5.2. The composite characteristic property of the antifriction layer obtained through FINPLAST technology

- After chip removing process in the superficial layer, a plastic layer (under D layer) is obtained. These represent the plastic matrix of composites material.
- In layer A, A', and B, there are a lot of particles like: oxides of components of antifriction materials, crystals, or crystal fragments and specific amorphous layer BEILBY;After finishing through FINPLAST technology, the layers A' and A will penetrate the Beilby layer, which will penetrate into the layers found beneath it (matrix of composites). All these will result in the enlargement of their dimensions.
- After this process, a metal (plastic) matrix which incorporates in its masse the hard particles is obtained;
- After incorporating in the masse the hard particles, a new superficial layer is obtained. This layer is more tensioned, and has better elastic properties;
- According to the theories of reinforcing (Orowan) of these materials, this irreversibility of structural transformation is very advantageous for antifriction layer of sliding bearings. In the normal case of using a sliding bearing, the material is stressed with variable forces. According the Orowan theory, these variable forces cause in the plastic material the aging and precipitation phenomenon.
- This solution of composites is also utilized in construction/ architecture, where the special composites with soft matrixes (different mortars) and different hard materials for special ornamental, mechanical characteristic or friction surfaces are obtained. Usually, in the soft matrix (mortar) it's loading of stock. Very seldom the hard and special components are pushing in soft matrices.
- In construction, this method gives the possibility to obtain special quality composites.

REFERENCES

[1] Dascălu D., New finishing process FINPLAST, The Publishing House Printech, Bucharest, 2004.

[2] Dascălu D., Contribution for up grating perform of sliding bearings, The Doctoral dissertation, 'TRANSILVANIA' University, Brasov, ROMANIA; 2004

[3] Dascălu D., FINPLAST, An Ecological Process, proceedings of SNOM, Brasov, pp 109-112, „TRANSILVANIA" University, Braşov, 9-10 June, 2005, Brasov, ROMANIA;

[4] Dascălu D., A New Concept In The Fabrication And The Design Of The Sliding Bearings, Proceeding of the 1st International Conference „Advanced engineering in mechanical systems ADEMS'07, Technical University of Cluj-Napoca, page. 389-392, 2007,
[5] Dascălu D., The Pitting Phenomena to Antifriction Alloys, in journal Meridian Ingineresc, Nr. 1, February, 2006, CHISINAU, MOLDOVA, page 51-55.
[6] Pavelescu, D., Tribotehnica, Techniques Publishing House, Bucharest, 1983.
[7] Groza, I. Pridvoriuc, M., Sechel, P., Drăgulin, M., The Cold Flow Of The Metals And Non-Ferrous Alloys, Techniques Publishing House, Bucharest, 1977

THE FAST ESTIMATION OF THE CUTTING FORCES AT TURNING AND DRILLING

Valentin Dițu
Transilvania University, Brașov, ROMÂNIA, vditu@unitbv.ro

Abstract: *The paper presents a fast experimental method to estimate the cutting forces at turning and drilling, method that is based on the electric current at cutting. The method can be used at the cutting processes of metal with edges that are good conductors of electricity. In the first part the paper presents in detail the fast estimation of the cutting force at turning steel OL37 with a cutting tool that has the edge made of P20 metallic carbide and in the second part it is shown the possibility of estimation of the axial force at drilling steel OLC45 with a drill from fast steel Rp3.*
Keywords: *electric current at cutting, cutting forces, turning, drilling*

1. INTRODUCTION

It is known the fact that the measuring of the cutting forces, for any processing method, has a great practical importance because besides the dimensioning and checking that are done with their help, these can be also used to diagnose the cutting process.
The new systems for measuring the cutting forces are based on piezoelectric transducers and are reasonable accurate, but they are modifying the stiffness characteristics of the technological system. So, in the tables 1 and 2 are presented the types of piezoelectric dynamometers and their characteristics, that can be used at turning, milling and drilling, produced by the company „Kestler". The cost of these measuring systems is reasonable high (more than 20000$) and in consequence there are few that can afford these systems.

Table 1. Piezoelectric dynamometers for turning and milling produced by the company „Kistler"

Crt. No.	Type	F_x, F_y		F_z	
		Domain	Work Freq.	Domain	Work Freq.
1	9255A - milling -	± 20 KN Er.<0,01 N	1,5 KHz	-10÷40 KN Er.<0,01 N	1,5 KHz
2	9257A milling/turning	± 5 KN Er.<0,01 N	4 KHz	± 5 KN Er.<0,01 N	4 KHz
3	9265A1 - turning -	± 15 KN Er.<0,01 N	1,5 KHz	0÷30 KN Er.<0,01 N	2 KHz
4	9265A2 - milling -	± 15 KN Er.<0,01 N	1,5 KHz	-10÷30 KN Er.<0,01 N	2 KHz
5	9281B22 - milling -	± 20 KN Er.<0,01 N	0,6 KHz	-20÷40 KN Er.<0,01 N	0,6 KHz
6	9281B23 - milling (different plate) -	± 20 KN Er.<0,01 N	0,6 KHz	-20÷40 KN Er.<0,01 N	0,6 KHz

From the tables can be observed that the produced systems are accurate but they modify the stiffness of the technological system and in consequence there must be searched other solutions to avoid this fact but also to avoid the high aquisition cost. In the next rows it is proposed a method to diagnose the cutting forces, method that mostly eliminates the negative facts previous presented.

Table 2. Piezoelectric dynamometers for drilling produced by the company „Kistler"

Crt. No.	Type	F_x, F_y		F_z		M_z	
		Domain	Work Freq.	Domain	Work Freq.	Domain	Work Freq.
1	9271A drilling	-	-	-5÷20 KN Er.<0,02 N	3 KHz	± 100 Nm Er.<0,02 Ncm	3 KHz
2	9273 drilling	± 5 KN Er.<0,02 N	1,5 KHz	-5÷20 KN Er.<0,02 N	3 KHz	± 100 Nm Er.<0,02 Ncm	3 KHz
3	9291 drilling	-	-	-100÷200 KN Er.<0,02 N	5 KHz	± 2000 Nm Er.<0,05 Ncm	2,5KHz
4	9293 drilling	± 20 KN Er.<0,01 N	4,5 KHz	-100÷200 KN Er.<0,02 N	5 KHz	± 2000 Nm Er.<0,05 Ncm	2,5KHz

2. THE FAST ESTIMATION OF THE CUTTING FORCES AT TURNING

2.1. The strategy of the fast estimation of the cutting forces

At turning of the metallic materials with edges that are good electricity conductors appears an electric current, in consequence, mainly, of the heat in the cutting process. As the heat in the cutting process depends on the cutting forces, results that we have data about the cutting electric current and we can appreciate the value of the cutting forces.

The measuring of the cutting electric current is more simple and more accurate than the measuring of the forces with rezistive systems or even with piezoelectric systems, and, in addition, it doesn't modify the stiffness of the technological system of processing.

In order to use the value of the cutting electric current tension at the appreciation of the cutting forces it must be determinated the relation between force and electric current. This can be done using two methods.

The first method consists in the determination of the relation that gives the value of the electric current depending on the cutting parameters, and based on the data from the specialty literature it can be determinated the relation that gives the value of the cutting force depending on the value of the cutting parameters, and finally, through the cutting parameters results the relation between the cutting force and the measured cutting electric current tension. The method is simple, it is necessary minimum of equipment, so it is also cheap, but the weak point it is represented by the data from the specialty literature.

The second approach, more accurate, consists in simultaneous using of a system to measure the tension of the electric current with another system to measure the cutting forces, the data being collected by an acquisition card and saved in the memory of a computer, and then it will be determinated the relation between the force and the tension of the electric current. This method is the best but it also needs an accurate system to measure the cutting forces, fact that is not available for anyone.

From economical reasons it will be used the first approach.

2.2. The elaboration of the methodology and experimental data

To determinate the cutting force is used the relation (1).

$$F_z = C_{Fz}.t^{xFz}.s^{yFz}.(HB)^{nz}.K_{Fz} \quad [daN] \tag{1}$$

The experiments were done at turning OL37 steel with a P20 metallic carbide plate, that has the next geometry: $\alpha = 8^0$; $\gamma = 10^0$; $\lambda = 0^0$; $K= 45^0$; r = 0,5 mm. According [4], OL37 steel, having HB = 164, the next data are found: $C_{Fz} = 27{,}9$; $x_{Fz} = 1$; $y_{Fz} = 0{,}75$; $n_z = 0{,}35$; $K_{Fz} = 0{,}33$.

Introducing the below data in the relation (1) it results:

$$F_z = 54{,}87 \,.\, t \,.\, s^{0,75} \quad [daN] \tag{2}$$

To determinate the relation for the tension of the electric current depending on the cutting parameters it is used the installation presented in papers [1,2,3]. The obtained experimental data are centralized in table 3.

The processing of the experimental data has led to relation (3).

$$U = 1,615 \,.\, v^{0,506} \,.\, s^{0,323} \,.\, t^{0,129} \quad [mV] \tag{3}$$

Table 3. Experimental data, for tension of the electric current at cutting, obtained at turning OL37 steel with CMP20 plate.

Exp. No.	n [rot/min]	d [mm]	v [m/min]	s [mm/rot]	t [mm]	U [mV]
1	200	49,7	31,22	0,106	0,5	4,13
2	250	49,7	39,03	0,106	0,5	4,65
3	315	49,7	49,18	0,106	0,5	5,05
4	400	49,7	62,45	0,106	0,5	5,7
5	500	49,7	78,06	0,106	0,5	6,65
6	630	49,7	98,36	0,106	0,5	7,36
7	400	49,2	61,82	0,053	0,5	4,18
8	400	49,2	61,82	0,106	0,5	5,66
9	400	49,2	61,82	0,151	0,5	6,40
10	400	49,2	61,82	0,208	0,5	6,90
11	400	49,2	61,82	0,25	0,5	7,12
12	400	49,2	61,82	0,302	0,5	7,33
13	400	49,5	62,2	0,106	1,5	6,93
14	400	49,5	62,2	0,106	1,25	6,71
15	400	49,5	62,2	0,106	1,00	6,56
16	400	49,5	62,2	0,106	0,75	6,28
17	400	49,5	62,2	0,106	0,5	6,00

Relation (2) can be written:

$$F_z = 54,87 \,.\, t \,.\, s^{0,75} \,.\, U(v,s,t)/U(v,s,t) \quad [daN] \tag{4}$$

Using also the relation (3) is obtained the connection relation between the force and the tension of the electric current:

$$F_z = 33,98 \,.\, v^{-0,506} \,.\, s^{0,427} \,.\, t^{0,871} \,.\, U \quad [daN] \tag{5}$$

The verifying of the relation (5) it is done for the next example: having v = 62,2 m/min; s = 0,106 mm/rot; t = 1,5 mm experimental resulting U = 6,9 mV, with relation (2) is obtained F_z = 15,29 daN and with relation (5) is obtained F_z = 15,83 daN.

3. THE FAST ESTIMATION OF CUTTING FORCES AT DRILLING

To estimate the cutting forces at drilling it is used the same method as for turning. In addition, it will approach the cutting moment.

In order to determinate the force and the cutting moment at drilling OLC45 steel, with a drill made from Rp3 fast steel, that has the next geometry, $\alpha = 8^0$; $\gamma = 25^0$; $2K = 118^0$, there are used the relations (6) and (7).

$$F = C_F.t^{xF}.s^{yF}.K_F \quad [daN] \tag{6}$$

$$M = C_M.t^{xM}.s^{yM}.K_{FzM} \; [daN.cm] \tag{7}$$

According to [4] results the next practical relations:

$$F = 78,75 \,.\, d^{1,07} \,.\, s^{0,72} \quad [daN] \tag{8}$$

$$M = 7,705 \,.\, d^{1,71} \,.\, s^{0,84} \quad [daN] \tag{9}$$

To determinate the relation that gives the tension of the thermocurrent depending on the cutting parameters it is used the installation presented in the papers [1,2]. The obtained experimental data are centralized in table 4.

The processing of the experimental data has led to relation (10).

$$U = 0{,}222 \,.\, v^{0,723} .\, s^{0,462} .\, d^{0,396} \quad [mV] \qquad (10)$$

The relation (8) can be written:

$$F = 78{,}75 \,.\, d^{1,07} \,.\, s^{0,72} . U(v,s,t)/U(v,s,t) \quad [daN] \qquad (11)$$

Table 4. Experimental data, for the tension of the cutting thermocurrent, obtained at drilling the OLC45 steel with drills with edges from Rp3 fast steel.

Exp. No.	n [rot/min]	d [mm]	v [m/min]	s [mm/rot]	U [mV]
1	160	9	4,52	0,13	0,63
2	315	9	8,90	0,13	0,93
3	450	9	12,72	0,13	1,30
4	630	9	17,81	0,13	1,80
5	900	9	25,44	0,13	2,06
6	630	9	98,36	0,10	1,4
7	630	9	61,82	0,13	1,7
8	630	9	61,82	0,19	2,0
9	630	9	61,82	0,27	2,4
10	630	9	61,82	0,38	2,6
11	1250	6	23,56	0,13	1,6
12	900	7	19,79	0,13	1,6
13	630	9	17,81	0,13	1,7
14	630	11	21,77	0,13	1,9
15	630	12	23,75	0,13	2,2

Using also the relation (10) is obtained the connection relation between the axial force and the tension of the electric current at cutting:

$$F = 354{,}73 \,.\, v^{-0,723} .\, s^{0,258} .\, d^{0,684} .\, U(v,s,t) \quad [daN] \qquad (12)$$

In a similar way is obtained the relation (13).

$$M = 34{,}71 \,.\, v^{-0,723} .\, s^{0,378} .\, d^{1,324} .\, U(v,s,t) \quad [daN] \qquad (13)$$

4. CONCLUSION

The estimation method of the cutting forces at turning and drilling, presented in this paper, is using the connection relation between the tension of the electric current at cutting and the force and/or the cutting moment. The main advantage of this method consists in it's simplicity and it's accuracy and shows that using few technical equipment, very good results can be obtained. Also, the estimation it is done in a short time, processing by cutting almost 20 seconds and measuring the tension of the resulted electric current. The tension of the resulted electric current is measured by milivolts fact that means that it can easily be measured with the actual technical equipment. Comparing turning with drilling it can be observed that the tension of the electric current at turning is greater than the tension of the electric current at drilling, the differences resulted because of the processed material, because of the different geometry of the tools, because of the used cutting parameters and most of all because of the material of the cutting edge.

REFERENCES

[1] Diţu, V., Theoretical and experimental research about the diagnose of the cutting process. PhD Thesis, Transilvania |University of Braşov, 1997.

[2] Diţu, V., The basis of surface generation and the cutting tool, Edit. „Transilvania" University of Braşov, 1999.

[3] Diţu, V., The Measurement of Cutting Forces at Turning Steel OLC 45 Without Modification of Technological System's Characteristic, 10-th National Scientific Symposium with international participation "Metrology and Metrology Assurance '2000", sept. 2000, Sozopol, Bulgaria.
[4] Picoş, C., ş.a., Calculus of processing allowance and of the cutting parameters, Edit. Tech., Bucharest, 1974.

COMPOZITES ADHESIVES BASED ON SYNTHETIC AND LIGNOCELLULOSIC MONOMERS

L. Dumitrescu[1], I. Manciulea[1], A. Matei[2]

[1] Transilvania University of Brasov, Research Centre Renewable Energy Systems and Recycling, Brasov, ROMANIA, lucia.d@unitbv.ro

[1] Transilvania University of Brasov, Research Centre Renewable Energy Systems and Recycling, Brasov, ROMANIA, i.manciulea@unitbv.ro

[2] National Institute for Research and Development in Microtehnologies, Bucharest, ROMANIA, alina.matei@imt.ro

Abstract: *Important progress has been made nowadays in development of new composites engineering materials from renewable, large available lignocellulosic waste materials, as a more sustainable alternative to raw materials from fossil sources. Our research has been focused on synthesis and characterization of some new composites adhesives based on woody wastes fillers ammonium, aluminium and iron and chromium lignosulfonates, as partial substitutes for toxic synthetic monomers phenol and formaldehyde in synthesis of phenol-formaldehyde resins (matrix). The improvements achieved, by using the lignosulfonates, in the properties of the new composites resins phenol-formaldehyde-lignin derivatives consist on decrease in adhesive viscosity, better water resistance of finished boards, increasing of the pot life, resistance to biodegradation and low costs.*

Keywords: *recycling biomass waste, composites, lignosulfonates.*

1. INTRODUCTION

Lignocellulosic materials are important natural renewable resources. The processes for chemical reutilization of lignocellulosic materials offer opportunities to produce a new generation of high-performance, high quality products. As a natural product of biological origin, lignocellulosic materials are characterized by high degree of diversity and variability in their properties. Lignocellulosic materials contain natural polymers cellulose, hemicellulose and lignins which possess many active functional groups susceptible to reaction [1, 2]. Lignin, the most abundant organic polymer in the plant world, is a three dimensional aromatic polymer with a phenylpropane unit, and have in structure phenolic and alcoholic hydroxyl groups. Lignin has a complex and heterogeneous structure caused by variations in its composition, size, cross-linking and functional groups. This heterogeneity depends on the species of plant from which it is obtained, the pulping process used to separate it from cellulose, and the methods by which it is recovered from the pulping liquor [3, 4].

The presence of lignin as a waste product in pulp mills has made it an attractive raw material for adhesives. Lignin is insoluble in water, but during the technical sulphite pulping, lignin becomes soluble in water, due to the partial degradation and introduction of sulfonic groups [5,6].

At elevated temperatures, when lignosulfonate is treated with strong mineral acid, condensation reactions leading to diphenylmethanes and sulfones take place. Hydroxybenzyl alcohol groups, as well as sulfonic acid groups on the carbon alpha to the aromatic rings of the phenylpropane units of the random polymer, react with the aromatic nuclei of other phenylpropane units in the presence of the strong mineral acids. This reaction, leading to diphenylmethane, is of the same type as the formation of phenolic resins from phenol and formaldehyde. Lignin also reacts with formaldehyde and can be cross-linked by it, in the same manner of synthetic polyfenolic resins [7, 8, 9].

The production of phenol-formaldehyde (PF) resins has not undergone significant changes for decades. However, the oscillations in crude oil price have generated a great interest in development of alternatives to oil-derived binders. Thus, naturally occurring raw materials such as lignin and furfuryl alcohol offer interesting possibilities as substitutes for phenol. On the basis of its structure, lignin may be considered a type of phenolic condensate in the broadest sense of the term, making it logical to use it in the production of phenolic resins.
Moreover, the structural changes suffered by lignin during the pulping process can exert a great influence on its characteristics [10, 11, 12, 13]. Lignosulfonates represent a type of technical lignin whose utilization in polymer synthesis is still insufficiently developed. There are certain properties of lignosulfonates which make them a potential substitutes for phenol in phenol-formaldehyde resins. The most important is their macromolecular structure which allows a quick gelation process [14, 15]. Phenol-formaldehyde (PF) resins have been intensively

studied to find economically and environmentally applicable natural raw materials as substituents of toxic and costly phenol and formaldehyde

Phenolic resins are considered an attractive area for commercial lignin applications because of its chemical reactivity. Lignin is available, less toxic and less expensive raw material than phenol. Due to the increase of phenol cost, research have been made to partially substitute monomer phenol with natural products, having phenolic hydroxyl groups such as lignin/lignosulfonates [16].

Our research has been focused on the obtaining of some new composites adhesives, based on metal complexed lignosulfonates (fillers), as partial substitutes for formaldehyde and phenol in the phenol-formaldehyde adhesives (the matrix of composites).

2. EXPERIMENTAL

2.1 Materials and Methods

Phenol (99%), 37% formaldehyde aqueous solution and sodium hydroxide (98%) were used. All the chemicals were supplied by Merck. The lignosulfonates were obtained by modifying ammonium lignosulfonate ($LSNH_4$) (from wood and paper industry) with different salts.

The aluminium lignosulfonate (LSAl) was obtained from ammonium lignosulfonate with $AlCl_3$ and the iron and chromium lignosulfonate ($LSFe^{3+}Cr$) was obtained from ammonium lignosulfonate with $Fe(NO_3)_3$ and $Na_2Cr_2O_7$. The lignosulfonates were analysed conforming to the specific methodology for lignins [17, 18, 19].

In order to partially substitute the phenol and formaldehyde, ammonium, aluminum and iron and chromium lignosulfonates were used in the synthesis of the new composites adhesives phenol-formaldehyde resins. The metal complexed lignosulfonates have also been used as comonomers in the polycondensation reaction because they possess in their structure reactive functional groups: phenolic and alcoholic hydroxyl, carbonyl, carboxyl, sulfonic groups, as can be seen from the data in Table 1 and from the FT-IR spectra (Figure 1) performed with a FTIR-Spectrometer model BX II (Perkin Elmer, 2005).

Table 1. The physical-chemical characteristics of the lignosulfonates

Characteristic	$LSNH_4$	LSAl	LSFeCr
pH- value	4.85	3.07	3.80
Solids, %	44.96	38.99	42,50
Density at 20 C, g/cm	1.0214	1.1400	1.1850
Viscosity at 20 C, cP	70.00	66.00	68.00
Ash, %	0.96	2.69	5.28
Cation, %	7.39	7.62	6.70 (Fe), 6.50 (Cr)
Functional groups:			
- OH phenolic, %	14.16	18.74	22.90
- OH alcoholic, %	13.74	16.50	19.40
- carbonyl, %	1.71	2.66	11.31
- carboxyl, %	0.60	0.72	3.93

The FTIR spectra of ammonium, aluminum and iron and chromium lignosulfonates show characteristic absorption bands for specific chemical functional groups, able to participate as comonomers to the polycondensation process of phenol-formaldehyde-lignosulfonates resins:

- 3039.71-3208.16 cm^{-1} - absorption bands assigned to aliphatic and aromatic hydroxyl (-OH phenolic) groups from chemical monomer (phenol) and lignosulfonates
- 2359.77- 2363 cm^{-1} intense aromatic C–H absorption bands from phenol and lignin structure
- 1630-1675 cm^{-1} - absorption bands corresponding to aromatic carbonyl bonds (C=O) from lignosulfonate, which also can be involved in substitution of formaldehyde monome
- 1419.41-1507.43 cm^{-1} - absorption bands specific for aromatic $-CH_2-$ skeletal vibration and aromatic nuclei
- 1029.00 cm^{-1} absorption bands specific to methoxy group ($-OCH_3$) from lignin and lignosulfonates
- 1109.79-1155.00 cm^{-1} - absorption bands assigned to aromatic methylene groups and to $-SO_3H$ groups from lignosulfonates

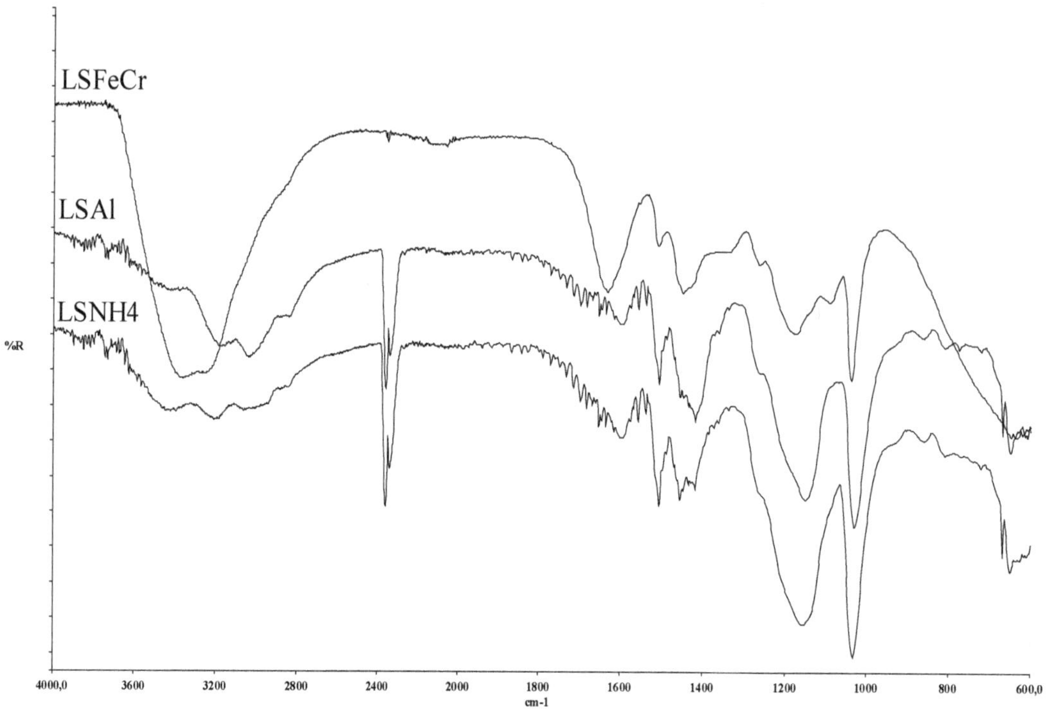

Figure. 1: The FTIR spectra of ammonium, aluminum and iron and chromium lignosulfonates

2.2 Synthesis of the composites based on phenol-formaldehyde adhesives with ammonium, aluminum and iron and chromium lignosulfonates

Phenolic resins were the first true synthetic polymers developed commercially, obtained by polycondensation of phenol with formaldehyde. The characteristic that renders these resins invaluable as adhesives is their ability to deliver, at relatively low cost, water, weather, a and high-temperature resistance to the cured glue line of a joint bonded with phenolic adhesives [5,20].

Phenols condense initially with formaldehyde at pH either acid or alkaline, to form a methylol phenol or phenolic alcohol, and then, dimethylol phenol. The initial attack may be at 2-, 4-, or 6-position of the phenol molecules. The second stage of the reaction involves the reaction of the methylol groups with other phenol or methylol phenol, leading first to the formation of linear polymers and then to the formation of hard-cured, highly branched structures. Resols are obtained as a result of alkaline catalysis and an excess of formaldehyde. A resol molecule contains reactive methylol groups. Heating causes the reactive resol molecules to condense to form large molecules without the addition of a hardener [21].

A typical phenolic resin was made in a glass reactor equipped with a turbine-blade agitator, a reflux condenser and heating and cooling facilities. Molten phenol, formalin (37% formaldehyde) and water are charged into the reactor in molar proportions between: 1:1:1 under mechanical stirring.

Quantities of: 10%, 15%, 20% of ammonium, aluminium and iron and chromium lignosulfonates were also added to the mixture of above mentioned monomers.

To make a resol-type resin, such as used in wood adhesive manufacture, an alkaline catalyst, such as sodium hydroxide was added, and the reaction mixture was heated to 80-90^{0}C for about 2-3 hours. Since the resol can gels into the reactor, the temperature was kept below 100^{0}C. Tests have to be done in order to determine first the degree of advancement of the resin, and second, when the batch should be discharged. Such tests consist on the measurements of the gel time/reactivity of the resin on a hot plate or at 100^{0}C in a water bath.

The resins are water soluble and of low molecular weight and are finished at a low temperature, usually around 40^0 to 60^{0}C. It is important that the liquid, water soluble resols, retain their ability to mix with water easily, since when they are used as wood adhesives they often require the addition of water to counterbalance the effect of the fillers added.

The characteristics of the new composites adhesives, based on phenol, formaldehyde (the matrix) and ammonium, aluminum and iron and chromium lignosulfonates (as fillers) are presented in Table 2.

Table 2. The characteristics of the new adhesives phenol-formaldehyde-lignosulfonates

Chemical characteristic	PF resin Standard	PF resin with $LSNH_4$			PF resin with LSAl			PF resin with LSFeCr		
		10%	**15%**	**20%**	**10%**	**15%**	**20%**	**10%**	**15%**	**20%**
Aspect	White liquid	Brown liquid			Brown liquid			Brown liquid		
Density, g/cm^3	1.1700	1.13	1,16	1.19	1.10	1.15	1.20	1.15	1.18	1.20
Solids, %	45 – 50	47.0	48.0	49.0	48.3	49.0	50.0	48.5	49.5	50.0
PH-value	8 - 11	9.5	10.0	10.0	8.5	9.5	10.0	9.5	10.0	10.5
Viscosity, at 20^0C, cP	120 – 180	160	156	150	150	145	140	140	134	128
Miscibility with water	2 : 1	2:1	2:1	2:1	2:1	2:1	2:1	2:1	2:1	2:1
Gel time/Reactivity at 160^0C, s, max.	180	155	135	115	140	132	120	135	115	110

The new type of composites based on phenol-formaldehyde resin with lignosulfonates present similar physical and chemical characteristics comparing with the commercial phenol-formaldehyde resin. The improvements achieved by using the lignosulfonates as comonomers can be explained by the polyphenolic structure of lignin, respectively, the presence of carbonyl and phenolic hydroxyl groups which improve the polycondensation process. Better properties are obtained for the lignosulfonate-phenol-formaldehyde composites resins, consisting in decreased adhesive viscosity (which will insure better wettability of wood particles), better water resistance of finished boards, the decreasing of the reactivity, and the increasing of the pot life of these new composites adhesives. Due to the chemical structure richer in functional groups carbonyl and phenolic hydroxyl, the iron and chromium lignosulfonate presents an increased reactivity which conducted to superior lignosulfonate-phenol-formaldehyde resins.

The FT-IR spectra of the new ecological phenol-formaldehyde lignosulfonates composites resins (PFR) (Figure 2) were performed with a FTIR- Spectrometer model BX II (Perkin Elmer, 2005).

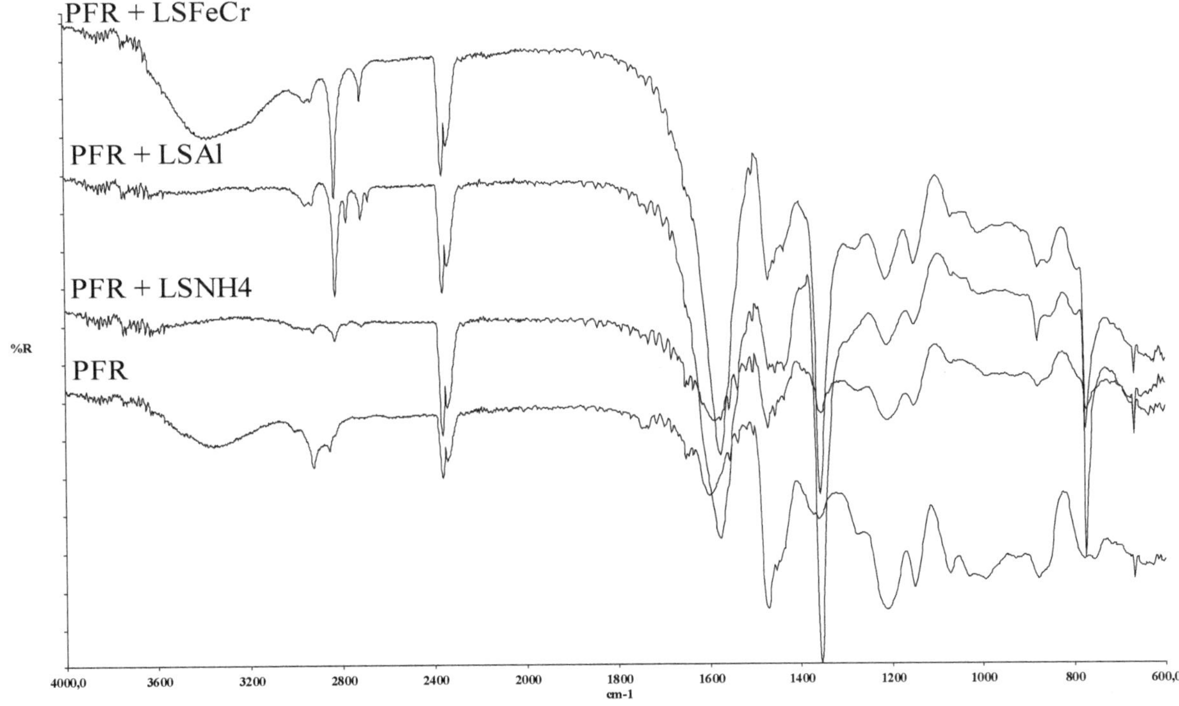

Figure 2. The FTIR spectra of the lignosulfonate-phenol-fromaldehyde resins synthesized

The condensation reaction of phenol and formaldehyde with lignosulfonates is certified by characteristic absorption bands in FTIR spectra:

- The characteristic absorption bands of monomers formaldehyde (911 cm^{-1}) and phenol (3039-3208 cm^{-1}, 1578 cm^{-1} and 773 cm^{-1}) diminished with increasing the time of polycondensation reaction, while the absorbance bands of hydroxymethyl and methylene groups, characteristic for the new adhesives increased
- Absorption bands are registered at 1595.80 cm^{-1} and 1474.05 cm^{-1} corresponding to the C=C aromatic ring vibration
- The absorption band at 1070 cm^{-1} corresponds to single bond C-O stretching vibrations of $-CH_2OH$ groups formed in resins
- Absorption band at 1451 cm^{-1} indicates the presence of benzene ring obscured by $-CH_2-$ methylene bridge formed between chemical and lignosulfonates monomers
- In the phenol-formaldehyde resins spectra, additional characteristic signals of methylene bridge $-CH_2-$ at 1478 cm^{-1} and methylene-ether bridge C-O-C at 1116 cm^{-1} are present, certifying the reaction with lignosulfonates
- Absorption bands at 880 cm^{-1} and 762.62 cm^{-1} correspond to 1,2,4- substituted benzene ring, assigning the formation of 2,3,6-trihydroxymethylphenol, as precursor of the new adhesives [6, 7, 8, 9].
- Stretching vibration of aromatic lignosulfonates compounds can be seen in a wide absorption band at 1578.02 – 1586.36 cm^{-1} for all resins. The presence of these functional groups certify the polycondensation products of the reaction between phenol-lignin adduct and formaldehyde [10, 22].

3. CONCLUSION

Our research was focused on the synthesis of some new composites adhesives based on phenol-formaldehyde resins with ammonium, aluminium and iron and chromium lignosulfonates, as an ecological alternative to the existing production of commercial- phenol-formaldehyde resin.

Improved properties, respectively water resistance, decreasing of the gel time/reactivity and increasing of the pot life of the new ecological composites adhesives (especially with iron and chromium lignosulfonate) have been obtained.

The recycling of biomass wastes, lignosulfonates, had as benefits the lowering of costs, resulting from the difference in cost between chemical monomers and lignosulfonates and reducing the toxicity by partially replacing monomers phenol and formaldehyde with aqueous solutions of lignosulfonates, biomass raw materials renewed by photosynthesis.

In the future, taking into account both, the need for recycling biomass waste as chemical reactants for organic synthesis, and the environmental protection, it will be encouraging to use lignocellulosic materials to obtain ecological, low costs composites materials.

REFERENCES

[1] Hon, D.N.S., and N. Shiraishi, (Eds), Wood and Cellulosic Chemistry. 2nd Edition. Marcel Dekker Publishers, New York, NY, 2001.

[2] Dence C.W., Lin, S.Y., General structure features of lignin. In: Methods in Lignin Chemistry, Springer Verlag, Berlin, pp. 3-6, 1992.

[3] Derek, S., Lignin as a base material for materials applications: Chemistry, application and economics. Industrial Crops and Products, 27(2), 202–207, 1998.

[4] Moore K.J., Jung, G., Lignin and fiber digestion. In:Journal Range Management, 54, pp. 420, 2001.

[5] Pizzi, A., Advanced Wood Adhesives Technology, Marcel Dekker, Inc., 1994.

[6] Hon, D. N. S. Chemical Modification of Lignocellulosic Materials,Mark Dekker, Publish,, N.Y., 1996.

[7] Vezquez-Tourres, H., Canchi-Escamilla, G., and Gruz-Ramas, P., Influence of biomass on the curing of novolac-composites.In: Journal Applied Polymer Science, 46, pp. 646, 1992.

[8] Mansouri, N., E.,. Salvado J., Structural characterization of technical lignin for the production of adhesives: Application to lignosulfonate, kraft, soda-anthraquinone, organosolv and ethanol process lignins. Industrial, Crops and Products, 24, 8–16, 2006.

[9] Muzaffar, A. K; Sayed, M. A and Ved, P. M., Development and Characterization of Wood Adhesive using Bagasse Lignin. In: International Journal of Adhesion & Adhesives, 24, 6, pp. 485-493, 2004.

[10] Mohamad I., M.N; Ghani, A. Md. and Nen, N., Formulation of lignin phenol formaldehyde resins as a wood adhesive. In: The Malaysian Journal of Analytical Sciences, 11, 1, pp. 213-218, 2004.

[11] Turunen, M., Alvila, L., Pakkanen, T., and Rainio, J., Modification of phenol-formaldehyde resol resins by lignin, starch, and urea. In: Journal Applied Polymer Science, 88, pp. 582–588, 2003.
[12] Forss, K.G., Fuhrmann, A., Finnish plywood, particleboard, and fiberboard made with a lignin-base adhesive. In: Forest Prod. J., 29, pp. 39-43, 1979.
[13] Gargulak, J.G., Lebo, S.E., Commercial use of lignin-based materials. In: Lignin: Historical, Biological, and Materials Perspectives. ACS Symposium Series. American Chemical Society, pp. 304–320, 2000.
[14] Nihat S. C. and Nilgul O, Use of Organosolv Lignin in Phenol-Formaldehyde Resin for Particleboard Production, International Journal of Adhesion & Adhesives, 22 : 477-480, 2002.
[15] Perez, H., M., Rodriguez, F., Alonso, V., Echeverria, J., M., Characterization of a novolac resin substituting phenol by ammonium lignosulfonate as filler or extender. Bioresources 2(2), 270-283, 2007.
[16] Gardziella, A., Pilato, A., Knop A., Phenolic Resins. Chemistry, Applications, Standardization, Safety and Ecology, Springer-Verlag, Berling, Heidelberg, 2000.
[17] Dumitrescu L., Ph.D.Thesis, "Gheorghe Asachi" Technical University of Iassy, Roumania, 1999.
[18] Zakis G., F., Functional analysis of lignins and their derivatives, TAPPI Press, Atlanta, Ga, 1994.
[19] Zaha C., Dumitrescu, L., Sauciuc A., Recycling biomass waste as copolymerization partners, Bulletin of the Transilvania University of Brasov, Series I, vol 3, (52), 215-220, 2010.
[20] Bousoulas, J., Tarantili, P., and Andreopoulos, A., Resole resin as sizing agent for aramid fibres. In: Advanced Composites Letters, 10, 5, pp. 249-255, 2001.
[21] Patsch, H., Schrod, M., Preparation and Analysis of Phenol-Formaldehyde Resins by High-Throughput Techniques, Macromolecular Rapid Communications, 25, 224-230, 2004.
[22] Poljansek, I., Krajnc, M., Characterization of Phenol-Formaldehyde Prepolymer Resin by In Line FT-IR Spectroscopy. In: Acta Chimica. Slov., 52, No. 3, pp. 238-244, 2005.

CONTROL OF ACTIVE VEHICLE SUSPENSION USING DUAL OBJECTIV ANALYSIS

Dumitru D. Nicoara
Transilvania University, Brasov, ROMANIA, tnicoara@unitbv.ro

Abstract: *When designing vehicle suspensions, the dual objective is to minimize the vertical forces transmitted to the passengers (i.e., to minimize vertical car body acceleration) for passenger comfort, and to maximize the tire-to-road contact for handling and safety. While traditional passive suspensions can negotiate this tradeoff effectively, active suspension systems have the potential to improve both ride quality and handling performance, with the important secondary benefits of better braking and cornering because of reduced weight transfer. This improvement, of course, is conditional upon the use of feedback to control the hydraulic actuators. In this paper we propose two methods for analyzing this tradeoff. First, we use frequency domain analysis of passive quarter- car suspension system and second we design a backstepping controller for analyze a parallel active suspension in witch the hydraulic actuator force is viewed as the control input.*
Keywords: *active suspension, control, transfer functions, backstepping controller*

1. INTRODUCTION

Development of control methods for passive and active suspension systems is a major topic of automotive industries. In general, ride comfort, road handling, and stability are the most important factors in evaluating suspension performance. Ride comfort is proportional to the absolute acceleration of the vehicle body, while road handling is linked to the relative displacement between vehicle body and the tires.

On the other hand, stability of vehicles is related to the tire-ground contact. The main concern in suspension design and control is the fact that currently, achieving improvement in these three objectives poses a challenge because these objectives will likely conflict with each other in the vehicle operating domain [1], [4].

A good suspension system shall improve ride quality and passenger comfort simultaneously. For ride quality improvement vertical acceleration that caused by road profile shall be limited. This means that suspension system shall absorb road disturbances. In the other word, contact of tire with road surface shall decrease. In the other side, for increasing the controllability of vehicle, tire shall contact to road more. Therefore, reach to a suitable suspension system is difficult, Because a tradeoff between ride quality and vehicle controllability exists.

In this paper we present two methods to analyze the tradeoff between ride quality and suspension travel of automotive suspensions: modal analysis to obtain approximate transfer functions and backstepping methodology to design control force generated by the actuator.

2. DYNAMICAL MODEL OF THE SUSPENSION SYSTEM

Since many of the proposed electronic suspension being considered today are independent, i.e. using local sensor information and control law, the completely active suspension system of a quarter car model, with two degrees of freedom, show in Fig. 1, has been considered in this paper. We used the following notation: m_{us} is the equivalent unsprung mass consisting of the wheel and its moving parts; m_s is the sprung mass, i.e., the part of the whole body mass and the load mass pertaining to only one wheel; k_t is the elastic constant of the tire, whose damping characteristics have been neglected.

If assume that the tire does not leave the ground the liniarized equations of the motions are

$$M\ddot{x} + C\dot{x} + Kx = D_1 x_r + D_2 u \tag{1}$$

$$\begin{bmatrix} m_s & 0 \\ 0 & m_u \end{bmatrix}\begin{Bmatrix} \ddot{x}_s \\ \ddot{x}_u \end{Bmatrix} + \begin{bmatrix} c_s & -c_s \\ -c_s & c_s \end{bmatrix}\begin{Bmatrix} \dot{x}_s \\ \dot{x}_u \end{Bmatrix} + \begin{bmatrix} k_s & -k_s \\ -k_s & k_s + k_t \end{bmatrix}\begin{Bmatrix} x_s \\ x_u \end{Bmatrix} = \begin{Bmatrix} 0 \\ k_t \end{Bmatrix} x_0 + \begin{Bmatrix} 1 \\ -1 \end{Bmatrix} u \tag{2}$$

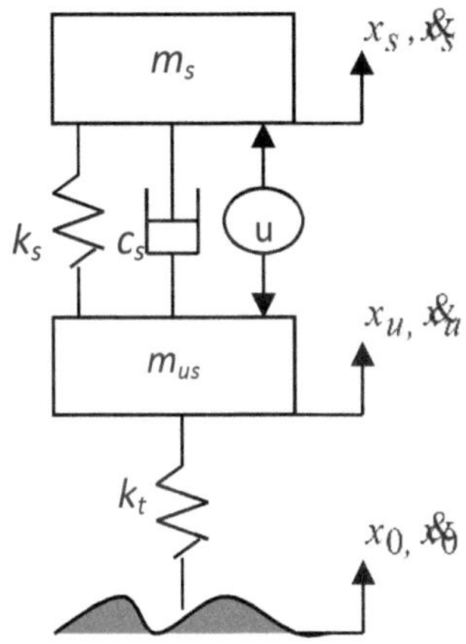

Figure 1: Quarter car model

Using this the state variables

$x_1 = x_s - x_u$, - suspension deflection (rattle space);

$x_2 = \dot{x}_s$, - the vertical absolute velocity of the sprung mass m_s;

$x_3 = x_u - x_0$, - tire deflection;

$x_4 = \dot{x}_u$, - the vertical absolute velocity of the unsprung mass m_u;

$u(t)$ - the control force produced by the actuator;

$x_0(t)$ - represents the disturbance, it coincides with the absolute vertical velocity of the point of contact of the tire with the road;

we can rewrite (1) in state space as

$$\dot{x} = Ax + Bu + L\dot{x}_0 \tag{3}$$

where

$$x = \begin{Bmatrix} x_1 \\ x_2 \\ x_3 \\ x_4 \end{Bmatrix}; A = \begin{bmatrix} 0 & 1 & 0 & -1 \\ -\frac{k_s}{m_s} & -\frac{c_s}{m_s} & 0 & \frac{c_s}{m_s} \\ 0 & 0 & 0 & 0 \\ \frac{k_s}{m_u} & \frac{c_s}{m_u} & -\frac{k_t}{m_u} & -\frac{c_s}{m_u} \end{bmatrix}; B = \begin{Bmatrix} 0 \\ \frac{1}{m_s} \\ 0 \\ -\frac{1}{m_u} \end{Bmatrix}; L = \begin{Bmatrix} 0 \\ 0 \\ -1 \\ 0 \end{Bmatrix} \tag{4}$$

3. PASSIVE SUSPENSION

In the case of passive suspension the control force $u(t)$ is set to zero. Since the system obtained from (2) is linear, we can use frequency domain analysis. For judge the effectiveness of the suspension system we are looking at the location of system poles and zeros, and the response of the vehicle outputs to road disturbances. System zeros can be obtained by the transfer function for the control input to the outputs.
The primary concern is acceleration transfer function:

$$H_A(s) = \frac{\ddot{x}_s(s)}{\dot{x}_0(s)} = \frac{k_t s(c_s s + k_s)}{d(s)} \tag{5}$$

where $d(s)$ is characteristic polynom

$$d(s) = m_u m_s s^4 + (m_u + m_s)c_s s^3 + ((m_u + m_s)k_s + m_s k_t)s^2 + c_s k_t s + k_s k_t \tag{6}$$

Similarly, we define the following two transfer functions: rattle space transfer function (suspension deflections) [4]

$$H_{RS}(s) = \frac{x_s(s) - x_u(s)}{\dot{x}_0(s)} = -\frac{k_t m_s s}{d(s)} \tag{7}$$

and tire deflection transfer function

$$H_{TD}(s) = \frac{x_u(s) - x_r(s)}{\dot{x}_0(s)} = -\frac{m_u m_s s^3 + (m_u + m_s)c_s s^2 + (m_u + m_s)k_s s}{d(s)} \tag{8}$$

We define the transfer functions with respect to the road input velocity ($\dot{x}_0(s)$), so all frequencies contribute equally to their mean square values. The system is observable for all three outputs and all states are controllable. This transfer functions will be used for comparison purposes later on. In this paper the numerical simulations were made for these values of parameters:

$$\begin{aligned} &m_s = 240\,Kg, \quad m_u = 36\,Kg, \\ &c_s = 1000\,N \cdot sec/m, \; c_t = 0 \\ &k_s = 16000\,N/m, \; k_t = 160000\,N/m. \end{aligned} \tag{9}$$

4. ANALYSIS OF PASSIVE SUSPENSION USING APPROXIMATE TRANSFER FUNCTIONS

Modal decoupling will be used to study the influence of different suspension parameters on the properties of the automotive suspension.

In order to study the effects of specific suspension parameters on the suspension performance, we calculate the natural frequencies and mode shapes of the suspension system and then transform to a new set of coordinates in which the two equations of motion are approximately decoupled [4]. For the numerical parameters (9), the two approximate decoupled equations are:

- sprung mass mode approximation

$$m_s\ddot{x}_s + c_s\dot{x}_s + k_s x_s = c_s\dot{x}_0 + k_s x_0, \text{ for } |x_s| >> |x_u| \tag{10}$$

- unsprung mass mode approximation

$$m_u\ddot{x}_u + c_s\dot{x}_u + k_t x_u = k_t x_0, \text{ for } |x_u| >> |x_t|\,. \tag{11}$$

So, we obtain the following approximate transfer functions:

- acceleration transfer function

$$\frac{1}{s}H_A(s) \approx \frac{x_s}{x_0} = \frac{c_s s + k_s}{m_s s^2 + c_s s + k_s} \tag{12}$$

- rattle space transfer function

$$sH_{Rs}(s) \approx \frac{x_s - x_0}{x_0} = -\frac{m_s s^2}{m_s s^2 + c_s s + k_s} \tag{13}$$

- tire deflection transfer function

$$H_{TD}(s) \approx \frac{x_u - x_0}{x_0} \approx \frac{-m_u s^2}{m_u s^2 + c_s s + k_t} \tag{14}$$

To evaluate the acuracy of the approximate transfer functions of equations (12) and (14), Figures 2 and 3 show a comparison between the original and approximate transfer functions.

It is clear that the approximate transfer function (12) matches the original transfer function (6) well for frequency range $\omega < \omega_1$ and the approximate transfer function (14) matches the original transfer function $H_{TD}(s) = (x_u - x_0)/x_0$ well for the frequency range $\omega > 0.5\omega_2$

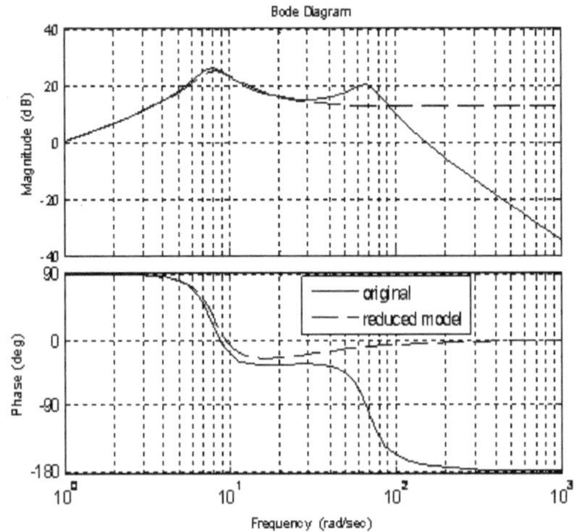

Figure 2: Bode for $H_A(s)$ sprung mass approximate mode

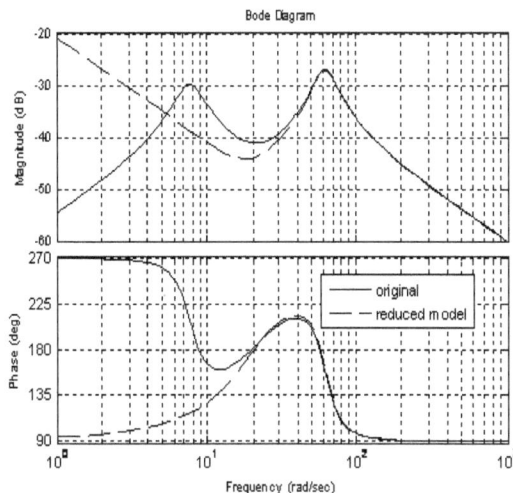

Figure 3: Bode for $H_{TD}(s)$ unsprung mass approximate mode

In order to improve passenger comfort the transfer function $H_A(s)$ from the road disturbance to the car body acceleration should be small in the frequency range from 0–65 rad/s. At the same time it is necessary to ensure that the transfer function $H_{RS}(s)$ from the road disturbance to the suspension deflection is small enough to ensure that even very rough road profiles do not cause the deflection limits to be reached.

The fact that the actuator force u is applied between the two masses places fundamental limitations on the transfer functions $H_A(s)$ and $H_{RS}(s)$. As shown in [2], [4] and [5] the acceleration transfer function has a zero at the "tyrehop frequency," $\omega_1 = \sqrt{k_t/m_u}$ .For the parameter values listed in (9), $\omega_1 = 60{,}7$ rad/s. Similarly, the suspension deflection transfer function has a zero at the "rattle space frequency," $\omega_1 = \sqrt{k_t/(m_u + m_s)}$. For the parameter values listed in (9), $\omega_2 = 24{,}07$ rad/s. The tradeoff between passenger comfort and suspension deflection is captured by the fact that is not possible to simultaneously keep both the above transfer functions small around the tyrehop frequency and in the low-frequency range.

5. ANALYSIS OF ACTIVE SUSPENSION USING BACKSTEPPING METHOD

Active suspension systems add actuators to the passive components. Active suspension systems have the potential to improve both ride quality and handling performance. This improvement is conditional upon the use of feedback to control the hydraulic actuators. In this paper we analyze the tradeoff between ride quality and suspension travel by backstepping design methodology [3].

The first step in the design of a backstepping controller is the choice of a quantity to be regulated. The choice of this variable is crucial to the performance of the closed-loop system, and one of the goals of this paper will be to show how to exploit the design flexibility built into this choice in order to achieve the desired closed-loop behavior.

If we want to minimize the car body acceleration (i.e. our objective control are the forces transmitted to passengers) then the desired value of actuator force is

$$u = k_s x_1 + c_s(x_2 - x_4) \tag{15}$$

to yield $\dot{x}_2 = \ddot{x}_s = 0$.

Substituting this expression into (4) yields the closed-loop system

$$\begin{Bmatrix} \dot{x}_1 \\ \dot{x}_2 \\ \dot{x}_3 \\ \dot{x}_4 \end{Bmatrix} = \begin{bmatrix} 0 & 1 & 0 & -1 \\ 0 & 0 & 0 & 0 \\ 0 & 0 & 0 & 1 \\ 0 & 0 & -\dfrac{k_t}{m_u} & 0 \end{bmatrix} \begin{Bmatrix} x_1 \\ x_2 \\ x_3 \\ x_4 \end{Bmatrix} + \begin{Bmatrix} 0 \\ 0 \\ -1 \\ 0 \end{Bmatrix} \dot{x}_0 \tag{16}$$

Now, if we want to minimize the suspension travel, $x_s - x_u$, then the regulated variable becomes x_1. So, we obtain the zero dynamics of the closed-loop systems, which consist of an unstable subsystem

$$\begin{cases} \dot{x}_3 = x_4 - \dot{x}_0 \\ \dot{x}_4 = -\dfrac{k_t}{m_u} x_3 \end{cases} \tag{17}$$

However, the zero dynamics are again oscillatory and hence this design is still not acceptable. We must therefore choose the regulated variable so as to avoid the oscillatory zero dynamics. One such choice is the variable

$$y_1 = x_s - \tilde{x}_u \tag{18}$$

where $\tilde{x}_u$ is a filtered version of the wheel displacement x_u

$$\tilde{x}_u = \frac{e}{s+e} x_u \tag{19}$$

This choice represents the first step towards the design, backstepping design of a controller which will accommodate the inherent tradeoff between ride quality and rattle space usage. The choice of the positive constant e affects the properties of our active suspension [3].

For small values of e, (19) is a low-pass filter. Hence, the regulated variable y_1 is essentially equal to the car body displacement x_u as long as the road input contains only high-frequency components which are rejected; however, at very low frequencies (constant or slowly changing road elevations) and in steady state, y_1 becomes almost identical to the suspension travel $x_s - x_u$. Thus, as we will see later on, the sustained oscillations are eliminated, and the active suspension rejects only high-frequency road disturbances, namely the ones which generate large vertical accelerations and cause passenger discomfort.

As the value of e becomes larger, more high-frequency components of the road input are allowed to pass through the filter (19). Hence, the regulated variable $y_1$approximates the suspension travel $x_s - x_u$: the high filter bandwidth renders $\tilde{x}_u \approx x_u$. As a result, the active suspension becomes stiffer and reduces its rattlespace use, at the price of significantly reduced passenger comfort [3]. With this choice of variable y_1, defined in (18) and using backstepping techniques we get a new control law for the calculation of transfer functions relating the road input x_0 to the car body displacement x_s and wheel travel x_u.

So, for the closed-loop systems, we obtain the transfer functions

$$H_{1A} = \frac{X_s(s)}{X_0(s)} = \frac{k_t e}{d_1(s)} \tag{20}$$

$$H_{1TD}(s) = \frac{X_u(s)}{X_0(s)} = \frac{s+e}{d_1(s)} \tag{21}$$

where

$$d_1(s) = m_u s^3 + e(m_s + m_u)s^2 + k_t s + k_t e \tag{22}$$

Furthermore we compare the transfer functions (5)-(8), of the passive suspension and the transfer functions (20)-(21) of active suspension.

These transfer functions are plotted in Figs.4-6 for the numerical parameters given by relations (9). As shown in Figs. 4 -6, the frequency response plots for any real suspension must pass through certain invariant point. These invariant properties are a result of the fact that the suspension forces are applied only between a wheel and the car body, and they place insurmountable limitations to what can be achieved by active suspension designs. With small $e(e=2)$, the active suspension design reduces both car body displacement and acceleration compared to the passive one, Figs. 4 and 5, but increases the suspension travel as seen in Fig. 6. On the other hand, if e is increased to 9, then the suspension travel can be significantly reduced, as seen in Fig. 5, but then the car body displacement and acceleration are increased.

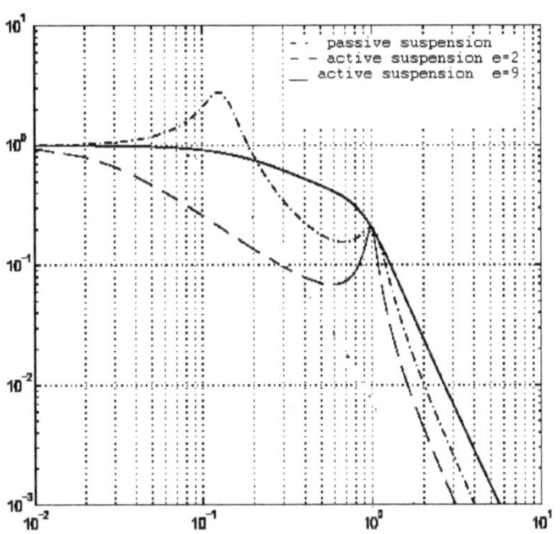

Figure 4: Car body displacement

Frequency response plot of $\left|\frac{X_s(j\omega)}{X_o(j\omega)}\right|$

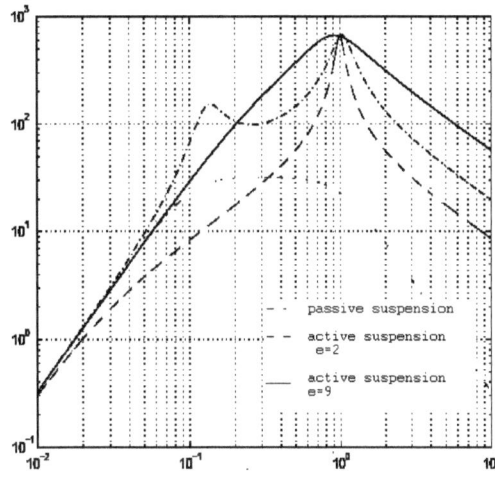

Figure 5: Car body acceleration

Frequency response plot of $\left|\frac{-\omega^2 X_s(j\omega)}{X_o(j\omega)}\right|$

6. CONCLUSION

Using the approximately decoupled models and backstepping method the following conclusions on suspension design were obtained:

- Decreasing suspension stiffness improves ride quality and road holding. However, it increases rattle space requirements.
- Increased suspension damping reduces resonant vibrations at the sprung mass frequency. However, it also results in increased high frequency harshness.
- Increased tire stiffness provides better road holding but leads to harsher ride at frequencies above the unsprung mass frequency.
- It was shown that considering ride quality and road holding trade-offs that both can be improved at low frequencies and at the sprung natural frequency.

Nevertheless, it is also clear that with the appropriate choice of the filter bandwidth e our active suspension design is superior not only to the passive suspension but also to the ideal one in some frequency ranges.

REFERENCES

[1] Hedrick, J. K., Butsuen, T., Invariant properties of automotive suspensions, Proceedings of the Institution of Mechanical Engineers, vol. 204, pp. 21–27, 1990.
[2] D. Karnopp, d., Theoretical limitations in active vehicle suspensions, Vehicle System Dynamics, vol. 15, pp. 41–54, 1986.
[3] Lin J. S., Kanellakopoulos, I., Nonlinear design of active suspension, IEEE Control Systems Magazin, vol. 17, pp.45-69, 2001.
[4] Rajamani, R., Vehicle dynamics and control, ISBN 0-387-26396 -9, Springer, 2006.
[5] Yue, C., Butsuen, T., Hedrick, J. K., Alternative control laws for automotive active suspensions, Transactions of the ASME, Journal of Dynamic Systems, Measurement and Control, vol. 111, pp. 286–291, 1989.

ESTIMATING THE TORQUE DISTRIBUTION IN A PLANETARY GEARBOX

Alexa Octavian[1], Ilie Constantin Ovidiu[2], Radu Vilău[3]
[1]Military Technical Academy, Bucharest, Romania, alexa.octavian@gmail.com
[2]Military Technical Academy, Bucharest, Romania, ovidiuilie66@yahoo.com
[3]Military Technical Academy, Bucharest, Romania, radu.vilau@yahoo.com

Abstract: *In the case where one intends to reveal the torque distribution onto the elements of a planetary gearbox, issuing a virtual model is previously needed. This model should replicate both the physical structure and the real working behavior. The torque is transmitted along the gearbox through the intermediary of three simple planetary mechanisms by successively locking different components with the aid of some multi-plate clutches. The simulation of the friction element's engagement is achieved by introducing a physical signal that replaces the pressure of the hydraulic command system. The pressure evolution has been previously experimentally determined, during the vehicle's taking-off procedure. The friction elements are modeled considering their three working stages: fully disengaged, partially engaged and fully engaged. The torque distribution is modeled starting from the energy conservation law as well as the torque balance law.*
Keywords: *simulation, torque, powertrain, planetary gearbox.*

1. INTRODUCTION

The planetary gearbox subjected to the present study consists of a train of planetary mechanisms and a set of multiple-disc clutches and brakes. Different scientists have previously described their working behavior by mathematical modeling and simulation. Jungang Wang [1] proposed a mathematical model able to describe the working process of the planetary mechanisms in transient modes. When issuing the dynamic equations, he has taken into account the losses due to the torsion vibrations within the meshing area of the gears. The loss assessment has been achieved by supplying the model with elasticity and friction coefficients. In another study [2], the authors proposed a theoretical approach to reveal the distribution torque within an automotive driveline. A better model would describe the real working mode if developing virtual simulation models. These models are featured by the fact that the system's elements are treated as physical subsystems, linked together by a net of physical connectors (shafts, flanges etc.). Otter et al. provided such a model in their paper [3]. When developing the model, its creator used a part of the predefined blocks of the Dymola Library to model the physical components of the planetary gears and multiple-disc clutches of the gearbox. All the blocks take advantage of predefined connecting physical ports; thus one can model the rigid linkages of the gearbox. Within the study, the gear shifting simulation was achieved by using a signal that successively locks the multiple-disc clutches.

2. DEVELOPMENT OF THE SIMULATION MODEL

For the present study, the Matlab programming environment was employed to model the planetary gearbox. The effects due to the torsion vibrations, the rotational inertia and the shafts' elasticity were taken into account. The basic components of the virtual simulation model were the three planetary gears and the four friction elements (three clutches and a brake), as it can be noticed in Figure 1. The basic components of the gearbox have been built using the SIMSCAPE module, by imposing the specific conditions of the real working environment. A simple planetary gear consists of three external elements (the sun gear, the crown and the carrier). The working behavior of the planetary gear has been achieved by writing within its source code (Figure 2) the so-called *Willis equation* (kinematic equation), the energy conservation law and the torque balance law [4]. The carrier locking generates these three equations and the gear's ratio K. The physical connection between the external elements is achieved by the (R, S, C) nodes, the „*through*" variables (M_s, M_r, M_c), the „*across*" variables (ω_s, ω_r, ω_c) and by the physical conserving ports. The power running through the three physical ports is directed, by means of the connection lines, towards the other components of the physical net.

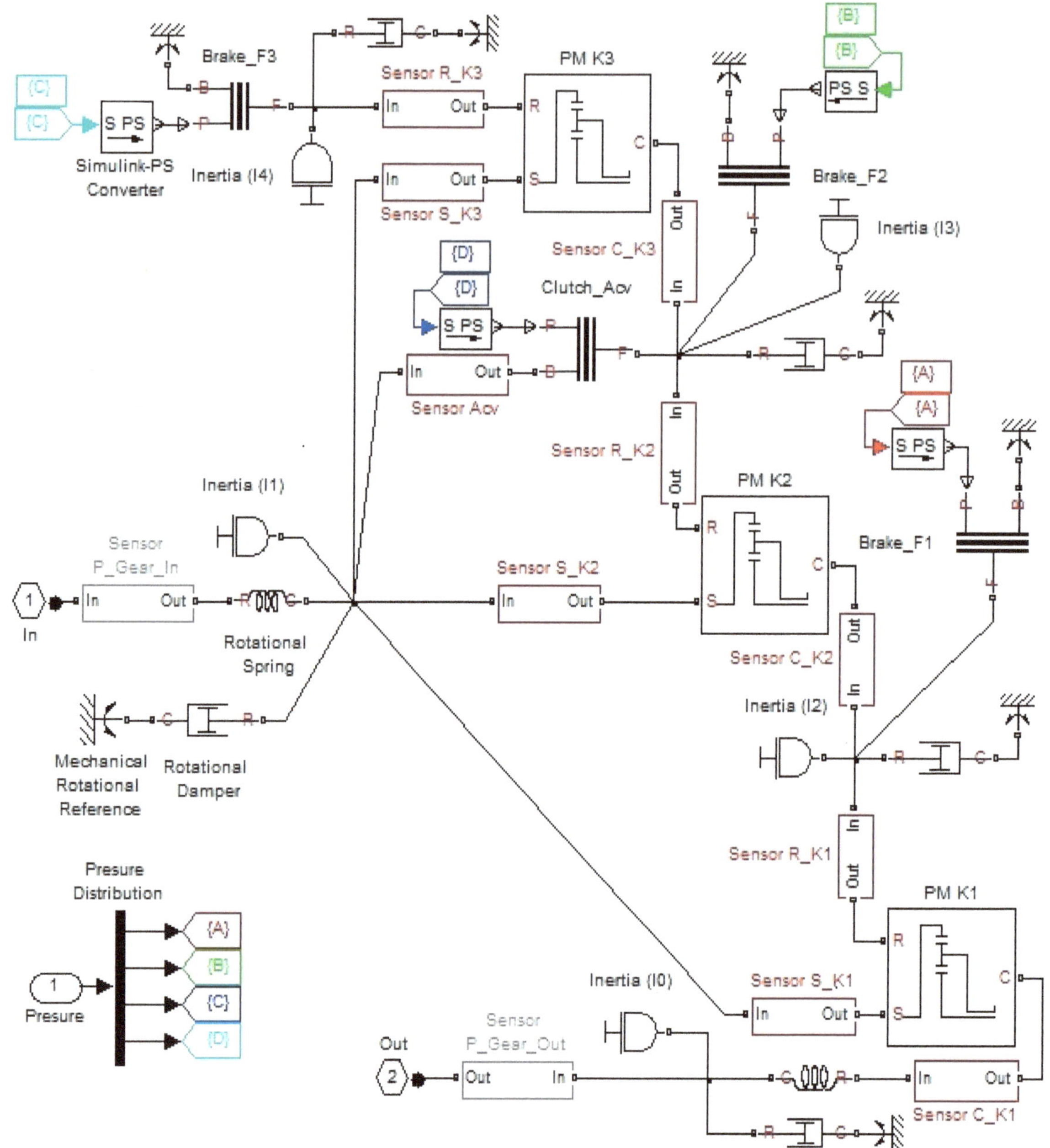

Figure 1: Virtual model to simulate the planetary gearbox

```
equations
  ws + K*wr == (1 + K)*wc; % Willis equation and torque balance law;
  Mr == K*eta*Ms; Ms + Mr + Mc ==0; % eta - efficiency of the planetary mechanisms;
end
```

Figure 2: Source code of a planetary gear

The multiple-disc clutches and brakes block on of the external elements. When using a brake, the element's angular speed drops to null. When using a clutch, the connected elements have the same angular speed. Physically speaking, the friction elements are featured by three working modes: fully disengaged, partially engaged and fully engaged In order to model the work of the friction elements when in partially engaged mode and fully engaged mode we used the *Stribeck model* [5]. Compared to the *classic friction model*, the Stribeck model (Figure 3) changes the variation mode of the friction force. It eliminates the discontinuity that might occur due to the simulating software, within the vicinity of the area where the relative angular speed (ω) between the friction surfaces closes to null. Therefore, the friction force has a single value when the relative angular speed (ω) is null. The friction force is approximated by the gain between the pressing force (F_n) and either the dynamic (cdfr) or the static friction coefficient (csfr).

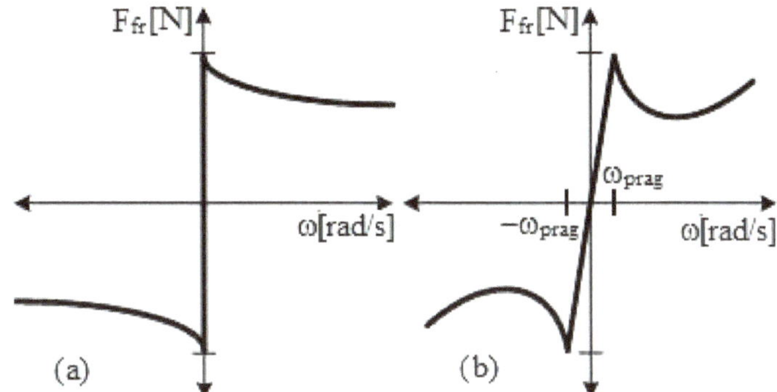

Figure 3: Friction force vs. angular speed: a) classic model; b) Stribeck model

The dynamic friction coefficient is used when slipping. In this situation the relative angular speed isn't null. The static friction coefficient is used when the relative angular speed is very close to null. The equations that estimate the physical behavior of the friction elements both in the slipping and locking stage are written within the source code of the friction element (Figure 4) according to the relative angular speed, for different variation intervals.

```
equations
 if ((w>=w_prag) || (w<=-w_prag))
   M ==2/3*p*A*cdfr*R*sign(w);  % sliding regime;
 elseif ((-w_prag<w<0) || (0<w<w_prag))
   M ==2/3*p*A*csfr*R*sign(w)*(w/w_prag);
 else (w=0)
   B.t ==F.t; B.w ==F.w;        % locking regime;
 end
```

Figure 4: Part of the friction element's source code

During the slipping stage, when the dynamic friction torque transfers at high rates, the discs' angular speeds are different; hence the relative angular speed has non-null values. It also tends to higher values than its limit value ω_prag. Modeling this situation implied the use of the „if" condition of SIMSCAPE followed by the equivalent equation of the friction torque. During time, the transfer rate of the friction torque is low and the friction element is still within the friction stage. The discs still have different angular speeds, but the relative angular speed rapidly decreases to zero. The situation is described by the „elseif" condition of the simulating environment. In the locking stage, described by „else", the discs of the friction element have the same angular speed and ω is zero.

Besides the basic components, in the virtual simulation model other blocks of the SIMSCAPE library occur. They were used to model the pressing force of the pistons that act inside the clutches of brakes, to visualize the torque distribution throughout the planetary mechanism or to model the inertia moments, elastic or viscous forces, viscous and dry damping forces. The time histories of the hydraulic system's parameters are obtained experimentally [6].

3. RESULTS

The simulation model presented in this paper is able to completely describe the work of the physical components of the analyzed gearbox for various working situations. For this study, we simulated the working conditions and gearbox's behavior during the vehicle's taking-off procedure. The torque diagrams obtained by simulation (Figure 5) underline the distribution of the torque for all the planetary mechanisms elements (K_1, K_2 and K_3).

Analyzing the diagrams, we noticed that the power flow runs only via the planetary mechanism K_1 in the first gear of the gearbox. Moreover, the other planetary mechanisms (K_2 and K_3) are consuming power due to the acceleration process and internal friction, although they do not actively take part in the power flux transit. When engaging the second and the fourth gears of the planetary gearbox, the power runs through K_1 and K_2 planetary gears. When engaging the third gear, the power uses all the three planetary gears to pass through the gearbox.

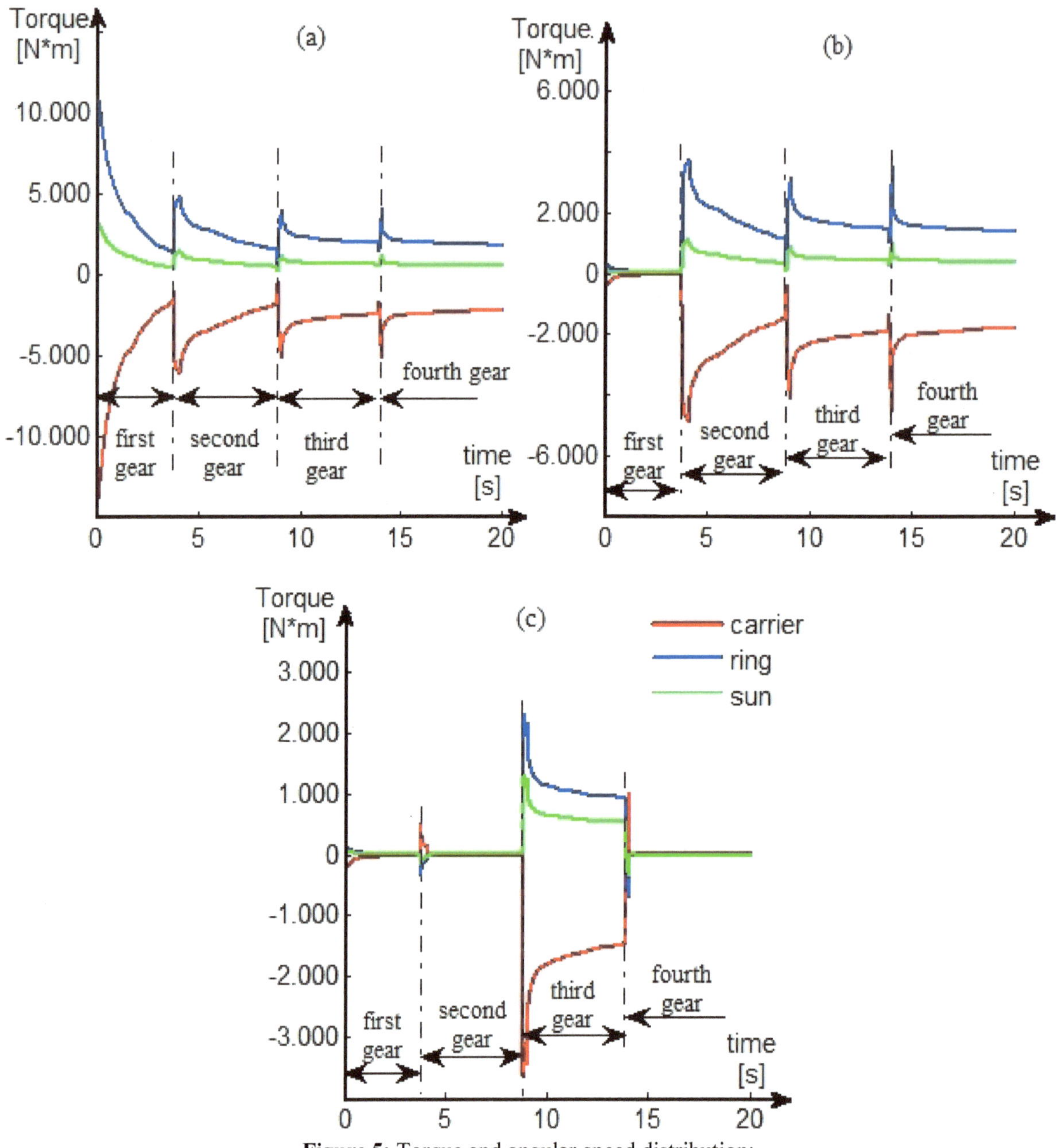

Figure 5: Torque and angular speed distribution:
a) K1 planetary gear; b) K2 planetary gear; c) K3 planetary gear;

4. CONCLUSIONS

The simulation model was conceived to analyze the modes of the power flow for all the four gears of the planetary gearbox. Analyzing the presented data, we noticed that the determination of the time histories of the torque for all the elements of the planetary gears, clutches and brakes are made of is possible. Moreover, the model can be used as a working platform for developing further an automating algorithm to shift the gears of the gearbox. As a result, we were able to draw some major lines in the improvement of the vehicle's working process efficiency that also leads to an improvement in the dynamic features of the vehicle.

Acknowledgements

This paper has been financially supported within the project entitled ***"Horizon 2020 - Doctoral and Postdoctoral Studies: Promoting the National Interest through Excellence, Competitiveness and Responsibility in the Field of Romanian Fundamental and Applied Scientific Research"***, contract number POSDRU/159/1.5/S/140106. This project is co-financed by European Social Fund through Sectoral Operational Programme for Human Resources Development 2007-2013. **Investing in people!**

REFERENCES

[1] J. Wang, Y. Wang and Z. Huo, Analysis of Dynamic Behavior of Multiple-Stage Planetary Gear Train Used in Wind Driven Generator, The Scientific World Journal, vol. 2014, Article ID 627045, 11 pages, 2014.

[2] Truţă M., Marinescu M, Vilău R, Alexa O., Ilie C., Self-Generated Torque Induced by the Lockable and Self-Locking Differentials within the 4wd Drivetrain, Applied Mechanics and Materials, Vol. 659 (2014), pp 268-273, ISBN 978-3-03835-272-3.

[3] Otter M., Schlegel C. and Elmqvist H., Modeling and Realtime Simulation of an Automatic Gearbox using Modelica, 9th European Simulation Symposium-ESS'97, pp. 115-121, 1997.

[4] Ciobotaru T., Grigore L., VÎNTURIŞ V. and LOGHIN L., Transmisii planetare pentru autovehicule militare, Publishing House of the Military Technical Academy, Bucharest, 2005.

[5] Samanuhut P., Modeling and control of automatic transmission with planetary gears for shift quality, PhD Thesis, The University of Texas at Arlington, 2011.

[6] Loghin L., Contribuţii privind studiul proceselor ce au loc la schimbarea etajelor în transmisiile hidromecanice ale autovehiculelor militare cu şenile, PhD Thesis, Publishing House of the Military Technical Academy, Bucharest, 2005.

THE MECHANICAL BEHAVIOR AND THE MATHEMATICAL MODELING IN THE CASE OF THE VIBRATIONS INDUCED TO THE SPINE

Olivia Ana Florea[1], Ileana Constanta Rosca[2]

[1] Faculty of Mathematics and Computer Science, Transilvania University of Brasov, email: olivia.florea@unitbv.ro
[2]Faculty of Product Design, Mechatronics and Environment, Dept. of Biomedical Engineering, Transilvania University of Brasov, email: ilcrosca@unitbv.ro

Abstract: *Back injuries, especially lower back pain, are amongst the most prevalent and costly non-lethal medical conditions affecting adults. An association between vehicle whole body vibration exposure and the development of lower back pain has been established through numerous epidemiological studies. In this paper we present a concept which could absorb the vibrations that appear at the driver or the passenger of a car or bus. To protect and prevent the lower back pains generated by the car vibrations and the rough road we construct a vibroisolator that can be inserted in the seats. Under those circumstances the following actions are required: the control, adjustment of the vibrations, their extinction, absorption or isolation – through vibro-isolation systems – environment protection, of real estate property and machines, the capture (absorption) of the vibrations, regularization and automated control of the sources by instruments of measure and control.*
Keywords: *spine injuries, vibrations, mathematical modeling, transfer function, stability*

1. INTRODUCTION

Intervertebral discs provide flexibility of the spine and transmit and distribute large loads through the spine. To carry out these tasks the intervertebral discs have a particularly complex structure consisting of a gelatinous nucleus pulpous (NP) and the annulus fibrous (AF). However, many people show degenerative changes in the intervertebral discs due to aging or pathological process. These changes affect the composition and structure of the intervertebral discs, and their mechanical functions too. Back pain is often a clinical consequence of disc degeneration.

The intervertebral disc is a complex structure, and its behavior is governed by its biochemical as well as mechanical composition. Simulation of the disc function is therefore challenging and has led to the development of a number of different approaches to represent its behavior, i.e. the NP has often been modeled as a non-linear incompressible solid governed by a Mooney-Rivlin law or a fluid , while the AF was modeled as a homogeneous, isotropic, linear-elastic solid. The highly layered and oriented structure of the AF suggests that its material behavior may be significantly anisotropic. The anisotropic behavior of the AF can be taken into account through discrete representation of the collagen fibers embedded within a homogeneous or hyperelastic matrix (the ground substance).

Back injuries, especially lower back pain, are amongst the most prevalent and costly non-lethal medical conditions affecting adults [1]. An association between vehicle whole body vibration exposure and the development of lower back pain has been established through numerous epidemiological studies [2].
In this paper we present a concept which could absorb the vibrations that appear at the driver or the passenger of a car or bus. To protect and prevent the lower back pains generated by the car vibrations and the rough road we construct a vibroisolator that can be inserted in the seats.

Under those circumstances the following actions are required: the control, adjustment of the vibrations, their extinction, absorption or isolation – through vibro-isolation systems – environment protection, of real estate property and machines, the capture (absorption) of the vibrations, regularization and automated control of the sources by instruments of measure and control.

2. MATHEMATICAL VIBROISOLATOR

The absorptions of the vibrations or their control, required ecologically or admitted for reliability, can be achieved through constructing mechanical, hydro-pneumatic or hydro-electric systems. These vibroisolators are implemented in the mechanical system, in general it is considered a large mass M (platforms) which vibrates and near by smaller masses (m) are installed – dynamic absorbers which are disrupted by M, or the masses m are connected at the amortization, resistance, friction systems (fig. 1) [4].

Usually, the vibration source of mass M operated by the disturbing force $F(t)$ and the amortized mass m must form a system of forces which acts on the same axis (usually the vertical) because the gravitation force is in most of the cases included in the component of the $F(t)$ force.

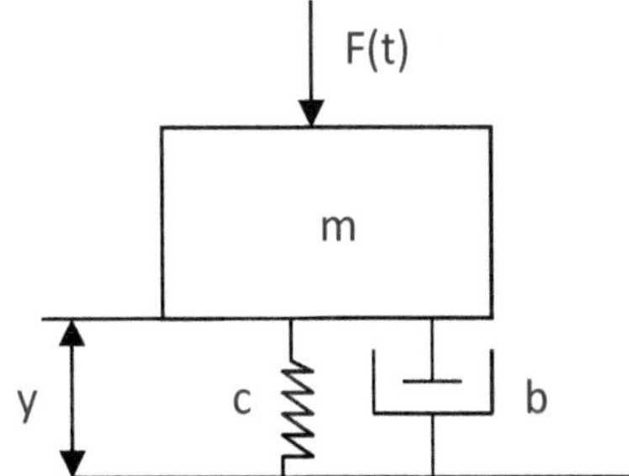

Figure 1: Mass-spring-damper mechanical model

The mass M is the fundament, and the mass m is the improvement-absorption source. The vibroisolator is located between M and m, has a resistance coefficient b and a vibro-isolation coefficient c (vibro-isolation rigidity). The rigidity coefficient c can be determined through equality conditions between the potential energy of the vibro-isolator and the equivalent elastic spring. This can be a non-linear function of coordinate y (rate) calculated from the equilibrium position. The resistance coefficient b is determined from the mechanical labor (energy) through the friction in the vibroisolator (amortization) and, in general, it can be a nonlinear function of y and $\dot{y}$.

$$m\ddot{y} = F(t) + Q(y,\dot{y}) \qquad (1)$$

Here, Q is the vibroisolator reaction. The purpose of the system is to minimize the components from Q on the coaxial direction y, with respect to the perturbation $F(t)$.

We consider the linear vibroisolator:

$$Q(y,\dot{y}) = -cy - b\dot{y} \qquad (2)$$

The movement equation is:

$$\ddot{y} + 2\gamma\dot{y} + \lambda^2 y = \frac{F(t)}{m} = \frac{M}{m} sin\omega t \qquad (3)$$

with the solution

$$y = \frac{M}{m\sqrt{(\lambda^2 - \omega^2) + 4\gamma^2\omega^2}} sin(\omega t - \theta) \qquad (4)$$

where:

$$tg(\theta) = \frac{2\gamma\omega}{\lambda^2 - \omega^2} \qquad (5)$$

equivalent with:

$$\dot{y} = \frac{M\omega}{m\sqrt{(\lambda^2 - \omega^2) + 4\gamma^2\omega^2}} sin(\omega t - \theta) \qquad (6)$$

From (4) and (5) we have the following relation in Q:

$$Q = -K_{din} M \left[sin(\omega t - \theta) + \frac{\omega b}{c} cos(\omega t - \theta) \right] \qquad (7)$$

where K_{din} is the dynamic coefficient of the perturbation, equal with the report of the sustained amplitude of the oscillations and the maximal value of movement M/c for the static equilibrium of the force:

$$K_{din} = \frac{\lambda^2}{\sqrt{(\lambda^2 - \omega^2) + 4\gamma^2\omega^2}} \tag{8}$$

we represent (7) with the hint: $asin\alpha + bcos\alpha = \sqrt{a^2 + b^2} sin(\alpha + \varepsilon)$, with $\varepsilon = arctgb/a$. For $a = 1$ and $b = \frac{2\gamma\omega}{\lambda^2}$, we have:

$$Q = -K_{din} M \sqrt{1 + \frac{4\gamma^2\omega^2}{\lambda^4}} sin(\omega t - \theta + \varepsilon) \tag{9}$$

where $\varepsilon = arctg\frac{2\gamma\omega}{\lambda^2}$, therefore:

$$Q_{max} = K_{din} M \sqrt{1 + \frac{4\gamma^2\omega^2}{\lambda^4}} \tag{10}$$

The report Q_{max}/M is called the transfer coefficient of the K_c force and is the same as the dynamic coefficient:

$$K_c = \frac{Q_m ax}{M} = \sqrt{\frac{1 + 4\beta^2\gamma^2}{(1 - \gamma^2)^2 + 4\gamma^2\beta^2}} \tag{11}$$

where $\gamma = \omega/\lambda$ is the frequency coefficient, and $\beta = \frac{\gamma}{\lambda}$ is the amortization (coupling) report. The transfer coefficient K_c is characterizing the quality of the vibroisolation as follows: if the link between m and M is rigid, then $K_c = 1$; for $K_c < 1$ the vibroisolation is efficient; for $K_c > 1$ the vibroisolation becomes disturbing for foundation . Alongside K_c (the forces transfer coefficient) it is also used the effective vibroisolation coefficient $K_{ef} = \frac{M}{Q_{max}} = \frac{1}{K_c}$

2. THE HYDROELECTRIC VIBROISOLATOR SYSTEM

The regulator systems, also called active control systems, are those vibroisolation systems in which the effective isolation against the vibrations is obtained by compensating the disturbing forces (disturbance based compensation) (fig. 2) Vibroisolator for the movement of the m mass relative to the static position [4], [5].

In the case of the electrohydraulic vibroisolator we have the external perturbation is $F(t)$. The vibroisolator is composed from the rigidity c and the damper b. The interaction (control) force F_y is given by the reaction of the hydro cylinder 1 of which piston acts on the object m through the rod 2 and the complementary elastic spring c_y. The movement of the piston through the pneumatic hydro cylinder is controlled with the help of the sensor signals 3 relative to the movement of the m object and the piston. This signal is transmitted through the amplifier 4, with an electric supply 5. The convector (amplifier) processes the signal and transmits it to the regulator (the debit translator) 6 which adjusts the fluid movement from the convector (pump) 7 through the force cylinder. In the intermediate position both pipes are closed. Through the movement of the valve 6, the fluid under pressure in moving in the superior half of the cylinder and the piston goes down, if the valve goes down, the piston climbs, respectively.

The movement of the m_g piston is z and implies the variation of the elastic force $F_y = c_y(y - z)$ which acts on m through the complementary spring c_y. Through an optimal choice of the vibroisolator parameters, the force F_y acts opposed (reaction) to the force $F(t)$ and reduces the movement y. This force is called controlled (conducted) interaction force.

2.1. The non-holonomous link in the electro-hydrodynamic system

The generalized coordinates are chosen to be y and z. This will determine a link that expresses the dependency of the piston speed $\dot z$ on the movement y. This link is defined by the variation of the fluid debit Q_p which passes through a window of the drawer 6, [7].

$$Q_p = S_p \dot z \tag{12}$$

where S_p is the effective area of the piston. We assume that the profile of the drawer's windows is chosen so that the fluid flow that enters in the cylinder is directly proportional with the drawer's opening.

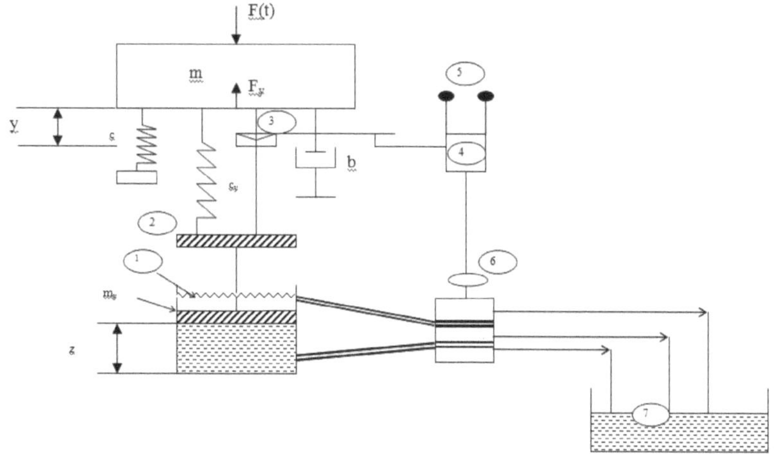

Figure 2. The electrohydraulic vibroisolator with the external perturbation is $F(t)$.

$$Q_p = K_{st} z_{st} \tag{13}$$

with K_{st}, the proportionality coefficient, depending on the drawer's parameters. Therefore, the electrical part of the system answers proportionally with the drawer's movement with respect to the movements $\varphi = y - z$ registered at 3:

$$z_{st} = K_\varphi (y - z) \tag{14}$$

K_φ is a proportional coefficient depending on the parameters of the electrical machine. Hence, from (12), (13), (14) we have:

$$\dot z = K(y - z) \tag{15}$$

where $K = \dfrac{K_{st} K_p}{S_p}$. The equation (15) can not be integrated since we have a non-holonomous link.

2.2. The movement equations of the system

We have two variables in the system, y, z but because of the link (15), through elimination we will have only one left degree of freedom left, y or z and therefore we'll obtain a differential equation. Further, we can write the movement equations, [3]:

$$\begin{aligned} &F(t) - cy - c_y(y - z) - b\dot y - m\ddot y = 0 \\ &c_y(y - z) - F_p - m_y \ddot z = 0 \end{aligned} \tag{16}$$

where m_y is the piston mass and, F_p the pressure force of the piston. We say that $\varphi = y - z$ and obtain the following system:

$$\dot y - \dot\varphi = K\varphi \qquad F(t) - cy - c_y\varphi - b\dot y - m\ddot y = 0 \qquad c_y\varphi - F_p - m_y K\dot\varphi = 0 \tag{17}$$

$$c_y\varphi = F(t) - cy - b\dot y - m\ddot y \tag{18}$$

$$c_y\dot\varphi = \dot F(t) - c\dot y - b\ddot y - m\dddot y \tag{19}$$

$$m\dddot y + (b + Km)\ddot y + (c + c_y + Kb)\dot y + cKy = KF(t) + \dot F(t) \tag{20}$$

Applying in (20) the Laplace transform with initial conditions we get:

$$[ms^3 + (b + Km)s^2 + (c + c_y + bK)s + cK]Y = (k + s)X \tag{21}$$

We define the transfer function W : $W = \frac{Y}{X}$

$$W = \frac{K+s}{ms^3 + (K+Km)s^2 + (c+c_y+bK)s + cK} = \frac{K+s}{P_3(s)} \tag{22}$$

If the function $F(t)$ is harmonic, meaning that $F = H\sin\omega t$ then $X = \frac{\omega}{s^2+\omega^2}$. In this situation we have:

$$W = \frac{\omega(K+s)}{(s^2+\omega^2)\left(ms^3 + (K+Km)s^2 + (c+c_y+bK)s + cK\right)} = \frac{\omega(K+s)}{P_5(s)} \tag{23}$$

In order to retrieve the original $y(t)$ we can discuss the roots of the polynomial $P_3(s)$ in the reduced form $P_3(s) = m(z^3 + pz + q)$ replacing the unknown function with $s = z - \frac{a_1}{a_0}$.

The effective command (control) coefficient for the vibroisolation K_{ef} will be defined as the report between the modulus $|y^o(i\omega)|$ which represents the complex amplitude without the vibroisolator interaction and $|y(i\omega)|$, the complex amplitude with the vibroisolator interaction.

$$K_{ef} = \left|\frac{y^o(i\omega)}{y(i\omega)}\right| \tag{24}$$

We choose in (23) $s = i\omega$ and we obtain the transfer function in the following form:

$$W(i\omega) = \frac{K+i\omega}{m(i\omega)^3 + (K+Km)(i\omega)^2 + (c+c_y+bK)(i\omega) + cK} \tag{25}$$

which can also be written as:

$$W(i\omega) = \frac{K+i\omega}{cK - (b+Km)(i\omega)^2 + i\omega(c+c_y+bK-m\omega^2)} \quad \textbf{or} \quad W(i\omega) = \frac{(K+i\omega)(s_1 - i\omega s_2)}{s_1^2 + \omega^2 s_2^2} \tag{26}$$

where we chose: $s_1 = cK - (b+Km)\omega^2$, $s_2 = c + c_y + bK - m\omega^2$

$$W = U + iV, \qquad U = \frac{Ks_1 + \omega^2 s_2}{s_1^2 + \omega^2 s_2^2}, \qquad V = \frac{\omega(s_1 - Ks_2)}{s_1^2 + \omega^2 s_2^2} \tag{27}$$

$$|W(i\omega)| = \sqrt{U^2+V^2} = \frac{\sqrt{(Ks_1^2 + \omega^2 s_2)^2 + \omega^2(s_1^2 - Ks_2)^2}}{s_1^2 + \omega^2 s_2^2} \tag{28}$$

If the hydraulic system is without the vibroisolator action, then $K = 0, c_y = 0, z = 0$. Hence, the movement equation of the damper becomes:

$$m\ddot{y}^o + b\dot{y}^o + cy^o = F(t) \tag{29}$$

getting the new transfer function:

$$W^o = \frac{1}{ms^2 + bs + c} \tag{30}$$

So the sequential transfer function is:

$$W^o(i\omega) = \frac{1}{-m\omega^2 + ib\omega + c}, W^o(i\omega) = U^o + iV^o, U^o = \frac{c - m\omega^2}{(c-m\omega^2)^2 + b^2\omega^2}, V^o = \frac{-b\omega}{(c-m\omega^2)^2 + b^2\omega^2} \tag{31}$$

$$|W^o(i\omega)| = \frac{1}{\sqrt{(c-m\omega^2)^2 + b^2\omega^2}} \tag{32}$$

$$K_{ef} = \left|\frac{W^o(i\omega)}{W(i\omega)}\right| = s_1^2 + \omega^2 s_2^2 \sqrt{\frac{(c - m\omega^2)^2 + b^2\omega^2}{(Ks_1 + \omega^2 s_2^2)^2 + \omega^2(s_1 - Ks_2)^2}} \tag{33}$$

To optimize the vibroisolation we have the condition: $K_{din} < 1 \Rightarrow K_{ef} = \frac{1}{K_{din}} > 1$, hence the parameters b, c, c_y, K are chosen so that we have: $K_{ef} > 1$. We can perform an analysis in the parameters space. For example: we settle b, c, K and we can study the variation of c_y; we settle c, K and we can study the variation of the parameters b, c_y; for $b = 0$ the damper is missing; for $c = 0$ the spring is missing.

3. THE STABILITY STUDY

In the movement equation (20) we consider that the free term $F(t) + K\dot{F}(t)$ is limited because for $F = H\sin\omega t$ the general solution is given by: $y(y) = y_g^o + y_p^n$, where y_p^n is limited. We are interested in the case of the homogenous equation when the free term is not in resonance with the frequency.

3.1. The stability with the Routh – Hurwitz criterion

$$a_0 r^3 + a_1 r^2 + a_2 r + a_3 = 0, \qquad a_0 = m, a_1 = b + Km, a_2 = c + c_y + bK, a_3 = cK \tag{34}$$

In order for the system to be stabile it is necessary to have:

$$a_i > 0, \qquad a_1 a_2 > a_0 a_3 \Leftrightarrow (b + Km)(c + c_y + bK) > mcK$$

If we consider $b = 0$ then the damper is missing, we have $c + c_y > c$ therefore in order to study the stability the spring must exist.

- If $c_y = 0$ we have full stability $bc + K^2 mb + b^2 K > 0$
- The general case is satisfied because the real part of the roots of the characteristic equation is negative $\Re(r_k) < 0$
- The resonance case: If the algebra equation would admit $\pm i\omega$ as roots, then we obtain an impossible condition: $-bc_y - bK^2 = \omega^2 b$. Therefore, the system can not have resonance.

3.2. The stability with the Nyquist criterion

This criterion is used for the analysis or the structure, functional and spectral analysis of the system.

The vibroisolator system is a linear system with a link of input-output type, where the report $\frac{y}{x}$ implies the analysis of $W = \frac{Y}{X}$. In this system the input quantity $x(t)$ is added to or subtracted from $x_y(t)$ (the piston – spring interaction F_y). The signal is transmitted to the output $y(t)$ through the transfer function W_i:

$$W_i = \frac{Y}{X \pm X_y} \tag{35}$$

The quantity $y(t)$, once found, is used to retrieve the value of F_y and so, the value of $x_y(t)$ that becomes X_y is determined by constructing the transfer function for the inverse link:

$$W_l = \frac{X_y}{Y} \tag{36}$$

The inverse link is called positive for $x + x_y$ and negative for $x - x_y$. The transfer function for the entire system is obtained after replacing Y from (36) and X from (35):

$$W = \frac{X_y / W_l}{(Y \mp X_y W_i)/W_i} = \frac{W_i}{1 \mp W_l W_i} \tag{38}$$

The signs are chosen in the following manner: - for a system with a positive link and + for a system with a negative link. Hence, the vibroisolator system looks like:

$$x = F(t), \qquad x_y = c_y(y-z) = c_y\varphi \tag{39}$$

The movement equation becomes:

$$F(t) - cy - c_y\varphi - b\dot{y} - m\ddot{y} = 0 \tag{40}$$

To this equation we apply the Laplace transform and we obtain:

$$(ms^2 + bs + c)Y = X - X_y \Rightarrow W_i = \frac{Y}{X - X_y} = \frac{1}{ms^2 + bs + c} \tag{41}$$

$$sY = K\frac{X_y}{cy} + s\frac{X_y}{c_y} \quad W_i = \frac{X_y}{Y} = \frac{c_y s}{s+k}$$

In conclusion, the transfer function has the following form:

$$W = \frac{s+K}{(s+K)(ms^2 + bs + c) + c_y s} \tag{42}$$

For systems without the vibroisolator action $W^o = W_i$ the following coefficient is obtained:

$$K_{ef} = \left|\frac{W_i}{W}\right| = \left|1 + W^o(i\omega)W_i(i\omega)\right| = \left|1 + \frac{c_y i\omega}{(i\omega + s)(-m\omega^2 + bi\omega + c)}\right|$$

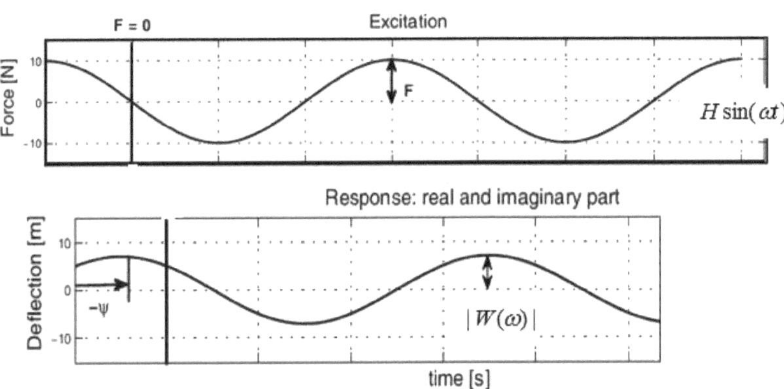

Figure 3. Excitation and response as a function of time.

4. CONCLUSION

In fig. 3 the excitation is a sinusoidal function with frequency ω and amplitude F. The bottom graph shows the complete response, which is a sinusoidal with the same frequency ω. The response characteristic has shifted compared to the excitation characteristic over the phase angle ψ. Here can be observed that the real part (Re, second panel) is in-phase with the excitation force, and that the imaginary part (Im, third panel) has a 90°-phase difference with the excitation force.

Back injuries, especially lower back pain, are amongst the most prevalent and costly non-lethal medical conditions affecting adults. An association between vehicle whole body vibration exposure and the development of lower back pain has been established through numerous epidemiological studies.

In this paper we have presented a concept which could absorb the vibrations that appear at the driver or the passenger of a car or bus. To protect and prevent the lower back pains generated by the car vibrations and the rough road we construct a vibroisolator that can be inserted in the seats.

ACKNOWLEDGEMENT

This paper is supported by the Sectorial Operational Programme Human Resources Development (SOP HRD), financed from the European Social Fund and by the Romanian Government under the project number POSDRU/159/1.5/S/134378.

REFERENCES

[1]. Deyo RA, Tsuiwu YJ, Descriptive epidemiology of low-back-pain and its related medical-care in the United States, Spine. 12, 264-268, 1987.

[2]. Bovenzi M, Hulshof CTJ, An updated review of epidemiologic studies on the relationship between exposure to whole-body vibration and low back pain (1986-1997), International Archives of Occupational and Environmental Health. 72, 351-365, 1999.

[3]. Rasvan, V., Theory of stability, (in rom), Ed. Stiintifica si Enciclopedica, Bucuresti, 1987

[4]. Florea, J., Petrovici, T., The dynamics of polyphasic fluids an its applications in technique , (in rom) Ed. Tehnica, Bucuresti, 1987

[5]. Florea, S., Dumitrache, C., Elemets of hydraulic and pneumatic execution, (in rom) EDP, Bucuresti, 1966

[6]. Chiriacescu, S., Balcu, I., Vibrations of the mechanical systems, (in rom) Ed. Univ. Transilvania, 2007

[7]. Vasiliu, N., Vasiliu, D., Hydraulics and pneumatic acting, (in rom), Ed. Tehnica, 2005, Vol. 1.

STRUCTURAL SYNTHESIS, ANALYSIS AND DESIGN OF THE MODULAR ANTHROPOMORPHIC GRIPPERS FOR INDUSTRIAL ROBOTS

I. Staretu

Transilvania University of Brasov, Brasov, ROMANIA, staretu@unitbv.ro

Abstract: *Anthropomorphic grippers for robots are similar to human hand and they can have two, three, four or more fingers, with two or three phalanxes. Anthropomorphic grippers for robots compared to other mechanical grippers have more advantages like: a higher degree of dexterity, a larger area of utility (more types of objects can be grasped) and a micro-movement of the grasped objects can be performed. This paper describes one group of modular mechanical anthropomorphic grippers with two versions, with three fingers and with four fingers, designed under coordinating of the author and there are shown more modular solutions. Synthetic, the stages of synthesis, analysis, design and functional simulation are shown too.*

Keywords: *anthropomorphic gripper, modular gripper, structural synthesis, cinematic analysis, functional simulation.*

1. INTRODUCTION

In general, the gripping systems are complex mechatronic systems used by robots, especially by industrial robots, for gripping operations on different pieces, to handle and transfer them from an initial position to a final one that is associated with a robotised action or technological process. In function of the gripping force type, the main categories of gripping systems are mechanical systems, vacuum systems and magnetic systems. In robotics field, mechanical gripping systems are also known as bilateral systems because the grasp is performed using at least two opposite forces onto the piece that is gripped.

Mechanical gripping systems have as main component a mechanical structure, a mechanism that provides the arrangement of the piece's contact elements towards the piece and enhances the contact force that is the necessary gripping force. In function of the constructive features of the mechanical structure, there are three main types of mechanical gripping systems: with jaws, with fingers (anthropomorphic) or with tentacles [1, 2, 3].

In present industrial robots use especially mechanical gripping systems with jaws, but anthropomorphic ones have become more and more popular, as simple shaped pieces grasped is replaced by grasp and micro handling of complex shaped pieces [4].

Current, anthropomorphic mechanical gripping systems with fingers can have two, three, four, five, or even six fingers with joints, having two or three phalanxes.

In this paper one category of anthropomorphic mechanical grippers, modular with jointed fingers, is described, and it was manufactured based on jointed bar mechanisms, more simple and with acceptable functionality like good alternative at very complex anthropomorphic mechanical hand with very high cost[5,6,7,8], what are in present on the market.

2. ANTHROPOMORPHIC MECHANICAL GRIPPERS

2.1. Main general aspects

Similar as in the mechanisms of prostheses, kinematic items most commonly used are the linkages[9]. They are found primarily in the construction of the fingers. For the rest of the mechanism, in addition to the mentioned elements, common mechanisms elements, of general mechanical transmission (gears, cams), usually smooth mechanical transmissions, are used[4]. Peculiarities of optimization for elements and couplings used in the robot gripping anthropomorphic mechanisms are arising from their structural features and construction, in the number of fingers, number, and relative position of the phalanges.In terms of optimizing the elements, because there may be more than three finger phalanxes, they must be very flexible but resistant, with similar or even identical forms. Results obtained by design optimization are used.

In the optimization of couplings, in addition to poly-couples use, it is envisaged the adoption of various structural forms that are no longer limited by the size of the model hand, as for the prostheses mechanisms.The phalanges that compose fingers can have the same size, different relative size or size that is proportional to the hand fingers size. Concerning the relative positioning of fingers, it can be similar to human hand; fingers can be placed in a plane, or in different planes. The relative position, depending on the number of fingers, at least two, must be chosen so that their access space is maxim. Several possible versions of relative positioning are illustrated in Fig. 1, of which it is very easy to obtain 3D fingers arrangement versions[2,4].

From the mobility degree point of view, all the fingers are usually actuated independently therefore, the mobility degree equals the number of fingers.

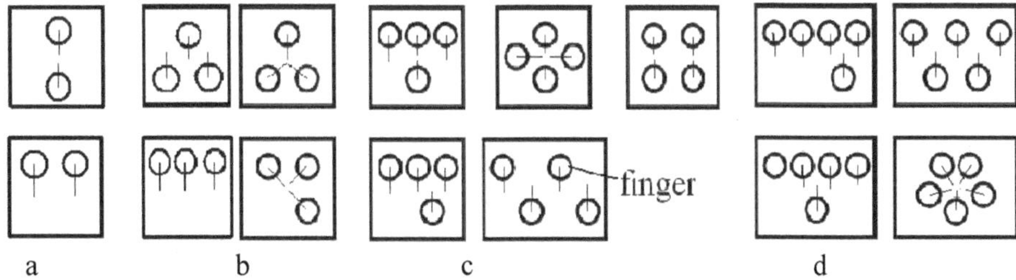

Figure 1: Relative positions of the fingers

In Fig. 2, there are fingers, of three, four and five phalanxes derived from the structural module mechanism. It is represented by the plane anti-parallelogram mechanism.

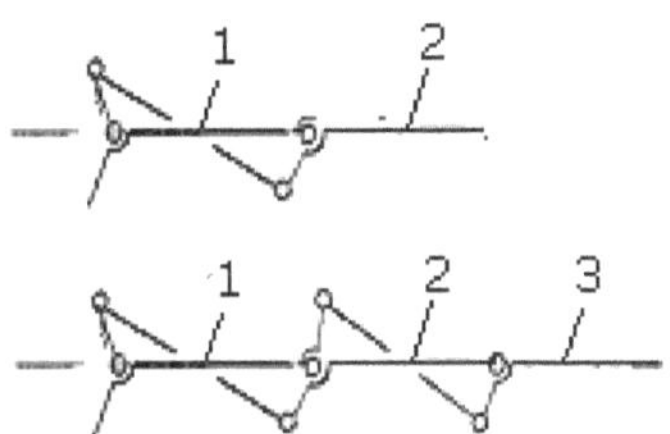

Figure 2: Modular fingers: with two or three phalanxes

2.2. Structural and cinematic synthesis and analysis

In the case of these mechanical grippers, the finger (Fig.3) is made by connecting more jointed bar mechanisms, in general anti parallelogram ones, according to the number of phalanxes (two or three)[11].

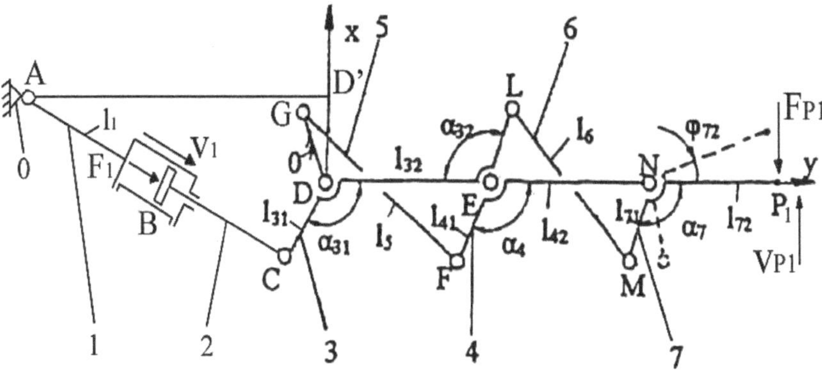

Figure 3: The structural scheme of one finger

Through **structural synthesis** the configuration of the finger is established that is the driving mechanism type and the number of phalanxes and the number of anti parallelogram mechanisms connected.
During the following stage, a **structural analysis** is performed in order to check if the mechanism is defined from the operational point of view (the mobility degree is determined, the cinematic and static parameters that are independent are identified, as well as the functions that convey external movements and forces).
For the mechanism in Fig.4, the mobility degree for each mono-contour mechanism is determined using the formula [3]: $M_k = \sum f_i - \chi_k$ (where $\sum f_i$ is the mobility degree of the couples - $f_i = 1$ and $\chi_k = 3$ is the cinematic rank of the mono-contour mechanism k=1, 2, 3). So, $M_1 = f_A + f_B + f_C + f_D - \chi_1 = 1+1+1+1-3=1$, $M_2 = f_D + f_E + f_F + f_G - \chi_2 = 1+1+1+1-3=1$, $M_3 = f_L + f_M + f_N + f_E - \chi_3 = 1+1+1+1-3 = 1$. For the multi-contour mechanism, the mobility degree is determined using the formula: $M = \sum M_k - \sum f_c$ (where M_k is the mobility degree for the mono-contour k and $\sum f_c$ is the mobility degree of the common couples $\sum f_c = f_D + f_E = 1+1=2$.). Therefore, $M = M_1 + M_2 + M_3 - \sum f_c = 1+1+1-2=1$.

M=1 represents an independent movement (independent speed): $v_1 = \dot{s}_1$ and a function that conveys external force: $F_m = F_m (M_7)$.
L-M =1 means a function that conveys external movement $\varphi_7 = \varphi_7 (s_1)$ or $\omega_7 = \omega_7 (v_1)$ and an independent momentum M7 (generated by the gripping force).

The cinematic synthesis means to adopt linear and angular dimensions necessary for the correct closing of the gripping system, and for the correct relative movements of the fingers, in order to grip the group of pieces given.
The cinematic analysis is performed using the method of the closed vector contour, applied successively to the vector contours corresponding to the mono-contour mechanisms underlined in Fig. 3.

3. DESIGN OF THE MODULAR MECHANICAL GRIPPERS

Under author's coordination the modular grippers designed are based on two modules, namely: a finger and a base (the palm). Two families that differ in structural features of the finger and platforms were designed. See Fig. 1, there are four gripping structures(a,b,c,d) obtained by the relative positioning of two, three, four or five identical fingers[12,13].
In Fig.1, the structural scheme of the finger is shown and the constructive form for this structural acheme is illustrated in Fig. 4a. Grippers in the family are based on a platform (palm) for three-finger versions (Fig. 4b) and another platform (Fig. 4c) for four-finger versions.

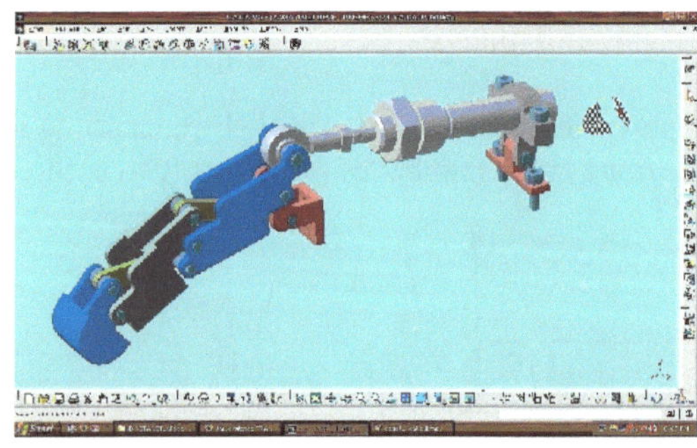

a

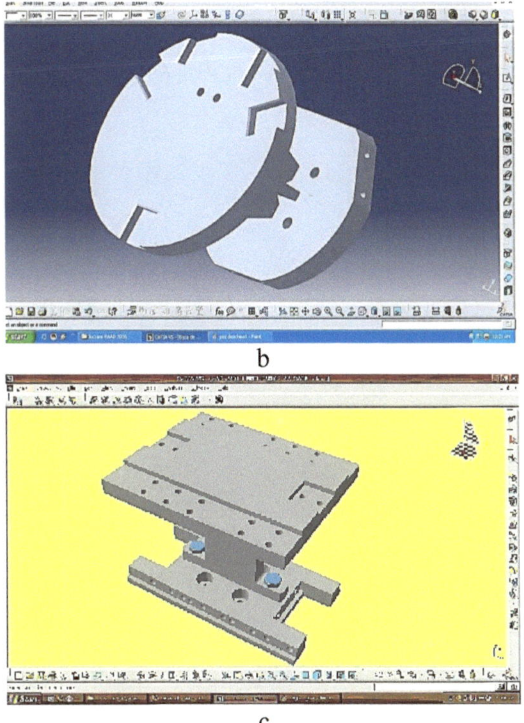

b

c

Figure 4: Constructive modules

Used these modules two main three-finger versions can be obtained (see Fig.1b and Fig. 4), the fingers having possible parallel (Fig. 5a) or concurrent movements (Fig. 5b). In Fig. 5c for the second situation, the gripper closing is simulated and in Fig. 5d a prototype, ready to be tested, is shown.

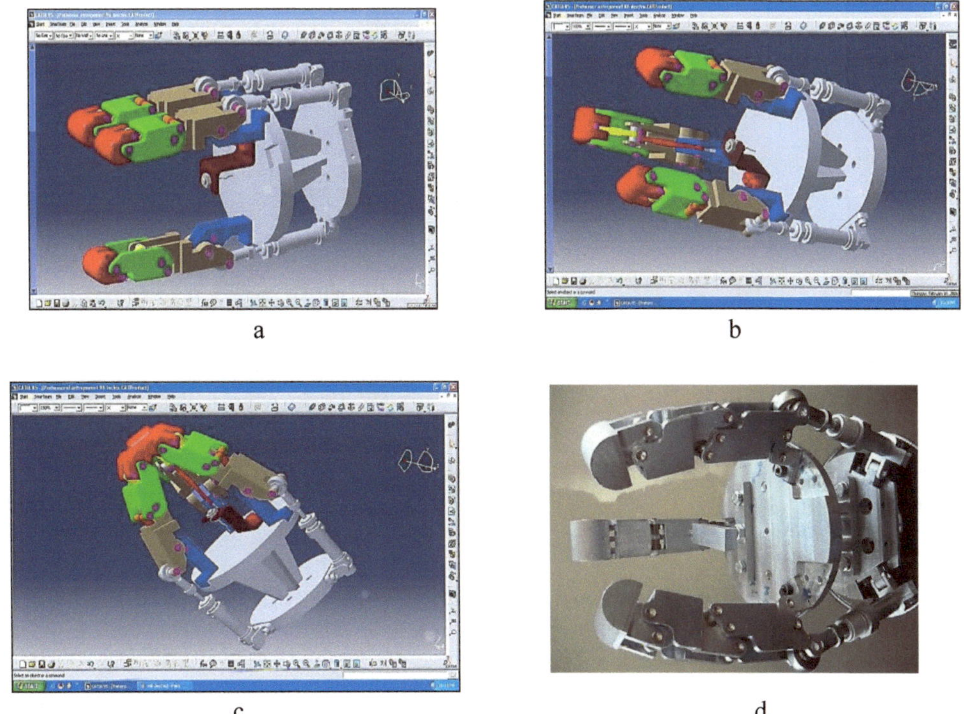

a

b

c

d

Figure 5: Modular anthropomorphic grippers' family with three fingers: the two main configurations(a and b), CAD simulation(c) and one prototype(d)

Technical characteristics of this prototype are: degree of freedom: M=3; weight hand: 12 N; payload: 40 N; gripping force: ~ 30 N/finger; dimensions: finger: 1:1 human fingers size and hand: 140x140x100 mm.
For four fingers modular mechanical anthropomophic gripper there are 5 variants (see also Fig. 1c), illustrated in Fig. 6.

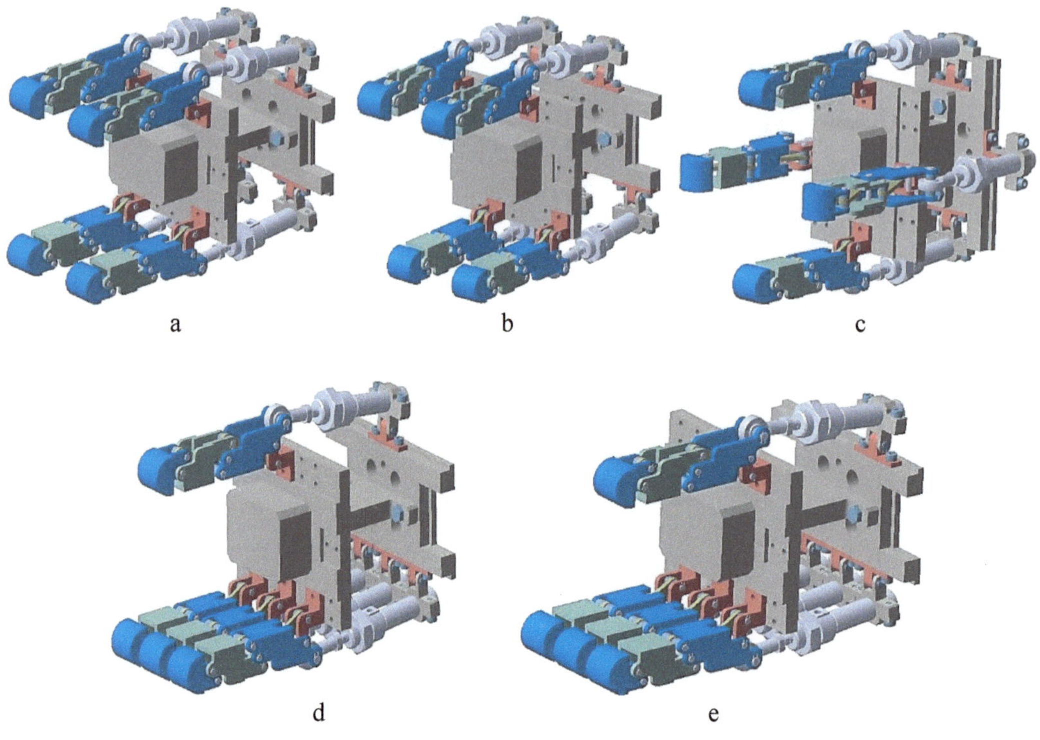

Figure 6: Modular anthropomorphic grippers' family , wth four fingers

In Fig. 7 is illustrated the gripper closing for the fingers intercalated parallel trajectories variants, without any piece to grip (Fig. 7a) and with a piece to grip (Fig. 7b).

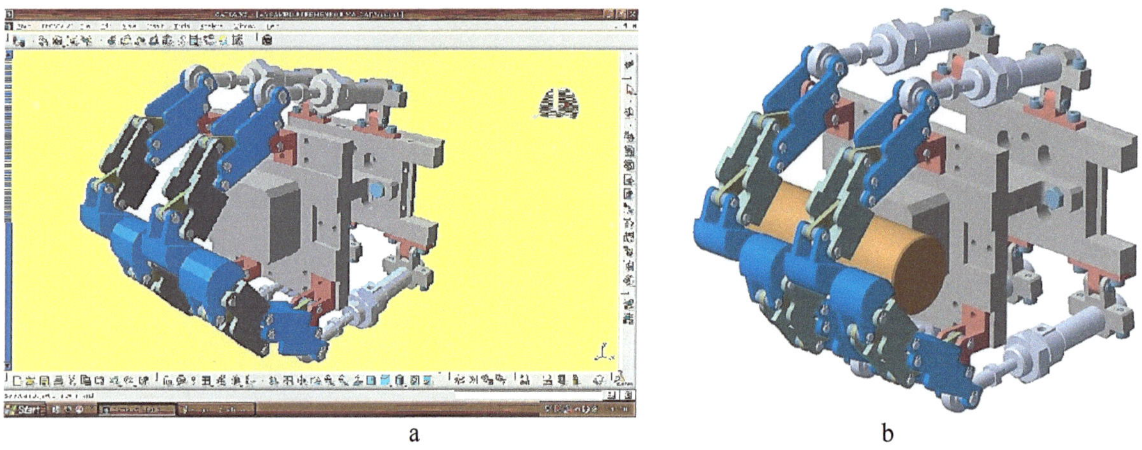

Figure 7: Modular versions closing CAD simulation,
four fingers, with no entity to grip (a), with entity to grip (b)

In this situation each finger is actuated by a pneumatic linear motor so that the degree of mobility of the grippers equals the number of fingers. Contact sensors are provided for the fingers, mounting them on the phalanges, and appropriate control equipment is used. Grippers can be mounted on the robot arm through the flange at the platforms base and they use the robot's sources of energy. For to change the gripper configuration (depending on the range of parts to grip) is possible without the gripper disassembling, only by changing the finger or the fingers position. With these grippers' families, a variety of parts can be gripped and they can successfully replace more sophisticated and highly expensive anthropomorphic grippers[6]. Similar, of the two basic versions, based on three, respectively, four fingers variants with two fingers, respectively, five and even six fingers can easily derive.

4. CONCLUSIONS

The main conclusions are as follows, In according with the ideas described in this paper:
- anthropomorphic gripping systems (with fingers) are used more and more frequently for industrial robots;
- there are one main type of anthropomorphic mechanical grippers (with fingers) according to their constructive elements : with jointed bars –linkages;
- there are two main types of anthropomorphic mechanical grippers, what can be classified in two groups: classical mechanical anthropomorphic grippers and modular mechanical anthropomorphic grippers;
- the synthesis and the structural and cinematic analysis of these gripping mechanisms can be done using classical methods that are popular in the theory of mechanisms, correspondingly adapted.
- functional simulation, in CAD software of these gripping mechanisms allows their constructive optimization and their use in order to perform the given gripping operations.

REFERENCES

1. Doroftei, I., Robotics(in Romanian), vol. 1 and 2 , Iasi, Technique, Scientific and Didactical Publishing House, CERMI, 2005-2006.
2. Staretu, I., Gripping Systems(in Romanian), Brasov, Lux Libris Publishing House, 2010.
3. Staretu, I., Gripping Systems, Tewksbury, Massachusetts, Derc Publishing House, 2011.
4. Staretu, I., Neagoe, M., Albu, N., Mechanical Hands. Anthropomorphic Gripping Mechanisms for Prostheses and Robots(in Romanian), Brasov, Lux Libris Publishing House, 2001.
5. Kawasaki, H., Komatsu, T., Uchiyama, K., Dexterous anthropomorphic robot hand with distributed tactile sensor: Gifu hand II, IEEE/ASME Transactions on Mechatronics, 2002, 7 (3), pp. 296-303.
6. http://www.barrett.com.
7. http:// www.shadowrobot.com /products/, 2013.
8. http://www.bebionic.com/wp-content/uploads/bebionic-Product-Brochure-Final.pdf.
9. Belter, J. T., Segil, J. L., Dollar, A.M., Weir, R. F., Mechanical design and performance specifications of anthropomorphic prosthetic hands: A review, Journal Rehabil Res. Dev.,2013, 50 (5), pp.9-618.
10. Staretu, I., Anthropomorphic Gripping Systems with Jointed Bars or Wheels and Wires for Industrial Robots-Constructive Synthesis, Analysis and Design, New Trend in Mechanism, Editors: S.-M.Cretu and N. Dumitru, Academica-Greifswald PH,2008, Germany, pp.133-144.
11. Bolboe, M., Staretu, I., Itu, A., Design, CAD Model and Functional Simulation for an Anthropomorphic Gripper for Robots. The 8th International Conference on Mechatronics and Precision Engineering, Cluj – Napoca, 2006, pp. 15-20.
12. Staretu, I., Ionescu, M., Runcan, V., Family of Mechanical Anthropomorphic Poly-Mobile Grippers for Robots – Synthesis, Analysis, Design and Functional Simulation. The 15th International Workshop on Robotics in Alpe- Adria- Danube Region, Balatonfured, June 15-17, 2006, pp. 271-277.

MATHEMATICAL MODEL OF A SPECIAL VEHICLE CLUTCH SERVOMECHANISM

Vilău Radu[1,a] , Alexa Octavian [1,b]

[1]Military Technical Academy, Faculty of Mechatronics and Integrated Systems for Armament, Dept. of Military Automotive Engineering and Transportation, Bd. George Cosbuc nr. 39-49, sector 5, Bucharest 050141, Romania

[a] radu.vilau@yahoo.com; [b] alexa.octavian@gmail.com

Abstract. *The working characteristic features of a process can be synthetically expressed as a mathematical model. For the technical systems, this model can be obtained based on the mathematical expressions that describe the working way of the system's components. The mathematical model can be used to analyse the dynamic performances of the system or a proper way to improve them, as far as the actual evolution of the system is accurately enough described. The paper makes an analysis using modelling and simulating techniques of a special clutch servomechanism. The modern simulation software allows studying the complex automotive systems; at the same time allows the fast variation of the system's parameters and the study of all modifications that appear within the working system.Some of dynamic charts are presented and improvement of the dynamic performances methods are given.*

Keywords: *Mathematical model, simulation, servomechanism, dynamic performances.*

1. INTRODUCTION

This paper deals with the analysis of a servomechanism working on the main clutch of a tracked vehicle's transmission. The main clutch ensures a smooth and progressive coupling of the engine to the gearbox as well as a fast de-coupling of the previously mentioned components. It also works as a safety device [1], starting to slip whenever the drag at the vehicle's driving wheels generates a coasting torque that exceeds the engine's output. The clutch is placed on the input shaft of the gearbox and it is referred as "the main clutch". The main clutch is a multi-disc type (it consists of two driving discs and three driven ones). It is also a "dry clutch" and its clutching (engaging/disengaging) mechanism is either hydraulic or pneumatic (when the hydraulic system fails). The main components of this clutch can be seen in fig. 1. The main hydraulic distributor is depicted in fig. 2.

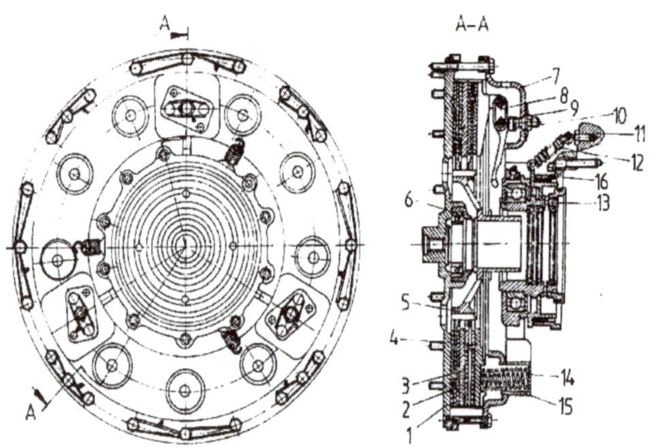

Figure 1 - The main clutch:

1-front plate; 2-friction discs; 3-intermediate plate; 4-plate; 5-friction discs' hub; 6-rolling bearing; 7-case; 8-clutching lever; 9-fork; 10-adjusting screw; 11-gearbox's case; 12-restoring spring; 13-piston; 14-springs; 15-cup; 16-cylinder.

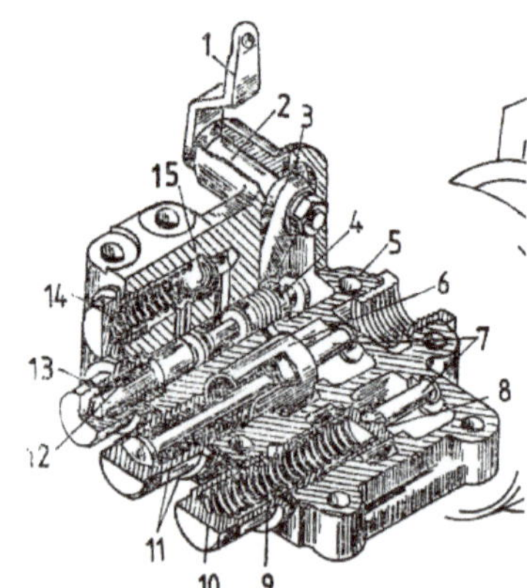

Figure 2 - The main distributor:

1–external command lever of the clutch; 2– shaft; 3–internal command lever; 4–spool valve; 5–body; 6–command system's valve; 7-guidance rod; 8–lubrication system's valve; 9–spring; 10–breaker; 11–gasket; 12–spring; 13–gasket; 14–spring; 15-ball;

The main distributor is in charge with distributing the working fluid, keeping the needed oil pressure within the hydraulic system, providing the power to clutching the main clutch, feeding the gear-shifting power system and providing a 0,2 MPa of oil pressure within the lubricating system of the gearbox. It is placed on the top of the gearbox's upper case.

To providing a smooth engagement of the main clutch, the driver should release the clutch pedal half-way of its whole travel and hold it in this position for about 4...5 seconds. In this situation, the plunger of the main distributor 2 reaches position II and allows the discharge of the oil pressure towards the oil tank via the smooth engaging valve. Hence, the oil pressure within the clutching system drops with a small gradient. When the clutch pedal is completely released, the plunger moves again, the distributor reaches position I, the oil is driven to the gearbox's case and the clutch is completely engaged under the springs' force.

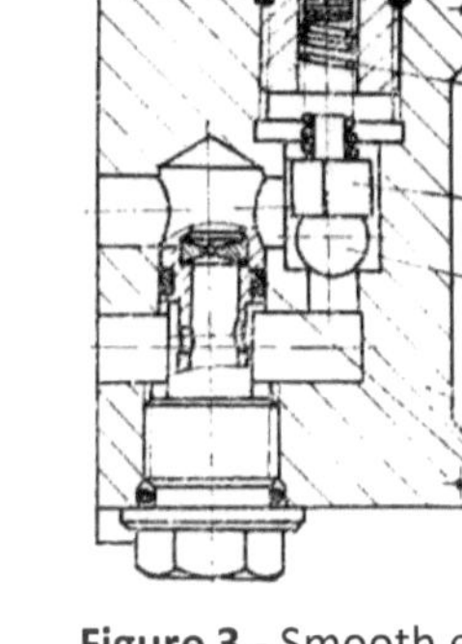

Figure 3 - Smooth engaging valve

The smooth engaging valve is depicted in figure 3, where: 2 - body, 1- throttle and ball valve, 4 – valve seat, 5 - spring. Within the oil circuit an additional flow-resistance is given by the throttle, which allows the command system's pressure smoothly drop.

2. MATHEMATICAL MODEL

The mathematical model of the hydraulic servomechanism has been issued taking into account the following equations

- the continuity equation, corresponding to the non-permanent motions within the acting hydraulic systems;
- the equation of the displacement - flow rate - pressure;
- the equation of the hydraulic engine piston's movement.

The hydraulic servomechanisms are mounted using assembling elements. The elasticity of these elements influences both the stability and the positioning accuracy of the servomechanism. If not considering the finite stiffness of these components, then the servomechanism is as good as ideal mounted (the stiffness of all the mounting elements to the supporting structure and to the actuated system is considered to be infinite). In real life, the stiffness is finite. Moreover, it slowly decreases due to the loads the mounting components have to bear.

Considering Q [2], [3], the flow through a stream tube, the continuity equation can be written as follows:

$$\frac{A\cdot\rho}{E}\cdot\frac{dp}{dt}+\rho\cdot\frac{\partial Q}{\partial s}=0 \tag{1}$$

If further processed, it leads to:

$$\frac{dp}{dt}=-\frac{E}{A}\cdot\frac{\partial Q}{\partial s}$$

$$\frac{dp}{dt}=\frac{E}{\vartheta}\cdot\left(Q_1-Q_2\right) \tag{2}$$

where:

- Q_1 – liquid flow through first (inlet) section;
- Q_2 - liquid flow through second (outlet) section;
- A – flow cross-section;
- υ - liquid system's volume;
- s – displacement (travel);
- t- time;
- p – liquid system's pressure;
- E – elasticity modulus;
- ρ - oil density

Liquid's elasticity modulus ranges between 7000 and 14000 bar. It reaches 14000 bar when the oil is free of dissolved gases

The continuity equation is applied to the "liquid system" which consists of both variable volumes of the hydraulic distributor and the hydraulic motor. It also subsumes the connection pipes between the distributor and the motor.

The flow Q_1, delivered by the motor's distributor [2], [3], generates the motor's piston displacement. It also completes internal and external losses as well as the extra fluid needed to compensate the liquid compression.

$$Q_1 = c_{ip} \cdot (p_1 - p_2) + c_{ep} \cdot p_1 + A_p \cdot \frac{d(z+u)}{dt} + \frac{\vartheta_1}{E_e} \cdot \frac{dp_1}{dt} \quad 3)$$

In the same time, the flow Q_2 runs out the distributor:

$$Q_2 = A_p \cdot \frac{d(z+u)}{dt} + c_{ip} \cdot (p_1 - p_2) - c_{ep} \cdot p_2 - \frac{\vartheta_2}{E_e} \cdot \frac{dp_2}{dt} \quad 4)$$

where:

- c_{ip}– internal leakage motor's coefficient;
- c_{ep} – external leakage motor's coefficient
- p_1 – pressure of the intake motor's chamber;
- p_2 - pressure of the exhaust motor's chamber;
- Ap - effective cross-section of the motor's piston;
- z – piston's rod displacement;
- u – servomechanism's body displacement;
- υ_1, υ_2 – flows through the motor's chambers and the coupling pipes;
- E_e - equivalent modulus of elasticity.

For a symmetrical hydraulic distributor, if using the hydraulic stiffness of the hydraulic motor, R_h, one could write:

$$R_h = \frac{2E_e}{\vartheta_0} \cdot A_p^2 \quad (5)$$

where: $2\upsilon_0 = \upsilon_1 + \upsilon_2$ – the initial volume of the chambers. Hence, the following continuity equation emerges:

$$Q = K_l \cdot P + A_p \cdot \frac{d(z+u)}{dt} + \frac{A_p^2}{R_h} \cdot \frac{dP}{dt} \quad 6)$$

where K_l – total flow coefficient throughout the distributor;
P - pressure drop on the hydraulic motor.

Next coefficients are defined:

a) amplifier flow factor of the distributor:

$$K_{Qx} = \frac{\partial Q_m}{\partial x} \quad (7)$$

b) flow – pressure coefficient:

$$K_{Qp} = -\frac{\partial Q_m}{\partial P} \quad (8)$$

c) pressure – displacement coefficient:

$$K_{px} = \frac{\partial P}{\partial x} = \frac{K_{Qx}}{K_{Qp}} \quad (9)$$

where Q_m is the flow taken by the hydraulic motor

The transfer function describing the dynamic behavior of the distributor is given by:

$$H_A(s) = \frac{X(s)}{I(s)} = \frac{K_A}{\frac{s^2}{\omega_n^2} + \frac{2\xi}{\omega_n} s + 1} \quad (10)$$

where:

- $\omega_n = \sqrt{\frac{K_s}{M_s}}$ - free oscillations frequency;
- $\xi = \frac{1}{2} \cdot \frac{f_s}{\sqrt{K_s \cdot M_s}}$ - damping factor;
- K_A – amplifying factor;
- M_s – spool valve mass;
- K_s – spool valve spring's coefficient of elasticity;
- f_s – fluid friction factor.

The pressure acting on the servomechanism's piston moves it with a motion law described by:

$$F = m_p \cdot \ddot{v} + K_{fv} \cdot (\dot{z} + \dot{v}) + R_c \cdot (z - v) + D_c \cdot (\dot{z} - \dot{v}) \quad (11)$$

where

- R_c – the stiffness of the connecting mechanism between the servomechanism and the load (actuated element)
- D_c – the damping coefficient of the connecting mechanism;
- z – displacement of the actuated element;
- K_{fv} – viscous friction coefficient;
- m_p – piston's mass.

3. SIMULATING THE SERVOMECHANISM'S WORK

The main distributor's work was simulated using a special simulating software, Simulink modulus from Matlab. The simulation diagram is depicted in fig. 4. The simulation results are given in the following pictures (fig. 5-9) that mainly analyze the time histories of the pressure within the hydraulic system [4], as well as the piston's displacement, speed and acceleration. The input, independent variable, was considered the driver's action upon the clutch pedal, on the second stage of the main distributor, when the oil flows via the hydraulic throttle.

When analyzing the distributor's working way, as well as the influence of various parameters on its functioning modes, useful conclusions emerge:

- The piston's speed and acceleration rapidly become steady. We can conclude that, after the first stage of clutch's engagement (the "smooth" stage) the friction discs slowly engage.
- Using throttling, the smaller flowing cross-section, the higher system's response time
- The higher the flowing cross-sections (as in second stage), the lower the system's response time; hence, the clutch's plate will quickly engage the discs and also the pressure gradient will increase.
- The check input signal was the ramp type (assuming the clutch pedal is steadily actuated, at constant speed)
- The higher the pressure coefficient (i.e. the pressure losses due the increased pressure within the distributor are higher) the larger the oscillations' amplitude of the piston's displacement while the system's response isn't quite affected. Nevertheless, high values of the flow coefficient lead to a destabilizing trend of the system (it is a common fact for the hydraulic system that need throttling to increase their stability)
- As the losses due to the pressure increase, the piston will have larger speed oscillations. This phenomenon leads to supplementary mechanical loads due to the inertia forces.
- A low hydraulic stiffness of the working oil leads to a pressure decrease within the system. On the other hand, we could notice an increase of the piston's travel during the first stage. It can be explained by a stronger compression of the oil; this phenomenon can also be met in terms of the speed variation.

On the contrary, a high hydraulic stiffness of the working oil leads to a pressure increase within the system. The piston's displacement during the first stage decreases, but the speed's oscillation frequency becomes higher.

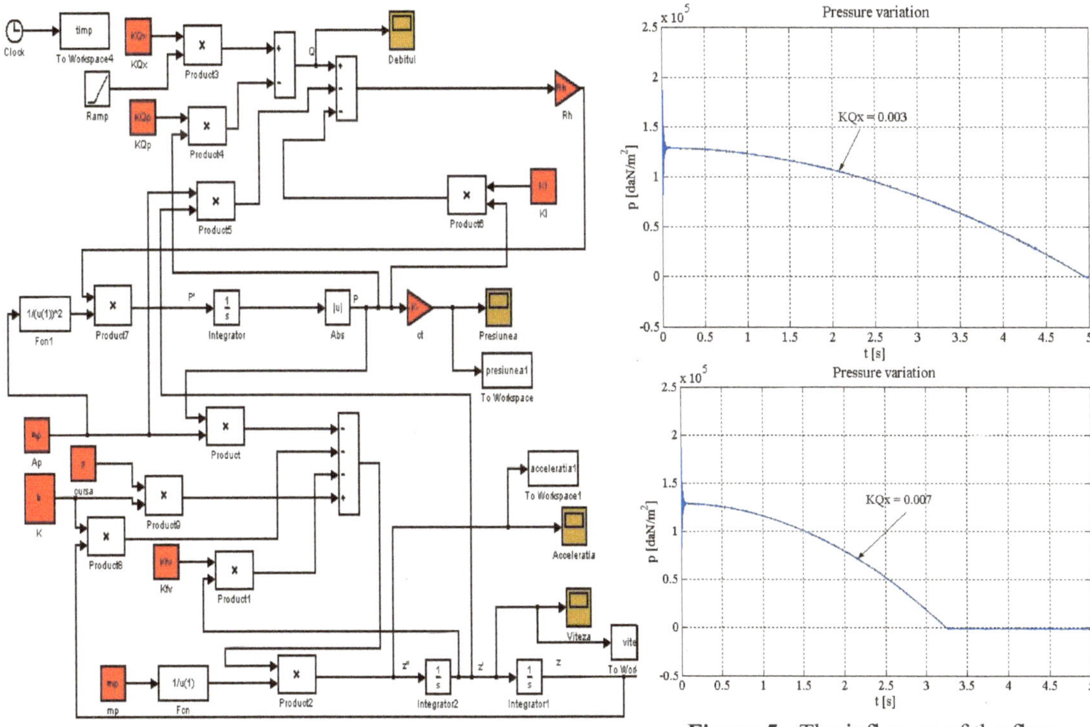

Figure 4 - Simulating chart

Figure 5 - The influence of the flow coefficient K_{Qx}, on the cylinder's pressure

- When the leaks on the piston are high, the pressure rapidly drops and the system' response time decreases; the systems behaves like taking advantage of a larger cross-section of the throttl
- If the restoring springs were stiffer, the response time decreases. On the other hand, the displacement (travel) of the piston in the first stage increases due to the strong compression of the working fluid. This leads to an increased loading of the piston due to the inertia effect (both by higher frequency and amplitude of the oscillation).
- If the friction of the piston-cylinder coupling increased, the system pressure decreases. Also, the speed increase of the piston is smoother.
- Eventually, when the piston's mass is changed, the pressure waves amplitude and frequency also vary.

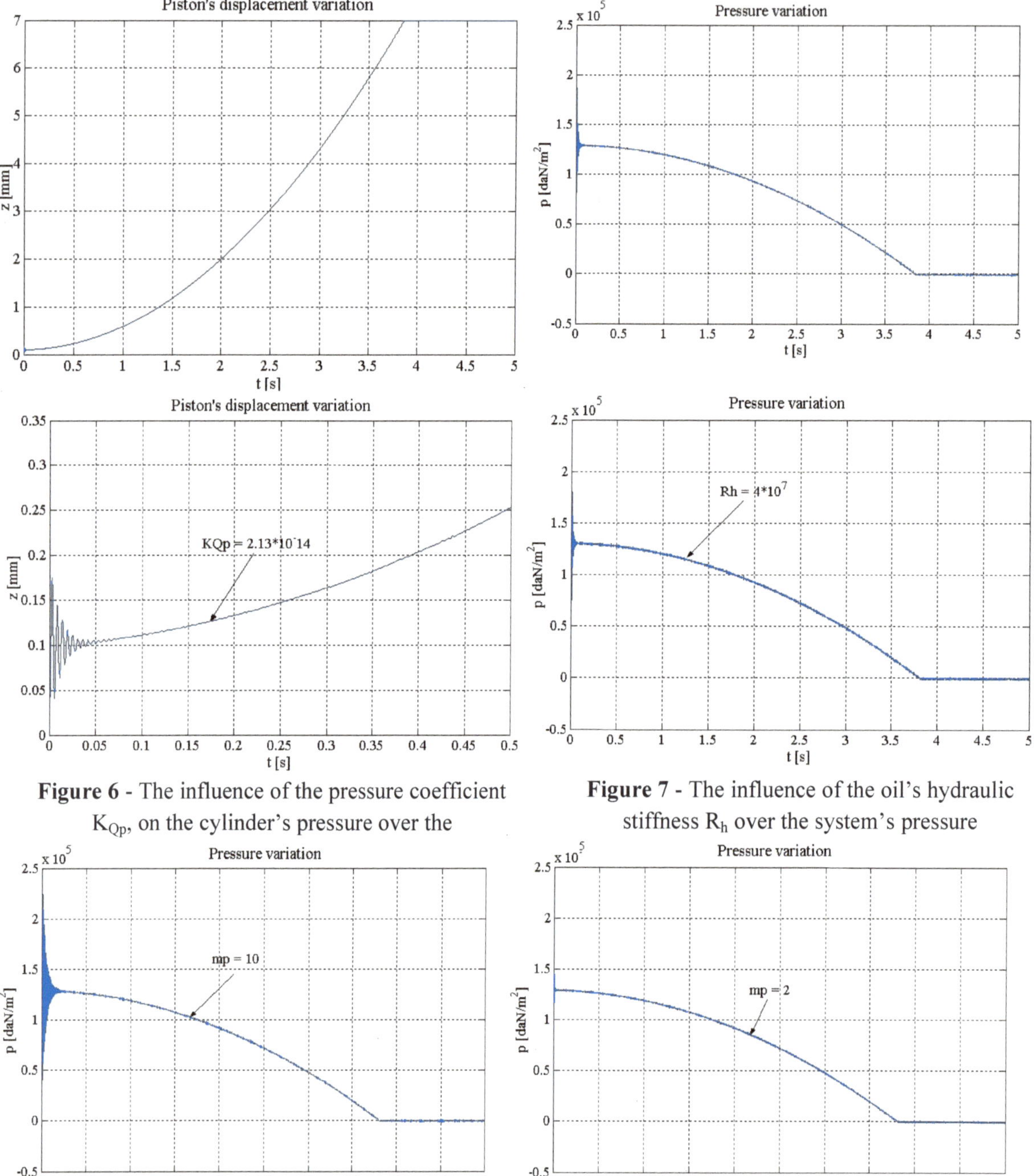

Figure 6 - The influence of the pressure coefficient K_{Qp}, on the cylinder's pressure over the

Figure 7 - The influence of the oil's hydraulic stiffness R_h over the system's pressure

Figure 8 - The influence of the actuated load m_p over the system's pressure

From the main distributor's work analysis we could notice that, when opening the oil path via the throttle, due to the restoring springs of the clutch, the piston is strongly pushed backwards. There is a small leap of displacement and but it is due to it to assuring the necessary flow that fills the distributor's bulks. Eventually, it could lead, according to the clutch's adjustment or to a fast-developing stage, to sudden engagements of the friction discs. After that, the flow becomes steady then drops slower. High masses of the actuated elements lead to the occurrence of the

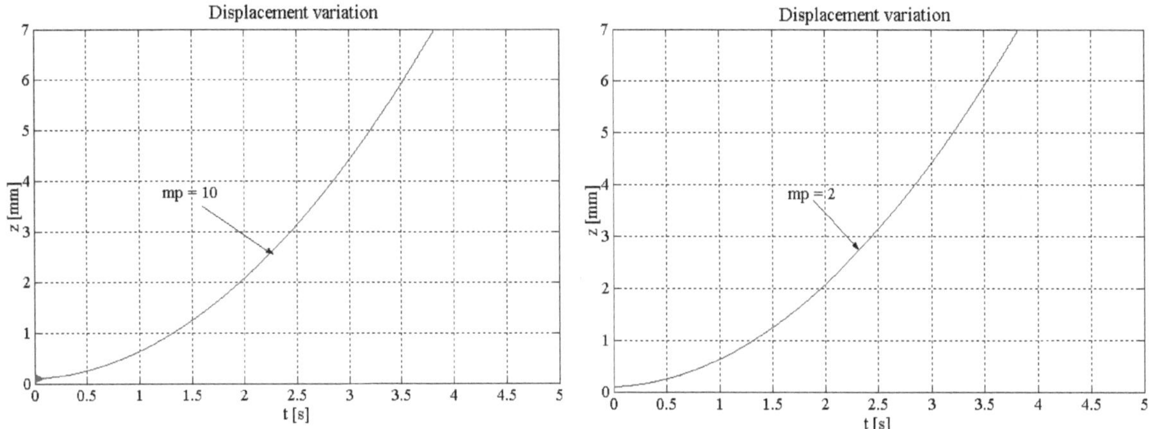

Figure 9 - The influence of the actuated load m_p over the system's r esponse

pressure oscillations from the very beginning of the process. The displacement leap occurs due the commuting the flowing process through one port to another. To avoiding this drawback, a continuously variable cross-section throttle can be used. The speed and acceleration of the piston rapidly stabilize; hence, after the first stage is completed, the rest of the engagement is gradually achieved.

REFERENCES

[1] Costache, D. - Actionari hidraulice si pneumatice la autovehicule - Editura Academiei Militare, Bucuresti, 1985.

[2] Oprean, A. ş.a. – Acţionări şi automatizări hidraulice. Modelare, simulare, încercare, Editura Tehnică, Bucureşti, 1989;

[3] Călinoiu, C. ş.a. – Modelarea, simularea şi identificarea experimentală a servomecanismelor hidraulice, Editura Tehnică, Bucureşti, 1998

[4] Marinescu, M. ş.a. - Theoretical and data-based mathematical model of a special vehicle braking system, Advanced Materials Research Vol. 837 (2014) pp 428-433, ISSN:1662-8985

[5] *** - Matlab toolbox and documentation.

ACKNOWLEDGMENT

This paper has been financially supported within the project entitled ***"Horizon 2020 - Doctoral and Postdoctoral Studies: Promoting the National Interest through Excellence, Competitiveness and Responsibility in the Field of Romanian Fundamental and Applied Scientific Research ",*** contract number POSDRU/159/1.5/S/140106. This project is co-financed by European Social Fund through Sectoral Operational Programme for Human Resources Development 2007-2013. **Investing in people!**"

BIOGAS PRODUCTION FROM MIXTURES OF ANIMAL MANURE AND FRESH BIOMASS WITH AND WITHOUT GLUCOSE ADDITION

Mirela Dincă[1], Mariana Ferdeş[1], Gheorghe Voicu[1], Gigel Paraschiv[1], Georgiana Moiceanu[1], Gabriel Muşuroi[1], Stela Ioniţă[1]

[1] University Politehnica of Bucharest, Bucharest, ROMANIA, e-mail: mirela_dilea@yahoo.com

Abstract: *Anaerobic digestion is a widely used technology that can process various kinds of organic wastes for biogas production by decomposing organic matter under oxygen-free conditions. Anaerobic digestion of organic matter has been considered as a suitable technology for organic wastes treatment and energy production in the form of biogas. The main aim of the paper is to present the results of different experiments carried out in order to provide data about biogas production from different substrates with and without glucose addition. The experimental data showed that the glucose addition was stimulatory for biochemical reactions and benefic for biogas production. We concluded that the biogas production resulted from anaerobic digestion of animal manure is substantially affected by the composition of feedstocks.*

Keywords: *biogas, anaerobic digestion, animal manure, glucose addition*

1. INTRODUCTION

The modern society generates large amounts of waste that represent a tremendous threat to the environment and human and animal health. To prevent and control this, a range of different waste treatment and disposal methods are used. The choice of method must always be based on maximum safety, minimum environmental impact and, as far as possible, on valorization of the waste and final recycling of the end products [1]. One of the most important processes of biomass conversion is the anaerobic digestion (methane fermentation) of organic matter to produce biogas, consisting mainly of methane and carbon dioxide [2].

Anaerobic digestion of energy crops, residues, and wastes is of increasing interest in order to reduce the greenhouse gas emissions and to facilitate a sustainable development of energy supply. Production of biogas provides a versatile carrier of renewable energy, as methane can be used for replacement of fossil fuels in both heat and power generation and as a vehicle fuel [3]. The main constituents of biogas are methane and carbon dioxide, but it can also contain, depending on the source's composition, traces or significant quantities of undesirable contaminants, such as hydrogen sulphide, ammonia and siloxanes, whose presence can cause corrosion, erosion, and fouling to the thermal or thermocatalytic device, and generate hazardous emissions. Therefore, biogas quality (purity and composition) is very important, and its purification represents a crucial final step of the overall production process in view of its final application [4].

The anaerobic digestion process can be divided into four major steps [5-7], namely: hydrolysis, acidogenesis, acetogenesis and methanogenesis. Firstly, high molecular materials and granular organic substrates (e.g., lipids and carbohydrates, protein) are hydrolyzed by fermentative bacteria into small molecular materials and soluble organic substrates (e.g., fatty acids and glucose, amino acids) [8]. Then, these small molecules are converted by fermentative bacteria (acidogens) to a mixture of volatile fatty acids and other minor products such as alcohol [9]. Thirdly, the acetogenic bacteria convert the volatile fatty acids to acetate, carbon dioxide and hydrogen, which provide the substrates for methanogenesis phase. Among the four microbial groups, methanogens bacteria have the slowest growth rate and are the most sensitive to changes of environmental conditions, such as temperature, pH, and the concentrations of inhibitors [10].

Crop residues and animal manure have recently been used together to produce biogas by anaerobic digestion process. Compared with the single digestion of feedstock, the co-digestion of crop residues and animal manures increases the rate of biogas production because of the greater balance between carbon and nitrogen [11]. Anaerobic co-digestion consists of the anaerobic digestion of a mixture of two or more substrates with complementary characteristics. It is well known that one of the main issues for the co-digestion process lies in balancing the C/N ratio. In fact, ideal co-substrates for manures, substrates with high nitrogen contents and high alkalinity, are wastes which have a high C/N ratio[12].

S. Astals *et. al* [13] tested the influence of glycerol on biogas production. Thus, they mixed fresh pig manure and digested pig manure with crude glycerol derived from biodiesel production. The researchers reported that an increase of about 400% in biogas production was obtained under mesophilic conditions when pig manure was co-digested with 4% of glycerol, on a wet-basis, compared to mono-digestion. Moreover, they found out that the digestate stability, evaluated through a respirometric assay, showed that co-substrate addition does not exert a negative impact on digestate quality.

Another study related to co-digestion is reported by K. Bulkowska *et. al* [14]. They tested the digestion process of crop silage (Zea mays L. and Miscanthus sacchariflorus) with 0%, 7.5%, 12.5% and 25% pig manure as co-substrate. The results indicated that the most stable anaerobic digestion was achieved using 7.5% and 12.5% pig manure. The authors concluded that compared to crop silage alone, pig manure favored the production of biogas and methane; the highest production rates were obtained with 12.5% pig manure.

Zhang T. *et al.* [15] investigated biogas production by co-digestion of goat manure with three crop residues, namely, wheat straw, corn stalks and rice straw, under different mixing ratios. Results showed that the combination of goat manure with corn stalks or rice straw significantly improved biogas production at all carbon-to-nitrogen (C/N) ratios. Goat manure(GM)/corn stalks (CS) (30:70), GM/CS (70:30), GM/rice straw (RS) (30:70) and GM/RS (50:50) produced the highest biogas yields from different after 55 days of fermentation.

The main aim of the paper is to present the results of different experiments carried out in order to provide data about biogas production from different substrates (horse and pig manure mixed with fresh biomass) with and without glucose addition. All the experiments were conducted on a small capacity pilot plant for obtaining biogas from biomass and the main parameters, temperature and pH, were kept constant. During the anaerobic fermentation process, the temperature was set in the mesophilic domain (35°C) and the pH value was set at 7. The composition of the obtained biogas was analyzed using a gas chromatography device.

2. MATERIAL AND METHODS

2.1. Feed material

For biogas production, animal manure (horse and pig manure) and fresh residual biomass were used as substrate. The residual biomass consisting of leaves, twigs and grass was chopped using an electric plant for grinding equipped with cutting knives. Pig manure and horse manure were obtained from the agriculture farm in Romania. The substrate subjected to the anaerobic fermentation process was mixed with tap water and the elements listed in Table 1.

Table 1: Substances used in the preparation of the feed substrate

	Substances g/100 L	**Symbol**	**Quantity** g
1	Glucose	$C_6H_{12}O_6$	3000
2	Ammonium phosphate	$(NH_4)_2HPO_4$	91.1
3	Ammonium chloride	NH_4Cl	56.6
4	Potassium chloride	KCl	8
5	Ferric chloride	$FeCl_3$	10
6	Magnesium chloride	$MgCl_2.6H_2O$	20
7	Aluminium chloride	$AlCl_3.6H_2O$	2.2
8	Calcium chloride	$CaCl_2.2H_2O$	2
9	Magnesium sulphate	$MgSO_4H_2O$	0.5
10	Zinc chloride	$ZnCl_2$	0.04
11	Ammonium molybdate	$(NH_4)6MoO_{24}4H_2O$	0.2

2.2. Biogas plant design

The experiments were conducted using a small capacity pilot plant for obtaining biogas from biomass. The plant used for the experiments is comprised of four main parts, namely [16]:

- a feed section, consisting of the feed preparation system and a pump to transfer the material into the reactor;
- the anaerobic digester, with all the instruments needed for feeding, measuring and control purposes;
- a gas line with thc relative treatment systems;
- a tank where the gas is stored before use.

The anaerobic digester made of plastic, has a working capacity of about 100 L. This is fitted with a sight window for viewing the content of the tank. Also, the digestor is equipped with temperature sensors, arranged lengthwise, that make it possible to asses the temperature variations. Moreover, inside there is a temperature control probe and a pH control probe. Mixing is obtained by circulating the content of the bioreactor through an external circuit by means of a pump.

The stainless steel tank has a working volume of 200 L and it is used to prepare the material to be fed into the digestor. This is equipped with a fixed-speed stirrer (100 rpm) and a level indicator.

2.3. Experimental set-up

The substrate subjected to the anaerobic fermentation mixed with 150 L of tap water and the substances in quantities listed in Table 1, were introduced into the stainless steel tank and for one hour the composition was stirred for one hour with 100 rpm using the stirrer. The concentration of substrate consisting of animal manure and fresh residual biomass is listed in Table 2.

In all the experiments, during the anaerobic fermentation process, the temperature was set in the mesophilic conditions (35°C) and the pH value was set at 7. After stirring the mass and having defined the operating conditions (pH and temperature) from the console, the mass is transfered into the anaerobic digester by means of a pump.

All the experiments, including the biomass preparation and the pilot start-up were carried out for a 7 days period.

The experiments were performed with and without glucose addition.

Table 2: Concentration of substrate used in anaerobic digestion

Experiment	Concentration of animal manure (% w/w)	Concentration of fresh residual biomass (% w/w)	Glucose
1	4	2	0
2	4	2	2%

3. RESULTS AND DISCUSSION

The biogas production obtained from mixture of animal manure and residual biomass with and without glucose addition was measured after 7 days of digestion using a gas chromatography device. Gas chromatography is an optimal analytical instrument for the analysis of components such as CH_4, CO_2, H_2S and siloxanes which are present in the gas.

In the table 3 and 4, are listed the concentrations of different components which are presented in the obtained gas, as a function of the used substrate. The data obtained from the experiments in which the substrate was mixed with glucose showed an improved biodegradability and biogas production compared with the substrate without glucose addition.

The graph in Figure 1 shows the main parameters taken into consideration in the anaerobic fermentation process and biogas production at different time intervals. In this section we present a selection of the graphs which we consider to be most representative for the results obtained in our experiments.

These results indicated that co-digestion with suitable animal manure, fresh residual biomass and glucose mixture is an effective way to improve biogas yield.

The chemical characterization of substrates used in the co-digestion experiment can be observed in Table 3.

Table 3: The chemical characterization of substrates, [15]

	Animal manure	Fresh residual biomass
pH	6,75	5,96
TSS%	25 - 35	75 - 85

VS%	75 -80	90 - 95
C/N	10 -25	80 - 90

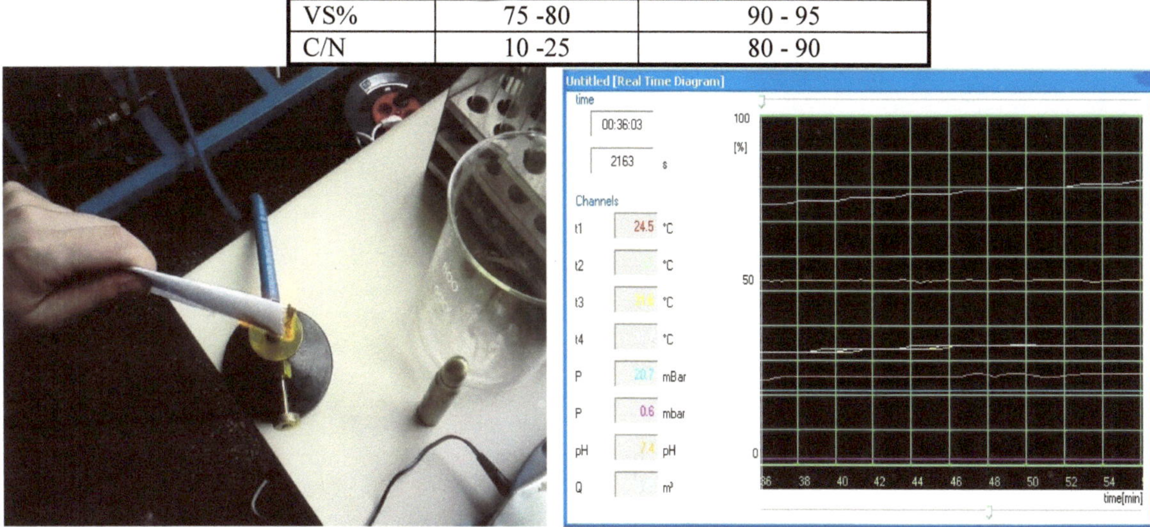

Figure 1: Biogas production at different time intervals

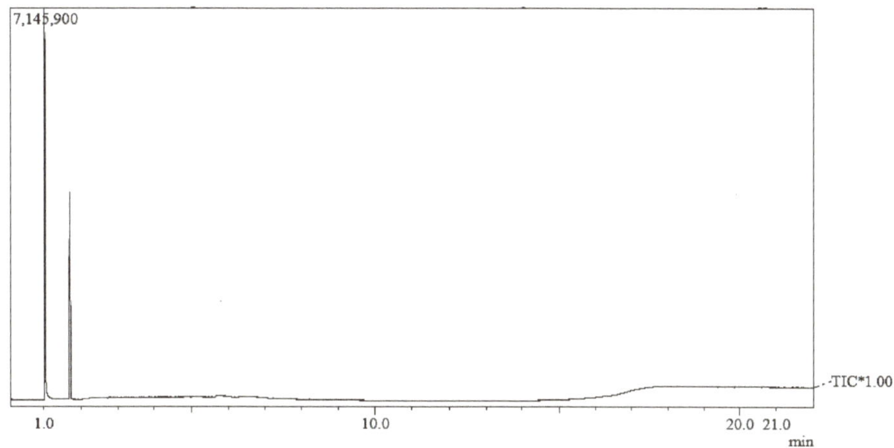

Figure 2: Gas chromatography diagram for the biogas obtained from the mixture without glucose addition

Table 4: The concentration of the biogas components

	Concentration (% v/v)	Proportion of substrate (% w/w)	Glucose addition
CO_2	38.537	animal manure 4 fresh biomass 2	0
CH_4	0.021		
CO	0.913		
N_2	42.723		
O_2	17.806		

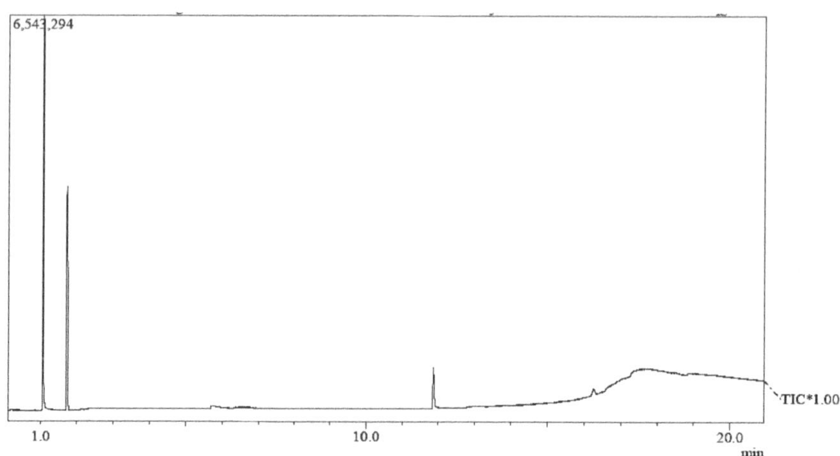

Figure 3: Gas chromatography diagram for the biogas obtained from the mixture with glucose addition

Table 5: The concentration of the biogas components

	Concentration (% v/v)	Proportion of substrate (% w/w)	Glucose addition
CO_2	43.030	animal manure 4 fresh biomass 2	2%
CH_4	0.024		
CO	0.001		
N_2	39.209		
O_2	16.046		

4. CONCLUSION

Biogas production by anaerobic fermentation is considered to be the optimal treatment for agricultural waste, manure and for a wide variety of organic waste, because these substrates are converted into renewable energy and organic fertilizer for agriculture. Moreover, anaerobic digestion is a most cost-effective bioconversion technology that has been implemented worldwide for commercial production of electricity, heat and compressed natural gas from organic material.

This study presented arguments for optimizing the anaerobic digestion process by glucose addition in the used substrate. Anaerobic digestion process efficiency is highly dependent on the type of used substrate.

The anaerobic co-digestion of animal manure, fresh biomass and glucose mixture, is a promising way for improving biogas production. Our results showed that the anaerobic co-digestion of the mixture mentioned above were efficient and produced more cumulative biogas compared with the mixture of substrate without glucose addition.

The future development of biogas from manure co-digestion includes the use of new feedstock types such as by-products from food processing industries, bio-slurries from biofuels processing industries as well as the biological degradation of toxic organic wastes from pharmaceutical industries.

ACKNOWLEDGEMENT

The work has been funded by the Sectoral Operational Programme Human Resources Development 2007-2013 of the Ministry of European Funds through the Financial Agreement POSDRU/159/1.5/S/134398.

REFERENCES

1. Ahring, B.K.., Perspectives for anaerobic digestion, Advances in Biochemical Engineering/Biotechnology 81, 2003, p. 1 – 30.

2. Vergara – Fernandez A., Vargas G., Alarcon N., Velasco A., Evaluation of marine algae as a source of biogas in a two-stage anaerobic reactor system, Biomass Bioenergy, 32, 2008, 338–344.
3. Weiland P., Biogas production: current state and perspectives, Appl Microbiol Biotechnol, 85, 2010, 849-860.
4. Abatzoglou N., BoivinS., A review of biogas purification processes, Biofuels, Bioprod. Bioref. 3, 2009, 42–7.
5. Cirne D.G., Lehtomaki A., Bjornsson L., Blackall LL., Hydrolysis and microbial community analyses in two-stage anaerobic digestion of energy crops, J Appl Microbiol, 103, 2007, 516–527.
6. Li Y., Park S., Zhu J., Solid – state anaerobic digestion for methane production from organic waste, Renewable and Sustainable Energy Reviews, 15, 2011, 821 – 826.
7. Ritari J., Koskinen K., Hultman J., Kurola J.M., Kymalainen M., Romantschuk M., Paulin L., Auvinen P., Molecular analysis of meso- and thermophilic microbiota associated with anaerobic biowaste degradation, BMC Microbiol. 12, 2012, 121.
8. Veeken A., Hamelers B., Effect of temperature on hydrolysis rates of selected biowaste components, BioresourTechnol, 69, 1999, 249–254.
9. Griffin M.E., McMahon K.D., Mackie R.I., Raskin L., Methanogenic population dynamics during start-up of anaerobic digesters treating municipal solid waste and biosolids, Biotechnol Bioeng, 57, 1998, 342-355.
10. Chen Y., Cheng J.J., Creamer K.S., Inhibition of anaerobic digestion process: a review. Bioresour Technol, 99 (10), 2008, 4044- 4064.
11. El-Mashad H.M., Zhang R, Biogas production from co-digestion of dairy manure and food waste, Bioresour Technol 101, 2010, 4021–4028.
12. Mata-Alvarez J., Dosta J., Mace S., Astals S., Codigestion of solid wastes: a review of its uses and perspectives including modelling, Crit. Rev. Biotechnol, 31 (2), 2011, 99–111.
13. Astals S., Nolla-Ardevol V., Mata-Alvarez J., Anaerobic co-digestion of pig manure and crude glycerol at mesophilic conditions: Biogas and digestate, Bioresource Technology 110, 2012, 63–70.
14. Bulkowska K.., Pokoj T., Klimiuk E., Gusiatin Z.M., Optimization of anaerobic digestion of a mixture of Zea mays and Miscanthus sacchariflorus silages with various pig manure dosages, Bioresource Technology 125, 2012, 208–216.
15. Tong Zhang, Linlin Liu, Zilin Song, Guangxin Ren, Yongzhong Feng, Xinhui Han, Gaihe Yang, Biogas production by co-digestion of goat manure with three crop residues, LoS ONE 8(6), 2013, e66845, doi:10.1371/journal.pone.0066845.
16. Didacta Italia, Re-Biomas, Pilot plant for the production of biogas from biomass, User's manual and exercise guide, Edition 01, 20008.

ON RESPONSE OF RANDOM VIBRATION FOR NONLINEAR SYSTEMS

Petre Stan[1], Marinică Stan [2]

[1]University of Pitesti, Romania, email: stan_mrn@yahoo.com

[2] University of Pitesti, Romania, email: petre_stan_marian@yahoo.com

Abstract: *In his paper, a nonlinear system under external excitation of white noise was replaced by an equivalent nonhysteretiv nonlinear system subject to the same excitation for the probality density of the stationary response was known. The basic idea of the method is proposed to obtain the approximate power spectral density for the stationary response. In general, and especially in random vibration analysis, it is difficult to obtain a closed form solution for dynamic response of a nonlinear system.*

Keywords: *Random oscillations, nonlinear damping, natural frequency,equivalent nonlinear system*

1. INTRODUCTION

A nonlinear system under external excitation of white noise was replaced by an equivalent nonhysteretic nonlinear system subject to the same excitation for the probality density of the stationary response was known. Nonlinear dynamic systems subject to random excitations are frequently met in engineering practice. Random differential equations appear in several different applications: study of random evolution of system with a spatial extension, study of stochastic models where the state variable is infinite dimensional, for example, a curve or surface. In this paper, a technique is proposed in order to evaluate the probality density function of the solution, based on the combination of the probalistic transformation methods.

2. SYSTEM MODEL

A nonlinear system will be considered with response $\eta(t)$ to an excitation $w(t)$ described by second differential equation

$$\ddot{\eta}(t)+2\xi p\dot{\eta}(t)+\alpha p^2\eta^5(t)=w(t), \tag{1}$$

where $\eta(t)$ is the displacement response of the system, c is the viscous damping coefficient, ξ is the critical damping factor, for the linear system, α is the nonlinear factor to control the type and degree of nonlinearity in the system.

The mechanical energy of the system is

$$H(\eta)=\int_0^\eta h(v)\,dv=\alpha\int_0^\eta p^2\eta^5(v)dv=\alpha p^2\eta^6, \qquad (2$$

The potential energy of the system [1,2,3] is

$$E_m(\eta,\dot{\eta})=\frac{1}{2}\dot{\eta}^2+\alpha p^2\eta^6. \tag{3}$$

Obtain for the mechanical energy

$$2E_m=\dot{\eta}^2+2\alpha p^2\eta^6, \tag{4}$$

$$\omega=\frac{d\varphi}{dt}=\frac{du_1}{d\eta}=\frac{1}{\sqrt{2H(\eta)}}H'(\eta)=\frac{\sqrt{2\alpha}}{2\alpha p\eta} \tag{5}$$

We have

$$\dot{u}_1 = \omega u_2 \tag{6}$$

and equation of motion becomes

$$\dot{u}_2 + 2\xi p u_2 + \omega u_1 = w(t). \tag{7}$$

Obtain

$$\ddot{u}_2 + 2\xi p \dot{u}_2 + \omega_1 \dot{u}_2 = w(t). \tag{8}$$

These two first order differential equation are conveniently combined into the matrix format

$$\frac{d}{dt}\begin{bmatrix} u_1 \\ u_2 \end{bmatrix} = \begin{bmatrix} 0 & \omega \\ -\omega & 2\xi p \end{bmatrix}\begin{bmatrix} u_1 \\ u_2 \end{bmatrix} + \begin{bmatrix} 0 \\ w(t) \end{bmatrix}. \tag{9}$$

and this format is similar to that of a linear oscillator
The energy relation is written in the form

$$\frac{d\eta}{dt} = \sqrt{2(E_m - H(\eta)}, \quad (10$$

and the period for the system is

$$T(E_m) = \int_0^T dt, \tag{11}$$

or

$$T(E_m) = 2\int_{\eta_{\min}}^{\eta_{\max}} \frac{d\eta}{\sqrt{2(E_m - H(\eta))}}. \tag{12}$$

The integral vanishes in the lower limit and therefore differential with respect to the energylevel gives

$$T(E_m) = 2\int_{\eta_{\min}}^{\eta_{\max}} \frac{d\eta}{\sqrt{2\{E_m - \alpha p^2 \eta^6\}}}, \tag{13}$$

The function $h(\eta) = \alpha p^2 \eta^5$ is odd function, $h(-\eta) = -h(\eta)$, , and the point $\eta = 0$ is an equilibrium position, so $h(0) = 0$. We consider the initial conditions $t = 0, \eta = 0, \dot{\eta} = 0$.

If the expression (12) we do $\eta_{\min} = 0$, $\eta_{\max} = A$ is get the time to walk the distance MO and how the function $h(\eta)$ is symmetric, that this time period is a quarter. There was thus obtained between

$$T(E_m) = \frac{4\eta}{\sqrt{2E_m}} \eta \Big|_0^A + \frac{\omega_0^2 \sqrt{2E_m}}{3E_m^2} \eta^3 \Big|_0^A + \frac{\omega_0^2 \beta^2}{191 E_m^4} \eta^4 \Big|_0^A, \tag{14}$$

relationship that writes

$$T(E_m) = \frac{4A}{\sqrt{2E_m}} + \frac{\omega_0^2 A^3 \sqrt{2E_m}}{3E_m^2} + \frac{\omega_0^2 A^4 \beta^2}{191 E_m^4} \tag{15}$$

Therefore, if the period nonlinear vibration depends on the initial conditions.
If differentiating the polar representation [2,3] in the (4) and the (6) in relation to the time, we obtain the system of equations

$$\dot{u}_1 = \frac{\dot{E}_m}{\sqrt{2E_m}} \sin\varphi + \dot{\varphi}\sqrt{2E_m}\cos\varphi, \tag{16}$$

$$\dot{u}_2 = \frac{\dot{E}_m}{\sqrt{2E_m}} \cos\varphi - \dot{\varphi}\sqrt{2E_m}\sin\varphi, \tag{17}$$

Obtain

$$\frac{\dot{E}_m}{\sqrt{2E_m}} = \sin\varphi \dot{u}_1 + \cos\varphi \dot{u}_2, \tag{18}$$

$$\dot{\varphi}\sqrt{2E_m} = \cos\varphi \dot{u}_1 - \sin\varphi \dot{u}_2 , \tag{19}$$

or

$$\dot{E}_m = -4E_m \xi p \cos^2\varphi + \sqrt{2E_m}\, w(t)\cos\varphi \tag{20}$$

$$\dot{\varphi} = \omega + 2\xi p \cos\varphi \sin\varphi - \frac{1}{\sqrt{2E_m}} w(t)\sin\varphi \tag{21}$$

The equations Fookker-Plank-Kolmogorov to determine the probability density [2,3] are

$$\frac{\partial p}{\partial t} = \frac{\partial}{\partial z_j}(-\alpha_j p + \frac{1}{2}\frac{\partial}{\partial z_k}(\beta_{jk} p)) . \tag{22}$$

The differential equations (20) can be placed by the approximation to a set of diffusion equations [1,2,3] in the form:

$$\frac{d}{dt} Z_j(t) = a_j(Z_k, t) + b_j(Z_k, t) y(t), \quad j = 1,2 \tag{23}$$

where

$$Z_1 = E_m \text{ şi } Z_2 = \phi . \tag{24}$$

For a broad-band excitation, we obtain Ito's equations [3,4] in the form

$$dZ_j = \alpha_j(Z_k, t)dt + \sigma_j(Z_k, t)d\tilde{W} , \tag{25}$$

The Fokker-Planck-Kolmogorov equations associated with equations (25) are

$$\frac{\partial}{\partial E}\left[-\alpha_{E_m} p_{E_m,\varphi} + \frac{1}{2}\frac{\partial}{\partial E_m}(\beta_{E_m,E_m} p_{E_m,\varphi}) + \frac{1}{2}\frac{\partial}{\partial \varphi}(\beta_{E_m,\varphi} p_{E_m,\varphi})\right] + \\ + \frac{\partial}{\partial \varphi}\left[-\alpha_{\varphi} p_{E_m,\varphi} + \frac{1}{2}\frac{\partial}{\partial E_m}(\beta_{\varphi,E_m} p_{E_m,E_m}) + \frac{1}{2}\frac{\partial}{\partial \varphi}(\beta_{\varphi,\varphi} p_{E_m,\varphi})\right] = 0. \tag{26}$$

3. THE PROBALITY DENSITY FUNCTION

The probality density function of simultaneous values of the response of a system of the form (1) to ideal white noise excitation satisfies a two dimensional Fokker-Planck-Kolmogorov equations, partial differential equation

$$p_{E_m}(E_m) = \frac{CT(E_m)}{\pi S_0} e^{-\int_0^{E_m} \frac{f_{eq}(r)dr}{\pi S_0}} , \tag{27}$$

where

$$f_{eq}(E_m) = \frac{\int_0^T 2\xi p dt}{\int_0^T \dot{\varphi}^2 dt} . \tag{28}$$

Introducting equation (27) into (26), obtain the probability density of response is

$$p_{E_m}(E_m) = \frac{CT(E_m)}{\pi S_0} e^{-\frac{2\xi\omega_0}{\pi S_0}\left(E_m + \frac{3}{4}\alpha E_m^2\right)} , \tag{29}$$

where C is the constant of normalization.

The probability density of response becomes

$$p_{E_m}\left(E_m\right)=\frac{C\left(\frac{4A}{\sqrt{2E_m}}+\frac{\omega_0{}^2 A^3\sqrt{2E_m}}{3E_m{}^2}+\frac{\omega_0{}^2 A^4\beta^2}{191E_m{}^4}\right)}{\pi S_0}e^{-\frac{2\xi\omega_0}{\pi S_0}\left(E_m+\frac{3}{4}\alpha E_m{}^2\right)}. \tag{30}$$

4. THE POWER SPECTRAL DENSITY OF THE RESPONSE FOR THE SYSTEM

The operator $s(E_m)$ is determined [3,4,] by the expression

$$s(E_m)=\frac{<\dot{x}^2>_t}{E_m}=\frac{1}{T(E_m)}\frac{1}{E_m}\int_0^{E_m}T(r)dr\,, \tag{31}$$

in which, if the period taken to introduce the expression (11)

$$s(E_m)=s_{E_m}=\frac{1}{E_m}\frac{4A\sqrt{2E_m}-\frac{2\sqrt{2}}{3}\frac{\omega_0{}^2A^2}{\sqrt{E_m}}-\frac{\omega_0{}^2A^4\beta^2}{573E_m{}^3}}{2A\sqrt{2E_m}+\frac{\omega_0{}^2A^3\sqrt{2E_m}}{3E_m{}^2}+\frac{\omega_0{}^2A^4\beta^2}{191E_m{}^4}}. \tag{32}$$

The power spectral density of response [4,5,6] is

$$S_\eta(\omega)=\frac{1}{\pi}\int_0^\infty\frac{s_{E_m}{}^2 f_{E_m}E_m}{(\frac{s_{E_m}{}^2 f_{E_m}{}^2}{4\xi^2}-\omega^2)^2+s_{E_m}{}^2 f_{E_m}{}^2\omega^2}p_{E_m}(E_m)dE_m, \tag{33}$$

or

$$S_\eta(\omega)=\frac{C}{\pi}\int_0^\infty\frac{s_{E_m}{}^2 f_{E_m}E_mT\left(E_m\right)}{(\frac{s_{E_m}{}^2 f_{E_m}{}^2}{4\xi^2}-\omega^2)^2+s_{E_m}{}^2 f_{E_m}{}^2\omega^2}\frac{1}{\pi S_0}e^{-\frac{2\xi\omega_0}{\pi S_0}\left(E_m+\frac{3}{4}\alpha E_m{}^2\right)}dE_m\,. \tag{34}$$

For the linear case we have

$$s(E_m)=\frac{<\dot{x}^2>_t}{E_m}=\frac{1}{T(E_m)}\frac{1}{E_m}\int_0^{E_m}T(r)dr=\frac{p}{2\pi E_m}\int_0^{E_m}\frac{2\pi}{p}dr=1\,. \tag{35}$$

Because nonlinearity factor is $\alpha=1$,

$$f_{eq}\left(E_m\right)=2\xi p \tag{36}$$

and the probability density of response is

$$p_{E_m}\left(E_m\right)=\frac{2C}{pS_0}e^{-\frac{2\xi p}{\pi S_0}E_m}. \tag{37}$$

Introducing all these parameters in response spectral density expression given by (33.), we get

$$S_\eta(\omega)=\frac{S_0}{(p^2-\omega^2)^2+(2\xi p)^2\omega^2}, \tag{38}$$

that formula known in the linear case.

5. NUMERICAL RESULTS

In this example, $m = 1kg$, $k = 36\frac{N}{m}$ $c = 4\frac{Ns}{m}$, $\alpha = 3m^{-2}$.

Obtain:

$$p = \sqrt{\frac{k}{m}} = 6s^{-1},\ \frac{c}{m} = 2\xi p \Rightarrow \xi = 0{,}33. \qquad (39)$$

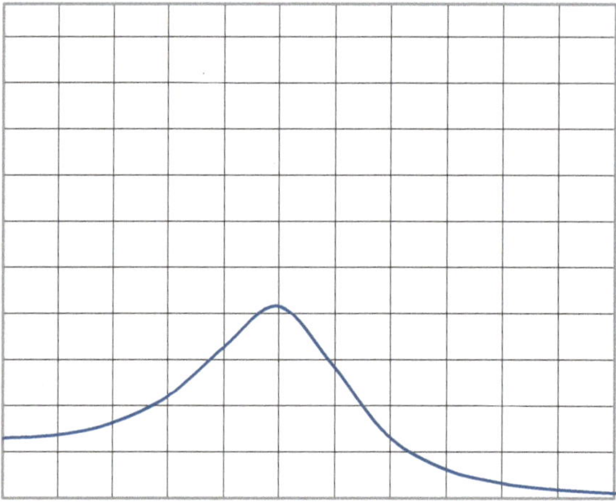

$\omega / 2\pi$

Figure 1. The power spectral density of the response $S_\eta(\omega)$ for $\xi = 0{,}33$, $\alpha = 3m^{-2}$.

6. CONCLUSION

In this article, we are analyzing a differential equation with random variable.Our new technique based on the combination of the transformation method with numerical method to evaluate the probability density function and the power spectral density of the response of the systems. The avantage of the method is that it yelds the approximate probality density for the stationary response of nonlinear stochastic systems rather than just a few statistical moments and that is may be applicable to nonlinear stochastic systems.

REFERENCES

1. Cai, G. Q., A new approximate solution technique for randomly excited nonlinear oscillator, International Jurnal of non-linear mechanics, 1988.
2. Zhu, W. Q., The equivalent non-linear system method, Jurnal of sound and vibration, 1989.
3. Zhu, W. Q., Equivalent non-linear system method for stochastically excited and dissipated integrable Hamiltonian systems, Journal of applied mechanics, 1997.
4. Zhu, W. Q., Stochastic averaging method in random vibration, Bulletin S.F.M, 5(1988)
5. Zhu, W. Q., Roberts, J.B., Spanos, P.D., Stochastic averaging:an approximate method of solving random vibration problems, Int. J. Non-Linear Mech. ,21(1986).
6. Pandrea, N., Parlac S.,-Mechanical vibrations, Pitesti University, 2000.

THE ABSOLUTE STABILIZATION OF THE SHIPS COURSE IN THE CASE OF ROLLING OSCILATIONS

Mircea Lupu[1], Olivia Ana Florea[2], Ciprian Lupu[3]

[1] Faculty of Mathematics and Computer Sciences, Transilvania University of Brasov, Romania, corresponding member of A.O.S.R., m.lupu@unitbv.ro

[2] Faculty of Mathematics and Computer Science, Transilvania University of Brasov, olivia.florea@unitbv.ro

[3] Faculty of Automatics and Computers, Politehnica University of Bucharest, Romania, cip@indinf.pub.ro.

Abstract: *In the first part of this paper it is presented the automatic regulation methods for the absolute stability (a.r.a.s) of the nonlinear dynamical systems that have applications on the stabilization of the rolling oscillations curse for aircraft or rackets. Two methods for the absolute stability are specified: a) the A.I. Lurie method with the effective determination of the Liapunov function; b) the frequency method of the Romanian researcher V.M. Popov who has used the transfer function in the critical cases. The authors develop a new sufficient criterion for (a.r.a.s.), with efficient technique of calculus.*

In the second part - there are obtained the analytic - numerical solutions and the conditions for the regulator parameters for the realization of the absolute stability for the airplane autopilot route in the case of rolling oscillations. At the end authors prove practically the theorem Kalman - Yakubovich - Popov for the equivalence of these methods. (Th. K-Y-P). In the last section of the paper it is presented the optimal control for the flight system in the case of rolling oscillations. The optimization is made using the maximum principle of Pontreaguine; the authors solve the problem of minimum time. It is determined the command function and the optimal trajectory for this system

Keywords: *nonlinear systems, automatic control system, absolute stability, optimal control., Pontreaguine principle, rolling perturbation flight.*

1. INTRODUCTION

The automatic regulation for the stability of dynamical systems occupies a fundamental position in science and technique, following the optimization of the technological process of the cutting tools, of the robots, of the movement vehicles regime or of some machines components, of energetic radioactive regimes, chemical, electromagnetic, thermal, hydro-aerodynamic regimes, etc.

The studies and the technical achievements are complex by mathematical models for closed circuits with input - output, following for the automatic regulation the integration of some mechanisms and devices with inverse reaction of response for the control and the fast and efficient elimination of the perturbations which can appears along these processes or dynamical regimes. Generally these dynamical regimes are nonlinear and it was necessarily some contributions and special achievements for automatic regulation, generating the automatic regulation of absolute stability (a.r.a.s.) for these classes of nonlinearities.

We highlight two special methods (a.r.a.s.): • Liapunov's function method discovered by A.I. Lurie [13,15,20] and developed into a series of studies by M.A Aizerman, V.A. Iakubovici, F.R. Gantmaher, R.E. Kalman, D.R. Merkin [14] and others [1,17].

• Frequency method developed by researcher VM Popov [18] generalizing the criterion of Nyquist, then developed in many studies [1,2,15].

We note the contributions of Romanian researchers recognized by the works and monographs on the stability and optimal control theory: C. Corduneanu, A. Halanay, V. Barbu, Th. Morozan, G. Dinca, M. Megan, Vl. Rasvan, V. Ionescu, M.E. Popescu, S. Chiriacescu, A. Georgescu and also who studied directly on (a.r.a.s.): I. Dumitrache [4] D. Popescu [16], C. Belea [2], V. Rasvan [19], S. Chiriacescu [3] and other recent works [6,...,12].

The research has shown that both methods are equivalent, and studies can be qualitatively or numerically. In this paper we presented the actual making methods in cases of singularity studies across applications.

2. (A.R.A.S.) USING THE LIAPUNOV'S FUNCTION METHOD

In this part we'll present the Lurie's ideas and the effective method for found the Liapunov's function [13,14,2,19]. Generally, the systems of automatic regulation are composed from the controlled processor system, and sensory elements of measurement, acquisition board, and the mechanism feedback controller. The regulator will mean all the sensors and the acquisition board, but the controller is included feedback mechanism. Parameters characterizing the object control system to control work mode are measured by sensors, and their records with the sensor response mechanism ζ is transmitted acquisition board. This processes the command σ, which is mechanically transmitted to the controller which, on its turn, distributes the object state and interact simultaneously adjusting the response mechanism. We highlight the dynamic system equations. We note by $x_1, x_2, \ldots, x_n$ the state parameters of the regime's subject which it must controlled, the coordinates and the sensorial speeds. We rename that the variation of these parameters if the open circuit (excluding the controller) system described by linear differential equations with constant coefficients: $\dot{x}_k = \sum_{j=1}^{n} a_{kj} x_j, k = 1, \ldots, n$. If the system is with closed loop then on the variables $x_1, x_2, \ldots, x_n$ will influence the regulation body, and we note by ξ its state. In this case for the autonomous closed system we have the equations:

$$\dot{x}_k = \sum_{j=1}^{n} a_{kj} x_j + b_k \xi, k = 1, \ldots, n \tag{1}$$

We'll consider that the mechanism or inverse reaction is determine on the output ζ with the rigidity connection on the input ξ :

$$\zeta = k\xi \tag{2}$$

The acquisition board collects the signals and transmits the input sensors in order to obtain the embedded system:

$$\sigma = \sum_{j=1}^{n} c_j x_j - r\xi \tag{3}$$

where c_j, r are transfer numbers, r is the transfer coefficient of the inverse rigid connection, $r > 0$ (the regulator characteristics) [13,14,15]. The connection between the output function σ (linear) of the controller and the nonlinear input φ in the case of automatic regulation is express by the relation:

$$\dot{\xi} = \varphi(\sigma) \tag{4}$$

The characteristic function of the controller $\varphi(\sigma)$, $\sigma \in (-\infty, +\infty)$ is continuous and verify the conditions [14,6,7]:

$$\begin{aligned} &a) \quad \varphi(0) = 0 \\ &b) \quad \sigma \cdot \varphi(\sigma) > 0, \quad \forall \sigma \neq 0 \\ &c) \quad \int_0^{\pm\infty} \varphi(\sigma) d\sigma = \infty \end{aligned} \tag{5}$$

Observe that $\varphi = \varphi(\sigma)$ is ascending in the quarters I, III where is graphically. The functions $\varphi(\sigma)$ are named admissible, and is verified the sector condition:

$$0 < \frac{\varphi(\sigma)}{\sigma} < k \tag{6}$$

where k is the amplification coefficient.

Example 1.

- $\varphi(\sigma) = sgn(\sigma) \cdot \ln(\sigma^2 + 1), k > 1$; • $\varphi(\sigma) = a(e^{\sigma} - 1), k \leq a$

The equations (1), (3), (4) model the perturbed system with the zeros $x(0,0,\ldots,0), \xi = 0$. Using the nonsingular square matrix $A = \|a_{kj}\|$ of degree $n > 1$, $B = \begin{pmatrix} b_1 \\ \vdots \\ b_n \end{pmatrix}, C = (c_1 \ \ldots \ c_n)$, C' the transpose matrix of C, this system can be:

$$\dot{X} = AX + B\xi, \quad \dot{\xi} = \varphi(\sigma), \quad \sigma = C'X - r\xi, \quad X = \begin{pmatrix} x_1 \\ \vdots \\ x_n \end{pmatrix} \tag{7}$$

Observation 1. It is known that for the linear system $\dot{X} = AX$, the second method of Liapunov for the null solution stability consists in determine a Liapunov function $V = V(x)$ fulfilled the regularity conditions associated of this system [1,20]. A simple technique is to search V like square form positive defined $V = X'PX$ and $\dot{V} = X'(A'P + PA)X$ associated of the autonomous system where $V(0) = 0, \dot{V}(0) = 0$. For the simple or asymptotic stability in the vicinity of the null solution must have negative sign (or negative defined). It must:

$$A'P + PA = -Q \tag{*}$$

Where the matrix $P, Q \in \mathbf{R}_{n \times n}$ Q are symmetrically and positives. So, practically it is choose Q randomize fixed and is determined the matrix P from the equation (*) with A nonsingular.
Bringing the system (7) to the canonical form and determine the Liapunov function:
Suppose that A with $\det A = \Delta_0 \neq 0$ is Hurwitz, that mean the characteristic polynomial $P(\lambda)$ has simple roots with $Re(\lambda_k) < 0, k = 1, \ldots, n$

$$P(\lambda) = (-1)^n \det(A - \lambda E) = 0 \tag{8}$$

The system (7) is bring to the canonical form if the matrix A is bring to the Jordan form $J = diagA = \begin{pmatrix} \lambda_1 & & 0 \\ & \ddots & \\ 0 & & \lambda_n \end{pmatrix}$. It is determine a non degenerate matrix $T = (t_{kj})$ for the diagonalization of matrix A with the relation:

$$T^{-1}AT = J, \quad AT = TJ, \quad \det T \neq 0 \tag{9}$$

We make the linear transform:

$$X = TY, Y = \begin{pmatrix} y_1 \\ \vdots \\ y_n \end{pmatrix} \tag{10}$$

Obtaining from (7):

$T\dot{Y} = ATY + B\xi, \quad \dot{\xi} = \varphi(\sigma), \sigma = C'TY - r\xi$ that mean:

$$\begin{gathered} \dot{Y} = JY + B_1\xi, \quad \dot{\xi} = \varphi(\sigma), \\ \sigma = C_1'Y - r\xi, B_1 = T^{-1}B, C_1' = C'T \end{gathered} \tag{11}$$

Reducing the system (1) with the linear transform:

$$Z = JY + B_1\xi, \sigma = C_1'Y - r\xi, Z = \begin{pmatrix} z_1 \\ \vdots \\ z_n \end{pmatrix} \tag{12}$$

$$\begin{cases} \dot{Z} = JZ + B_1\varphi(\sigma) \\ \dot{\sigma} = C_1'Z - r\varphi(\sigma) \end{cases} \tag{13}$$

The disturbed system (13) with the equilibrium solution ($z_k = 0, \sigma = 0$) will be equivalent with the system (7) with the equilibrium solution $(x_k = 0, \xi = 0)$ and the transform (12) will be non degenerate if the determinant of the system (13) is non null.

$$\Delta = \begin{vmatrix} J & B_1 \\ C_1' & -r \end{vmatrix} \neq 0, r + C_1'J^{-1}B_1 \neq 0 \tag{14}$$

Retuning to $J^{-1} = T^{-1}AB, B_1 = T^{-1}B, C_1' = C'T$ transforms we obtain from (14) the final condition:

$$r + C'A^{-1}B \neq 0 \tag{15}$$

The Lurie's problem consists in calculus the asymptotic stability conditions of the (7) equivalent with (13) with the null solution respectively $(x_k = 0, \xi = 0)$, $(z_k = 0, \sigma = 0)$ for the initial perturbations and for any admissible functions $\varphi(\sigma)$ defined in (5), (6). This type of stability where the systems (7), (13) have a linear part which is the

A and a non linear part which is $\varphi(\sigma)$ is named the absolute stability (a.s), [1,16] It is observe that if $\varphi(\sigma)$ is linear, than the systems are linearized being asymptotic stable. The simplicity of system (13) entails immediate techniques for determining the Liapunov function $V=V(z_1,\ldots,z_n,\sigma)$ attach to the system (13). The function $V(z,\sigma)$ of class C^1 is Liapunov from the system (13) if $V(z=0,\sigma=0)=0$ and is positive defined $V(z,\sigma)>0$ radial unlimited to ∞ , with the absolute derivative $\dot{V}=\frac{dV}{dt}$ $\dot{V}(0,0)=0$ and $\dot{V}$ negative defined $\frac{dV}{dt}<0$ for $(z\neq 0,\sigma\neq 0)$ in vicinity of the equilibrium point for have than absolute stability. Here, for the case of automatic regulation we choose $V,\dot{V}$ have the special form which verify these conditions. So we search the function $V=V(z,\sigma)$ compose by a square form z_k corresponding to the linear block A and an integral term corresponding to the non linear part.

$$V(z,\sigma)=Z'PZ+\int_0^{\sigma}\varphi(\sigma)d\sigma=V_1(z,\sigma)+\int_0^{\sigma}\varphi(\sigma)d\sigma \tag{16}$$

From theory [1,4] $Z'PZ$ is the square form defined strictly positive if the matrix P is symmetric $(P=P')$ and we have $A'P+PA=-Q$ where Q is symmetric and positive (with the eigenvalues positive). The integral term from (16) is strictly positive from the conditions (5) with $\sigma\neq 0$ and $V(z=0,\sigma=0)=0$. Next are verify the regularity conditions with $\dot{V}$ attach to (13) and with (15) will obtain the conditions for parameters c_k,r to obtain (a.r.a.s.). From (16) using (13) and: $Q=Q',P=P',B_1'PZ+Z'PB_1=B_1'PZ+(PB_1)'Z=2(PB_1)'Z$, for:

$$\frac{dV(z,\sigma)}{dt}=Z'(J'P+PJ)Z-r\varphi^2(\sigma)+\varphi(\sigma)(B_1'PZ+Z'PB_1)+\varphi(\sigma)C_1$$

we obtain:

$$\frac{dV}{dt}=-Z'QZ-r\varphi^2(\sigma)+2\varphi(\sigma)\left(PB_1+\frac{1}{2}C_1\right)Z;\dot{V}(z=0,\sigma=0)=0 \tag{17}$$

It can be see the connection from the matrix components $P(p_{ij}),Q(q_{ij})$ from $\lambda_i+\lambda_j\neq 0,i,j=1,\ldots,n,P=P',J=\text{diag}A$ than from $Q=Q'$ we have $q_{ij}=-(\lambda_i p_{ij}+\lambda_j p_{ij})$ that mean:

$$p_{ij}=-\frac{q_{ij}}{\lambda_i+\lambda_j} \tag{18}$$

Observation 2. The matrix A is stable with $\lambda_i+\lambda_j\neq 0$ if Q is a square form positive defined.

Example 2. If choose $Q=E$ the unit matrix and P obtain from (18) than the below observation is valid. Because $\dot{V}<0$ we prove that $(-\dot{V})$ is positive defined. Apply in (17) the Silvester criterion demanding that all diagonal minors of (17) to be positive. Because Q is positive like square form, than the first n inequalities are verify; it rest the last inequality from (17) after the square form in Z and which is:

$$r>\left(PB_1+\frac{1}{2}C_1\right)'Q^{-1}\left(PB_1+\frac{1}{2}C_1\right);\ Q=E,\sqrt{r}>\left\|PB_1+\frac{1}{2}C_1\right\| \tag{19}$$

If the regulator parameters verify the conditions (15), (19) there are sufficient conditions for the asymptotic stability of the system (1), (3), (4) for the solution $(x=0,\xi=0)$. [13,19,11].

Remark1. A choice technique of the square form $V_1(z)$ for p_{ij} according Lurie is:

$$V_1(z)=\varepsilon\sum_{k=1}^{s}z_{2k-1}z_{2k}+\frac{\varepsilon}{2}\sum_{k=1}^{n-2s}z_{2s+k}^2-\sum_{k=1}^{n}\sum_{j=1}^{n}\frac{a_k z_k a_j z_j}{\lambda_k+\lambda_j},\varepsilon>0$$

where $a_1,a_2,\ldots,a_{2s}$ are complex conjugated, $a_{2s+1},\ldots,a_n$ are real corresponding to roots λ_k determining the coefficients a_k .

Remark2.The two transforms for the diagonal system (1), (3), (4) to obtain (13) can be replacing directly with the transform [15]:

$$x_k = -\sum_{i=1}^{n} \frac{N_k(\lambda_i)}{D'(\lambda_i)} z_i \tag{20}$$

where from (7) it is obtained $P(\lambda) = (-1)^n D(\lambda), N_k(\lambda) = \sum_{i=1}^{n} b_i D_{ik}(\lambda)$, D_{ik} are the corresponding algebraic complements of (i,k) from $D(\lambda) = A - \lambda E$. In this case the simplified system analogous (13):

$$\dot{z}_k = \lambda_k z_k + \varphi(\sigma), \dot{\sigma} = \sum_{i=1}^{n} f_i z_i - r\varphi(\sigma), k = 1,...,n \tag{21}$$

for which we will build easier $V(z,\varphi)$.

Determining of $V(z,\varphi)$ with a new efficient method for (13) or (21)

Following the form of $V_1(z)$ we choose the function $V(z,\sigma)$ for (21).

$$V(z,\sigma) = \frac{1}{2}\sum_{j=1}^{n} A_j z_j^2 + F(\alpha_1 z_1, \alpha_2 z_2, ..., \alpha_n z_n) + \int_0^{\sigma} \varphi(\sigma) d\sigma \tag{22}$$

$$F(z_1, z_2, ..., z_n) = -\sum_{j.k=1}^{n} \frac{1}{\lambda_j + \lambda_k} z_j z_k, \lambda_{k<0} \tag{23}$$

where, $A_j > 0, \alpha_j \in \mathbf{R}$ will be determined. From:

$$-\frac{1}{\lambda_j + \lambda_k} = \int_0^{\infty} e^{(\lambda_j+\lambda_k)s} ds > 0 \qquad F(z_1, z_2, ..., z_n) = \int_0^{\infty} \sum_{j,k} z_j z_k e^{(\lambda_j+\lambda_k)s} ds = \int_0^{\infty} \left(\sum_{j=1}^{\infty} z_j e^{\lambda_j s} \right)^2 ds \geq 0$$

results that F is nullify just for $F(z_1 = 0, z_2 = 0, ..., z_n = 0) = 0$ and $\int_0^{\sigma} \varphi(\sigma) d\sigma > 0$.

So, $V(z,\sigma)$ has the positive sign defined and $V(z = 0, \sigma = 0) = 0$. Compute $\frac{dV}{dt}$ associate to the system (21) and it must be $(-\dot{V})$ of positive sign defined.

$$-\frac{dV}{dt} = -\sum_{j=1}^{n} A_j \lambda_j z_j^2 - 2\sum_{j,k=1}^{n} \frac{\lambda_j \alpha_j \alpha_k}{\lambda_j + \lambda_k} z_j z_k + r\varphi^2(\sigma) + \sum_{j=1}^{n} z_j \left[A_j + f_j - 2\alpha_j \sum_{k=1}^{n} \frac{\alpha_k}{\lambda_j + \lambda_k} \right] \varphi$$

From $2\sum_{j,k=1}^{n} \frac{\lambda_j \alpha_j \alpha_k}{\lambda_j + \lambda_k} z_j z_k = \left(\sum_{k=1}^{n} \alpha_k z_k \right)^2, r > 0, \lambda_j > 0$

We obtain the first three terms positives and must nullifying the coefficient of φ :

$$A_j + f_j - 2\alpha_j \sum_{k=1}^{n} \frac{\alpha_k}{\lambda_j + \lambda_k} = 0, j = 1..n \tag{24}$$

In this quadratic algebraic system (24) we can take $A_j = -\frac{1}{\lambda_j}$, and f_j, λ_j known, we determine the coefficients $\alpha_j, j = 1..n$ and other conditions from (19). If in (24) divide with λ_j and summing we obtain

$$\left(\sum_{j=1}^{n} \frac{\alpha_j}{\lambda_j} \right)^2 = -\sum_{j=1}^{n} \frac{A_j + f_j}{\lambda_j} \equiv \Gamma^2, \sum_{j=1}^{n} \frac{\alpha_j}{\lambda_j} = \pm\Gamma \tag{25}$$

So, must have $\sum_{j=1}^{n} \frac{A_j + f_j}{\lambda_j} < 0$, and the solution of the system (24) $(\alpha_1, \alpha_2, ..., \alpha_n)$ is in this hyper-plane (25).

For the case when a root is null $P(0)=0$ and the others have $Re(\lambda_k)<0, k=1,...,n-1$ than the system (13) with $Z=\begin{pmatrix}\tilde{z}\\ z_n\end{pmatrix}$ becomes:

$$\dot{\tilde{z}}=\tilde{J}\tilde{Z}+\tilde{B}_1\varphi, \dot{z}_n=b_0\varphi, \dot{\sigma}=\tilde{C}_1'\tilde{Z}+C_0z_n-r\varphi \tag{26}$$

where for $\tilde{z}$ we have the matrix $\tilde{Z}$ and $\tilde{J}$ of degree $(n-1)$, $\tilde{B}_1, \tilde{C}_1$ row, column matrix $(n-1,1),(1,n-1)$. In this case the Liapunov function search form:

$$V(\tilde{z},z_1,\sigma)=az_1^2+\left\{\tilde{z}'P\tilde{z}+\int_0^\sigma \varphi(\sigma)d\sigma\right\} \tag{27}$$

For proofs and recently applications we recommend the bibliography [2,15,14,11,12].

2.1. The frequency method for (a.r.a.s.)

This method obtained by V.M. Popov [18] is applied to the dynamical system with continuous nonlinearity. We present in this section the method with criterions given by Aizerman, Kalman, Jakubovici [19,14]. Let be the dynamical, autonomous, non homogeneous system:

$$\dot{x}_i=\sum_{l=1}^{n}a_{il}x_l+b_iu, i=1,...,n; \dot{x}=\frac{dx}{dt}; \sigma=\sum_{l=1}^{n}c_lx_l, u=-\varphi(\sigma) \tag{28}$$

where a_{il}, b_i, c_l are real constants, u is the arbitrary function of input , continuous, nonlinear with $\varphi(\sigma)$ and σ is the output function. Using the Laplace transform, replacing the operator $\frac{d}{dt}$ with s we obtain from (2):

$$sx_i=\sum_{l=1}^{n}a_{il}x_l+b_iu, \sigma=\sum_{l=1}^{n}c_lx_l, i=1,...,n \tag{29}$$

Eliminating from (21) the characteristic parameters of the regulator is obtained:

$$\sigma=W(s)u, \sigma=W(s)(-\varphi) \tag{30}$$

where $W(s)=\frac{Q_m(s)}{Q_n(s)}$ is the transfer function and $Q(s)$ are polynomials $m<n$. [4,6,16] The transfer function connect σ and φ ; the function φ verify the conditions (5) and the sector condition (6) $0<\frac{\varphi(\sigma)}{\sigma}<k\le\infty$ - the plot $\varphi=\varphi(\sigma)$ in the plane (σ,φ) will be the sector $0\le\varphi(\sigma)\le k\sigma$. The sector condition and the nonlinearity of φ determine the system (σ,φ) with closed loop through the impulse function φ . We study the absolute stability of the perturbed system (29) from the null solution $(x=0,u=0)$. Because the system is closed and nonlinear we can't applied directly the Nyquist criterion, [4,6,18]. If $\varphi\equiv k\sigma$ then the system is linear and it cab be applied this criterion. It observe that the block $\sum a_{il}x_l$ is linear and b_iu is nonlinear and result that the roots of characteristic polynomial $P(\lambda)=(-1)(A-\lambda E)=0, P(\lambda_i)=0,$ the poles of $W(s)$ and k will influence the determination of the absolute stability criteria. From $W(s=j\omega)=U(\omega)+jV(\omega), j=\sqrt{-1}$ we have the hodograph for the axis (U,V) [2,4,6,7,15]:

$$U=U(\omega), V=V(\omega), 0\le\omega\le\infty \tag{31}$$

If all poles of $W(s)$ have $Re(s_i)<0$ then the system is uncritically; if through the poles of $W(s)$ are a part null or on the imaginary axis and the rest have $Re(s_i)<0$ then the system is in the critical case. We enunciate the criteria for absolute stability of automatic control (a.r.a.s.) by the frequency method.

Criterion1. *(the uncritically case). Let be the conditions:*

a) The function $\varphi(\sigma)$ *verifies (5), (6)*

b) All poles of $W(s)$ *have* $Re(s_i)<0$

c) If there exists a real number $q\in R$ *that* $\forall\omega\ge0$ *is satisfied the condition:*

$$\frac{1}{k}+Re[(1+j\omega q)W(j\omega)]\ge0 \tag{32}$$

Then the system (20) is automatic regulated and absolute stable for the null solution $(x=0,u=0)$.
From (32) is obtained:

$$\frac{1}{k}+U(\omega)-q\omega V(\omega)\geq 0 \tag{33}$$

The criterion (32) geometrically shows that in the plane geometric $U_1=U, V_1=\omega V$ exists the line (33) passing through $\left(-\frac{1}{k},0\right)$ and the plot of the hodograph is under this line for $\omega\geq 0, k>0$.

Criterion2. *(the critical case when there are a simple null pole* $s_0=0$ *). Let be satisfied the conditions:*

a) The function φ *verify (5), (6).*

b) $W(s)$ *has a simple null pole, and the others poles* s_i *have* $Re(s_i)<0$.

c) We have $\rho=\lim_{s\to 0} sW(s)>0$ *and exists* $q\in R$ *for* $\forall\omega\geq 0$ *verifying the condition (33). Then for the system (28) for the null solution we have (a.r.a.s.).*

Criterion3. *(the critical case when* $s=0$ *is a double pole). Let be the conditions:*

a) The function $\varphi(\sigma)$ *verify (5), (6) and the sector condition for* $k=\infty$ *in the quarters I, III.*

b) $W(s)$ *has a double pole in s=0 and the others poles has* $Re(s_i)<0$.

c) Is verifying $\rho=\lim_{s\to 0} s^2W(s)>0$, $\mu=\lim_{s\to 0}\frac{d}{ds}\left[s^2W(s)\right]>0$, $\pi(\omega)=\omega ImW(j\omega)<0$ *for* $\forall\omega\geq 0$ *then for the system (28) we have (a.r.a.s.) for the null solution.*

Observation3. The shape of these criteria (I, II, III) has an analytical character and their verification is required for construction of hodograph values of the coefficients by numbers. For special cases the recommended monographs are [2,4,15,19].

3. THE STUDY OF THE ABSOLUTE STABILITY OF SOME AIRCRAFT COURSE WITH THE AUTOMATIC PILOT

We'll consider the airplane fly in the vertical plane xOy, the longitudinal axis of the aircraft is parallel with the horizontal axis Ox and the vertical plane is symmetry plane for the aircraft. In the longitudinal fly course (horizontal) can appear some perturbations with angular variations for: the pitch (tangage) angle ψ, between the longitudinal axis and Ox, the speed angle on the trajectory of fly θ, with the axis Ox compared with the considered system $\psi-\theta=\alpha$, represents the attack angle [17].
Considering these 3 angles without yaw and roll, it is written the system of disturbed differential equations compared with the mass centre, corresponding to ψ,θ,α, the coefficients are linearized, depend of the gyroscopic momentums created by the stability gyroscopes and the automatic regulations mechanisms for the pitch(tangage) stability [5,17]. Eliminating θ,α from the system we'll study the equation for ψ in concordance with the regulator characteristics. The object of automatic regulation is the horizontal course of the plane. The important elements of the measurement, control, sensors and with response with inverse reaction to the perturbations that compose the regulator are considered: a gyroscope that measure the pitch(tangage) speed ψ and a gyrotachometer that measure the angular speed $\dot{\psi}$, [5,17]. With sensors and potentiometers help these values are transmitted on the collector plate and transducers and amplifiers are turned into electrical signals, by summary they are transmitted through the input function φ for the output command function to the server $\sigma=-C_1\psi-C_2\dot{\psi}-r\xi$. By mechanical, electromagnetic, hydroelectric and gyroscopic effects, with the reaction parameter ξ determined, conform with the conditions from §2, it is obtain the stability for the null solution.

3.1. The method of the Liapunov solution for (a.r.a.s.).

We'll write the reduce system of equations dimensionless [17], corresponding to the pitch(tangage) perturbation $\psi=x$ in concordance with the functions and characteristics of the regulator connections.

$$\dddot{x}+a_1\ddot{x}+a_2\dot{x}=l\dot{y}+lmy;\sigma=-c_1x-c_2\dot{x}-r\xi;\psi=x;\dot{\psi}=\frac{dx}{dt} \tag{34}$$

Here, in the constants that appear have been included mass moments, moments of inertia, gyroscopic moments $a_1, a_2, l, m > 0, a_1^2 > 4a_2$ and the characteristic parameters of regulator $c_1, c_2, r > 0, b_2 = l, b_3 = l(m - a_1)$. The right side of the equation is actually the expression of server represented by the nonlinear function $\varphi(\sigma)$. Will write the system (34) with (1)-(4) using the next notations: $x_1 = x = \psi$, $x_2 = \dot{x} = \dot{x}_1 - \dot{\psi}$, $x_3 = \dot{x}_2 - ly$, $y = \xi$, $\dot{y} = \dot{\xi} = \varphi(\sigma)$.

$$\dot{x} = Ax + By, \dot{y} = \dot{\xi} = \varphi(\sigma), \sigma = c'x - r\xi \tag{35}$$

The matrix from (35) are:

$$x = \begin{pmatrix} x_1 \\ x_2 \\ x_3 \end{pmatrix}, A = \begin{pmatrix} 0 & 1 & 0 \\ 0 & 0 & 1 \\ 0 & -a_2 & -a_1 \end{pmatrix}, B = \begin{pmatrix} 0 \\ b_2 \\ b_3 \end{pmatrix}, C = \begin{pmatrix} -c_1 \\ -c_2 \\ 0 \end{pmatrix} \tag{35'}$$

Using the linear transform:

$$u = AX + B\xi, \dot{\sigma} = C'x - r\xi, u = \begin{pmatrix} u_1 \\ u_2 \\ u_3 \end{pmatrix} \tag{36}$$

Obtain the simplify system, by derivation:

$$\dot{U} = AU + B\varphi(\sigma), \dot{\sigma} = C'U - r\varphi(\sigma) \tag{37}$$

The system (35) has the unique solution $(x = 0, \xi = 0)$ and (37) $(U = 0, \sigma = 0)$. The absolute stability will be realize compare with these null solutions. The characteristic polynomial $P(\lambda) = \det(A - \lambda E) = 0$, $\lambda(\lambda^2 + a_1\lambda + a_2) = 0$ with the notations: $a_1 = 2p, a_2 = q$ has the roots:

$$\lambda_1 = -p + \sqrt{p^2 - q}, \lambda_2 = -p - \sqrt{p^2 - q}; \lambda_1 < 0, \lambda_2 < 0, \lambda_3 = 0 \tag{38}$$

After the diagonalization method (9) – (13), will transform the system (37) with $U = Tz$, $T(t_{ij}), i, j = 1,2,3$, determining the matrix T with (9) $AT = TJ, J = diagA$, obtaining :

$$T = \begin{pmatrix} \frac{1}{\lambda_1(\lambda_1 - \lambda_2)} & -\frac{1}{\lambda_2(\lambda_1 - \lambda_2)} & \frac{1}{\lambda_1\lambda_2} \\ \frac{1}{\lambda_1 - \lambda_2} & -\frac{1}{\lambda_1 - \lambda_2} & 0 \\ \frac{\lambda_1}{\lambda_1 - \lambda_2} & -\frac{\lambda_2}{\lambda_1 - \lambda_2} & 0 \end{pmatrix} \quad T^{-1} = \begin{pmatrix} 0 & -\lambda_2 & 1 \\ 0 & -\lambda_1 & 1 \\ \lambda_1\lambda_2 & -(\lambda_1 + \lambda_2) & 1 \end{pmatrix}, z = \begin{pmatrix} z_1 \\ z_2 \\ z_3 \end{pmatrix} \tag{39}$$

$$\dot{z} = Jz + T^{-1}B\varphi(\sigma), \qquad \dot{\sigma} = C'Tz - r\varphi(\sigma) \tag{40}$$

The system (40) is equivalent with (35) (36) and has the unique solution $(z = 0, \sigma = 0)$ and for this solution we study (a.r.a.s), determining the Liapunov function. To build the Liapunov function corresponding to the transformed system (40) $V = V(z, \varphi(\sigma))$, apply the calculus technique presented in (22) – (25) for the special case $\mathrm{Re}(\lambda_{1,2}) < 0, \lambda_3 = 0$ at (26), (27). The system (40) became:

$$\dot{z}_1 = \lambda_1 z_1 + b_1'\varphi(\sigma); \dot{z}_2 = \lambda_2 z_2 + b_1'\varphi(\sigma), \dot{z}_3 = b_3'\varphi(\sigma); \dot{\sigma} = f_1 z_1 + f_2 z_2 + f_3 z_3 - r\varphi(\sigma) \tag{41}$$

$b_1' = b_3 - \lambda_2 b_2, b_2' = b_3 - \lambda_1 b_2, b_3' = b_3 - (\lambda_1 + \lambda_2)b_2$; $f_1 = -\frac{c_1 + \lambda_1 c_2}{\lambda_1(\lambda_1 - \lambda_2)}, f_2 = \frac{c_1 + \lambda_2 c_2}{\lambda_2(\lambda_1 - \lambda_2)}, f_3 = -\frac{c_1}{\lambda_1\lambda_2}$.

In this case we choose the Liapunov function conforms with (22), (27)

$$V(z, \sigma) = \frac{1}{2}A_1 z_1^2 + \frac{1}{2}A_2 z_2^2 + \frac{1}{2}A\, z_3^2 + \int_0^{\sigma} \varphi(\sigma)d\sigma \tag{42}$$

where $A_1, A_2, A > 0$ are fixed, $V(z = 0, \sigma = 0) = 0$ and $V(z, \sigma)$ is positive defined. Compute the derivative $\dot{V}$ associated to the system (41)

$$\dot{V} = \sum_{j=1}^{2} A_j \lambda_j z_j^2 - r\varphi^2 + \sum_{j=1}^{2} (A_j \lambda_j b_j' + f_j) z_j \varphi + (Ab_3' + f_3) z_3 \varphi(\sigma) \tag{43}$$

We observe that taking $A_j = -\frac{1}{\lambda_j} > 0$ the negativity of this form is ensured from the first terms, forcing the cancellation of the last term: $Ab_3' + f_3 = 0$, that means:

$A = -\frac{f_3}{b_3'} = \frac{c_1}{a_1(b_3 + a_1 b_2)} = \frac{c_1}{a_1 lm} > 0$. From

$$\dot{V} = -(z_1^2 + z_2^2) - r\varphi^2 + \sum_{j=1}^{2} \varphi z_j \left(\frac{b_j'}{\lambda_j} - f_j \right) \tag{44}$$

The quadratic form is positive defined for $(-\dot{V})$ in relation with z_1, z_2, φ, with the system (41) or (9). From the Silvester determinant is obtained the necessary and sufficient condition (41) for the rigidity coefficient.

$$r > \left(\frac{b_1'}{\lambda_1} - f_1 \right)^2 + \left(\frac{b_2'}{\lambda_2} - f_2 \right)^2 \tag{45}$$

In this way the characteristic parameters of the regulator r, c_1, c_2 verify the condition (45), ensure the absolute stability of the horizontal fly course of the aircraft. It is observe that in conditions do not appear the function φ, so the nonlinear control function can be choose arbitrary from the admissible class (5), (6).

3.2. The frequency method for (a.r.a.s.).

For this study will applied the frequency method used in §3. because the system (35) is equivalent with (37) and (41), the function $u = -\varphi(\sigma)$ verify the sector condition. By replacing the operator $\frac{d}{dt}$ with the factors is found the transfer function $W(s)$. For simplicity we choose the system (37) with (35), we deduce the transfer function $W(s)$ that is the same for (35) and (41). Applying the Laplace operator in (37) we have:

$$U_1 s = U_2, U_2 s = U_3 + b_2\varphi, U_3 s = -a_2 U_2 - a_1 U_3 + b_3\varphi; \quad \sigma s = -c_1 U_1 - c_2 U_2 - r\varphi \tag{46}$$

Eliminating from these relations U_1, U_2, U_3 it is found the connection $\sigma = W(s)(-\varphi)$:

$$W(s) = \frac{1}{s^2}\left(rs + \frac{[b_2(s + a_1) + b_3](c_2 s + c_1)}{s^2 + a_1 s + a_2} \right) \tag{47}$$

We observe that $W(s)$ has a double pole in $s_0 = 0$ and $s_1 = \lambda_1 < 0, s_2 = \lambda_2 < 0$, being in the special case of the frequency method, Criterion3 (a.r.a.s) from §3. next, we verify the conditions from Criterion3.

$$\rho = \lim_{s \to 0} s^2 W(s) = \frac{lmc_1}{a_2} > 0, b_2 = l > 0, b_3 = l(m - a_1) > 0, a_1 > 0, a_2 > 0, c_1 > 0 \tag{48}$$

$$\mu = \lim_{s \to 0} \frac{d}{ds}(s^2 W(s)) = r + \frac{l}{a_2^2}\left[c_1(a_1^2 + a_2) - m(a_1 c_1 - a_2 c_2)\right] > 0 \tag{49}$$

From (49) we obtain conditions for r, m, c_2

$$r > \frac{l}{a_2^2}\left[m(a_1 c_1 - a_2 c_2) - c_1(a_1^2 + a_2)\right] > 0; \qquad m > \frac{c_1(a_1^2 + a_2)}{a_1 c_1 - a_2 c_2} > 0, \frac{a_1 c_1}{a_2} > c_2 > 0 \tag{50}$$

$$\pi(\omega) = \omega \operatorname{Im} W(j\omega) = -r - l\frac{\omega^2[a_1 c_2 - (c_1 + mc_2)] + [a_2(c_1 + mc_2) - a_1 c_1(m - a_1)]}{(a_2 - \omega^2)^2 + a_1^2\omega^2} = -r + g(\omega) \tag{51}$$

$$\lim_{\omega \to \infty} \pi(\omega) = -r < 0, \lim_{\omega \to 0} \pi(\omega) = -r + g(0) < 0 \tag{52}$$

From (52) we observe that $r > g(0)$ is from (50) condition. For the rigidity coefficient r we obtain the equivalence with (45). It is observe that by this qualitative criterion are necessary and numerical data in the space of parameters for regulator. The condition $\pi(\omega) = -r + g(\omega) < 0, \forall \omega \geq 0$ because $g(0) > 0$ is the right member from (50), $g(\omega)$ is derivable, $g'(\omega) < 0$, $\lim_{\omega \to \infty} g(\omega) = 0$ ($g = g(\omega)$ is an even function on $(-\infty, \infty)$ with $g(0)$ maximal.

4. THE ABSOLUTE STABILITY IN THE AUTOMATIC REGULATION OF THE WOOD CUTTING

The high precision of the tools wood cutting with tools machine, implies an automatic regulation of the processes. Here, are modeling and are studying the nonlinear dynamics of the cutting processes (CP) witch tools inside of the wood blocks, the composite materials blocks or hardwood. [3]
These (CP) are: CP of drilling, CP of milling, CP of grinding, screw machine, spindle bearing. Machine tool bar is provided with an inner elastic hard wood cutting, cutting inside to run the required geometric rotation and advancing to step slow. Because of the variation in hardness, density, coefficient of elasticity, material composition manufactured by the process disturbances will occur in work mode: transverse vibration due to shaft rotation or longitudinal vibrations to advance. Automatic controller is equipped with sensors, micrometers, tensiometers, rigid response mechanisms of signals output power amplifiers and accelerators. Their purpose is to adjust the characteristics to obtain asymptotic stability of the system work, resulting in high precision components. We will study the two methods described above in §2, §3.

4.1. The (a.r.a.s.) method by Liapunov function

Consider the dynamic system modeled mathematically, brought to a canonical form of Cauchy, autonomous, with features automatic adjustment for absolute stability of dynamic cutting machining processes. [3] [14]

$$\begin{cases} \dot{x}_1 = a_{11}x_1 + b_1\xi \\ \dot{x}_2 = a_{23}x_3 \\ \dot{x}_3 = a_{31}x_1 + a_{32}x_2 + a_{33}x_3 \\ \dot{x}_4 = a_{44}x_4 + b_4\xi \end{cases}, \sigma = c_2x_2 + c_4x_4 - r\xi, \dot{\xi} = \varphi(\sigma) \tag{53}$$

where a_{ij}, b_i, r, ξ are constants $i, j = 1,2,3,4$.

$$\begin{aligned} &a_{11} = -m < 0, a_{31} = n > 0, a_{32} = -\varepsilon n < 0, a_{33} = -p < 0, a_{44} = -l < 0 \\ &a_{23} = 1. c_2 = 1, c_4 = c < 0, b_1 = b > 0, b_4 = d - r > 0, r > 0 \end{aligned} \tag{54}$$

These, according to Lagrange's equations of the parameters are mass produced; mass inertia, elastic constants, strain or pressure coefficients, and σ, r, ξ are the characteristics of the server. We assume that the input function φ is generally nonlinear and check conditions (5) (6). We observe that the linear response function σ of the server control the elements x_2 - the speed of rotation of the cutting bar and x_4 - the speed of advancing its material.

We check the absolute stability of the system solution from zero $(x = 0, \xi = 0)$. Suppose that the block linear system (XA) is asymptotically stable as follows from relations: $\det A \neq 0, \operatorname{Re}(\lambda_i) < 0, i = 1,2,3,4$.

$$P(\lambda) = D(\lambda) = \begin{vmatrix} a_{11} - \lambda & 0 & 0 & 0 \\ 0 & -\lambda & a_{23} & 0 \\ a_{31} & a_{32} & a_{33} - \lambda & 0 \\ 0 & 0 & 0 & a_{44} - \lambda \end{vmatrix} = (a_{11} - \lambda)(a_{44} - \lambda)(\lambda^2 - \lambda a_{33} - a_{23}a_{32}) = 0 \tag{55}$$

$$\lambda_1 = a_{11} = -m < 0, \lambda_4 = a_{44} = -l < 0, \lambda_{2,3} = \frac{1}{2}\left(-p \pm \sqrt{p^2 - 4\varepsilon n}\right) < 0, \lambda_i \in \mathbf{R} \tag{56}$$

In this case, following the diagonalization method §2 with the formulas (9) - (13) or directly choose the option remark (R2) we get the diagonal system in z_i and $\dot{\sigma}$ (12), (13):

$$\dot{z}_i = \lambda_i z_i + \varphi(\sigma); \dot{\sigma} = \sum_{i=1}^{4} f_i z_i - r\varphi(\sigma), i = 1,\dots,4 \tag{55}$$

$$f_1 = \frac{b_1 a_{31}}{(\lambda_1 - \lambda_2)(\lambda_1 - \lambda_3)}, f_2 = \frac{-b_1 a_{31}}{(\lambda_1 - \lambda_2)(\lambda_2 - \lambda_3)}, f_1 = \frac{b_1 a_{31}}{(\lambda_1 - \lambda_3)(\lambda_2 - \lambda_3)}, f_4 = b_4 c_4 < 0 \tag{56}$$

We observe that $f_1 + f_2 + f_3 = 0$ and whatever is the choice of order quantities $\lambda_1, \lambda_2, \lambda_3$ are strictly negative, always two of the functions $f_i, i = 1,2,3$ have the same sign and the third function takes opposite sign using relation

(26). In this case the stability of the following (29) from the null solution ($z_i = 0, \sigma = 0$) and that $f_{4<0}$ we can construct such Lyapunov function:

$$V(z,\sigma) = -\frac{1}{2} f_4 z_4^2 - \frac{1}{2}\sum_{i=1}^{3} \frac{a_i^2 z_i^2}{\lambda_i} - \sum_{i=1}^{3}\sum_{j=1}^{3} \frac{a_i a_j}{\lambda_i + \lambda_j} z_i z_j + \int_0^{\sigma} \varphi(\sigma) d\sigma \tag{55}$$

where the real coefficients a_1, a_2, a_3 will be determined.

From $\lambda_i < 0, \lambda_i + \lambda_j < 0, V(0) = 0$, the summative terms after $i = 1,2,3$ determine a positive quadratic form positive definite and the integral positive term, we have $V(z,\sigma) > 0$ allowed in the vicinity. We calculate $\dot{V}(z,\sigma)$ attach to the system (29), and obtaining:

$$\dot{V}(z,\sigma) = -f_4 \lambda_4 z_4^2 - (a_1 z_1 + a_2 z_2 + a_3 z_3)^2 - \varphi \sum_{i=1}^{3} z_i \left(\frac{a_i^2}{\lambda_i} + \sum_{j \neq i} \frac{2 a_i a_j}{\lambda_i + \lambda_j} - f_i \right) \tag{56}$$

Observe that $\dot{V}(z = 0, \sigma = 0) = 0$ and to have the strict negativity it must that the parenthesis from the term φ to be null

$$\frac{a_i^2}{\lambda_i} + \sum_{j \neq i} \frac{2 a_i a_j}{\lambda_i + \lambda_j} - f_i = 0, i, i = 1,2,3 \qquad \text{(S*)} \tag{57}$$

The system (S*) $F_i(a_1, a_2, a_3) = 0, i = 1,2,3$ is implicit with three equations with three unknowns and the existence of solutions is provided by the system Jacobian $J = \dfrac{D(F_1, F_2, F_3)}{D(a_1, a_2, a_3)} \neq 0$. A helping calculation prove that if each equation from (33) is multiplying respectively by $\dfrac{1}{\lambda_i}$ and summing, it is obtain:

$$\Gamma^2 = \left(\sum_{i=1}^{3} \frac{a_i}{\lambda_i} \right)^2 = \sum_{i=1}^{3} \frac{f_i}{\lambda_i} = S > 0, \Gamma = \pm\sqrt{S} \tag{58}$$

A conditions that indicates that the knowing sum (S) is strictly positive and in the parametric space (a_1, a_2, a_3) the symmetrical plane (π_{12}) , $\sum_{i=1}^{3} \frac{a_i}{\lambda_i} = \pm\sqrt{S}$ where exists a solution, don't admit the null solution because $f_i \neq 0$. Is obtained as:

$$J = \pm 8 \frac{\sqrt{S}(a_1 + a_2 + a_3)[a_1(\lambda_2 + \lambda_3) + a_2(\lambda_1 + \lambda_3) + a_3(\lambda_1 + \lambda_2)]}{\lambda_1 \lambda_2 \lambda_3 (\lambda_1 + \lambda_2)(\lambda_1 + \lambda_3)(\lambda_2 + \lambda_3)} \neq 0 \tag{59}$$

Because the two factors from the numerator parenthesis are planes passing through the origin, the solution is contained in the planes (π_{12}). Analyzing the system (S*) after the sign of $f_i, (a_1, a_2, a_3)$ these solutions from (π_{12}) are not in the I,V octant. If all would have the same sign than $f_i > 0$. We proved that exists solutions of the system (33) and $\dot{V} < 0$; results that the Liapunov function provide the automatic regulation of the absolute stability. For this application and sufficient conditions of type (15), (16) with numerical data, are obtain.

4.2. The frequency method for (a.r.a.s.)

For this study will applied the frequency method used in §3. Because the system (53) is equivalent with (39) the function $u = -\varphi(s)$ verify the sector conditions. By replacing the operator $\dfrac{d}{dt}$ with the factor s is found the transfer function $W(s)$. So, from (29) is obtain for $\sigma = W(s)(-\varphi)$

$$s z_i = \lambda_i z_i + \varphi; s\sigma = \sum_{i=1}^{4} f_i z_i - r\varphi \tag{60}$$

Eliminating from (60) z_i we obtain the transfer function from $\sigma = W(s)(-\varphi)$

$$W(s)=\frac{1}{s}\left(r-\sum_{i=1}^{4}\frac{\lambda_i f_i}{s-\lambda_i}\right) \qquad (61)$$

Because the real roots $s_i=\lambda_i$ verify $\mathrm{Re}(\lambda_i)<0$ the transfer function has a simple pole in $s=0$ and the rest of real roots with $\mathrm{Re}(s_i)<0$. In this case we have the Criterion II of critical singularity from §3 for (a.r.a.s.). here, the conditions (15), (19) and II a), b) are verified fro the method A and must verified the condition c). So, $\rho=\lim_{s\to 0} sW(s)=r+\sum_{i=1}^{4} f_i=r+f_4=r+b_4c_4>0$ implies $r+c(d-r)>0$ that mean $r>\frac{cd}{1-c}>0, d<0$. From $W(s=j\omega)=U(\omega)+jV(\omega)$, we have: $U(\omega)=\sum_{i=1}^{4}\frac{f_i\lambda_i}{\lambda_i^2+\omega^2}$, $V(\omega)=-\frac{r}{\omega}+\frac{1}{\omega}\sum_{i=1}^{4}\frac{\lambda_i^2 f_i}{\lambda_i^2+\omega^2}$. For given $k>0$, from the condition $0<\varphi(s)<k\sigma$ with $\varphi(\sigma)$ specified, it can be determine $q\in\mathbf{R}$ verifying the condition (24'). The parameters λ_i, f_i are knowing from (56), (60), the nonlinear function φ is chosen with σ from (53) and for specified numerical data determine k and the delimitation of q. The existence of these conditions can be perform hodographically for (a.r.a.s.) at this application.

5. CONCLUSIONS

The importance of this paper is evident in the fact that the problem of absolute stability is systematized by the two methods. It is remark that fact that the application regarding (a.r.a.s.) for the horizontal fly course with automatic pilot is studied for the critical difficult cases, when the roots of characteristic polynomial or the pole of transfer function is in origin (on the imaginary axis). For the Liapunov function building we applied an original method. For another studies are recommend the published results of the researchers [1,15,19,20,11].

ACKNOWLEDGEMENT

The second author of this paper is supported by the Sectorial Operational Programme Human Resources Development (SOP HRD), financed from the European Social Fund and by the Romanian Government under the project number POSDRU/159/1.5/S/134378.

REFERENCES

References

[1] E. Barbasin, Liapunov's function, Ed. Nauka, Moscov, 1970 (in Russian).

[2] C. Belea, The system theory - nonlinear system, Ed. Did. Ped., Bucuresti, 1985 (in Romanian).

[3] S. Chiriacescu, Stability in the Dynamics of Metal Cutting, Ed. Elsevier, 1990; Dynamics of cutting machines, Ed. Tehnica, Bucuresti, 2004 (in Romanian).

[4] I. Dumitrache, The engineering of automatic regulation, Ed. Politehnica Press, Bucuresti, 2005.

[5] R. Lungu, Gyroscopic equipment and systems , Ed. Press Universitaria, Craiova, 1997 (in Romanian)

[6] C. Lupu, and co., Industrial process control systems, Ed. Printech, Bucuresti, 2004, (in Romanian).

[7] C. Lupu, and co., Practical solution for nonlinear process control, Ed. Politehnica Press, Bucuresti, 2010 (in Romanian).

[8] M. Lupu, O.Florea, C. Lupu, Theoretical and practical methods regarding the absorbitors of oscillations and the multi-model automatic regulation of systems, Proceedings of The 5th International Conference Dynamical Systems and Applications, Ovidius University Annals Series: Civil Engineering, Volume 1, Special Issue 11, pg. 63-72, 2009.

[9] M. Lupu, F. Isaia, The study of some nonlinear dynamical systems modeled by a more general Reylegh - Van der Pol equation, J. Creative Math, 16(2007), Univ. Nord Baia Mare, pp. 81-90.

[10] M. Lupu, F. Isaia, The mathematical modeling and the stability study of some speed regulators for nonlinear oscilating systems, Analele Univ. Bucuresti, LV(2006), pp 203-212.

[11] M. Lupu, O. Florea, C. Lupu, Studies and applications of absolute stability in the automatic regulation case of the nonlinear dynamical systems, Romai Journal, Vol 6, 2(2010), pp. 185-198

[12] M. Lupu, E. Scheiber, O. Florea, C. Lupu, Studies And Applications For The Automatic Regulation Of Absolute Stability For The Dynamics Tools Cutting Machine, Journal of Computational and Applied Mechanics, Vol. 12., No. 1., (2011), (will appear)

[13] A. Y. Lurie, Nonlinear problems from the automatic control, Ed. Gostehizdat Moskow, 1951 (in Russian).

[14] D. R. Merkin, Introduction in the movement stability theory, Ed. Nauka, Moskow, 1987 (in Russian).

[15] R. A. Nalepin, and co.: Exact methods for the nonlinear system control in case of automatic regulation, Ed. M.S. Moskva, 1971.

[16] D. Popescu, and co.: Modéllisation, Identication et Commande des Systemes, Ed.Acad. Romaine, Bucuresti, 2004.

[17] E. P. Popov, Applied theory of control of nonlinear systems, Ed. Nauka, 1973, Verlag Technik, Berlin 1964.

[18] V. M. Popov, The hypersensitivity of automatic systems, Ed. Acad Romane, 1966, Ed. Springer -Verlag, 1973, Ed. Nauka 1970

[19] V. Rasvan, Theory of Stability, Ed. St. Enciclopedica, Bucuresti, 1987, (in Romanian)

[20] N. Rouche, P. Habets, M. Leloy, Stability theory by Liapunov's Direct Method, Springer Verlag, 1977

ANALYTICAL AND EXPERIMENTAL DETERMINATION OF STRESSES IN THE PLANE COMPOSITE PLATES USED TO BUILD CRAFTS

Corneliu Moroianu

"Mircea cel Bătrân" Naval Academy, Constanța, Romania, e-mail: cmoroianu2000@yahoo.com

Abstract: *This work presents the mathematical method for solving the differential equations by means of which we can determine the stresses in the plane composite plates used to build crafts (the impregnation resin is NESTRAPOL 450). The results analytically determined are compared with the experimental ones.*
Keywords: *composite plates, stresses, mathematical method, build crafts.*

1. INTRODUCTION

The fundamental researches on the composite materials (with material orthotropy) are in process of development. By applying the elements of "The elastic theory" to the composite plates normally and in median plane stressed, the differential equations of strained median surfaces for deflected plate ($0,5<w<5h$) are:

$$E_x \frac{\partial^4 F}{\partial x^4} + 2E_{xy} \frac{\partial^4 F}{\partial x^2 \partial y^2} + E \frac{\partial^4 F}{\partial y^4} = E_x E_y \left[\left(\frac{\partial^2 w}{\partial x \partial y} \right)^2 - \frac{\partial^2 w}{\partial x^2} \frac{\partial^2 w}{\partial y^2} \right]$$
$$D_x \frac{\partial^4 w}{\partial x^4} + 2H \frac{\partial^4 w}{\partial x^2 \partial y^2} + D_y \frac{\partial^4 w}{\partial y^4} = h \left[\frac{p(x,y)}{h} + \frac{\partial^2 F}{\partial y^2} \frac{\partial^2 w}{\partial x^2} - 2 \frac{\partial^2 F}{\partial x \partial y} \frac{\partial^2 w}{\partial x \partial y} + \frac{\partial^2 F}{\partial x^2} \frac{\partial^2 w}{\partial y^2} \right] \quad (1)$$

By analyzing the equation system (1) and comparing it with the equation system for isotropic plate (steel or aluminum) we note the appearance of stiffness on the two directions which changes the structure of solutions.
The equation system (1) has as unknowns the stress function $F(x,y)$ and the deflection $w(x,y)$, which can be determined by means of the boundary conditions for various supporting ways (rigid fixing, simple or free side suspension). In the previous work, the special forms of equation system (1) have been presented, from which we are interested in particular, in the rigid plate with small deflection ($w \le 0.2h$) of the following form:

$$D_x \frac{\partial^4 w}{\partial x^4} + 2H \frac{\partial^4 w}{\partial x^2 \partial y^2} + D_y \frac{\partial^4 w}{\partial y^4} = p(x,y) \quad (2)$$

The presence of a small strain of the plate means that the ship's shape changes very little due to the water action, the stream lines don't change very much and the heading resistance doesn't increase very much due to the hull 's strain.

The mathematical resolution of differential equation system (1) is possible only in particular cases. So, it is necessary a careful analysis of strength structure of ship and her skin. Taking into account only the local loading, the ship's strength structure is formed both by keelsons, girders and lines alongside and floors, frames and beams athwart wise forming a network on which the ship's skin is fixed. I consider the plate mesh, between the stiffening members, stressed by water pressure, being rigid with a small deflection ($w \le 0.2h$) where the sectional stresses N_x, N_y, N_{xy} don't influence the bending. In this case, the equation system (1) under the form of (2) represents a linear differential equation with partial derivates and constant coefficients.
To determine the stress and strain conditions in the plate mesh resulted from the local loading, means to find a function w which to check the differential equation (2) and in the same time the boundary conditions depending on the supporting pattern.

2. THE ANALYTICAL RESOLUTION OF DIFFERENTIAL EQUATIONS

We consider the general case when the plate is of *a x b x h*, simply supported on the contour line, normally loaded with *p(x,y)* varying on both directions. The system of axes is like in Figure 1.

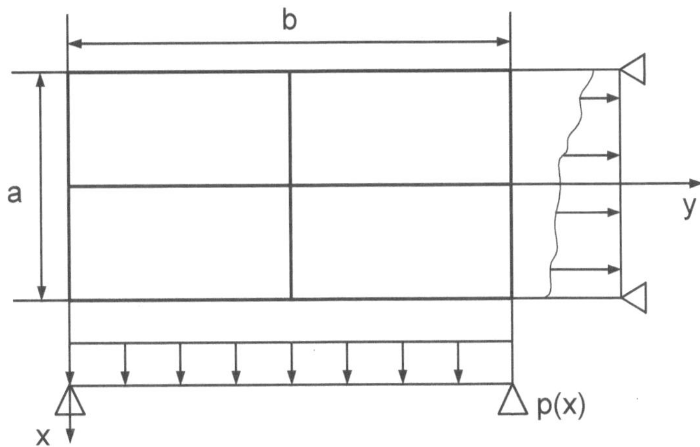

Figure 1. Plate loaded with a load distributed on the surface varying on both directions.

It is developed the normal load *p(x,y)* in double Fourier's series:

$$p(x,y)=\sum_m\sum_n p_{mn}\sin\frac{m\pi x}{a}\sin\frac{n\pi y}{b} \qquad \text{m, n} = 1,2,3, \tag{3}$$

The parameters p_{mn} are determined by Euler's method and become:

$$p_{mn}=\frac{4}{ab}\int_0^a\int_0^b p(x,y)\sin\frac{m\pi x}{a}\sin\frac{n\pi y}{b} \qquad \text{m, n} = 1,3,5,... \tag{4}$$

For the load uniformly distributed p_o the parameters p_{mn} become:

$$p_{mn}=\frac{16p_0}{\pi^2 mn} \qquad \text{m, n}=1,3,5,\ ... \tag{5}$$

The strain function (deflection) is also developed in Fourier's series under the form of:

$$w(x,y)=\sum_m\sum_n w_{mn}\sin\frac{m\pi x}{a}\sin\frac{n\pi y}{b} \tag{6}$$

The coefficients w_{mn} are determined from the condition that the expression (6) to satisfy the differential equation of the plate (2) for any values *x, y* and the boundary conditions on the contour line (for the plate simply supported $w = 0$ and $M_x = M_y = 0$) and it is obtained:

$$w_{mn}=\frac{16p_0}{\pi^6 mn}\frac{1}{D_x\frac{m^4}{a^4}+2H\frac{m^2n^2}{a^2b^2}+D_y\frac{n^4}{b^4}} \tag{7}$$

The expression of deflection (6) for the particular case when the load $p(x,y) = p_0$, the case of bottom plates of the ship, becomes:

$$w(x,y)=\frac{16p_0}{\pi^6}\sum_m\sum_n\frac{\sin\frac{m\pi x}{a}\sin\frac{n\pi y}{b}}{mn\left(D_x\frac{m^4}{a^4}+2H\frac{m^2n^2}{a^2b^2}+D_y\frac{n^4}{b^4}\right)} \tag{8}$$

The expressions of sectional moments become:

$$M_x = \frac{16p_0}{\pi^4}\sum_m\sum_n \frac{\left(D_x\frac{m^2}{a^2}+D_1\frac{n^2}{b^2}\right)\sin\frac{m\pi x}{a}\sin\frac{n\pi y}{b}}{mn\left(D_x\frac{m^4}{a^4}+2H\frac{m^2n^2}{a^2b^2}+D_y\frac{n^4}{b^4}\right)} \qquad (9)$$

$$M_y = \frac{16p_0}{\pi^4}\sum_m\sum_n \frac{\left(D_y\frac{n^2}{b^2}+D_1\frac{m^2}{a^2}\right)\sin\frac{m\pi x}{a}\sin\frac{n\pi y}{b}}{mn\left(D_x\frac{m^4}{a^4}+2H\frac{m^2n^2}{a^2b^2}+D_y\frac{n^4}{b^4}\right)} \quad \text{m, n} = 1,3,5,\ldots \qquad (10)$$

$$M_{xy} = -\frac{32p_0}{\pi^4 ab}\sum_m\sum_n \frac{D_{xy}\cos\frac{m\pi x}{a}\cos\frac{n\pi y}{b}}{\left(D_x\frac{m^4}{a^4}+2H\frac{m^2n^2}{a^2b^2}+D_y\frac{n^4}{b^4}\right)} \qquad (11)$$

The expressions of shearing forces become:

$$T_x = \frac{16p_0}{\pi^3}\sum_m\sum_n \frac{\left[D_x\left(\frac{m}{a}\right)^3+H\frac{m}{a}\left(\frac{n}{b}\right)^3\right]\cos\frac{m\pi x}{a}\sin\frac{n\pi y}{b}}{mn\left(D_x\frac{m^4}{a^4}+2H\frac{m^2n^2}{a^2b^2}+D_y\frac{n^4}{b^4}\right)} \qquad (12)$$

$$T_y = \frac{16p_0}{\pi^3}\sum_m\sum_n \frac{\left[D_y\left(\frac{n}{b}\right)^3+H\frac{n}{b}\left(\frac{m}{a}\right)^3\right]\sin\frac{m\pi x}{a}\cos\frac{n\pi y}{b}}{mn\left(D_x\frac{m^4}{a^4}+2H\frac{m^2n^2}{a^2b^2}+D_y\frac{n^4}{b^4}\right)} \qquad (13)$$

The maximum deflection is produced at the middle of the plate, that is, in the coordinate point x = $a/2$ and $y = b/2$, and in this case the relation becomes:

$$w_{\max} = \frac{16p_0}{\pi^6}\sum_m\sum_n \frac{(-1)^{\frac{m+n}{2}-1}}{mn\left(D_x\frac{m^4}{a^4}+2H\frac{m^2n^2}{a^2b^2}+D_y\frac{n^4}{b^4}\right)} \quad \text{m,n} = 1,3,5,\ldots \qquad (14)$$

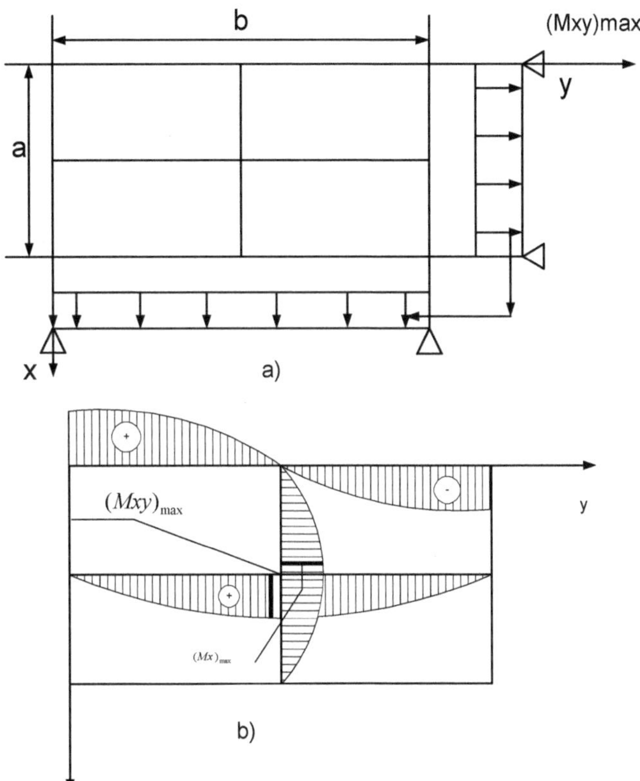

Figure 2. The rectangular orthotropic plate loaded with a normal load uniformly distributed:
a) The plate simply supported loaded with $p(x,y) = p_0 = ct.$
b) The diagram of stresses in the rectangular plate simply supported.

The numerical results are listed in Table 1. The calculus was performed for the first three terms.

3. EXPERIMENTAL RESULTS AND CONCLUSION

To check the value of maximum deflection obtained by the theoretical methods mentioned above, we built a device by means of which we measured the maximum deflection in the middle of the plate. The device is formed of two rigid angle bar frames by means of which we performed the fixing and with only one frame we made the support on the sides. The loading was made with fine, dry sand with a density of $\rho = 1.3\ kg/dm^2$. The thickness of sand layer was calculated from the condition of loading with a load uniformly distributed $p = 3000 N/m^2$. The deflection was measured in the middle of the plate by a comparator. The comparation between the calculated values and the measured ones for the five plied laminar is shown in Table 1.

We conclude that the methods of resolution can be divided into:

- approximative analytical methods (energetically) : when the unknown function, w, is approximated, from energetically reasons, satisfying both the system and the supporting conditions on the contour line. They are: the orthogonally method, Ritz, Rayleigh-Ritz, Bubnov-Galerkin, Treftz, etc.
- approximative numerical methods: the finite element method or the finite difference method.

Both methods offer the possibility of determining the efforts and good results with acceptable approximations.

Table 1. Experimental results.

Method of determination	Maximum analyzed values					
	w_{max} [mm]	M_x [mm/mm]	My[mm/ mm]	M_{xy} [mm/mm]	σ_x[mm/mm]	Σ_y [mm/mm]
The plate simply supported						
Experimental	4,2	43,601	11,91	-6,91	14,15	3,86
Double trigonometric	4,383	-	-	-	-	-
Simple Trigonometric	1,413	44,157	16,31	-5,44	-	-
Ritz method	4,18	44,8	13,2	-8,31	14,57	4,278
MEF (COSMOS M program)	4,424	-	-	-	-	-
MEF (ALGOR program)	4,205	-	-	-	-	-
The fixed plate						
Experimental	1,19	-	-	-	-	-
MEF (COSMOS M program)	0,961	16	3,44	-	5,91	1,12

REFERENCES

[1] Beschia N., Strength of Materials, Ed. Didactic and Pedagogic Bucharest, 1971.

[2] Caracosea A., Theory of Elasticity, University of Galati, 1980.

[3] Modiga M., Mechanical shipbuilding, University of Galati, 1980.

[4] Panait M., Topa N., Ieremia M., Theories of elasticity and plate applications in construction calculation, Ed. Technique, Bucharest, 1986.

DESIGN, ANALYSIS AND IMPROVEMENT OF SEATBELT COMPONENTS

Imre P.
Technical University of Cluj-Napoca, Baia Mare, ROMANIA, imrepaul@yahoo.com

Abstract: *This paper presents the design challenges form car manufacturer request to concept and concept validation, examining the customer's needs and assimilating them in the design goal document used by component suppliers for the automotive industry. The rules and regulations that apply to the to the safety systems components bought as international law, customer specifications and internal specifications. The design challenges for this are to integrate all the design goals into the product as sometimes there is a thin line between product failures and over-engineering as designers and engineers must not only build a product that is robust, but it must also be cost effective bought in used terms of materials and component design.*
Keywords: *seat belt, design, simulation, analysis*

1. INTRODUCTION

When a car crashes due to collision with an obstacle, the seatbelt is there to protect your body from severe injury. NTF states that seatbelts are the most important safety device in automotive, both in the front seats and in the back seats. According to NTF a three-point seatbelt decrease the risk of severe injury in a collision by 50 % [5].

When a car collides with an obstacle at high speeds, the occupant's inertial speed is very high and it takes a high force to stop the occupants. If the car collides with an obstacle during low speed, the force to stop the occupant is not as high. This means that the seatbelt will have to withstand a higher force to stop you when colliding at a higher speed. The result of this is that the seatbelt will apply a higher chest compression to the occupant.

2. FROM CUSTOMER REQUEST TO MANUFACTURING CONCEPT

For any new car platform, the same platform could be used form multiple car models, the manufacturer will make a design goal document (DGD) in which he defines and lays-out all the important information for the component manufacturers. The important information for seatbelt design is defining the safety and comfort requirements, this will include: positioning of each seatbelt component; the vehicle destination market, as each market has their own set of standards and directives or regulations for car safety systems (ex: ECE-R16 for Europe, FMVSS 208 and 210 for North America, CCC for China, etc.) [1]. Also a note must be made that a standard is different from a directive or a regulation.

A standard is considered by the general public or by the authorities as a basis for comparisons between different products or just to ensure that a product meets a certain level but in most cases only serve as a guideline and are not included in national legislation [2].

A directive in the EU is defined as a direction from authorities or as an instruction, such as a rule or specific order, this can over time be adopted as a national law by member states over time [2].

Based on the documents mentioned above if there is no current serial part that can fulfill the customer's request a multi-department team is formed in order to assure all needed factors will be taken into account. This factors impact departments such as: development, sales, purchasing, quality, manufacturing and logistics.

In the concept phase the department leading the team should be the development department, but it has to take into consideration all the requests and limitations imposed by the other team members.

3. DESIGN CHALANGES

A design methodology provides two important characteristics. First, it acts as a check list to ensure all design steps have been completed. Second, it provides focus on what the design must accomplish, based on the user's needs [4].
The design challenges for seat belt suppliers are in meeting the quality, innovation, time to market, global standardization and pricing requirements from the car manufacturers as they became stricter in all global region.
The quality aspect is higher as new revisions for the safety regulations of seat belt systems in a car become more demanding in order to allow car manufactures to have the same New Car Assessment Program (NCAP) rating. An evolution of the Euro NCAP ratings obtained from 1997 to 2007 is shown in *figure 1.*

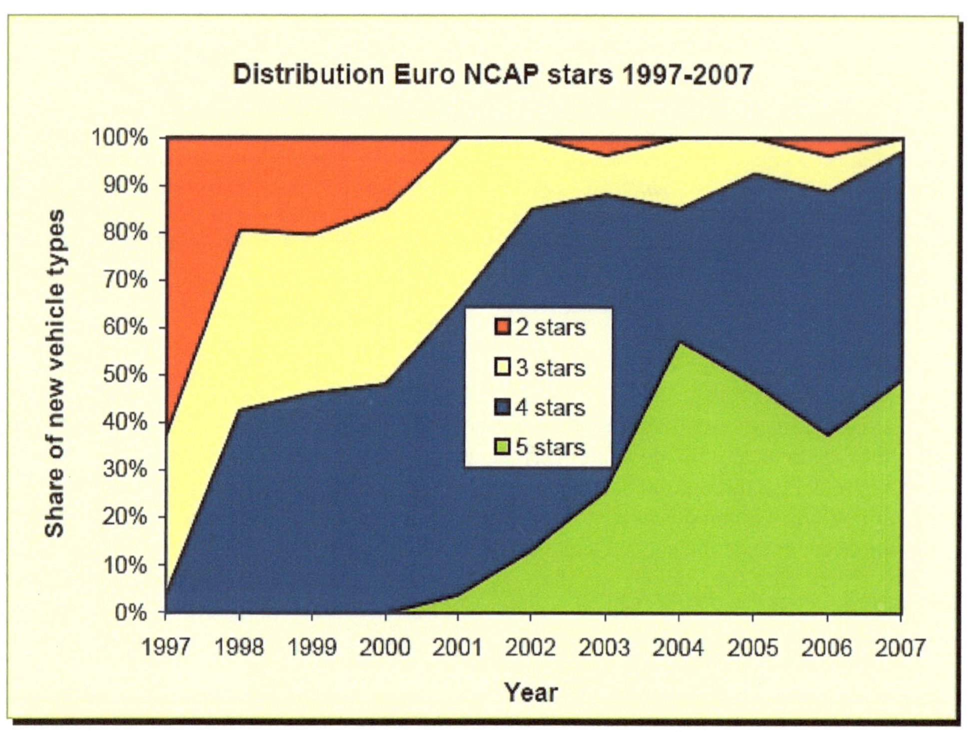

Figure 1: Share of Euro NCAP stars for new types of vehicle on the European market for the safety of adult occupants in the old Euro NCAP system [7]

Quality issues are raised as more pressure is put on the part supplier and given the high volumes of common parts between Original Equipment Manufacturers (OEMs) and this can result in potential "lethal" penalty fees [5].
Innovation is a key aspect in large seat belt manufacturing companies as they always try to stay ahead of the market, in regards of product performance, durability, weight, applications, size and last but not least cost. The innovation factor is important as it gives the company an edge over their competitor. This allows the part manufacturer to set a more favorable price could translate into a higher Earning Before Interest and Taxes (EBIT), resulting in more capital for future development investment.
Time to market for a product gets shorter every time, as the OEMs request products to be developed quicker for them to be able to maintain or gain a lead over the competition with the proposed innovation. So the time that you can best save is in the concept development stage as the tests and tooling manufacturing time tend to get longer due to high demand.

Global standardization is important for any part manufacturer as this will allow them to reduce costs due to needing les development time (one product) and les time to build and set up the manufacturing process (one process). Also the lessons learned could be shared more effectively and if the product can meet standards and regulations from different regions than after it is developed it could be sold as a standard product.
Pricing is a delicate balance between just seeing one project or also the future business it might result from an innovation. With this in mind the cost of a part is influenced by factors such as the impact it will have in terms of being a regional project versus a global one, or impacting only one customer versus multiple customers.

The trend in the automotive industry is to reduce the cost/part as the project life moves towards end of production. Also a careful consideration has to be mad in regards to the other project that may result from the same OEM as they can have a big difference in volumes, thus impacting the cost/part.

4. CONCEPT ANALYSIS AND IMPROVMENT

The first concept id built in a Computer Aided Design (CAD) interface to allow for easier future adjustments to the model, also in this case an initial check of the system integration form the component could be performed. After an optimization of the model is performed in order to better allow for easier real live part production and assembly a simulation should be performed using programs for Finite Element Analysis (FEA) studies on the part as shown in *figure 2*, also if the assembly has any moving component this could also be simulated to better understand the forces in the system.

This initial CAD model and simulation should allow for a quick and effective concept verification and improvement. Based on the simulation results the geometry or material of the part could change.

A prototype part could then be build using 3D printers as shown in *figure 3*. This will be used to verify product or part dimensions, any potential assembly problems and also possible manufacturing issues or difficulties.

After the development team decides that they have the best concept based on available data that prototype parts from a more robust material and closer to the real life product could be built for testing. This testing will be performed with only a few samples and it has the purpose to uncover any possible design flaws in the product durability or performance and at the same time to give an indication if the sample could be over engineered.

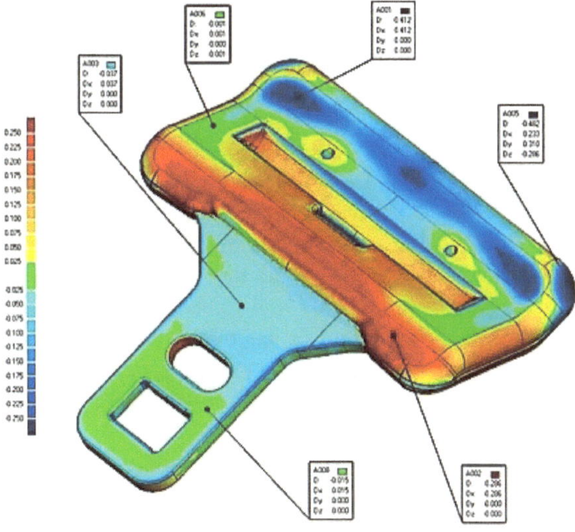

Figure 2: FEA analysis on seatbelt tongue [6]

Figure 3: Tongue produced using a 3D printer

Failure Modes and Effects Analysis (FMEA) is a method used for predict what errors that can occur and how serious they can be. A table is used where all the possible failures are presented. They are then rated on a scale from one to ten, the higher number they get the greater influence they have [5].

The method is about identification of errors in the components on system level and the consequences of them. An evaluation of how severe damage they can do, how likely it occurs and the probability that the failure is detected before it occurs. This will result in a greater chance to fix the problem before it occurs.

In this case a Design Failure Modes and Effects Analysis (DFMEA) will suit the needs and is better used when doing a concept evaluation for a new design.

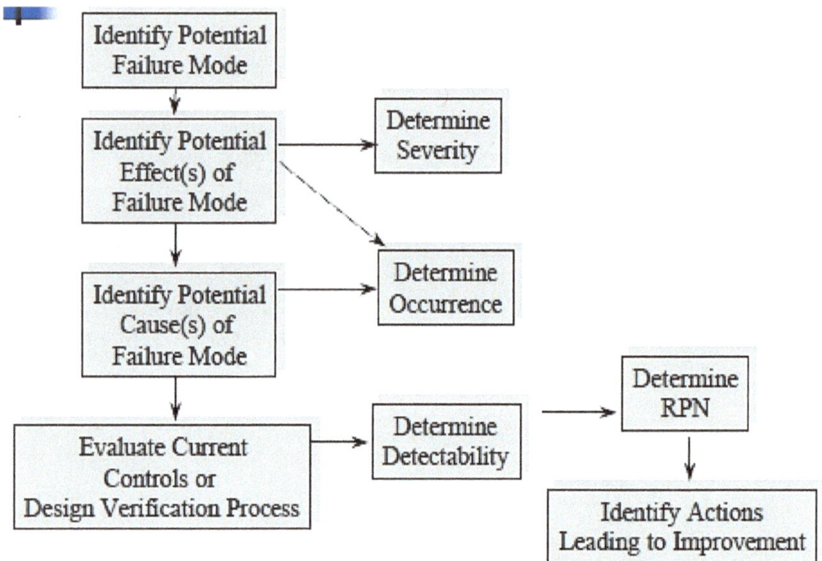

Figure 4: FMEA roadmap [6]

Using the FMEA analysis a product can be improved in all the phases from its life cycle, from concept to end of productions, and the lessons learned could be applied in the next concept design.

To aid this structure and because most durability and robustness test are destructive test, it is important to use statistic tools and guidelines to track the product performance over time and to better optimize it.

5. CONCLUSION

For the future of car safety the seat belt system will always be needed and it will always need improving as to reduce the risk of injury or death as much as possible.

That is why it is so important to have the improvements start using lessons learned, know-how and keeping an open mind for new innovation. The analysis of the concept should be done using 6 Sigma tools to better optimize the resources needed bought in terms of numbers of samples tested and also result interpretation.

Seat belt components can be and must be improved in any number of categories, from materials used for their parts to radical new concepts and designs, but this has to be controlled as to assure a product that it is not mature or lacks enough performance data is introduced to market.

REFERENCES

[1] Imre, P., Cotetiu, R., Contribution to Validation and Testing of Seatbelt Components, The International Conference of the Carpathian Euro-region Specialists in Industrial Systems CEurSIS 2014, 10 h Edition. Proceedings Baia Mare, September, 11st – 13rd , 2014. Editura U.T.PRESS Cluj-Napoca.

[2] Pedersen C.D., Safety standards and test procedures, Dhal Engineering, Denmark, 2010.

[3] Pettersson S., Andersson M., VEVA Multifunctional Spindlering, Maskiningenjor – Teknisk Design, Halmstad, 2013.

[4] ***, 3D software for inspection and reverse engineering. Retrived on 18.05.2014, from www.otto-jena.de/3D_software.html.

[5] ***, Driving on thin ice, Roland Berger and Lazard, 2013.
[6] ***, Potential Failure mode and effect analysis reference manual. Automotive Industry Action Group (AIAG), 1995.
[7] ***, SWOV Fact sheet, Leidschendam, Netherlands, 2010.

ON THE MODELING AND SIMULATION OF AN ACOUSTIC CLOAK

Veturia Chiroiu[1], Rodica Ioan[1,2], Cornel Brişan[3], Ruxandra Ilie[4]

[1] Institute of Solid Mechanics, Romanian Academy, Bucharest, e-mail veturiachiroiu@yahoo.com
[2] University Spiru Haret, Bucharest, e-mail rodicaioan08@yahoo.com
[3] Technical University of Cluj-Napoca, e-mail cornel.brisan@mmfm.utcluj.ro
[4] Technical University of Civil Engineering, Bucharest, e-mail rux_i@yahoo.com

Abstract: *Transformation acoustics opens a new avenue towards the modeling and simulation of acoustic cloaks. The design of acoustic cloaks is based on the property of Helmoltz equations to be invariant under a geometric transformation. In this paper, a spherical shell cloak made of auxetic material (material with negative Poisson's ratio) is discussed. The original domain consists of spheres made from conventional foam with positive Poisson's ratio. The spatial compression leads to an equivalent domain filled with auxetic material.*
Keywords: *Helmoltz equation, transformation acoustics, acoustic cloak, auxetic material*

1. FORMULATION OF THE PROBLEM

Transformation acoustics is the key for the design of acoustic cloaks. Recent works show that such materials could cloak regions of space, making them invisible to sound [1-3]. We refer to acoustic cloaking which occurs when a medium contains a region in which noisy objects can be acoustically hidden. It is easy to imagine an object invisible to sound by building a box around it to prevent the wave from reaching the object.
The principle how to cloak a region of space to make its contents invisible or transparent to waves was discussed in [4, 5]. The geometric invariance of Helmoltz equation shows how a region of the space can be made inaccessible to acoustic waves by surrounding it with a suitable shield. As an alternative to a box made from a metamaterial, sonic composites exhibit the full band-gaps, where the sound is not allowed to propagate due to complete reflections [6-9].

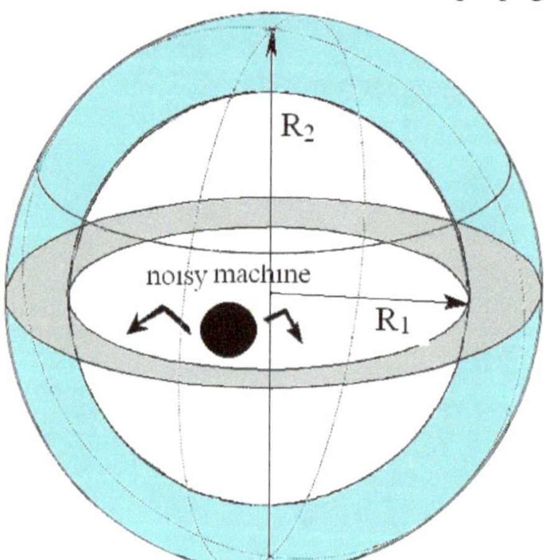

Figure 1: Sketch of the spherical cloak surrounding a noisy machine.

In this paper, we apply the 3D concave-down transformation to design a spherical cloak which surrounds a noisy machine, see Figure 1. The original domain is a sphere of radius R_2, consisting of a traditional foam with positive Poisson's ratio. After the transformation, the cloak contains a region $r < R_1$ which contains the noisy source, filled with air, while the shell $R_1 < r < R_2$ is filled by auxetic material.

2. TRASFORMATION ACOUSTICS

A finite size object surrounded by a coating consisting of a specially designed material would become invisible for acoustic waves at any frequency. In acoustics, the idea of the invisibility cloak is that the sound sees the space differently. For the sound, the concept of distance is modified by the acoustic properties of the regions through which the sound travels. In geometrical acoustics, we are used to the idea of the acoustical path; when travelling an infinitesimal distance $\mathrm{d}s$, the corresponding acoustical path length is $c^{-1}\mathrm{d}s$, where $c^{-1}=\sqrt{\rho/\kappa}$ with ρ is the fluid density and κ is the compression modulus of the fluid. To understand the problem, we consider the 3D acoustic equation for the pressure waves propagating in a bounded fluid region $\Omega \subset \mathrm{R}^3$

$$\nabla\cdot(\underline{\underline{\rho}}^{-1}\nabla p)+\frac{\omega^2}{\kappa}p=0, \tag{1}$$

where p is the pressure, $\underline{\underline{\rho}}$ is the rank-2 tensor of the fluid density, κ is the compression modulus of the fluid, and ω is the wave frequency. Let us consider the geometric transformation from the coordinate system (x',y',z') of the compressed space to the original coordinate system (x,y,z), given by $x(x',y',z')$, $y(x',y',z')$ and $z(x',y',z')$. The change of coordinates is characterized by the transformation of the differentials through the Jacobian $J_{xx'}$ of this transformation, i.e.

$$\begin{pmatrix}\mathrm{d}x\\ \mathrm{d}y\\ \mathrm{d}z\end{pmatrix}=J_{xx'}\begin{pmatrix}\mathrm{d}x'\\ \mathrm{d}y'\\ \mathrm{d}z'\end{pmatrix},\quad J_{xx'}=\frac{\partial(x,y,z)}{\partial(x',y',z')}. \tag{2}$$

From the geometrical point of view, the change of coordinates implies that, in the transformed region, one can work with an associated metric tensor [10]

$$T=\frac{J_{xx'}^{\mathrm{T}}J_{xx'}}{\det(J_{xx'})}. \tag{3}$$

In terms of the acoustic parameters, one can replace the material from the original domain (homogeneous and isotropic) by an equivalent compressed one that is inhomogeneous (its characteristics depend on the spherical (r',θ',ϕ') coordinates) and anisotropic (described by a tensor), and whose properties, in terms of $J_{x'x}$, are given by

$$\underline{\underline{\rho}}'=J_{x'x}^{-\mathrm{T}}\cdot\rho\cdot J_{x'x}^{-1}\cdot\det(J_{x'x}),\quad \kappa'=\kappa\det(J_{x'x}), \tag{4}$$

or, equivalently, in terms of $J_{xx'}$

$$\underline{\underline{\rho}}'=\frac{J_{xx'}^{\mathrm{T}}\cdot\rho\cdot J_{x'x}}{\det(J_{xx'})},\quad \kappa'=\frac{\kappa}{\det(J_{xx'})}. \tag{5}$$

Here, $\underline{\underline{\rho}}'$ is a second order tensor. When the Jacobian matrix is diagonal, (4) and (5) can be more easily written. The geometric transformation may be linear or nonlinear. Qiu [11] classified the geometric transformation functions in terms of the negative (i.e., concave-down) or positive (i.e., concave-up) sign of the second order derivative of this function. All transformations, i.e. linear, concave-up and concave-down transformations, are perfect cloaks for the exact inhomogeneous design.

The concave-down nonlinear transformation compresses a sphere of radius R_2 in the original space Ω into a shell region $R_1<r'<R_2$ in the compressed space Ω' as

$$r(\beta)=\frac{R_2^{\beta+1}}{R_2^{\beta}-R_1^{\beta}}\left(1-\left(\frac{R_1}{r'}\right)^{\beta}\right), \tag{6}$$

where β denotes the degree of the nonlinearity in the transformation. By taking $\beta\to 0$ in (6), the linear case is obtained, namely

$$r(\beta)=\frac{R_2\mathrm{Ln}(r'/R_1)}{\mathrm{Ln}(R_2/R_1)}. \qquad (7)$$

All curves belonging to (6) have negative second order derivative with respect to the physical space r'. This class of transformations is termed as the *concave-down* transformation. The transformation function (6) depends on the radial component r' in the spherical coordinate system (r',θ',ϕ').

The concave-up nonlinear transformation compresses a sphere of the radius R_2 in the original space Ω into a shell region $R_1<r'<R_2$ in the compressed space Ω' as

$$r(\beta)=\frac{R_2R_1^{\beta}}{R_2^{\beta}-R_1^{\beta}}\left(\left(\frac{r'}{R_1}\right)^{\beta}-1\right). \qquad (8)$$

As $\beta\to 0$, one obtains again the linear case (7). This class of transformations is termed as the *concave-up* transformation because (8) has positive second order derivatives.

All curves belonging to (6) have negative second order derivative with respect to the physical space r'. This class of transformations is termed as the *concave-down* transformation. The nonlinear transformation function in (6) only depends on the radial component r' in the spherical coordinate system (r',θ',ϕ'). The cloak properties in the both transformed coordinates are given by (4) and (5) where $J_{r'r}=\partial r'/\partial r$.

Indeed, the equations governing the propagation of elastodynamic waves with a time harmonic dependence are written, in a weak sense, as

$$\nabla\cdot C:\nabla u+\rho\omega^2u=0, \qquad (9)$$

where ρ is the scalar density of an isotropic heterogeneous elastic medium, C is the fourth-order elasticity tensor, ω is the wave angular frequency, and $u(x_1,x_2,x_3,t)=u(x_1,x_2,x_3)\exp(-i\omega t)$ is the vector displacement. It is easy to show that under a change of coordinates (x',y',z') to (x,y,z) such that $u'(x')=J_{x'x}^{-T}u(x)$, $J_{x'x}=\dfrac{\partial(x',y',z')}{\partial(x,y,z)}$, Eq. (9) takes the form

$$\nabla''\cdot(C'+S'):\nabla'u'+\underline{\underline{\rho}}'\omega^2u'=D':\nabla'u', \qquad (10)$$

which preserves the symmetry of the new elasticity tensor $C'+S'$. Equation (10) contains two third-order symmetric tensors S' and D' with $D'_{pqr}=S'_{qrp}$, and a second-order tensor ρ'_{pq}.

3. ACOUSTIC CLOAK

We write the Helmholtz equation which governs the behavior of the sphere filled with traditional foam in the coordinate system (x_1,x_2,x_3) as

$$\nabla\cdot(\zeta^{-1}\nabla\Theta)+\omega^2\Lambda^{-1}\Theta=0. \qquad (11)$$

where $\zeta=(2k(1-2\nu))^{-1/2}$, k is the bulk modulus of the foam, and $\nu>0$ is the Poisson's ratio of the foam, ρ its effective density, $\Lambda=\rho^{-1}$ and ω the frequency. Let us apply the concave-down transformation (6) to (11), which compresses the original domain Ω occupied by a sphere of radius R_2 into a shell region $R_1<r'<R_2$ in the compressed space Ω', characterized by

$$\underline{\underline{\zeta}}'^{-1}_{p,e}(r')=J_{rr'}^{\mathrm{T}}\zeta_{p,r}^{-1}(r)J_{rr'}/\det(J_{rr'}),\ \underline{\underline{\Lambda}}'^{-1}(r')=J_{rr'}^{\mathrm{T}}\Lambda^{-1}(r)J_{rr'}/\det(J_{rr'}),\ J_{rr'}=\partial r/\partial r', \qquad (12)$$

In the new coordinates, the transformed equation (11) now reads as

$$\nabla\cdot\underline{\underline{\zeta}}_{p,e}^{-1}\nabla(\Delta_{33}\nabla\cdot\underline{\underline{\zeta}}_{p,e}^{-1}\nabla\Theta')-\Lambda_{33}^{-1}\gamma_0^4\Theta'=0, \quad (13)$$

where $\underline{\underline{\zeta}}_{p,e}^{-1}$ is the upper diagonal part of the inverse of $\underline{\underline{\zeta}}$ and Λ_{33}^{-1} is the third diagonal entry of $\underline{\underline{\Lambda}}^{-1}$, and $\nu<0$.

The cloak has the inner radius $R_1=0.5\text{m}$ and outer radius $R_2=1\text{m}$.

We must say that the condition of $-1<\nu<0.5$ corresponds to the usual range of properties for stability of the material. The Poisson's ratio $\nu=\nu_{yx}$ (for tensile and compressive tests) was calculated as the negative ratio between the radial and longitudinal strains using a best fit to the strain-strain graph $\nu_{yx}=-\frac{\varepsilon_x}{\varepsilon_y}$. The most important physical parameter to dominate the negative Poisson's ratio transformation is the compression ratio $\vartheta=\frac{(R_2'^2-R_1'^2)l'}{R_2^2 l}$, where prime denotes the final parameters. Figure 2 shows the transformed annulus domains for $\vartheta=0.25$, 0.26, 0.3 and 0.4. . We observe that the conventional foam becomes auxetic ($-0.15\le\nu<0$) for $0.55\le 1-\vartheta\le 0.77$, or $0.23\le\vartheta\le 0.45$. It is very interesting to see that the auxetic foam is changing the sign for its Poisson's ratio for $0.46\ge 1-\vartheta$. It is of interest to underline that the results provides an overall agreement with the experimental values for the auxetic foam [11].

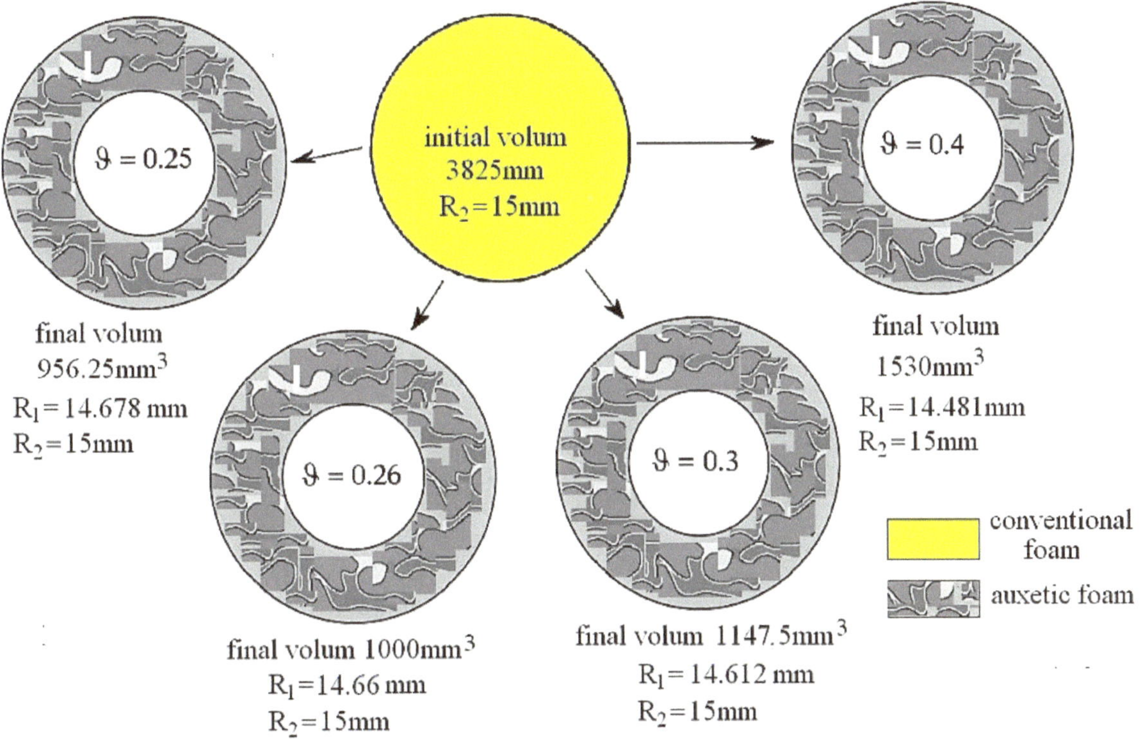

Figure 2: Transformed domains.

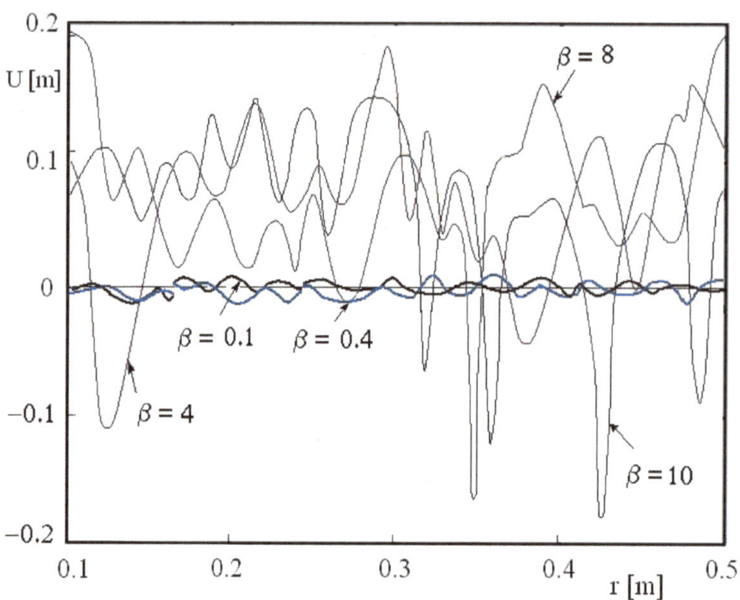

Figure 3: Variation of the displacement amplitude with respect to β in the region $r \le R_1$.

The concave-down transformation presents an overlapping for all mapping curves for $\beta < 0.1$, which means the same results in applications. The effect of β on the amplitude of displacements, which vary from $-U$ to U ($U = \sqrt{u_1^2 + u_2^2 + u_3^2}$) inside the cloak $r \le R_1$, is illustrated in Figure 3. It can be seen that when β increases, the amplitude increases significantly inside the region $r \le R_1$ of the cloak. This is due to the fact that more energy is guided towards the inner boundary $r = R_1$, which in turn makes the cloaked object more *acoustically visible* to external incidences. For $\beta = 0.1$ and 0.4, the acoustically invisibility is good. The effect of β on the amplitude of displacements in the shell region $R_1 < r < R_2$ is illustrated in Figure 4. In a similar manner, when β increases, the amplitude increases significantly in the shell region of the cloak.

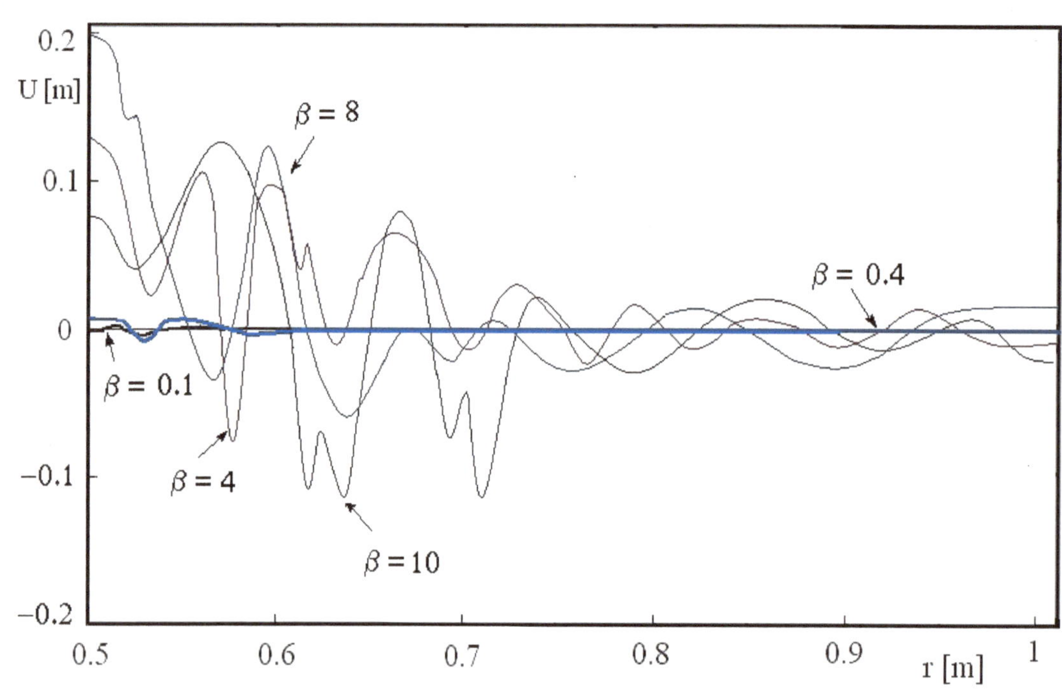

Figure 4: Variation of the displacement amplitude with respect to β in the region $R_1 < r < R_2$.

The absence of the scattering of waves generated by an external source outside the cloak is observed in Figure 5 for $\beta = 0.1$ and $\nu = -0.15$. The waves are smoothly bent around the central region inside the cloak. The results reported in Figure 5 show that the wave field inside the cloak, i.e. the inner region of radius R_1 which surrounds the noisy machine, is completely isolated from the region situated outside the cloak. The waves generated by a noisy source are smoothly confined inside the inner region of the cloak, and the sound invisibility detected from the observer is proportional to β. The inner region is acoustically isolated and the sound is not detectable by an exterior observer because the amplitudes on the boundary vanish. The domain $r < R_1$ is an acoustic invisible domain for exterior observers. The waves generated by the exterior source outside the cloak do not interact with the interior field of waves. A possible interaction or coupling between the internal and external wave fields is cancelled out by the presence of the shell region $R_1 < r < R_2$ filled with auxetic material.

Hence we can conclude that for the concave-down spherical cloaks, smaller values for β lead to a smaller disturbance in the acoustic fields in both the inner and the outer spaces $r < R_2$ and $r > R_2$, respectively

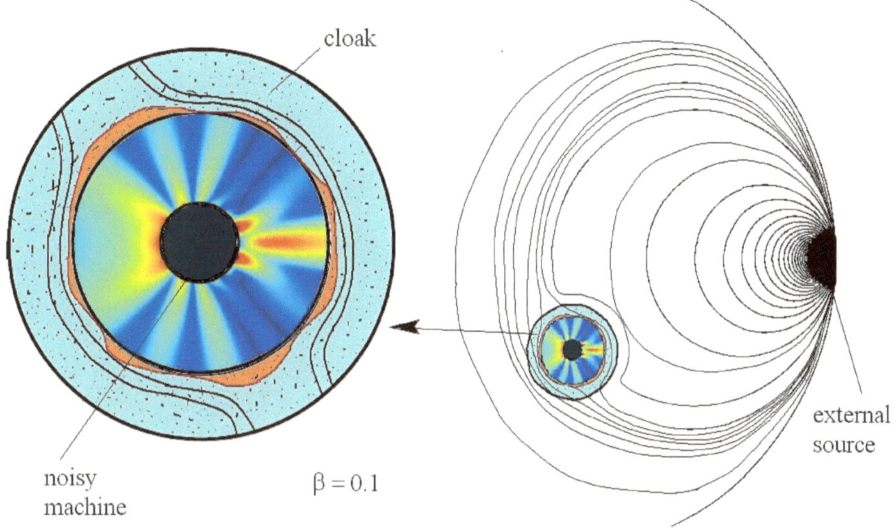

Figure 5: The wave fields inside and outside the cloak for $\beta = 0.1$ and $\nu = -0.15$.

4. CONCLUSION

In this paper, we have identified new aspects in the 3D spherical cloaking related to new reflectionless solutions which may exist for cloaking systems that are not isomorphic to electromagnetism. The original domain consists of traditional foam. The spatial compression obtained by applying the concave-down transformation has led to an equivalent domain of auxetic material. However, the present study represents an application of the aforementioned analytical results, in the sense that a numerical implementation, which treated a new kind of material that might be useful in the design of elastic cloaking devices, was developed.

Acknowledgement. The authors gratefully acknowledge the financial support of the National Authority for Scientific Research ANCS/UEFISCDI through the through the project PN-II-ID-PCE-2012-4-0023, Contract nr.3/2013. The authors acknowledge the similar and equal contributions to this article.

REFERENCES

1. Munteanu, L., Chiroiu, V., *On the three-dimensional spherical acoustic cloaking,* New Journal of Physics, 13 (8), 083031, 1-12, 2011.
2. Munteanu, L., Brisan, C., Donescu, St., Chiroiu, V., *On the compression viewed as a geometric transformation*, CMC: Computers, Materials & Continua, 31(2), 127-146, 2012.
3. Nicorovici, N.A., McPhedran, R.C., Milton, G.W., *Optical and dielectric properties of partially resonant composites*, Phys. Rev. B, 490, 8479–8482, 1994.

4. Miller, D.A.B., *On perfect cloaking*, Optical Society of America, 14,25, Optics Express 12465, 2006.
5. Leonhardt, U., *Optical conformal mapping*, Science, 312, 1777–1780, 2006.
6. Munteanu, L., Chiroiu, V., *On the dynamics of locally resonant sonic composites*, European Journal of Mechanics-A/Solids, 29(5), 871–878, 2010.
7. Chiroiu, V., Brişan, C., Popescu, M.A., Girip, I., Munteanu, L., *On the sonic composites without/with defects,* Journal of Applied Physics, 114 (16), 164909-1-10, 2013.
8. Munteanu, L., Chiroiu, V., Donescu, St., Brişan, C., *A new class of sonic composites*, Journal of Applied Physics, 115, 104904, 2014.
9. Hirsekorn, M., Delsanto, P.P., Batra, N.K., Matic, P., *Modelling and simulation of acoustic wave propagation in locally resonant sonic materials*, Ultrasonics, 42, 231–235, 2004.
10. Guenneau, S., McPhedran, R.C., Enoch, S., Movchan, A.B., Farhat, M., Nicorovici, N.A., *The colours of cloaks*, Journal of Optics, 13, 2, 024014, 2011.
11. Qiu CW, Hu L, Zhang B., Wu BI, Johnson SG and Joannopoulos JD 2009 Spherical cloaking using nonlinear transformations for improved segmentation into concentric isotropic coatings *Optics Express* 17(16) 13467–13478.
12. Bezazi, A., Scarpa, F., *Mechanical behaviour of conventional and negative Poisson's ratio thermoplastic polyurethane foams under compressive cyclic loading,* International Journal of Fatique, 29, 922–930, 2007.

EVALUATION OF STRAIN AND STRESSES STATES OF REINFORCED GUITAR NECK WITH FINITE ELEMENTS METHOD

Mariana Domnica Stanciu[1], Ioan Curtu, Dragos Apostol

[1] Transilvania University of Brasov, Brasov, Romania, mariana.stanciu@unitbv.ro, curtui@unitbv.ro

Abstract: *Many guitarists have found that guitar neck deforms due to tensions of the strings during the singing and the relaxation due to internal stress accumulated in the neck structure. This bending of the neck is a defect that affects the sound quality of the guitar. As a result, there are numerous methods of reinforcement of the neck with the metal rods which can be pre-tensioned or can be adjusted. The paper evaluated variants guitar neck reinforcement (rods of circular section, rectangular section, T section). The five alternatives were modeled and were determined the strain and stresses states of the guitar structure for different reinforcement bars applied in different areas of the neck of the guitar. It was found that both shape, material and position of bars from the fingerboard different and leads to smaller displacement of neck. The results may be useful to manufacturers of musical instruments.*

Keywords: *reinforced guitar neck, stress, finite element method, stiffness, strain*

1. INTRODUCTION

One of the most important stresses of mechanical engineering is bending produced both and shear forces bending moments. In case of mechanical structure of the guitar, guitar neck which can be considered a cantilever beam bending in the plastic domain conducting to fall upon quality acoustic musical instrument. The researchers focused on stiffening the neck by a reinforced system with metal rods, fixed or adjustable tension rods made from hardwood timber, steel, carbon fiber, glass fiber. Gordis (http://www.ukuleles.com/Technology/neck.html) conducted a study on the calculation of flexural strength of ukulele and guitar neck, causing the neutral axis of tension produced in the neck section. These rods have different sections and ways of handling. They are made of lightweight material that does not change the structure weight. They act as static: the position of the rods in cross section neutral axis position changes so that the balance of tension and compression stresses (Figure 1). From the mechanical point of view, most of the tension rod is adjustable by means of a screw-nut system (Figure 2) [1, 2].

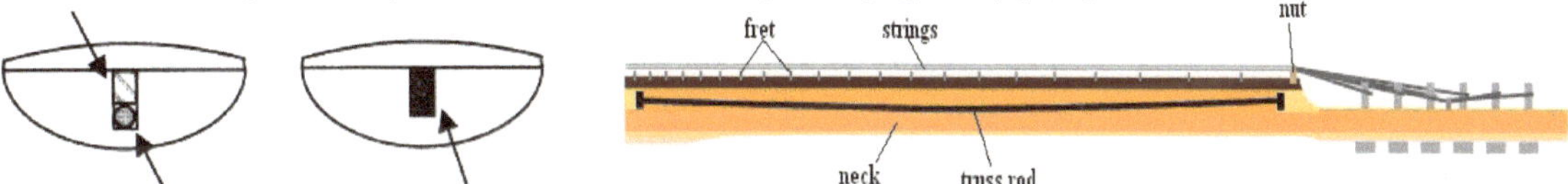

Figure 1. Transverse and longitudinal section of the neck of the guitar in the reinforced http://en.wikipedia.org/wiki/Truss_rod, http://www.warmoth.com/guitar/necks

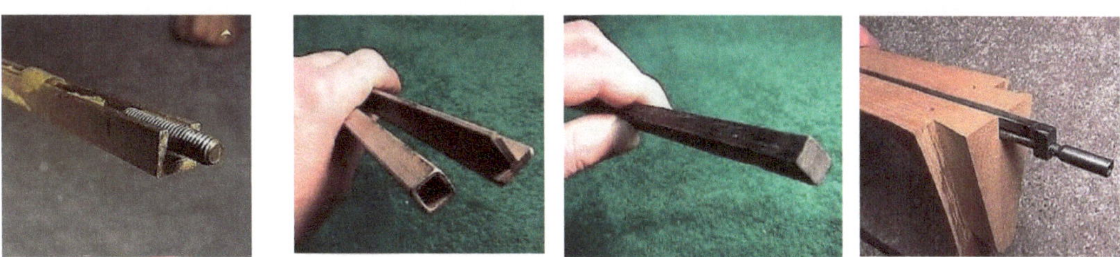

Figure 2. Types of reinforcing rods neck (http://www.frets.com)

This paper aims to assess the effectiveness of the reinforcement system of the neck of the guitar from the point of view of the loads and displacements size during bending.

2. EVALUATION OF DEFLECTION CURVE OF GUITAR NECK USING FINITE ELEMENT ANALYSIS

The neck of the guitar is a cantilever beam used both for strength and ergonomic, aesthetic and functional. These bars are effective when the variation law of cross-sectional is optimal choose such that in all cross-sections maximum normal stresses are equal. Only at rest but loaded with forces / moments of strings tension, the guitar neck meets the criteria of a bar of equal resistance. The neck of the guitar is stressed by normal forces (N_i) of tensioned strings which generate a bending moment (M_b) and torques moment (M_t). Depending on the bridge height, bending moments have higher or lower value. Torques moment is produced by variation of intensity of normal force related to symmetrical longitudinal axe. In previous articles, it was presented an analytical calculus to determine the sectional efforts from guitar structure due to tune of guitar strings [3, 4]. Geometry of guitar was modeled in Abaqus taking into account the real dimensions of guitar and the main parts of them – guitar body, neck, fret board and frets (Figure 3). For meshing were used tetrahedron elements.

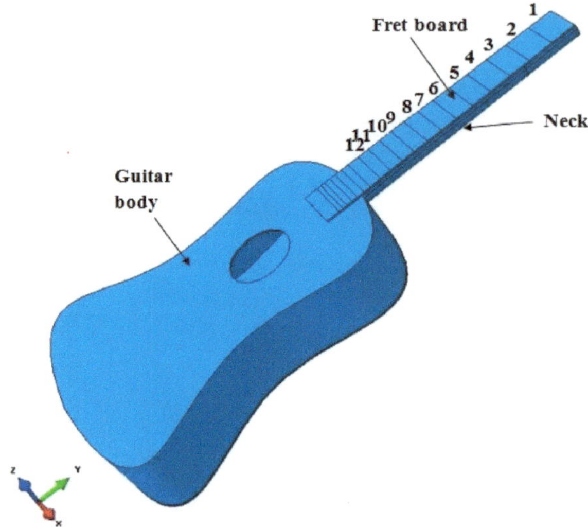

Figure 3. Main parts of guitar

The most important part of guitar in this study is neck which was reinforced successively with three types of rods from section point of view. Seven cases were analyzed: case 1 - neck without reinforcement system; case 2 – neck reinforced with rectangular rod; case 3 - neck reinforced with rectangular rod translated with 5 mm; case 4 - neck reinforced with T rod; case 5 - neck reinforced with T rod translated with 5 mm; case 6 - neck reinforced with circular rod; case 7 - neck reinforced with circular rod translated with 5 mm (Figure 4). The rod material was considered steel with elasticity modulus of $2x10^5$ MPa.

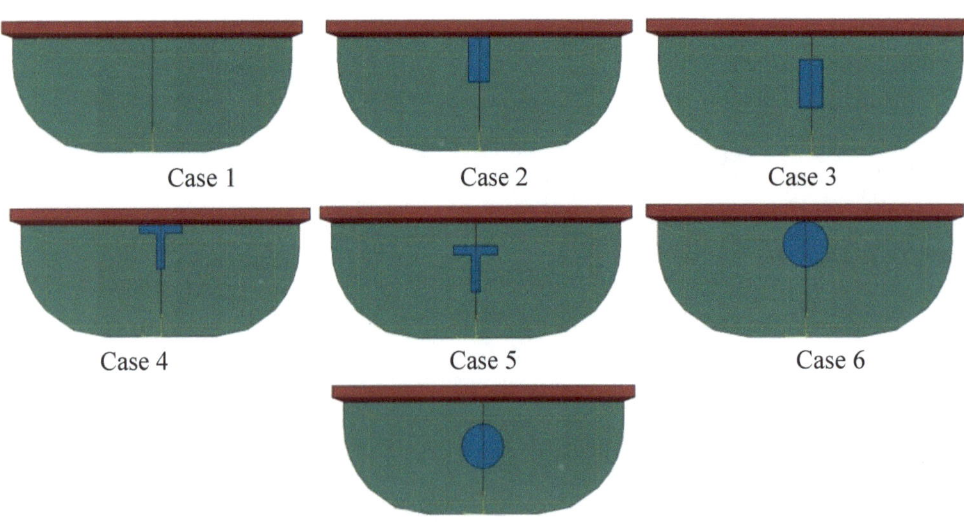

Case 7

Figure 4. Reinforced system of guitar neck

In the preprocessing stage, there were introduced the specific parameters of the guitar neck and the guitar body. Table 1 summarizes the elastic characteristics of guitar components taken from literature [5, 6, 7].

Table 1: Elastic characteristics of wood species used in guitar structure

Components	Wood species	Modulus of Longitudinal Elasticity *E [MPa]*	Modulus of Transversal Elasticity *G [MPa]*	Poisson's Coefficient ν
Neck	Beech *Fagus sylvatica*	E_L=14200 E_T= 1160 E_R= 2280	G_{LR}=1970 G_{LT}= 950 G_{RT}= 467	ν_{TR}=0,30; ν_{RT}=0,64 ν_{LT}=0,50; ν_{TL}=0,085 ν_{RL}=0,12; ν_{LR}=0,32
Fret board	Locust *Robinia pseudoacacia*	E_L =19900 E_T = 1650 E_R = 3040	G_{LR}=2100 G_{LT}= 980 G_{LR}= 530	ν_{TR}=0,31; ν_{RT}=0,66 ν_{LT}=0,63; ν_{TL}=0,090 ν_{RL}=0,14; ν_{LR}=0,40
Guitar Body	Spruce *(Picea abies (l.) Karst.)*	E_L =16225 E_T = 400 E_R = 700	G_{LR}= 650 G_{LT}= 416 G_{RT}= 347	ν_{TR}=0,25; ν_{RT}=0,42 ν_{LT}=0,33; ν_{TL}=0,013 ν_{RL}=0,028; ν_{LR}=0,40

It was established the boundary conditions: the head of the neck was pin connected and the end of guitar body was fixed. It was successively applied a distributed forces of 0,1 MPa on each fret.

3. RESULTS AND DISCUSSION

After running the program with finite element analyses, the maximum stresses and strains values obtained when the structure was loaded on each fret, being processed into graphical variation. In Figure 5 is represented the displacement and stresses von Mises. The results in terms of maximum stresses and displacements for all types of stiffening system are presented in Table 2 and 3. It can be noticed that variation of static behavior of guitar neck depends on type of reinforcement and position of them in transversal section. Position of load on fretboard influences the tension of guitar neck, So, with increasing the distance between force and supports conduct to increasing the bending moment and stresses. Also, the type of reinforce bar lead to different behavior of guitar neck related to stresses. The most efficient stiffening bar is T shape (case 4 and 5). Minimum value of stress is obtained in case of simple neck – without reinforcement (case 1) and in case of rectangular shape of bar section displaced with 5 mm below fretboard (case 3).

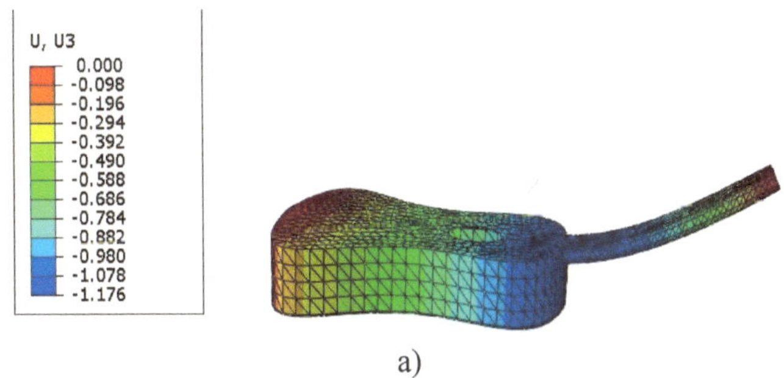

a)

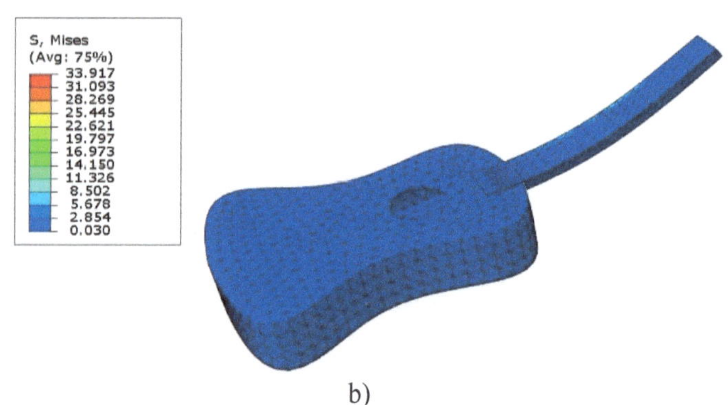

b)

Figure 5. Stress and strain of guitar: a) displacement; b) normal stress

Table 2: Maximum normal stresses obtained for load on each fret of guitar neck

	Fret 1	Fret 2	Fret 3	Fret 4	Fret 5	Fret 6	Fret 7	Fret 8	Fret 9	Fret 10	Fret 11	Fret 12
	Maximum normal streses σ [MPa]											
Case 1	13,82	22,59	23,15	26,69	27,61	32,89	31,01	31,1	36,39	32,02	26,82	19,51
Case 2	25,59	39,54	37,27	39,03	32,48	33,94	33,14	33,92	34,47	34,38	29,83	29,15
Case 3	6,7	10,99	11,33	13,11	13,19	13,85	13,75	13,93	13,88	13,88	14,24	11,12
Case 4	56,31	92,47	94,7	106,94	101,35	102,85	107,96	108,42	108,85	112,04	98,24	70,94
Case 5	51,04	83,84	86,5	98,13	92,91	94	97,74	102,89	100,1	97,13	84,46	63,37
Case 6	25,37	41,69	43,08	48,22	44,73	39,91	43,29	46,54	43,82	40,57	39,79	28,85
Case 7	25,85	42,94	43,6	50,07	47,39	52,42	50,62	49,88	53,78	53,25	44,37	31,83

Table 2: Maximum displacement obtained for load on each fret of guitar neck

	Fret 1	Fret 2	Fret 3	Fret 4	Fret 5	Fret 6	Fret 7	Fret 8	Fret 9	Fret 10	Fret 11	Fret 12
	Maximum displacement Δ [mm]											
Case 1	1,11	1,79	1,79	1,99	1,94	2,08	2,07	**2,11**	2,07	2	1,74	1,29
Case 2	0,5	0,81	0,79	0,86	0,83	0,87	0,85	0,86	0,83	0,79	0,69	0,51
Case 3	0,3	0,47	0,45	0,49	0,46	0,48	0,47	0,47	0,45	0,43	0,37	0,28
Case 4	0,7	1,05	1,04	1,14	1,1	1,17	1,16	1,17	1,14	1,1	0,96	0,71
Case 5	0,64	1,02	0,99	1,09	1,05	1,12	1,1	1,11	1,08	1,035	0,89	0,67
Case 6	0,73	1,15	1,14	1,26	1,22	1,29	1,28	1,29	1,26	1,22	1,06	0,79
Case 7	0,68	1,09	1,07	1,18	1,14	1,21	1,19	1,2	1,17	1,12	0,97	0,72

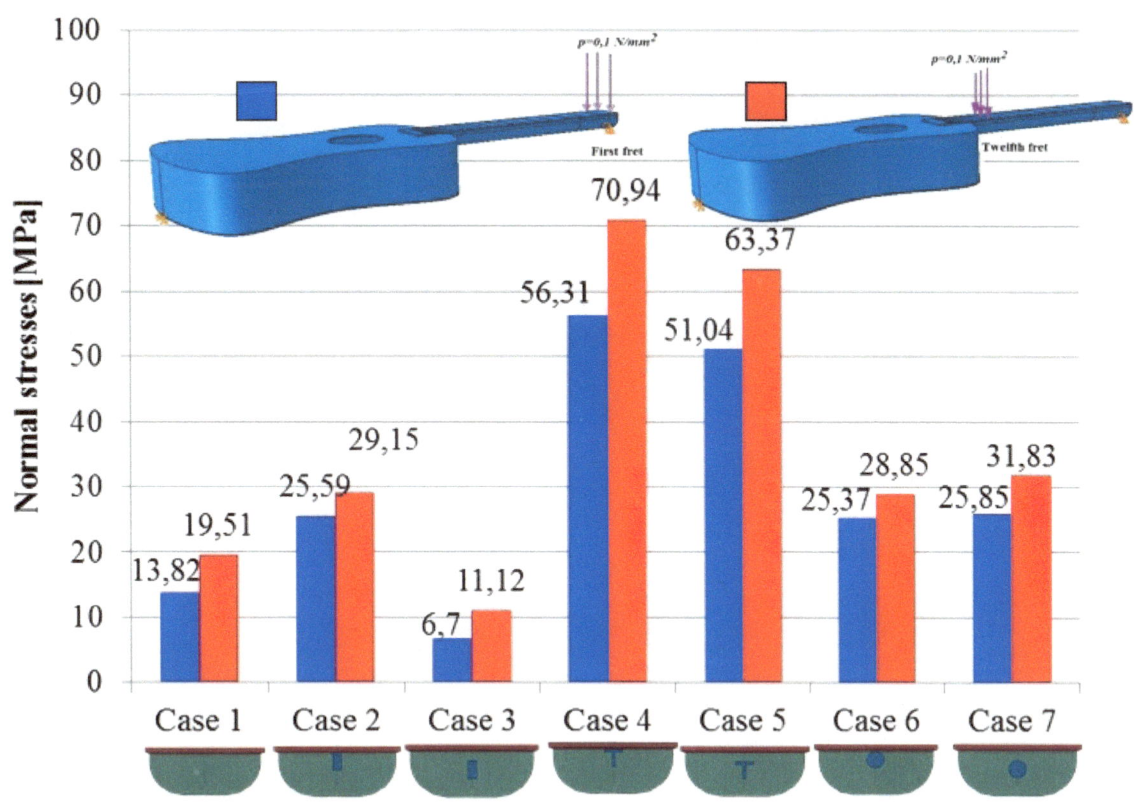

Figure 6. Variation of normal stresses with type of reinforcement and position of load on fretboard

Concerning of displacement variation, maximum values are obtained in case of simple guitar neck (case 1) which means that the neck is exposed to flexural bending and in time to residual displacements. This phenomenon leads to decreased acoustic quality of guitar. The lower values of displacement are recorded for rectangular shape of bar section translated with 5 mm below fret board. In this case (case 3), the position of load on fret board does not influence the size of flexural.

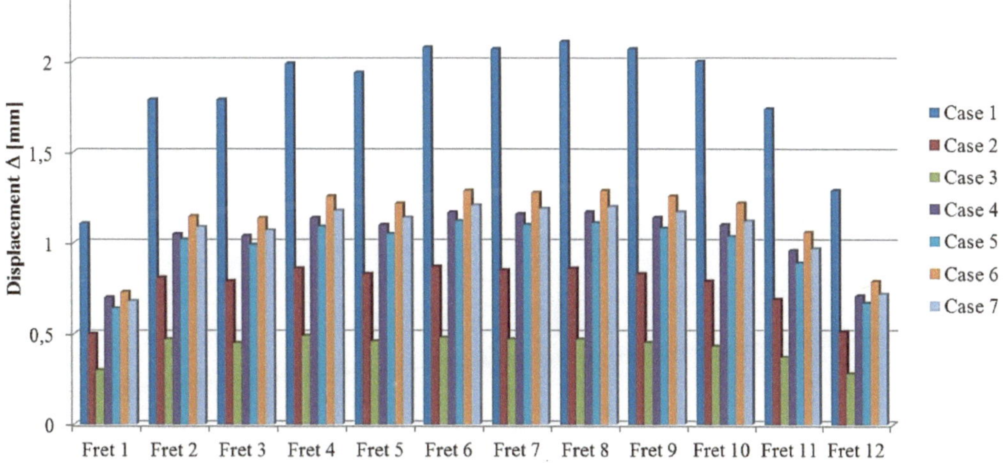

Figure 7. Variation of displacemen with type of reinforcement and position of load on fretboard

The reinforcement bars inserted in guitar neck modify the weight of guitar and thus the own frequency of guitar. So, it was analyzed the natural frequency and modal shapes for each type of reinforcement bar (Figure 8).

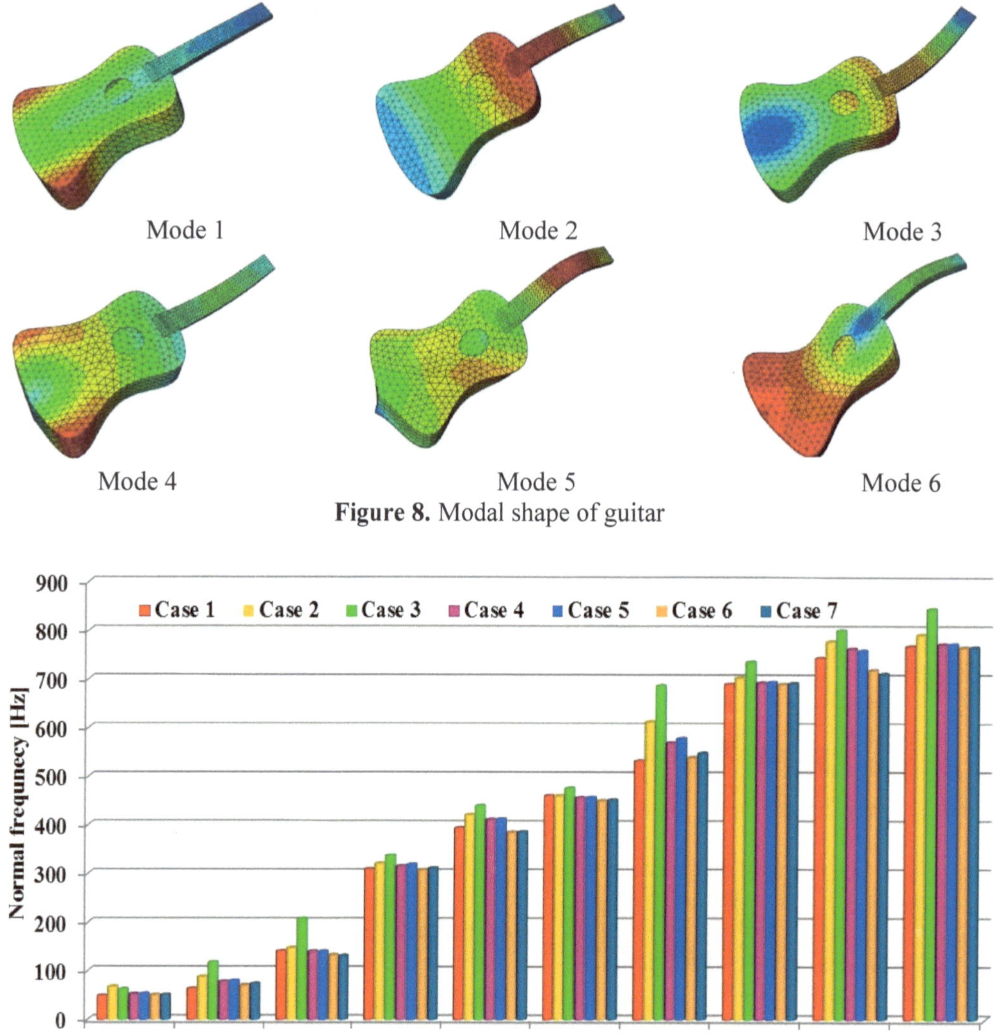

Figure 8. Modal shape of guitar

Figure 9. First eight natural frequencies for each case

The higher value of natural frequency is recorded in case of rectangular shape of bar section (case 2) and the lower value is obtained for not reinforced guitar neck. Similar values of natural frequency were obtained for the other types of cross section of bars (T and circular).

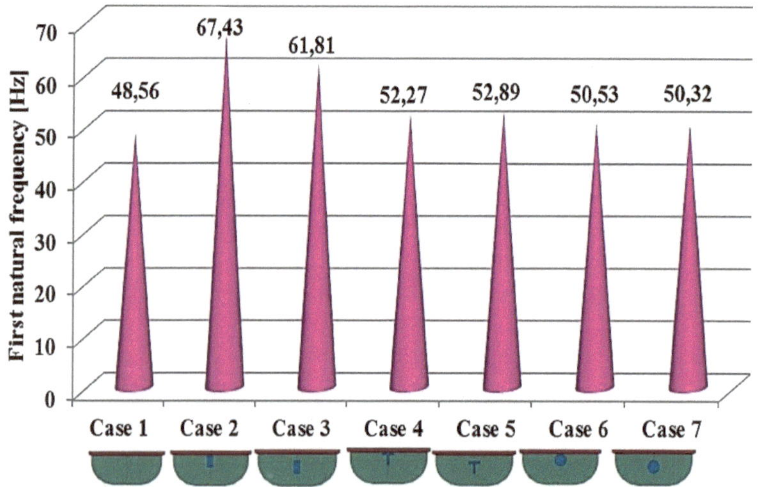

Figure 10. First natural frequencies for each case

4. CONCLUSION

In this paper was presented a numerical study of flexural bending with application of guitar neck which can be considered a cantilever beam. Due to dynamic loads during playing, the neck of guitar is subjected to flexural bending with residual displacement which leads to false tone. So, from mechanical point of view, is important to improve the rigidity of guitar neck using reinforce rods with different cross section shapes. It was analyzed seven cases and the conclusion is that regardless of type of cross section, the structure is stiffer compare to simple neck (case 1).

REFERENCES

[1] Stanciu M. D., Curtu I. Dinamica Structurii Chitarei Clasice, Ed . Universitatii Transilvania din Brasov, 2012, ISBN 978-606-19-0074-9.

[2] Curtu, I., Stanciu M.D., Baba, M., The Numerical Modeling of the Acoustic Plates on the Guitar Structures, in: Annals of the University of Petrosani – Mechanical Engineering, vol. 10 (XXXVII), Ed. Universitas Petrosani Romania, (2008) 41-46.

[3] French, R.M. Engineering the guitar, Springer Science, 2009.

[4] Stanciu, M. D., Curtu, I. Muzica Lemnului. in: PROLigno (CNCSIS B+), vol. 3, Nr. 2-2007, p. 61-68, ISSN 1841-4737.

[5] Curtu, I., Ghelmeziu, N. Mecanica lemnului şi a materialelor pe bază de lemn. Ed. Tehnică, Bucureşti, România, 1984.

[6] Bucur, V. Acoustic of wood. Springer-Verlag Berlin Heidelberg New York, 2006

[7] Stanciu M. D., Curtu I., Mihalache Daniel: Design of Experimental Test Bench for Determining the Stresses and Strains State of Guitar Neck, in Applied Mechanics and Materials, Vol. 658 (2014) pp 225-230

ON THE DYNAMIC CONTACT WITH FRICTION

Ligia Munteanu[1], Veturia Chiroiu[1], Cornel Brişan[2], Dan Dumitriu[1]
[1] Institute of Solid Mechanics, Romanian Academy, Bucharest,
e-mails ligia_munteanu@hotmail.com, veturiachiroiu@yahoo.com, dumitri04@yahoo.com,
[2] Technical University of Cluj-Napoca, e-mail cornel.brisan@mmfm.utcluj.ro

Abstract: *In this paper, the 3D normal vibro-contact problem with friction is investigated for the contact between the tire and the off-road. The road profiles are developed by using the image sonification technique. Direct and inverse approaches for the image sonification are developed in order to determine the contact domain, the natural frequencies and modes when the tire is in ground contact, and also, the pressure distribution of the surface of the half-space which defines the tire tread.*
Keywords: *Vibro-contact; Friction; Tire/off-road contact, Image sonification*

1. INTRODUCTION

In this work, the dynamic road concept is introduced in order to characterize a particular stretch of road by total longitudinal, lateral, and normal forces as well as their geometric distributions in the contact patches. These forces would give an equivalent driveline which if applied to a contact patch would have the same mechanical effect to the vehicle. For example, if we neglect any slip and if we consider only a point contact, the climbing on a curb or stone with a wheel could be reproduced with a translational couple which can vertically act to raise the wheel. The image sonification is used in this paper to explore the geometric properties of real off-roads with hardly detectable details, or different cross-sectional slices of the image by interpolating the sound parameters furnished by the transformation map.
By applying an inverse algorithm, the analytical representations of the off-roads curves can be developed by using a genetic algorithm. Sonification can be defined as the use of non-speech audio to convey information, possible in a bijective way [1, 2]. The aim of sonification is to turn some sort of data into some sort of sound [3]. General purpose of sonification is the facilitating communication or interpretation of data [4].

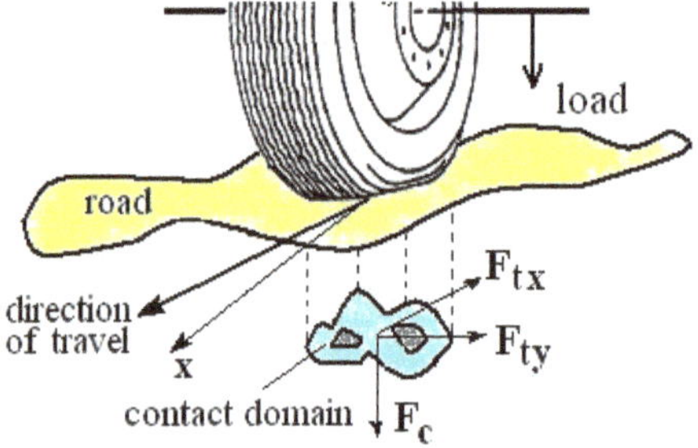

Figure 1: The contact tire/off-road.

Sonification is based on different disciplines ranging from the sciences to the arts, both of which are often linked with technology. As a result, no traditional methods, or canonical form exist for sonification [5].
The sound and vision are interesting and attractive subjects for application of sound in the playback of curve shape and curvature data [6]. In a particular way, they can be designated for driving simulator platforms
A scheme of contact between the tire and the off-road is represented in Figure 1. F_c is the contact force, F_{tx} the longitudinal component of the friction force acting in the X direction, F_{ty} the lateral component of the friction force acting in the Y direction.
In this paper, we propose an approach based on the direct and inverse image sonification algorithms, for solving the 3D normal vibro-contact problem with friction for the contact tire/off-road.

2. SONIFICATION ALGORITHM

The images chosen for sonification are those of various off-roads. These images depict abrupt changes in profile and can benefit from the accompanying sound by calculating of several cross-sectional slices of them, or even by providing additionally information to that gained from images alone. By using sonification, sounds will be generated directly from digital images, possible in a bijective way, and tested to see whether they aid in the study of them, or not. Cross-sectional slices of the image can be built by interpolating the sound parameters furnished by the transformation map. The aim of sonification is to create a road design algorithm as previously mentioned in [7-9].
A 3D sample of the virtual off-road is represented in Figure 2. The direction of travel is OX , while the depth of the road is OZ . Possible 2D (X,Y) cross-sectional slice of this virtual off-road, before and after deformation, are delimitated by two plane curve $s_1(Z,X)$ and $s_2(Y,X)$ that can be represented as series of cnoidal functions

$$s_1(Z,X)=\sum_{j=1}^{N}\alpha_j \mathrm{cn}^j(m_j,k_{1j}X+k_{3j}Z),\quad s_2(Y,X)=\sum_{j=1}^{N}\beta_j \mathrm{cn}^j(m_j,k_{1j}X+k_{2j}Y), \tag{1}$$

where N is the finite number of degrees of freedom of the curves, $0\le m_j\le 1$ is the modulus of the Jacobean elliptic function, and α,β,k are unknown parameters that are determinedf from an inverse technique.
Consider the set $D\{d_i, i=1,...,M\}$ of sound parameters obtained by applying the sonification operator S to the virtual off-road shown in Figure 2

$$D=S(p), \tag{2}$$

where $S(p):\Omega_1\to\Omega_2$ is a nonlinear differentiable operator, and Ω_1,Ω_2 are open and bounded subset of R^n. Ω_1 is an open bounded subset of R^n representing the set of image data, and Ω_2 is an open bounded subset of R^n representing the set of sound data. Consider the inverse nonlinear ill-posed operator equation

$$S^{-1}(D)=s, \tag{3}$$

where $S^{-1}(D):\Omega_2\to\Omega_1$, and s ($s_1(Z,X)$ or $s_2(Y,X)$). is the curve of the cross-sectional slice of the image. Suppose that (3) has at least one solution $s\{p_i, i=1,...,P\}$. The problem (3) is an ill-posed problem in the sense that the solution of (3) does not depend continuously on data. Almost always only noisy data are available in practice, so that a direct inversion of noise-contamined data D^δ would not lead to a meaningful solution. This article uses a genetic algorithm for recovering of the unknown solution s of (3) from a noisy version D^δ which verifies $S^{-1}(D^\delta)=s$. The noisy data D^δ with noise level δ verify $\|D^\delta-D\|\le\delta$.
The purpose of the inverse problem is to find the parameters

$$P=\{\alpha_j,\beta_j,m_j,k_{1j},k_{2j},k_{3j},\omega_n\},\ j=1,...,N, \tag{4}$$

by a least-squares optimization technique. The objective function $J(P)$ is defined as the distance between the computed data $p(u)$ for a fixed location of a point belonging to $s_1(Z,X)$ and/or $s_2(Y,X)$, and the corresponding sound data $D\{d_i, i=1,...,M\}$. The function $J(P)$ can be expressed as

$$J(P)=\int_C |p(u)-d(u)|^2\,\mathrm{ds}, \tag{5}$$

A genetic algorithm is used to minimize $J(P)$. The genetic algorithm is running until it is reached a non-trivial minimizer $s(P)$, which will be a point at which (5) admits a global minimum.

Figure 2: A 3D sample of the virtual off-road.

For a cross-sectional slice (Z,X) of the off-road displayed in Figure 3, the spectra plotted data is represented in the left-hand side of Figure 3, before deformation. It is shown each spectrometer pixel in the range 300 nm to 1000 nm, each pixel is 3.125 nm wide. The absolute amplitudes between plots are not significant. In the right-hand side of Figure 3, the corresponding curve $s(Z,X)$ before deformation is represented.

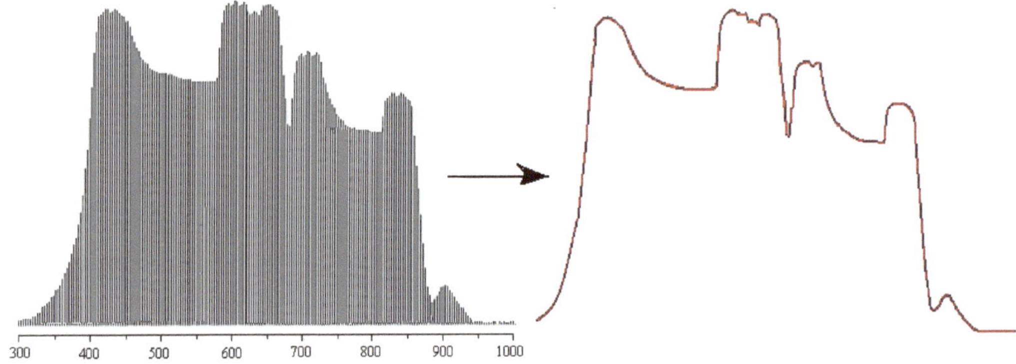

Figure 3: Spectra of the cross sectional slice (Z,X), and the corresponding curve $s(Z,X)$ before deformation.

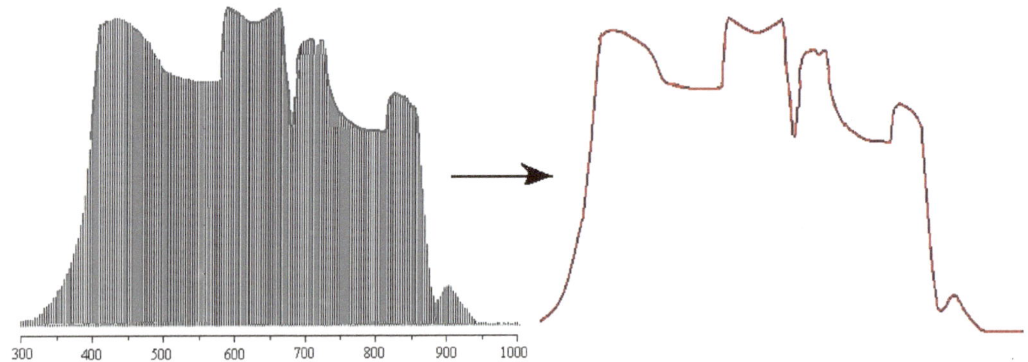

Figure 4: Spectra of the cross sectional slice (Z,X), and the corresponding curve $s(Z,X)$ after deformation.

Measurement noise is artificially introduced for determining the curve $s(Z,X)$ after deformation. It is done by multiplication of the data values for the curve $s(Z,X)$ before deformation by $1+r$, r being random numbers uniformly distributed in $[-\varepsilon,\varepsilon]$, with $\varepsilon = 10^{-1}, 10^{-2}, 10^{-3}$. Spectra of the cross sectional slice (Z,X), and the corresponding curve $s(Z,X)$ after deformation are shown in Figure 4 for $\varepsilon = 10^{-1}$.

To shape of the unknown contact domain is taken as a superellipse defined by a Lamé curve [10-13]

$$\left(\frac{x}{a(t)}\right)^n+\left(\frac{y}{b(t)}\right)^n=1,\ n>0, \tag{6}$$

where x and y define the envelope of the contact area, a is half of the contact length, and b is half of the contact width (radii of the oval shape are depending of time), and n the power of the ellipsoid.
A representation of in polar coordinates $x=r\cos\theta$, $y=r\sin\theta$, is given by

$$r(\theta)=\left[\left|\frac{\cos\theta}{a}\right|^n+\left|\frac{\sin\theta}{b}\right|^n\right]^{-1/n},\ 0\le\theta<\frac{\pi}{2}, \tag{7}$$

with $r=\frac{\tilde{r}}{\max(a,b)}$, $0\le r\le 1$, $\tilde{r}=\left[\frac{1}{n}(a^n+b^n)\right]^{1/n}$, $\lim_{n\to\infty}\tilde{r}=\max(a,b)$. By applying the Sneddon dual integral equations for elastic contact problem [11], the solution of the contact problem can be written as

$$p(r(\theta))=\left[\frac{3PH^2E}{2(1-\nu^2)}\right]^{1/3}\left(\frac{1}{2\bar{E}(\varphi,r\bar{k})}\right)^{2/3}\left(1-r^2\bar{k}^2\right)^{1/6}, \tag{8}$$

where $H=\frac{1}{n}\left(\frac{1}{\rho_1}+\frac{1}{\rho_2}\right)$ is the mean curvature of the contact domain $z=\frac{1}{n}\left(\frac{x^n}{\rho_1}+\frac{y^n}{\rho_2}\right)$, with ρ_1 and ρ_2 are the curvature radii in origine., $\varphi=\arcsin(1-r)$, $\varphi\le\frac{\pi}{2}$, $\bar{k}=\frac{\left(n\tilde{r}-2b^n\right)^{1/n}}{a}$, $a=\max(a,b)$, and $F(\varphi,\bar{k}r)$, $\bar{E}(\varphi,\bar{k}r)$ are the elliptic integrals of the first and of the second kind, respectively . The maximum value of $p(r)$ is obtained for $r\to 0$ $\left(\varphi=\frac{\pi}{2}\right)$

$$p_{\max}=p(0)=\left[\frac{3PH^2E}{(1-\nu)}\right]^{1/3}\left(\frac{1}{2\bar{E}(\pi/2,\bar{k})}\right)^{2/3}\left(1-\bar{k}^2\right)^{1/6}. \tag{9}$$

The displacement δ at the surface of the half-space is expressed as

$$\delta_z=\left[\left(\frac{3}{2}\right)^2\frac{(1-\nu)^2}{4\pi^2E^2}P^2H\right]^{1/3}\frac{\left(1-\bar{k}^2\right)^{1/3}K(\bar{k})}{\bar{E}^{1/3}(\bar{k})}. \tag{10}$$

where $F(\pi/2,\bar{k})=K(\bar{k})$ and $\bar{E}(\pi/2,\bar{k})=\bar{E}(\bar{k})$ are the complete elliptic integrals of the first and of the second kind, respectively, given by

$$c=\ln\left(\tan\left(\frac{3\pi}{8}\right)\right),\ d=\left(4^{1/4}c\frac{2\pi P}{l}\frac{\Gamma\left(\frac{1}{2}+\frac{1}{4}\right)}{\Gamma\left(1+\frac{1}{4}\right)}\right)^{1/3}. \tag{11}$$

We consider now the same3D sample of the virtual off-road represented in Figure 2. For arbitrary four cross-sectional slices (Z,X) of the off-road, the corresponding vertical displacements $s_j(t)$, $j=1,2,3,4$ of tires, are represented in Figure 5.

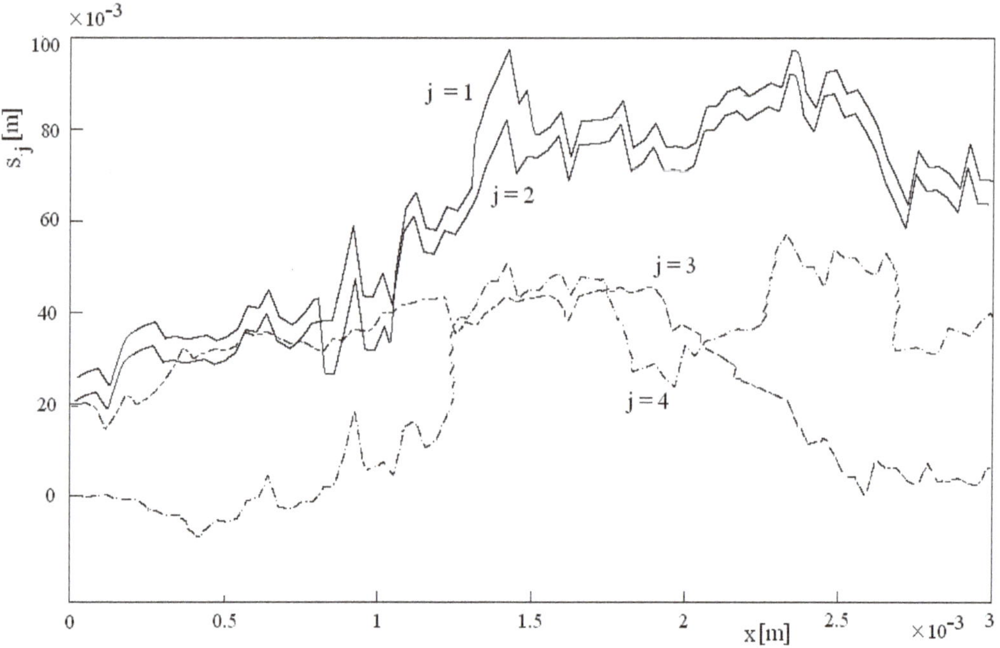

Figure 5: Functions s_j, $j = 1,2,3,4$ of vertical displacements of tires in the plane (Z, X).

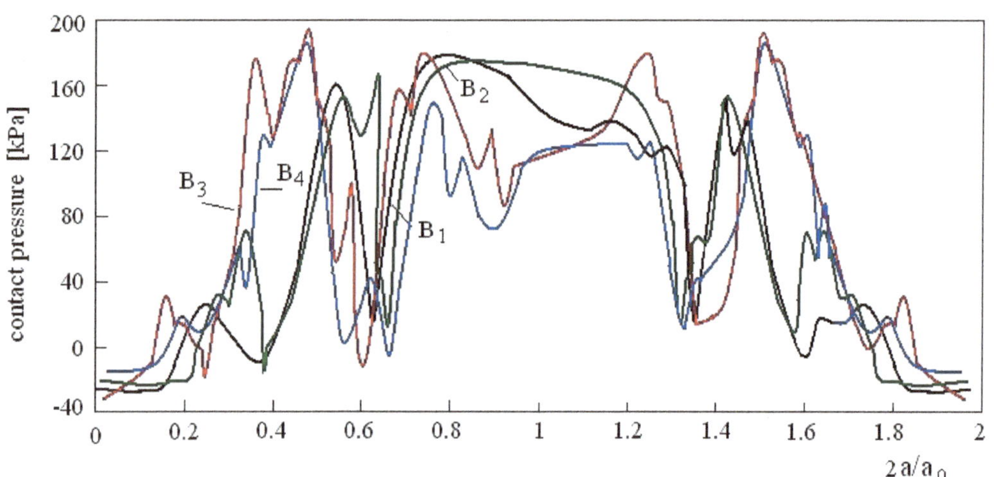

Figure 6: Maximum contact pressure in arbitrary contact patch.

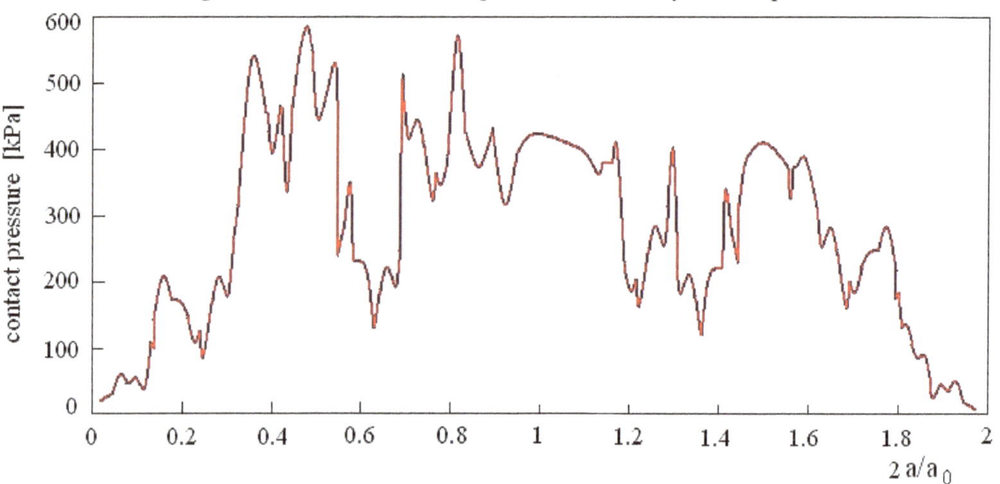

Figure 7: Maximum contact pressure in the contact patch.

To compare our results, consider the experiment done to the automotive laboratory of the Mechanical Engineering Faculty, Technische Universiteit Eindhoven with a real car wheel on the flat-plate [14]. After that the pressure is kept constant for approximately 5 seconds and next the pressure is released in 5 seconds. The maximum pressure is 4000 N. Maximum value of the contact pressures for this contact patch with respect to $2a/a_0$ are plotted in Figure 7, where a_0 is one reference radius of the contact domain. The results are in a perfect agreement with the pressure distribution obtained from experiment.

3. CONCLUSION

The aim of this work is to present a virtual experiment concerning the driving on the off-roads, via the building of the road profiles by direct and inverse sonification algorithms. The feasibility of communicating geometrical data through sonification can be useful to build a low-cost virtual reality environment with an increased degree of realism for driving simulators and higher user flexibility. The results are encouraging since they suggest that humans, previously unexposed to the system, are capable of establishing a clear connection between the sound and the underlying data.

Acknowledgement. This research was elaborated through the PN-II-PT-PCCA-2011-3.1-0190 Project nr. 149/2012 of the National Authority for Scientific Research (ANCS, UEFISCDI), Romania. The authors acknowledge the similar and equal contributions to this article.

REFERENCES

1. Munteanu, L., Chiroiu, V., Brişan, C., Dumitriu, D., Sireteanu, T., Petre, S., *On the 3D normal tire/off-road vibro-contact problem with friction,* Mechanical Systems and Signal Processing, 2014.
2. Kramer, G., *An introduction to auditory display*. In: Kramer G (eds) In auditory display. pp 1–79 Addison-Wesley, Boston, MA, 1994.
3. Hermann, T., *Taxonomy and definitions for sonification and auditory display*. Proceedings of the 14th International Conference on Auditory Display, Paris, France June 24 – 27, 2008.
4. Bonebright, T., Cook, P., Flowers, J. H., *Sonification Report: Status of the Field and Research Agenda*, Faculty Publications, Department of Psychology, Paper 444, 2010.
5. Licht, A., *Sound art, beyond music, between categories*. Rizzoli International Publications, Inc., New York, NY, 2007.
6. Shelley, S., Alonso, M., Hollowoof, J., Pettitt, M., Sharples, S., Hermes, D., Kohlrausch, A., *Interactive sonification of curve shape and curvature data*, In Lecture Notes in Computer Science 5763, Haptic and Audio Interaction Design, 4th International Conference, HAID2009, Dresden, Germany, Sept 10-11, 2009 (eds. M.Ercan Altinsoy, Ute Jekosch, Stephen Brewster) 51-60, 2009.
7. Brişan, C., Vasiu, R.V., Munteanu, L., *A modular road auto-generating algorithm for developing the road models for driving simulators.* Transportation Research part C: Emerging Technologies, 26, 269-284, 2013.
8. Vasiu, R.V., Brisan, C., *Aspects regarding modular road design in virtual reality*. In: Proceedings of WINVR, Milan, 27–29 June 2011.
9. Vasiu, R.V., Melinte, O., Vlădăreanu, V., Dumitriu, D., *On the responseof the car from road disturbances.* Romanian Journal of Technical Sciences – Applied Mechanics, 58(3), 2013.
10. Dumitriu, D., Munteanu, L., Brişan, C., Chiroiu, V., Vasiu, R. V., Melinte, O., Vlădăreanu, V., *On the continuum modeling of the tire/ road dynamic contact.* CMES: Computer Modeling in Engineering and Sciences, Materials & Continua, 94 (2), 159-173, 2013.
11. Dumitriu, D., Chiroiu, V., *On the dual equations in contact elasticity.* Revue Roumaine des Sciences Techniques – Série de Mécanique Appliquée, 2006.
12. Munteanu, L., Brişan, C., *On the modeling of contact interfaces with frictional slips*, Analele Universităţii "Eftimie Murgu" Reşiţa, Fascicola de Inginerie, XX(2), 17-24, 2013.

13. Munteanu, L., Brisan, C., Chiroiu, V., Donescu, Şt., *A 3D model for tire/road dynamic contact*. Acta Technica Napocensis, Series: Applied Mathematics and mechanics, 55(3), 611–614, 2012.
14. Backs, P.W., *Tyre/road contact measurements using pressure sensitive films*. Technische Universiteit Eindhoven, Report DCT 2007-026, 1-35, 2007.

SYNTHESIS AND CHARACTERIZATION OF SOME COMPOSITE MATERIALS OBTAINED FROM ELECTRONIC RECYCLING WASTE, WITH INTERSECTORIAL APPLICATIONS

A. R. Caramitu[1], S. Mitrea[1], D.Patroi[1], V. Tsakiris[1], V. Marinescu[1], G.A.Ursan[2], C. Tugui[2], C. Banciu[1]

[1]National Institute for R&D in Electrical Engineering, ICPE-CA, Bucharest, ROMANIA, e-mail: alina_caramitu@icpe-ca.ro

[2]SC ALL GREEN SRL, Iasi, ROMANIA, e-mail: andrei_urs@yahoo.com.

Abstract:*The main goal of the paper was to obtain and characterize from physical - mechanical, chemical and thermal point of view some new polymer matrix based composites. These composites consist in re-granulated low density polyethylene (LDPE), re-granulated polypropylene (PP) and high density polyethylene (HDPE) grinding, obtained by the recycling from electronic waste and Nano-conductive (NC) powder used as reinforcement material, in order to obtain new composites for inter-sectorial applications (automotive, electrical, etc.).Were been prepared and characterized different samples of composites by varying the percentage of the NC powder relative to the polymer matrix. (3%, 7% and 10%).*

Keywords: *composite materials, physical-mechanical and thermal characteristics.*

1. INTRODUCTION

The recent studies show that Polyethylene (PE), especially high density Polyethylene (HDPE), is the second most widely used material in civil constructions after Polyvinyl chloride (PVC). More, the trend is to diminish (up to complete replacement) the PVC in this sector (due mainly to the high degree of flammability and toxicity of combustion product). HDPE is also very convenient for automotive applications.

The same trend can be observed in the case of Polypropylene (PP), the interest being maxim in the filed of automotive applications.

Therefore, recycling of materials such as PE and PP from electronic waste and their use in construction/ automotive industry (like custom composites) represents a real interest because they can be easily identified and recovered (generally are not mixed with other thermoplastic compounds or contaminants).

2. TECHNICAL REQUIREMENTS

2.1. Materials

The raw materials used to obtain the composite materials are based on: low density Polyethylene (LDPE), high density Polyethylene (HDPE) and Polypropylene (PP), recycled from electronic waste. They were classified as follows: re-granulated LDPE from electronic waste, re-granulated PP from electronic waste and HDPE flakes from electronic waste. These materials were been mixed with different percents of NC powder (3%, 7% and 10%), obtained also from electronic waste

The samples were codified as follows: re-granulated LDPE from electronic waste- M_1; re-granulated LDPE from electronic waste/3NC- M_2; re-granulated LDPE from electronic waste/7NC- M_3; re-granulated LDPE from electronic waste/10NC- M_4; HDPE flakes from electronic waste- M_5; HDPE flakes from electronic waste/3NC- M_6; HDPE flakes from electronic waste/7NC- M_7; HDPE flakes from electronic waste/10NC- M_8; re-granulated PP from electronic waste-M_9; re-granulated PP from electronic waste/3NC- M_{10}; re-granulated PP from electronic waste/7NC- M_{11}; re-granulated PP from electronic waste/10NC- M_{12}.

2.2. Equiments, methods and procedures

2.2.1. Obtaining raw materials. Grinding was performed with a SPEX mill type, 8000M series. The milling time was 4 hours, with rotation speed of 875 cycles / min. Grinding was done in metallic or ceramic (tungsten carbide, alumina, zirconia, silicon nitride) mills. Dry milling was the simplest and succesfully to use.
2.2.2. Obtaining composite materials. In order to obtain a uniform distribution of the components of the mixtures in the whole structure, the reinforcing material (NC powder) and polymers were mixed for one hour in a cylindrical type TURBULA T2F mixer. After homogenization, the mixture was placed in a Dr. Boy 35A injection machine (Germany) where were obtained specimens according to the necessary types of tests to perform.
2.2.3. Identification of the crystalline phases by X-ray diffraction method were carried out on a X-ray diffractometer D8 Advance type.
2.2.4. Determination of mechanical properties (tensile strength and three points bending strength) was performed on the equipment for mechanical strength in static regime, LFM model 30kN, Walter & Sai AG Switzerland. Shore hardness was determined using a Shore A hardness tester (with measurement uncertainty of 0.6%).
2.2.5. Qualitative, semi quantitative and quantitative elemental chemical analyzes were performed on wave lengh X-ray fluorescence spectrometer S8 Tiger, having wide limits of quantification (from 10 ppm up to 100%).
2.2.6. Thermal analyzes were: termogavrimetrice analysis and analysis for determining conductivity and thermal diffusivity (measured between 25°C and 95°C). The analyzes were performed on a TG-DSC simultaneous thermal analyzer type STA 449 F3 Jupiter, and respectively a LFA-type device 447 NanoFlash – both from Netzsch (Germany).

2.3. Obtaining composite materials specimens by injection

In order to perfom the proposed test program, was neccessary to realize some dedicated specimens, using the injection method.
According to each composite material, were chosen the temperatures of the five heating zones of the injection machine cilinder, as follows:

- for re-granulated LDPE from electronic waste recipes : 190/190/200/210/220 (°C)
- for re-granulated PP from electronic waste recipes: 190/195/205/215/225(°C)
- for HDPE flakes from electronic waste recipes: 210/210/210/220/230(°C)

The specimens, having dedicated design, have been subjected to full program of tests, measurements and analyzes to determine the physico-mechanical, thermal and chemical properties, for possible intersectorial applications.

2.4. Identification of the crystalline phases by X-Ray diffraction method

In the Figures 1, 2 and 3 below, are presented the XRD spectra of the composite materials based on LDPE, HDE and PP (with different percents of NC powder). The XRD analyses were performed according to SR EN 13925-1, 2/2003 [1, 2], analysis identifying the network type of crystalline compounds and determining the unit cell parameters and crystallite size.
The XRD spectra of the samples have shown the follows:

- **in the case of re-granulated LDPE from electronic waste recipes** (figure 1): was identified and emphasized the specific structure of LDPE and the basic components of NC powder (Calcium/silicon/titanium oxides);
- **in the case of HDPE flakes from electronic waste recipes** (figure 2): was identified and emphasized the specific structure of a mix between HDPE and PP and also the basic components of NC powder (Calcium/silicon/titanium oxides);
- **in the case of re-granulated PP from electronic waste recipes** (figure 3): was identified and emphasized the specific structure of PP and the basic components of NC powder (Calcium/silicon/titanium oxides);

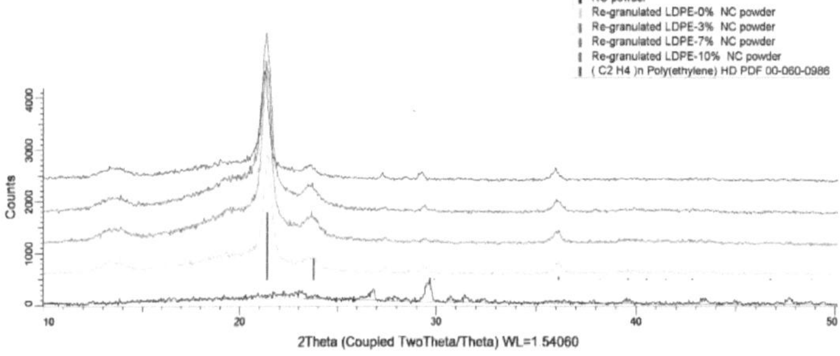

Figure 1 XRD spectra of re-granulated LDPE from electronic waste recipes

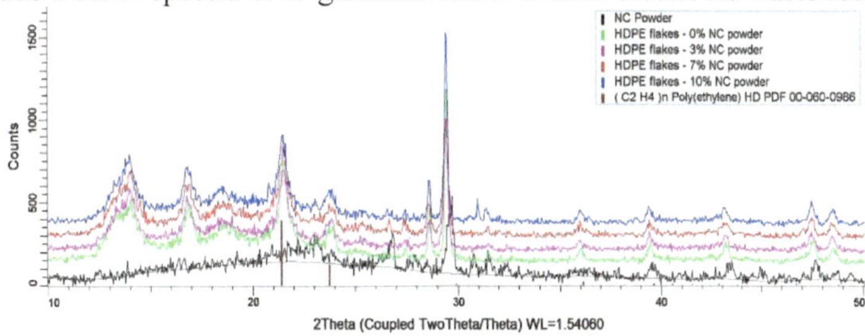

Figure 2 XRD spectra of **HDPE flakes** from electronic waste recipes

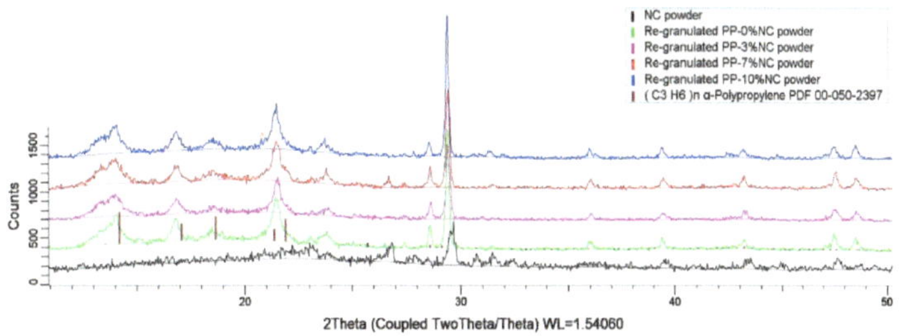

Figure 3: XRD spectra of re-granulated PP from electronic waste recipes

2.5. Determination of mechanical properties

The mechanical properties (tensile strength and three points bending strength were determined according to the SR EN ISO: 527-2:2000 standard. [3]

2.5.1. Tensile strength

The experimental results are summarized for each type of composite, in the Figure 4, below:

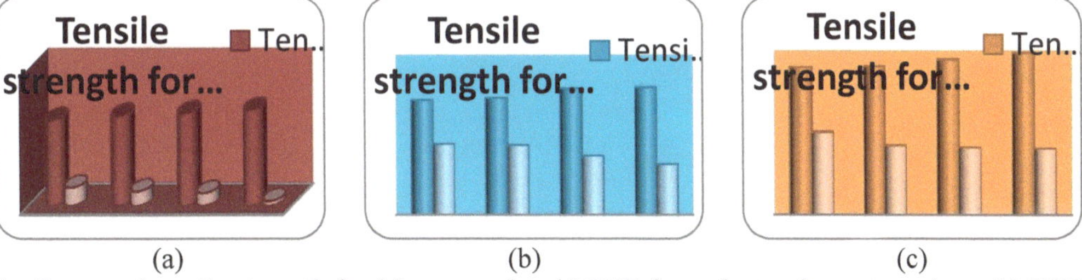

(a) (b) (c)

Figure 4: Compared tensile strength for (a) re-granulated LDPE from electronic waste recipes, (b) HDPE flakes from electronic waste recipes and (c) re-granulated PP from electronic waste recipes

The experimental results revealed the following:

- ***in the case of re-granulated LDPE from electronic waste recipes:*** the best values of the tensile strength were obtained for the sample M4- re-granulated LDPE from electronic waste/10NC. Increasing percentage addition of NC powder, leads to improved tensile strength as follows: Rm M_2> 5% Rm M_1; Rm M_3>7%Rm M_1 and Rm M_4>10%Rm M_1
- ***in the case of HDPE flakes from electronic waste recipes**:* the best values of the tensile strength were obtained for the sample HDPE flakes from electronic waste/10NC- M_8. Increasing percentage addition of NC powder, leads to improved tensile strength as follows: Rm M_6> 2% Rm M_5; Rm M_7>9%Rm M_5 and Rm M_8>10%Rm M_5
- ***in the case of re-granulated PP from electronic waste recipes:*** the best values of the tensile strength were obtained for the sample M12- re-granulated PP from electronic waste/10NC. Increasing percentage addition of NC powder, leads to improved tensile strength as follows: Rm M_{10}>0,5% Rm M_9 ; Rm M_{11}>5% Rm M_9 and Rm M_{12}>9% Rm M_9 .

2.5.2. Three points bending strength

2.5.3.

The experimental results are summarized for each type of composite, in the Figure 5, below:

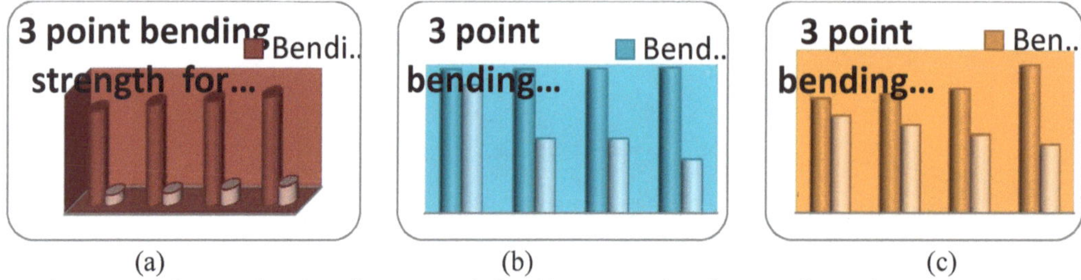

(a) (b) (c)

Figure 5: Compared three points bending strength for (a) re-granulated LDPE from electronic waste recipes, (b) HDPE flakes from electronic waste recipes and (c) re-granulated PP from electronic waste recipes

The experimental results revealed the following:

- ***in the case of re-granulated LDPE from electronic waste recipes:*** the best values of the bending strength were obtained for the sample M4- re-granulated LDPE from electronic waste/10NC. Increasing percentage addition of NC powder, leads to improved the bending strength as follows: Rm M_2> 6% Rm M_1; Rm M_3>9%Rm M_1 and Rm M_4>13%Rm M_1
- ***in the case of HDPE flakes from electronic waste recipes:*** the best values of the bending strength were obtained for the sample HDPE flakes from electronic waste/10NC- M_8. Increasing percentage addition of NC powder, leads to a very slight improved the bending strength as follows: Rm M_6> 0.03% Rm M_5; Rm M_7>0.4%Rm M_5 and Rm M_8>1%Rm M_5
- ***in the case of re-granulated PP from electronic waste recipes:*** the best values of the bending strength were obtained for the sample M12- re-granulated PP from electronic waste/10NC. Increasing percentage addition of NC powder, leads to improved the bending strength as follows: Rm M_{10}>4% Rm M_9 ; Rm M_{11}>8% Rm M_9 and Rm M_{12}>22% Rm M_9.

2.5.3. Shore hardness

Shore hardness was determined using a Shore A hardness tester (with measurement uncertainty of 0.6%), according to ASTM D 785 standard [4]. The values are represented as average of three measurements on each sample.
The statistical analysis leads to very closed values in terms of Shore hardness for all samples, whatever the nature of the composite (LPDE, HDPE or PP). More than, increasing percent of the NC powder, does not lead to modify the values of Shore hardness.

2.6. Chemical elemental analysis using X-Ray fluorescence spectrometry technique

Due the fact that the intermediate percents of NC powder (3% and respectively 7%) do not lead to suplimentary information in the case of such type of analysis, the chemical elemental analysis was performed only on the samples prepared without NC powder (M_5 and M_9) and with 10% NC powder (M_4 and M_{12}).
The analysis were been perform according to the internal procedure PI-18/2013 and the Operating Manual of the S8 TIGER device [5].
The figures 6-9 below, show the chemical composition identified for each composite analyzed:

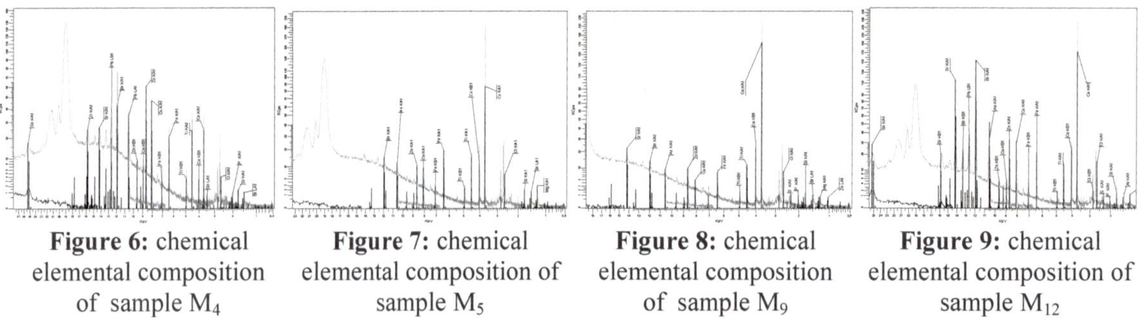

Figure 6: chemical elemental composition of sample M_4

Figure 7: chemical elemental composition of sample M_5

Figure 8: chemical elemental composition of sample M_9

Figure 9: chemical elemental composition of sample M_{12}

The results indicate the following:

- The NC powder used as reinforcing material, contains mainly elements: Si, Ca, Fe, Al, Cu, Pb, Ti, Sn, P, Sr (like oxides);
- The reference polymeric materials (LDPE, HDPE and PP) are chemically pure and is only emphasized organic matrix without any inclusions, impurities or additives

• The LDPE based composites have in composition additionally to polymeric matrix, elements such Ca, Ti, Si, and Fe (like oxides);

• The HDPE based composites have in composition additionally to polymeric matrix, the elements Ca, Ti and Si (like oxides);

• The PP based composites, have in composition additionally to polymeric matrix, the elements Ca, Ti, Si and Mg (like oxides);

• In all cases, the NC powder added in 10 percent into the polymeric matrix, leads both to the increasing of the common elements concentration and the appearance in the chemical composition of specific items only for NC powder (such as. Sn and Zr). This indicates an appropriate degree of mixing of polymers with NC powder during injection process to realize the testing samples.

2.7. Determination of thermal properties

2.7.1. *Thermal gravimetry and dinamic scanning calorimetry analysis (TG-DSC)*

The analysis was performed according to the ASTM E831-2006 standard [6].

In the Figure 10 below, are presented the thermograms of the analized samples:

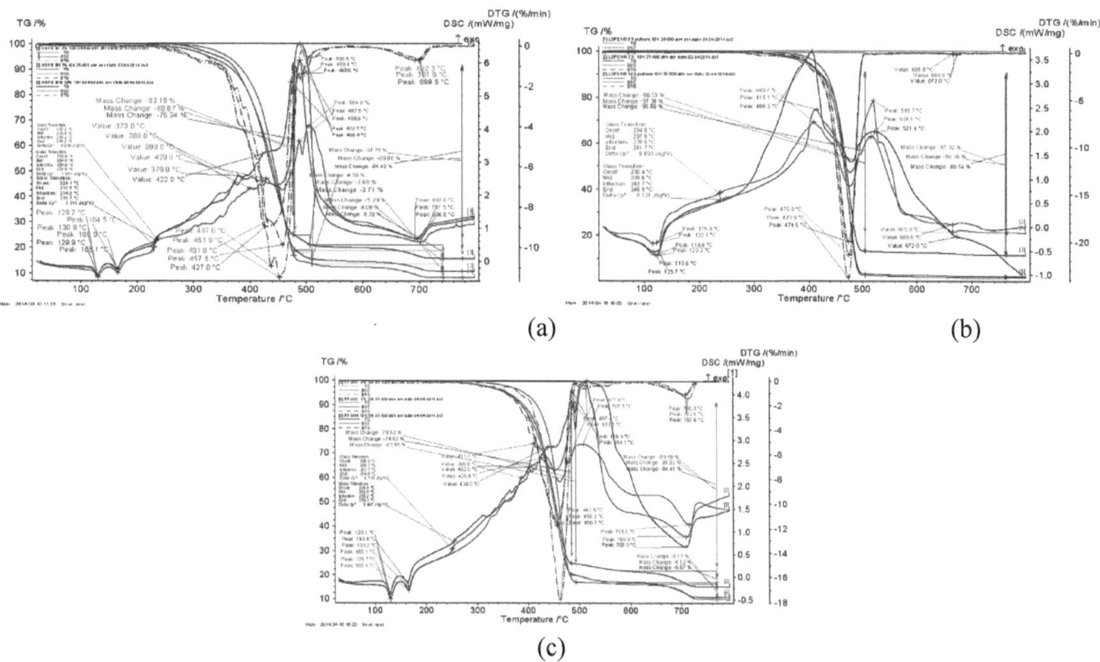

Figure10: TG- DSC curves for (a) HDPE flakes from electronic waste recipes, (b) re-granulated LDPE from electronic waste recipes and (c) re-granulated PP from electronic waste recipes

The thermograms represent an average result of determinations on three different samples, with a confidence level of 95% .

The materials present a similar behavior due the polymeric matrix. This behavior is limited to a series of phenomena, clearly identified in TG-thermal mass loss and their related derivative- DTG. DSC curves shows the absorbed or given heat by such materials in dynamic regime (differential scanning calorimetry).

♦ **Process I – Melting.** In all cases is distinguished the melting point (a glass transition of second kind**).**

▪ In the case of HDPE flakes from electronic waste recipes and re-granulated PP from electronic waste recipes , it shows a split of melting phenomenon, which may be due to re-granulation of polymeric material and to the addition of reinforcing powder.

▪ In the case of re-granulated LDPE from electronic waste recipes, it appears only the tendency to split the melting process, but without area widening.

- **Process II Oxidation .** In all cases can be observed the begining of the oxidation around 300°C. It is distinguished also the related mass loss TG-DTG.
- **Process III Thermo-oxidation.** This process is outlines in the TG – DTG curves.
- **Process IV Decomposition.** In the case of pure polymeric materials it observes a behavior consisting of a melting process, a glass transition (only for LDPE) and an oxidation-thermo-oxidation process. Compared to the pure polymeric materials, the analized composites shows a trend of translation of thermal phenomena to higher temperatures. This phenomenon is confirmed by the temperature differences found during glass transition process, where is observed increasing of starting process temperature, which may be due both to material re-granulation and NC powder addition.

2.7.2 Themal conductivity

The thermal conductivity measurements were been conducted between 25°C si 95°C, according to the ASTM E-1461:2007 standard, using "flash" method [7].
In the figure 11 below, is presented compared the variation of the thermal conductivity for each type of samples:

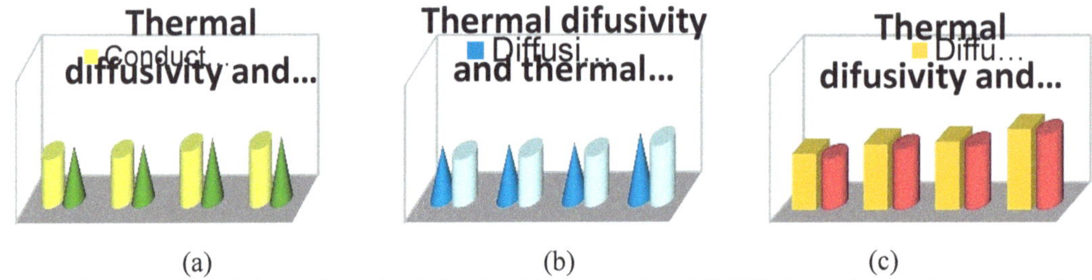

(a) (b) (c)

Figure 11: The variation of thermal conductivity for (a) re-granulated LDPE from electronic waste recipes, (b) HDPE flakes from electronic waste recipes and (c) re-granulated PP from electronic waste recipes

The experimental results show that the adding of increased percents of NC powder, has not important influence in terms of thermal conductivity. Anyway, the best values were obtained for the receipes containing the highest percent of NC powder (10%), respectively: re-granulated LDPE from electronic waste/10NC- M_4; HDPE flakes from electronic waste/10NC- M_8 and re-granulated PP from electronic waste/10NC- $M_{12.}$ Of theese, the best value was obtained for the re-granulated PP from electronic waste/10NC- M_{12} composite.

3. CONCLUSIONS

The tests conducted on the composites based on some materials from electronic waste, were able to draw the following conclusions:
-The composites based on different polymers (LDPE, HDPE and PP) from electronic waste, have basic polymer composition. The adding of the NC powder like reinforcing material, leads in all cases to the appearance of specific peaks in the existing compounds (oxides of calcium, titanium, silicon and / or combinations thereof);
-Mechanical -tests have emerged as the best option composite the re-granulated PP from electronic waste/10NC-M_{12};
- Chemical tests indicate that NC powder (used in all three types of composites) contains mainly elements Si, Ca, Fe, Al, Cu, Pb, Ti, Sn, P, Sr (like oxides). Reference materials (LDPE, HDPE and PP) are almost chemically pure and is only emphasized the organic matrix, without any inclusions, impurities or additives;
- The thermal analyses revealed the presence of melting, glass transition (only for LDPE) and oxidation-thermo-oxidation processes. Compared to the reference materials, the reinforced composites show a trend of translation of thermal phenomena to higher temperatures.
-Thermal conductivity highest value was found also for is the best option for re-granulated PP from electronic waste/10NC- M_{12};
All the tests indicate the re-granulated PP from electronic waste/10NC receipt as the best option for intersectorial applications.

4. REFERENCES

[1] SR EN 13925-1, 2/2003 Qualitative phase analysis
[2] SR EN 13925/3 -2005 and PI - 04
[3] SR EN ISO: 527-2:2000 – Plastic materials. Tensile strength. Part 2-Test conditions for injected and extruded plastic materials. Test Method for Rockwell Hardness of Plastics and Electrical Insulating Materials.
[4] PI-18/2013 and Operating Manual of the S8 TIGER device
[5] Workstation Auriga SmartSEM V05.04.
[6] ASTM E831-2006 Linear Thermal Expansion of Solid Materials by Thermomechanical Analysis.
[7] ASTM E-1461:2007 Standard Test Method for Thermal Diffusivity by the Flash Method

TEMPERATURE ADAPTIVE CONTROL USING THE ADDITIVE MANUFACTURING FOR INJECTION MOLDING POLYMERIC PRODUCTS

Teodorescu Drgahicescu Florin[1], Opran Constantin Gheorghe[2], Pascu Nicoleta Elisabeta[3]

[1] University Polytehnica, Bucharest, ROMANIA, florin.teodorescu@ltpc.pub.ro
[2] University Polytehnica, Bucharest, ROMANIA, opran.constantin@ltpc.pub.ro
[3] University Polytehnica, Bucharest, ROMANIA, nicoletaelisabeta_pascu@yahoo.ro

Abstract: *This paper presents the adaptive control of the temperature in the injection mold using additive manufacturing in order to achieve the controlled and specific cooling channels in the components parts of the injection molds of polymeric products. The increasing complexity of technical characteristics by construction and technology of the polymer products lead to increasingly stringent specifications for their dimensional stability and to complicated injection molds cost prices involving high reliability issues. The adaptive control of the temperature in the injection molds produced polymeric is the solution that solves these performance requirements. The main problems in the adaptive control of the temperature by injection molds is given by the thermodynamic characteristics of active plate material of the mold and by the distribution of the thermal fields and the cooling of the polymer product injected. The differential temperature control of the various areas of the active plaque cavity by injection mold using the conventional or assisted cooling systems lead to high manufacturing times and the bad quality of the product. This paper presents ways to realize of specific cooling channels so that they will be conform to the contour of the parts or integrated in the specific heat conductive material, embedded in the active plate for the injection polymer product. It is also presents the technologies used to achieve the specific cooling channels like direct laser metal forming, "direct metal laser sintering" (DMLS) and the method of introduction of high conductivity material in the active areas of the injection molds plate. Each type of product requires a specific construction technology to achieve specific.By creating these specific channels occurs adaptive temperature control in injection mold active areas resulting low manufacturing times and conformal quality.*

Keywords: *temperature adaptive control, additive manufacturing, injection molding,direct metal laser sintering*

1. INTRODUCTION

Of particular importance for injection molds is that they allow obtaining products injected for material losses to a minimum, in compliance with technical requirements. The injection parts of complex shapes, concave, must apply an injection temperature controlled by the surface of the part while maintaining pressure duration further. When opening the mold can occur in parts molded internal tensions that can cause cracking of parts (in case more rigid amorphous thermoplastics such as polystyrene shock-resistant PAS) or their deformation (in the case of semi-crystalline thermoplastics such as PE-polyethylene flexible . Temperature plays an important role influencing mold: internal stress, deformation, dimensional accuracy, weight, surface quality. Cooling time is determined by the temperature of the mold surface. The heating equipment - cooling must ensure a constant temperature of the mold with some limitations related to the cooling channels position closer to the nest mold. The characteristic cooling time of the cooling process during the longest part of the injection cycle representing approximately 68% of the total duration of the cycle. To achieve, in view of a high productivity, short cycle time, shall be provided reduction measures for the cooling. Injection technology, involves getting a time and cooling rate so as to ensure the quality prescribed injected part. By varying the cooling time and mold temperature (which determines the cooling rate) is depleted the technological possibilities of influencing the process of cooling in the mold. Although these technological possibilities seem easy to achieve, the cooling process involves many technical difficulties. This paper presents adaptive control temperature in the mold using manufacturing technologies through the generation of material to achieve controlled and specific cooling channels in injection molds components of polymeric products.

2. COOLING SYSTEMS OF INJECTION MOLDS

The higher the temperature of the thermoplastic material is higher, the more it is fluid filled lighter mold and injection time are reduced. Mold temperature is decisive phase cooling and solidification of the part.

Table 1: Temperatures of major thermoplastics and molds

Code	Material	The temperature of the molten material [°C]	Mould temperature [°C]
PS	Polystiren	190-220	40-50
ABS	Acrilonitril-butadien-styren Copolymer	210-240	60-70
P.A.6	Polyamid	230-250	100-110
P.A.6.6	Polyamid	270-290	100-110
PP	Polypropilene	240-270	30-50
PMMA	Polimetacrilat Metil (Plexiglas)	200-240	50-80
POM	Poliacetatpolioximetilen	200-220	60-100
PC	Policarbonat	290-310	80-120
PTFE	Politetrafluoretylene (Teflon)	340-370	140-230

In terms of technology, cooling channels position to the part, especially in parts of complex shapes is an intractable problem technologically. Such a classic mold LTPC existence in the laboratory of the University Politehnica of Bucharest, for a common part, as well as can be seen in Figure 1 where the cooling channels are driven on straight paths, trajectories relatively easily obtained in terms of technology. Rectilinear trajectory cooling channel can not be too close to a piece of complex size and different thickness. It is therefore very important that the cooling channels position to make the play closer or mold nest.

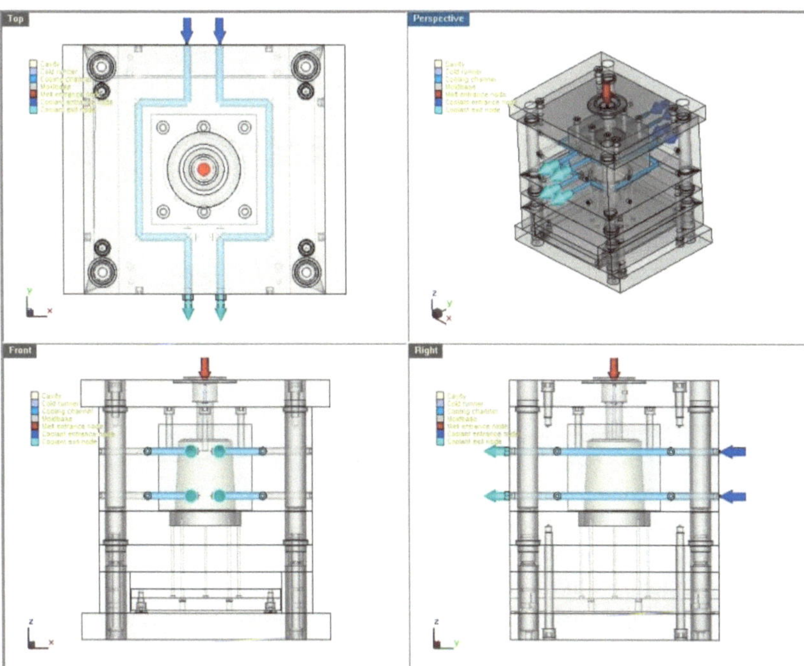

Figure 1: Making a mold cooling channels for a common parts (made in Moldex 3D)

Adaptive control temperature inside the mold must however be optimized relative to the temperatures of the injected material, mold temperature and cooling system after the injection mold piece. The main problem is the cooling of injection molds mold steel heat resistance. Depending on the quality of steel used typical thermal conductivity values are approximately 25 W/mk and down to about 12 to 15 W/mK, depending on the proportion of alloy additives. Tool steels are poor conductors of heat as compared to copper alloy or pure copper which has a thermal conductivity of approximately 390 W / mK. Figure 2 illustrates the situation of a conventional cooling heat the molds during the manufacturing process [1].

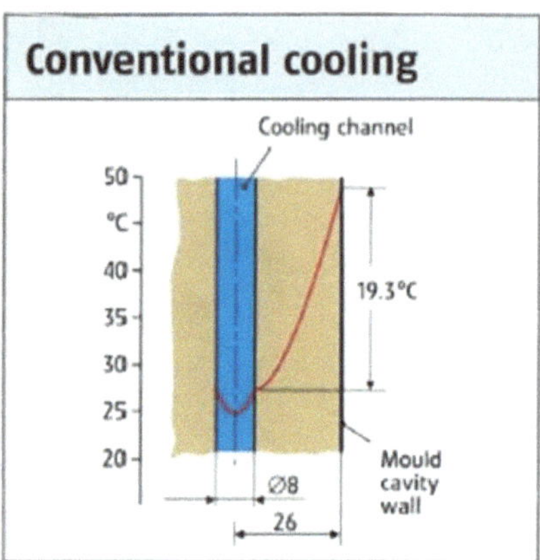

Figure 2: The temperature in the injection mold walls as compared to the reference temperature of the cooling medium

Although the distance from the cooling channel to the wall of the mold cavity is only 22 mm, the mold wall temperature increases only about 19°C to a core temperature of the cooling medium. If we consider now the situation in the mold installation channel, channel cooling situation that is far away from the mold wall, or even a region whose thin-walled cavity is completely encased in the polymer, the temperature difference between the cooling medium and the wall mould often reach values of 60 °C and more.

Such hot spots are due to the considerable extension of the time of closing of the mold and also may affect the quality of parts and the injection itself.

The problem can be addressed only by the arrangement of cooling channels so that they conform to the contour or integrated in highly thermally conductive material to be clamped tightly in tool steel workpiece regions.

The polymeric material of the nest during a cycle gives mold injection mold body, the amount of heat Q, which is calculated using the equation:

$Q=m(i_2-i_1)$ kcal

m – mass injected part, kg;

i_2 – enthalpy of the polymeric material from entering the mold;

i_1 – enthalpy of the polymeric material to mold release.

The enthalpy of the polymeric material is calculated using the equation:

$\Delta i= i_2- i_1=c_p(T_{Mp}-T_D)$ [kcal/kg]

c_p – specific heat of the polymeric material kcal/kg°C;

T_{Mp} – temperature of the material in the nest;

T_D – mold release temperature;

The amount of heat transfer of the piece is taken by mold and transported to the environment of moderation. The amount of heat Q is determined by the relationship:

$$Q = \frac{\lambda_M}{\delta} S(T_{pc} - T_{pT}) \text{ [W]}$$

where:

λ_M – thermal conductivity of the mold =0,197 [W/m·K]

δ – Channel Tempering distance from the mold surface,[m]

S – the active cross-sectional area of the mold ,[m²]

T_{pc} – The average temperature of the wall of the cavity =433 [k°]

T_{pT} – channel wall average temperature tempering =333 [K°]

Heat transfer from the mold (solid medium) to the tempering environment (liquid medium) is by convection and can be expressed by the relation:

$Q=\alpha_T S_T(T_{pT} - T_T)$ [W]

where:

α_T – heat transfer coefficient tempering environment =1310 [w/m²K]

S_T – active surface tempering channels, [m²]

T_{pT} - channel wall temperature tempering, =433 [k°]

T_T - ambient temperature tempering =393 [k°]

Heat transfer coefficient is calculated using the equation:

$$\alpha_T = \frac{\lambda}{d_c}[3,65 + \frac{0,0668 p_e \frac{d_c}{L_c}}{1 + 0,045 (p_e \frac{d_c}{L_c})^2}] \qquad [w/m^2K]$$

where:

d_c – Channel Tempering diameter , [m]

L_c – Channel Tempering length, [m]

3. MANUFACTURING TECHNOLOGY USE OF COOLING CHANNEL USING TECHNOLOGY GENERATION OF MATERIAL

The increasing complexity and geometry of injection molded parts and the more stringent specifications for dimensional stability increased the parts cost to put pressure in the injection molds and enhance feasibility problems already difficult. A recently launched service that has already gained attention is the direct laser metal forming (Direct Metal Laser Sintering-DMLS). This process generates an insert molded from a metal powder bed, which often is a material inserted in the original mold. A fine metal powder is placed in a heated room built on a platform that can be lowered. A direct laser melting the powder layer, which was initially leveled with a slide according to a 3D model, to directly form the contour of the part webs, and the powder is welded to a monolithic workpiece subsequent to the final hardness about 52 Rockwell C (fig. 3).

The conformal cooling, or the segmented mold cooling is a new technology that meets these chalenges.

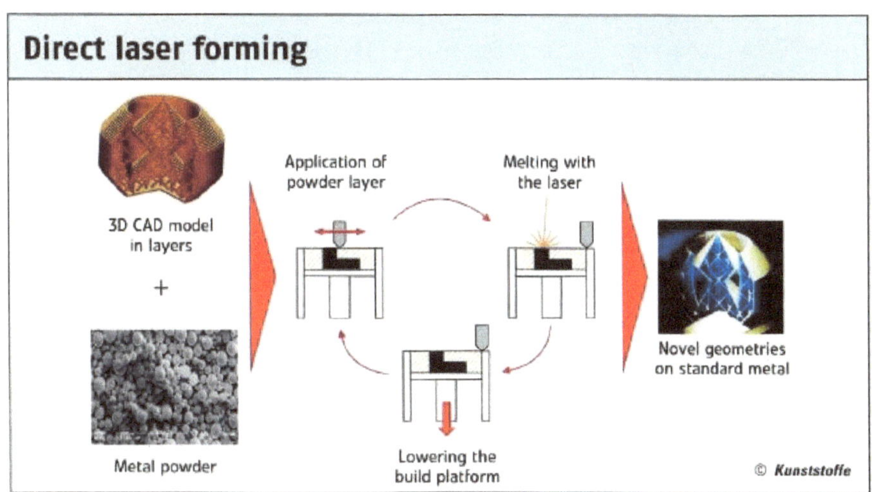

Figure 3: Laser Sintering using the method of 3D generating metal powder directly [2]

The steel in the mold has a density of about 100%. If the position of the cooling channel is in accordance with the shape of a part of the plastic material of the workpiece then is introduced into the 3D data set, which is then further insertion of the printed precisely on the three-dimensional matrix in the form of a channel. The advantage of the above-mentioned manufacturing process is complete freedom of design in creating channel. Although the method does not require cooling medium flowing through conformal channels, it provides adequate heat transfer due to high thermal conductivity of the material. Heat conductive needles, such as copper of high purity which can achieve a thermal conductivity of 390 W/mK can be inserted into the powder bed by taking the shape of the contour. [1]

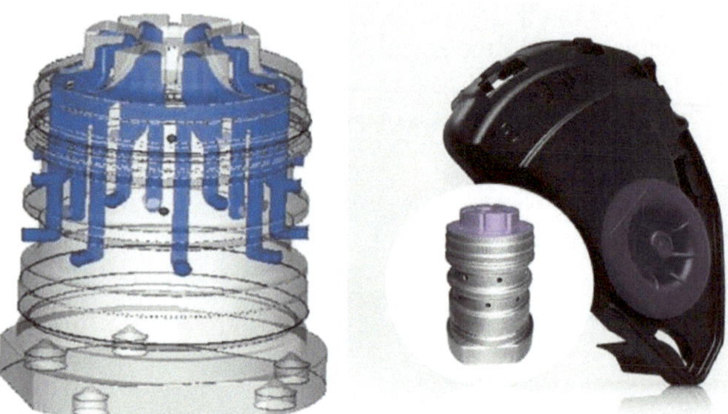

Figure 4: A 3D view of the inner cooling chanels of a tool insert, wich could not be manufactured using conventional machining [3]

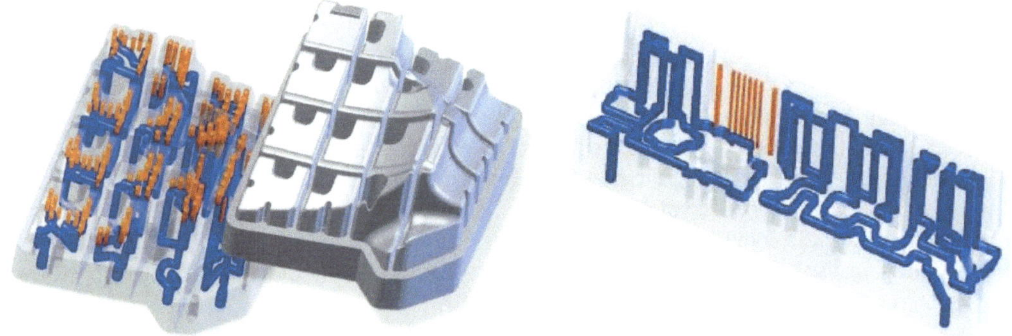

Figure 5: The mold for producing a liquid-cooled engine [3] ; **Figure 6** : Electronic Parts from ABS [3]

In the case where such copper elements are welded or sintered high-purity (by diffusion bonding) placed in the cavities prior to the insertion of the mold, average thermal conductivity of the new composite steel/copper is much higher than that of the steel used in the mold conventional manner. Depending on the design of the mold insert, such items steel/copper can reach or even exceed the thermal conductivity values, which are usually carried out only by copper alloys are also used in the manufacture of the mold .

Examples of building cooling channels to achieve a consistent cooling using modern additive manufacturing technologies for material generating are shown in Figures 4, 5 and 6.

The position and shape of the channels to be determined from the point of view of 3D additive manufacturing technologies being achieved by the design and proper segmentation of the individual channels. This segmentation is essential especially for large surface components because then the form is usually operated at different temperatures - in accordance with the various areas of the part. This means that mold temperatures are tailored to specific thermal balance of the piece. Such segmentation is a segmentation channels under different water temperatures and helps to influence and control the contraction of individual areas play actively. [4]

Figure 6 shows an example of an automotive cooling comply. The core of electronic parts ABS is about 300 mm long, 50 mm wide and about 100 mm in height. For the ABS-based core, defining the shape is controlled two separate cooling channels running beneath it. [5]

In the narrower base region, the heat transfer copper-composite steel pins provide for adequate heat removal. Placing the lower half of a mold core of a coolant tank for the engine shown in Figure 5 shows how impressive the different technologies interact to provide a viable solution. Space cooling channels in figure are projected on a technical level of production possible, being designed only to supply the coolant core regions and to allow application in channel cross sections. In further narrow base regions can be observed copper pins embedded in mold steel needles remove heat efficiently, even in areas with narrow contour regions.

Figure 7 and Figure 8 is trying to achieve cooling channels made by generating a 3D printer . These tests were obtained by the Polytechnic University of Bucharest.

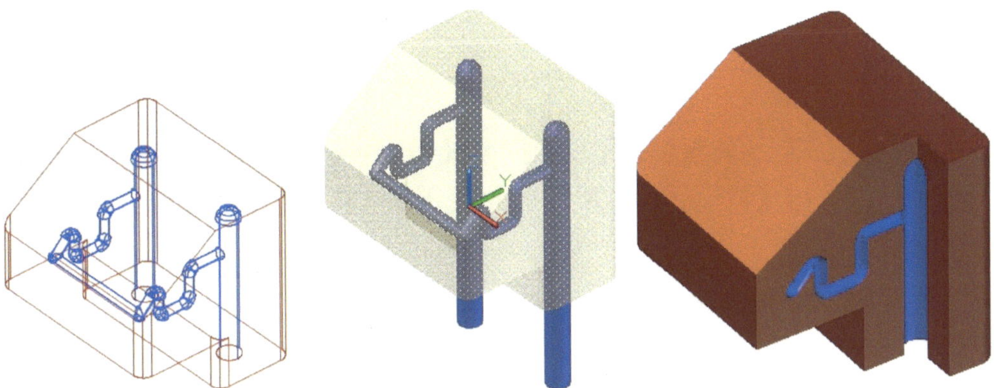

Figure 7: Cooling channels made for a sample generated by 3D printing through the additive manufacturing technologies

Figure 8: Physical part performed in a 3D printing machine

4. CONCLUSION

Mold productivity increases with the quality and efficiency of the cooling system. The ability to remove heat as quickly as the design of more efficient cooling channels and the choice of materials as appropriate and without making too complicated mold, mold will increase productivity. Copper-steel composite solutions are not subject to the disadvantages of the known copper-based alloys, such as reduced resistance to bending (due to the low modulus of elasticity), hardness and even less costly processing. A reduction of cycle time by an average of 30%, with lower rates for scrap, can quickly lead to a reduction in unit costs by 15% for injectable products. Differential temperature control of various areas of active plaque cavity injection mold using conventional cooling systems or assisted stroke lead to high manufacturing quality and conform.

New technologies used to produce specific cooling channels and direct laser metal forming "direct metal laser sintering" (DMLS) and the method of introduction of high conductivity material in the active areas of the plate as a cooling injection molds comply with contour.

By creating these specific channels trough additive manufacturing technologies occurs adaptive temperature control in injection mold active areas resulting in low and quality manufacturing times comply.

REFERENCES

[1] Westhoff R., Conformal Cooling on the Advance, Kunststoffe International nr. 8/2006, Carl Hanser Verlag, pp.24-25, 2006;

[2] Meyer R., Rapid Technologies in Automotive Engineering, ATZautotechnology nr. 11-12/2008, volume 8, pp. 34-37, 2008;

[3] Conformal Cooling Irons out Production Problems, European Tool&Mould Making, pp 33-35, May 2014;

[4] Metal Additive Manufacuring to Grain Ground in Die and Mould, European Tool&Mould Making, pp 38 -39, May 2014;

[5] SASAM Project, First European Standards at the European Level, tctmagazine, pp. 58-59, July 2014;

RENAULT VEHICLES AND THEIR PROGRESS REGARDING THE REDUCING OF ENVIRONMENTAL FOOTPRINT

Petrica Heroiu, Marius Ciprian Rusu, Oana Elena Gradin
Engineering & Material Characteristics Department, Renault Technologie Roumanie, Bucharest, Romania,
e-mail: petrica.heroiu@renault.com

Abstract: One of the major world ecological challenges is to reduce the environmental impact of the activity and products throughout their life cycle. Renault, 4th company in the automotive industry, has its environmental policy which makes the company more efficient and competitive. This policy is focused on following priorities: climate change and energy efficiency, resources and competitive circular economy, health and ecosystem, innovative mobility systems and environmental management and transparency.
Today, Renault vehicles are design to be at least 85% recyclable, integrating already parts based on recycled materials. For Europeans customers, the ECO2 label show that the Renault and Dacia vehicles have best environmental credentials and are the most economical to drive.
Keywords: *environment footprint, recycling materials, renewable sources, automotive*

1. INTRODUCTION

Automotive industrial processes and products present an important impact on health and environment. In order to manage these impacts, each vehicle producer makes big efforts to develop and to implement their own Environment Policy, according to worldwide environmental assessments and regulations [1].
Renault, 4th company in the automotive industry, has its environmental policy which makes the company more efficient and competitive. This policy is focused on following priorities: climate change and energy efficiency, resources and competitive circular economy, health and ecosystem, innovative mobility systems and environmental management and transparency. The worldwide network extended to all levels is used to implement and to monitor the respect of the Renault Environmental Policy.

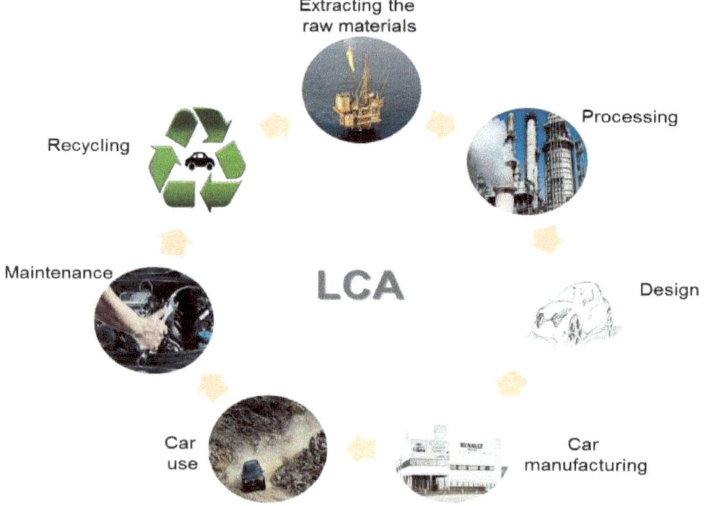

Figure 1. Schema of LCA method in automotive

The most worldwide used methodology to measure the environmental footprint of processes, products and services is Life Cycle Analysis (LCA). By this method, according to the ISO 14040 standard, are evaluated all manufacturing phases, from obtaining the raw materials to possible re-use or recycling of the product and its eventual disposal (Figure 1) [1].

Renault use Life Cycle Analysis (LCA) in order to measure our environmental progress from one vehicle generation to the next (e.g. Renault Clio - Figure 2).

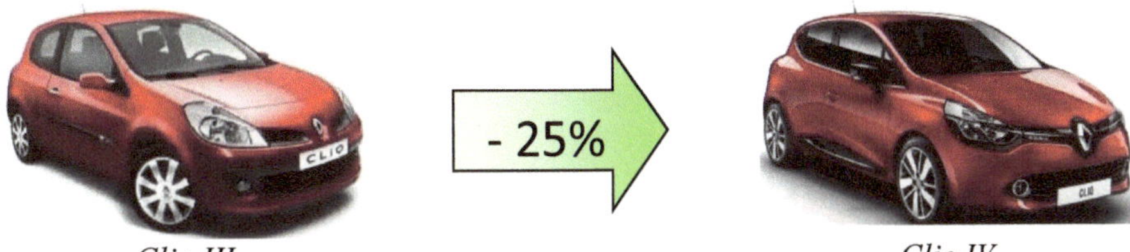

Clio III *Clio IV*

Figure 2. Decrease of the environmental impact during the car model evolution

In the present, Renault vehicles are design to be at least 85% recyclable, integrating already parts based on recycled materials (Figure 3). For Europeans customers, the ECO2 label show that the Renault and Dacia vehicles have best environmental credentials and are the most economical to drive.

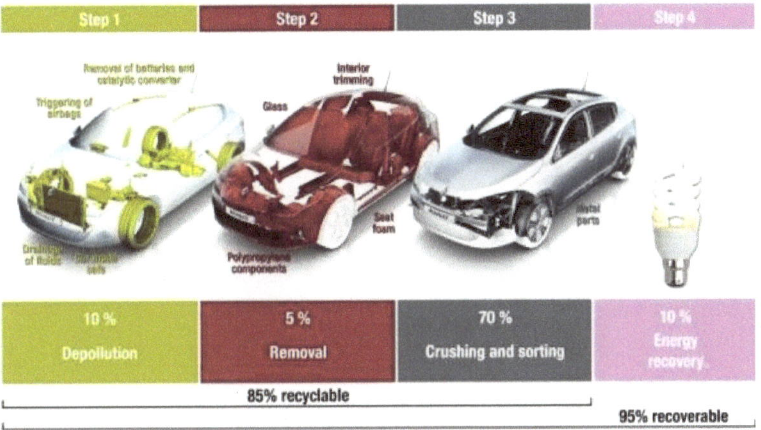

Figure 3. Recycling process at the car end life

2. RECYCLING PLASTIC MATERIALS USED IN RENAULT VEHICLES

As early as the 1990s, Renault led the way on the use of recycled plastics on its vehicles. Starting with launch of Clio in 1996 (first vehicle with recycled plastic materials parts – wheel housing liner), Renault has been carrying out an important recycling policy, increasing each year the amount of the used recycled plastic materials from 4 kg to more than 32 kg (Renault Laguna). The current objective of the company is to have 20 % of recycled plastic materials on new vehicles by 2016 [2].

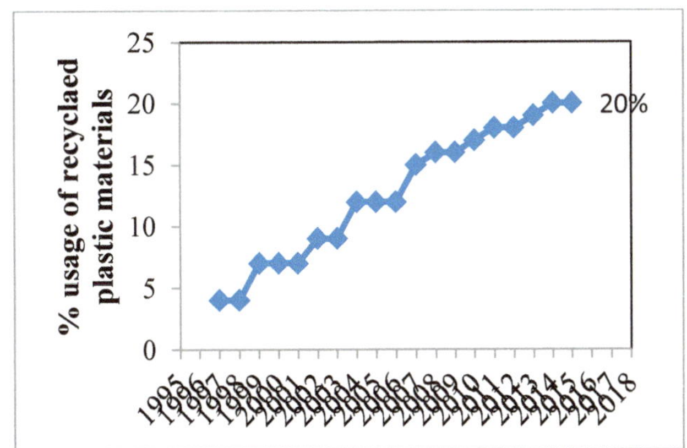

Figure 4. Evolution of the recycled materials used on Renault cars

Recycled materials are usually perceived as being a second choice when engineers start to design a certain vehicle part. However, it was showed that a well-recycled material can provide good results regarding the properties of the recycled material in comparison with original material. The potential of the recycled materials was highlighted when decorative parts were made from recycled plastic (Renault Megane), having no differences to the customers regarding the quality and functionality.

Figure 5. Parts in recyclable and renewable plastic material for Renault Megane vehicle

Renault Group funded research projects in order to develop recycling processes, especially for plastic materials. The most used plastic in Renault vehicles is polypropylene (PP). For this type of material Renault develop design guidance tools in order to prevent the pollution of the polypropylene with parasite polymers of equal density and incompatible materials present into the polypropylene process (PVC, glued textiles, metallic inserts etc.). A reference part is the door panel from Renault Laguna. To make it easier to recycle, engineers worked to the extent that all of its elements (speaker covers, shock absorbers etc.) located in the lower part of the panel be made entirely of polypropylene. Recyclers can cut along the visible markings to recover and recycle more than 1.5 kilograms of matter per panel. This innovation has been patented and is visible to our customers, thanks to the affixing of the conventional logo to symbolize this recycling [2].

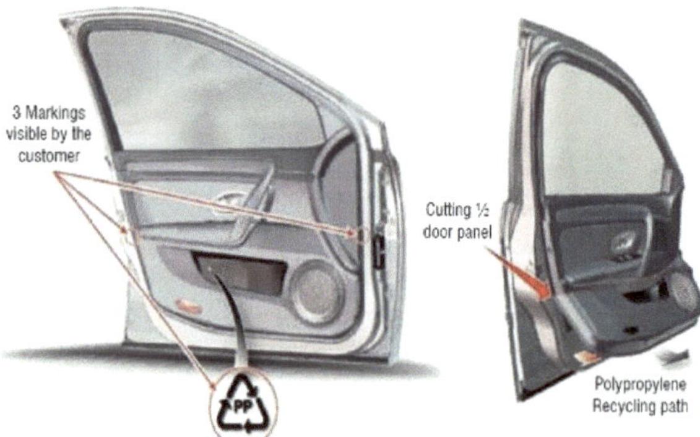

Figure 6. Renault Laguna III panel door trim

3. CONCLUSION

Looking beyond economic considerations, Renault uses the circular economy to shrink the environmental footprint of its products and activities. Working on the entire life cycle of its vehicles – and in particular by reducing CO_2 emissions – the Renault group leads an ambitious policy on the reduction of its environmental footprint.

The ECO2 label is based on Renault environmental policy and highlights the three key stages of the vehicle life cycle: manufacturing – vehicles produced in ISO 14001 certified plants, utilization – vehicles emitting less than 120g of CO_2 / km and recycling – at least 85% of the car must be recyclable.

REFERENCES

[1] Hashmi S. (Editor-in-chief), Comprehensive Materials Processing, Elsevier, 2014

[2] Abraham F., Design to recycling – The Renault way, Autofocus Asia, 2013

COMMENTS ON THE SOLITON-LIKE INTERACTIONS IN NONDISPERSIVE MEDIA

Valerica Moşneguţu[1], Veturia Chiroiu[1], Ruxandra Ilie[2]
[1]Institute of Solid Mechanics, Romanian Academy, Bucharest
e-mails veturiachiroiu@yahoo.com, valeriam732000@yahoo.com
[2]Technical University of Civil Engineering, Bucharest, e-mail rux_i@yahoo.com

Abstract: *Seymour and Varley [1] analyse certain media whose responses are governed by the nonlinear nondispersive wave equation. Any two pulses traveling in opposite directions interact nonlinearly for a finite time when they collide but then part unaffected by the interaction. More clearly, when any two pulses are traveling in opposite directions meet and interact they emerge from the interaction region unchanged by the interaction. This interaction is similar to those that occur when two solitons collide. The main difference is that solitons are represented by waves of permanent form whose profiles are specific.*
Keywords: *DRIP media, nondispersive waves*

1. INTRODUCTION

In this work we present some comments on the properties of a coupled system of wave equations discussed by Seymour and Varley in [1]

$$y_{tt} = A^2(y_x) y_{xx} \ , \tag{1}$$

$$A_{,y_x} = A^{3/2}(\mu + \nu A), \tag{2}$$

that govern the motion of an heterogeneous string, where $y(x,t)$ is the physical displacement, $A(y_x)$ a positive function representing the local speed of propagation, and μ, ν the material constants. If $y_x = y_x(x)$ and $A(y_x(x)) = A(x)$, the above system of equations becomes

$$y_{tt} = A^2(x) y_{xx} \ , \tag{3}$$

$$A_x = A^{3/2}(\mu + \nu A) y_{xx} \ . \tag{4}$$

We show that the waves described by Seymour and Varley are dispersive and dissipative. Single bounded solution of (2) may be written under the form

$$A(y_x) = \frac{\lambda[e_3 + (e_2 - e_3)\mathrm{sn}^2(\sqrt{e_1 - e_3}\, y_x + \delta')]}{1 + \rho[e_3 + (e_2 - e_3)\mathrm{sn}^2(\sqrt{e_1 - e_3}\, y_x + \delta')]} \ , \tag{5}$$

with λ, ρ constants depending on μ, ν, and e_i, $i = 1,2,3$ the solutions of the equation $4y^3 - g_2 y - g_3 = 0$ with constants g_2, g_3 depending on μ, ν. The solution (5) shows cnoidal dependence on y_x, being characterized by the dependence of the amplitude on the argument of sn. For a certain values of μ, ν, for that $m = 1$, $m = \dfrac{e_2 - e_3}{e_1 - e_3}$, the solution (5) becomes

$$A(y_x) = \frac{\lambda[e_1 - (e_1 - e_3)\mathrm{sech}^2(\sqrt{e_1 - e_3}\, y_x + \delta')]}{1 + \rho[e_1 - (e_1 - e_3)\mathrm{sech}^2(\sqrt{e_1 - e_3}\, y_x + \delta')]} \ . \tag{6}$$

The solutions of Seymour and Varley equations are also characterized by the dependence of the amplitude on the argument sech . The interaction (collision) of two solutions such (5) or (6) has solitonic properties: may propagate without change of form, being regarded as a local confinement of the energy of the wave field; at the collision each may come away with the same character as it had before the collision [2 -4]. Two solutions of (2) traveling in opposite directions interact nonlinearly and the collision is influenced by the properties of A.

2. LINEAR VIBRATING STRING

Consider 1D string motion equation

$$u_{tt} - c^2 u_{xx} = 0, \tag{7}$$

with c a real positive number. Let x range from $-\infty$ to ∞. For the transverse vibrations of a string $c^2 = T/\lambda$ where T is the constant tension and λ the mass per unit length at the position x. For the compressional vibrations of an isotropic elastic solid in which the density and elastic constants are functions of x only (laminated medium) $c^2 = (\lambda + 2\mu)/\rho$. For the transverse vibrations of such laminated solid $c^2 = \mu/\rho$. The characteristics are given by $\mathrm{d}x/\mathrm{d}t = \pm c$, that are straight lines inclined to the axis at $c = \tan\varphi$. The D'Alembert solution of (7) is

$$u(x,t) = f(x-ct) + g(x+ct), \tag{8}$$

where the functions $f, g : R \to R$ are determined from the initial conditions attached to (7)

$$u(x,0) = \Phi(x), \ u_{,t}(x,0) = \Psi(x). \tag{9}$$

The solution (8) describes two waves $f(x-ct)$ and respectively $g(x+ct)$. The observer sees, at any moment of time the unchanged profile $f(x)$ at the initial time $t_0 = t'_0 = 0$. This is way the function $f(x-ct)$ represents a right travelling wave or a forward-going wave with the velocity c. For a similar reason, $g(x+ct)$ represents a left travelling wave or a backward-going wave with the velocity c. As a consequence, both waves are not interacting between them and do not change their shape during the propagation. These waves can be superposed by a simple sum, because of the linearity of (7).

These waves can be called *solitary waves*, for the reason they are not changing their shapes during propagation process, and do not interact one with the other.

If (7) is not linear we have the case considered in [1] for which the superposition principle is not valid. If y_1, y_2 are two solutions of (1), the sum between them is not a solution of (1). The function $A(y_x(x))$ is a positive function that represents the local speed of the propagation.

3. NONLINEAR VIBRATING STRING

Consider the equation (1) that can governs the motion of an heterogeneous string, where y is the physical displacement, and $A(y_x)$ is a positive function representing the local speed of propagation and verifies

$$A_{,x} = A^{3/2}(\mu + \nu A) y_{xx}, \tag{10}$$

we obtain a coupled partial nonlinear differential equations for $y(x,t)$ and $A(x)$

$$y_{tt} = A^2(x) y_{xx}, \tag{11}$$

$$A_x = A^{3/2}(\mu + \nu A) y_{xx}. \tag{12}$$

The characteristics are given by $\mathrm{d}x/\mathrm{d}t = \pm A$, which equations define two congruences of curves in the (x,t) plane. First step is to straighten the characteristics of (11) for geometrical representation in a space-time plane. To do this we define the transformation [5, 6]

$$x \to u(x) = \int_0^x \frac{\mathrm{d}z}{A(z)}. \tag{13}$$

From (11) we obtain

$$y_{tt} - y_{uu} + \frac{c_u}{c} y_u = 0 \,,\ \ c(u(x)) = A(x) \,. \tag{14}$$

The function $c(u)$ is the transformed local speed. Strictly speaking we should not use the same symbols y in (12), because the function $y(x,t)$ of (11-12) is not the same function as the $y(u,t)$ of (14), but this is not likely to cause confusion if we remember that y may be regarded as a physical quantity in terms of (x,t) or (u,t).
The linearized form of (14) is

$$y_{tt} - y_{uu} + y_u = 0 \,. \tag{15}$$

Introducing the harmonic wave $y(u,t) = \tilde{A}\exp \mathrm{i}(ku - \omega t)$ into (15), we obtain the dispersion relation

$$\omega^2 = \mathrm{i}k + k^2 \,. \tag{16}$$

The phase velocity of the harmonic waves $y(u,t) = \tilde{A}\exp(tkb)\exp \mathrm{i}k(u - \frac{t}{2kb})$ is $c_p = \frac{1}{2kb} = \frac{-1}{\sqrt{2k(-k + \sqrt{1+k^2})}}$ and depends on k. In conclusion, the equation (11) is *dispersive and dissipative.* In a space-time plane in which u and t are Cartesian coordinates, the characteristics are $\mathrm{d}x / \mathrm{d}t = \pm 1$, and are straight lines inclined to the axis at 45^0.

4. BOUNDED SOLUTIONS

Consider the equation (1) written under the form

$$A'^2 = A^3 (\mu + \nu A)^2 \,, \tag{17}$$

where $A' = A_{,e}$ and $e = y_x$. Differentiating (17), we have

$$A'' = a_2 A^2 + a_3 A^3 + a_4 A^4 \,, \tag{18}$$

where

$$a_2 = 1.5\mu^2 \,,\ a_3 = 4\mu\nu \,,\ a_4 = 2.5\nu^2 \,, \tag{19}$$

and $a_i > 0$, $i = 2,3,4$. We assume the solution of (18) in the form [7]

$$A = \frac{\lambda P(e)}{1 + \rho P(e)} \,, \tag{20}$$

where $\lambda \neq 0$ and $\rho \neq 0$ are arbitrary real constants, and $P(e)$ is the Weierstrass elliptic function satisfying the differential equation

$$P'^2 = 4P^3 - g_2 P - g_3 \tag{21}$$

with two invariants g_2 and g_3 which are assumed to be real and satisfy

$$g_2^3 - 27 g_3^2 > 0 \,. \tag{22}$$

So, the exact periodic solutions can be written as

$$A(y_x) = \frac{\lambda P(y_x + \delta; g_2, g_3)}{1 + \rho P(y_x + \delta; g_2, g_3)} \,, \tag{23}$$

where δ is an integration constant with known quantities g_2, g_3, λ and ρ. The exact bounded periodic solution can be obtained by replacing the Weierstrass elliptic function by the Jacobean elliptic sine function using the formula

$$P(y_x + \delta; g_2, g_3) = e_3 + (e_2 - e_3)\mathrm{sn}^2(\sqrt{e_1 - e_3}\, y_x + \delta') \,, \tag{24}$$

where δ' is an arbitrary real constant, and e_i, $i=1,2,3$ are real roots of the equation $4y^3-g_1y-g_2=0$ with $e_1>e_2>e_3$. From $\mathrm{cn}^2+\mathrm{sn}^2=1$, we can express (24) in term of the cnoidal function cn.
Thus, it results that the exact bounded periodic solution of (18) is (5). The solitonic form of (5) is given by (6).
Let again consider the transformation (13). Equation (1) can be written in the form

$$\frac{d}{dt}v(t)=Lv(t), \tag{25}$$

where

$$v=\begin{pmatrix} y_u \\ y_t \end{pmatrix},\quad L=\begin{pmatrix} 0 & \partial_u \\ \partial_u-2\gamma & 0 \end{pmatrix},\quad 2\gamma(u)=\frac{c_u}{c}. \tag{26}$$

We calculate $v(t)$ by a sequence of linear transformations that reduce $v(t)$ to a perturbation of the pulse. For this we define the energy of the string

$$E=\frac{1}{2}\int_{-\infty}^{\infty}\frac{y_u^2+y_t^2}{c(u)}du, \tag{27}$$

and take a point v in the phase-space (y_u, y_t)

$$v=\begin{pmatrix} q \\ p \end{pmatrix},\quad q(u)=y_u(u,0),\quad p(u)=y_t(u,0) \tag{28}$$

The inner product is derived from the energy quadratic form (27)

$$<v_1,v_2>=(C^{-1}v_1,C^{-1}v_2),\quad C=\begin{pmatrix} \sqrt{c(x)} & 0 \\ 0 & \sqrt{c(x)} \end{pmatrix}, \tag{29}$$

$$(v_1,v_2)=\frac{1}{2}\int_{-\infty}^{\infty}[q_1(u)q_2(u)+p_1(u)p_2(u)]du. \tag{30}$$

The operator L is screw-symmetric with respect to the inner product (29) and so there exists a one-parameter group $V(t)$ of orthogonal transformation determined by

$$V_{,t}(t)=LV(t),\quad V(0)=1, \tag{31}$$

so that $v(t)=V(t)v$ is a solution of (31). The method is based on the decomposition of the phase-space (y_u, y_t) into a pair of complementary subspaces. This induces a decomposition of each initial datum into a forward propagating part and a backward-propagating part.
In the homogeneous case $(\gamma=0)$, (1) is reduced to $y_{tt}-y_{uu}=0$ and the solution are expressed as a sum of two waves $f(x-t)$ and $f(x+t)$ that propagate independently. In the heterogeneous case the both pulses are coupled by $\gamma\neq 0$ considered as a perturbation.
So, we take

$$V(t)=CR^{-1}\tilde{V}(t)RC^{-1},\quad \tilde{V}(t)=\exp(t\tilde{L}), \tag{32}$$

$$L=CR^{-1}\tilde{L}RC^{-1},\quad \tilde{L}=\begin{pmatrix} -\partial u & -\gamma \\ \gamma & \partial u \end{pmatrix}, \tag{33}$$

with $R=\begin{pmatrix} \frac{1}{\sqrt{2}} & -\sqrt{2} \\ \frac{1}{\sqrt{2}} & \sqrt{2} \end{pmatrix}$. It results that $\tilde{V}(t)=RC^{-1}V(t)CR^{-1}$, $\tilde{L}=RC^{-1}LCR^{-1}$. We see that $\tilde{L}$ can be written as a sum of two operators to separate the contribution of the coupling term $\gamma\neq 0$

$$\tilde{L}=\tilde{L}_0+\Gamma,\quad \tilde{L}_0=\begin{pmatrix} -\partial u & 0 \\ 0 & \partial u \end{pmatrix},\quad \Gamma=\begin{pmatrix} 0 & -\gamma \\ \gamma & 0 \end{pmatrix}. \tag{34}$$

In the homogeneous case we have $\Gamma = 0$ and

$$\tilde{V}(t) = \tilde{U}(t), \quad \tilde{U}(t) = \begin{pmatrix} T(t) & 0 \\ 0 & T'(t) \end{pmatrix}, \tag{35}$$

where $T(t)$ is right translation by t

$$[T(t)f](s) = f(s-t) \ , \tag{36}$$

and $T'(t) = T(-t)$ is left translation by t. The initial conditions (29) can be written under the form

$$y(u,0) = \varphi(u) \ , \ y_t(u,0) = \psi(u) \ . \tag{37}$$

For $\Gamma = 0$ the solution of $y_{tt} - y_{uu} = 0$ is written as the D'Alembert formula

$$y(u,t) = \frac{1}{2}[\varphi(u+t) + \varphi(u-t)] + \frac{1}{2}\int_{u-t}^{u+t} \psi(z)\mathrm{d}z \ . \tag{38}$$

Our aim is to obtain a similar formula for the inhomogeneous case $\Gamma \neq 0$.
For this we use the well-known perturbation formula [8]

$$\tilde{V}(t) = \tilde{U}(t) + \int_0^t \tilde{U}(t-s)\Gamma \tilde{V}(s)\mathrm{d}s \ . \tag{39}$$

From this we can obtain an infinite series for $\tilde{V}(t)$ by an iteration scheme

$$\tilde{V}^{(p+1)}(t) = \tilde{U}(t) + \int_0^t \tilde{U}(t-s)\Gamma \tilde{V}^{(p)}(s)ds \ , \qquad \tilde{V}^{(0)}(t) = \tilde{U}(t) \ . \tag{40}$$

We take account that $\tilde{V}(t) = \exp(tL)$ from (32) maps forward-going data into forward-going data and backward-going data. So, we write

$$\tilde{V}(t) = \begin{pmatrix} \tilde{V}_{FF}(t) & \tilde{V}_{FB}(t) \\ \tilde{V}_{BF}(t) & \tilde{V}_{BB}(t) \end{pmatrix}. \tag{41}$$

Here, $\tilde{V}_{FF}$ maps forward-going data into forward-going data, $\tilde{V}_{FB}$ maps forward-going data into backward-going data, $\tilde{V}_{BF}$ maps backward-going data into forward-going data and $\tilde{V}_{BB}$ maps backward-going data into backward-going data. From $(35)_2$ and (40) we obtain for $\tilde{V}_{FF}$

$$\tilde{V}_{FF}(t) = T(t) - \int_0^t\int_0^{t_1} T(t-t_1)\gamma T'(t_1 - t_2)\gamma T(t_2) dt_1 dt_2 + \ldots.. \tag{42}$$

The first term in (42) is simply translating a forward-going datum into a forward direction. The integrant $T(t-t_1)\gamma T'(t_1-t_2)\gamma T(t_2)$ translate a forward-going datum in the forward direction from time zero to time t_2 when it is reflected. On reflection it is multiplied by the local reflection coefficient γ, then translated backwards from time t_2 to time t_1, when it is reflected again, multiplied by γ and translated forwards from time t_1 to time t. So, the second term in (42) represents the contribution to the forward-going disturbance from all possible double reflections. The following terms in (42) consider third reflections and so on. Knowing this, we have

$$V(t) = \begin{pmatrix} V_{11}(t) & V_{12}(t) \\ V_{21}(t) & V_{22}(t) \end{pmatrix}, \tag{43}$$

with

$$V_{11}(t) = 0.5[V_{FF}(t) + V_{BB}(t) + V_{FB}(t) + V_{BF}(t)] \ , \quad V_{12}(t) = 0.5[V_{BB}(t) - V_{FF}(t) + V_{FB}(t) - V_{BF}(t)] \ ,$$
$$V_{21}(t) = 0.5[V_{BB}(t) - V_{FF}(t) + V_{BF}(t) - V_{FB}(t)] \ , \quad V_{22}(t) = 0.5[V_{FF}(t) + V_{BB}(t) - V_{FB}(t) - V_{BF}(t)] \ . \tag{44}$$

Taking account of the initial data (37) we have

$$\begin{aligned} y_t(u,t) = {} & 0.5\sqrt{c(u)}[V_{BB}(t) - V_{FF}(t) + V_{BF}(t) - V_{FB}(t)]c^{-1/2}\varphi'(u) + \\ & + 0.5\sqrt{c(u)}[V_{BB}(t) + V_{FF}(t) - V_{BF}(t) - V_{FB}(t)]c^{-1/2}\psi(u), \end{aligned} \tag{45}$$

and

$$y(u,t) = \varphi(u) + 0.5\sqrt{c(u)}\int_0^t [V_{BB}(t) - V_{FF}(t) + V_{BF}(t) - V_{FB}(t)]c^{-1/2}(u)\varphi'(u)dt_1 + \\ +0.5\sqrt{c(u)}\int_0^t [V_{BB}(t) + V_{FF}(t) - V_{BF}(t) - V_{FB}(t)]\psi(u)dt_1. \quad (46)$$

When $c'(u) = 0$, we have

$$y(u,t) = \varphi(u) + \frac{1}{2}\int_0^t [T'(t_1) - T(t_1)]\varphi'(u)dt_1 + \frac{1}{2}\int_0^t [T'(t_1) + T(t_1)]\psi(u)dt_1. \quad (47)$$

After integration by parts and a change of variable (47) yields to D'Alembert formula (40).

5. CONCLUSION

As conclusion, from (46) we see that the solution of $y_{tt} = A^2(x)y_{xx}$ is expressed in term of the $\sqrt{c(u)} = \sqrt{A(x)}$, where $x \to u(x) = \int_0^x \frac{dz}{A(z)}$, and $A(x) = A(y_x(x))$ verifying the solutions (5) and (6). These solutions are expressed in term of cnoidal (or soliton) solutions. Therefore, we can emphasis that the collisions between the Seymour and Varley waves have cnoidal or soliton characteristics. These waves distort as they propagate, and are of arbitrary shape and amplitude. Since such media transmit waves that *do not remember the interaction process* they are called *DRIP media.*

Acknowledgement. The authors gratefully acknowledge the financial support of the National Authority for Scientific Research ANCS/UEFISCDI through the through the project PN-II-ID-PCE-2012-4-0023, Contract nr.3/2013. The authors acknowledge the similar and equal contributions to this article.

REFERENCES

1. Seymour, B.R., Varley, E., *Exact solutions describing soliton-like interactions in a nondispersive medium*, Soc. Ind. Appl. Math..J. Applied Mathematics 42, 804-821, 1982.
2. Munteanu, L., Donescu, St., *Introduction to Soliton Theory: Applications to Mechanics,* Book Series Fundamental Theories of Physics, vol.143, Kluwer Academic Publishers, Dordrecht, Boston (Springer Netherlands) 2004.
3. Ionescu, M.F., Chiroiu, V., Dumitriu, D., Munteanu, L., *A special class of DRIP media with hysteresis,* Revue Roumaine des Sciences Techniques, série de Mecaniqué Appliquée, 56(1), 3-10, 2011.
4. Zabusky and Kruskal, *Interactions of solitons in a collisionless plasma and the recurrence of initial states*, Phys, Rev. Lett, vol 15, 1965.
5. Lewis. J. T., *The heterogeneous string: coupled helices in Hilbert space*, Quarterly of Applied Mathematics, XXXVIII, nr. 4, 1980.
6. Synge, J. L., *On the vibrations of a heterogeneous string*, Quarterly of Applied Mathematics, XXXIX, nr. 2, 1981.
7. Krishnan E. V.,*On the Ito-Type Coupled Nonlinear Wave Equation*, J. of the Physical Society of Japan, 55, 11, 1986, 3753-3755.
8. Kato T., *Perturbation theory for the linear operators*, Springer, New York, 1961.

THE USE OF ADSORPTION ISOTHERMS FOR MODELLING THE NITRITE AND SULPHITE ANIONS REMOVAL FROM AQUEOUS SOLUTIONS ON A COMPOSITE MEMBRANE

D.E. Pascu[1*)], M. Neagu(Pascu)[1,2], G.A. Traistaru[3], A.C. Nechifor[1], A.R. Miron[1]

[1]Faculty of Applied Chemistry and Materials Sciences, Politehnica University of Bucharest, 1-7 Gheorghe Polizu Street, Bucharest,011061, ROMANIA,dpascu@yahoo.com; nechiforus@yahoo.com; andra3005@yahoo.com;

[2]S.C. HOFIGAL S.A., Analytical Research Department, 2 Intr. Serelor, Bucharest, 042124, ROMANIA, mihhaela_neagu@yahoo.com;

[3]S.C. ENECO Consulting S.R.L, sos. Pantelimon, no. 247, sector 2, Bucharest, ROMANIA, traistaru_ginaalina@yahoo.com;

Abstract: *In this study was investigated the adsorption of nitrite and sulphite anions on a composite membrane. Initial anion concentration, pH and adsorbent dosage were investigated in this study, as well as the adsorption kinetics and isotherms for the composite membrane (10µm) used. Kinetics of adsorption was examined by means of three kinetic models, namely: pseudo-first-order, pseudo-second-order and intraparticle diffusion models and Langmuir, Freundlich, Temkin, Dubinin-Radushkevich isotherms. The results obtained revealed that the adsorption isotherms fitted the data in the following order: Temkin>Langmuir>Freundlich>Dubinin-Radushkevich models. The results indicated that the composite membrane exhibited potential application for removal of nitrite and sulphite from aqueous solutions. Mathematical modelling by means of adsorption isotherms shows that the interaction of sulphite and nitrite anions with membrane surface is localized in the monolayer adsorption, meaning that the adsorbed molecules are adsorbed at definite, localized sites.*

Keywords: *adsorption, isotherm, nitrite, sulphite, mathematical modelling*

1. INTRODUCTION

The increasing levels of nitrite and sulphite anions in drinking and groundwater due to natural and anthropogenic activities has been recognized as one of the major problems worldwide [1–5]. This imposes a serious threat to human health and environmental issues. Thus, a renewed interest in the anions removal from domestic and industrial wastewaters has been greatly increased [6–9]. Moreover, technologies for anions removal from water are of great relevance in waste treatment processes, which aim to achieve efficient strategies to obtain acceptable levels for disposal of the aforesaid anions.

Adsorption studies were conducted under various experimental conditions, such as pH, contact time, initial nitrite and sulphite concentrations, temperature and the presence of competing anions. The data from the experiments were fitted with different models to identify the adsorption mechanism. The results have been thoroughly discussed which would help in the better understanding of nitrite and sulphite anions adsorption mechanism by composite membrane [10-15].

Adsorption isotherms and anion exchange modelling equilibrium are widely studied to predict the relative affinities of anions and their distribution in the adsorbent-solution system during the purification process.Thus, we use adsorption isotherms to investigate the removal of nitrite and sulphite anions from aqueous solutions by a membrane composite adsorbent through the Langmuir, Freundlich, Temkin and Dubinin–Radushkevich adsorption isotherms. The adsorption parameters of the models are determined by nonlinear regression. The high anion - absorbing capacity of this product apparently suggests that it can be widely used in the treatment of wastewater.

2. EXPERIMENTAL

2.1.Materials

There was used one composite membranes with pore size 10 μm. The composite membranes were prepared in the laboratory of Analytical Chemistry and Environmental Engineering Department from University Politehnica Bucharest. Synthesis and characterization of these membranes has been the subject of another article [1,11]. The degree of selectivity of a membrane depends on the membrane pore size. Depending on the pore size, they can be classified as microfiltration (MF), ultrafiltration (UF), nanofiltration (NF) and reverse osmosis (RO) membranes. Membranes can also be of various thickness, with homogeneous or heterogeneous structure. Membranes can be neutral or charged, and particle transport can be active or passive.
The nitrite anion was determined according (STAS 3048/2-96 and SR ISO 6777/96) and the sulphite anion was determined according (STAS 7661/1989).

2.2. Theories of the kinetic models

Several models can be used to describe and analyze the adsorption process of a given analyte as a function of time. It should be noted that these models can describe adequately the kinetic behavior of a solute in a given substrate, but the parameters obtained from each model cannot necessarily have a coherent physical meaning because the theoretical development of these models is usually empirical. For that reason it is important to know the border conditions that describe the different kinetic models used and how the experimental factors can significantly affect the adsorption kinetics on the membrane (Ho and McKay 1998; Madden et al., 2006; Pyrzynska, 2010; Boparai et al., 2011). It is certainly feasible to select *a priori* a kinetic model that considers, as correctly as possible, the conditions under which the adsorption of some given analytes takes place, considering the characteristics of the substrate.

3. RESULTS AND DISCUSSION

3.1. Adsorption isotherms

The isotherm results were analyzed using the Langmuir, Freundlich, Temkin and Dubinin–Radushkevich isotherms. The Langmuir adsorption model is based on the assumption that maximum adsorption corresponds to a saturated monolayer of solute molecules on the adsorbent surface, with no lateral interaction between the adsorbed molecules. The Temkin adsorption isotherm has been successfully used to explain the adsorption of the two anions.
Temkin Isotherm:
Temkin and Pyzhev considered the effects of some indirect sorbate/adsorbate interactions on adsorption isotherms and suggested that because of these interactions the heat of adsorption of all the molecules in the layer would decrease linearly with coverage. The model is given by the following equation:

$$q_e = \frac{RT}{b_T} \ln(K_T \cdot C_e) \tag{1}$$

The linear form of Temkin equation is:

$$q_e = \frac{RT}{b_T} \ln K_T + \frac{RT}{b_T} \ln C_e \tag{2}$$

where:

- b_T is the Temkin constant related to heat of sorption (J/mg)
- K_T is the binding constant corresponding to the maximum binding energy (L/g).

The Temkin constants b_T and K_T are calculated from the slope and intercept of the plot of q_e versus $ln\ C_e$ respectively figure 1(a and b).

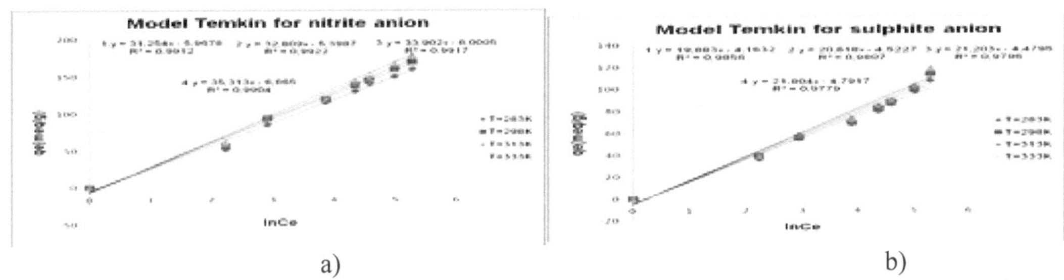

a) b)

Figure 1. Linear representation of Temkin isotherm for adsorption of nitrite (a), Sulphite (b) on composite 10 μm membrane

The high values in this study indicates a strong interaction between adsorbate and adsorbent, supporting an ion-exchange mechanism for the present study.

Langmuir Adsorption Isotherm

The expression of the Langmuir model is given by equation:

$$q = \frac{K_L q_m Ce}{1 + K_L Ce} \quad (3)$$

where:

- q_m and K_L are the Langmuir constants, related to the adsorption capacity and energy of adsorption, respectively.

The linear form of the Langmuir isotherm is given by the following equation:

$$\frac{Ce}{q_e} = \frac{C_e}{q_m} + \frac{1}{K_L q_m} \quad (4)$$

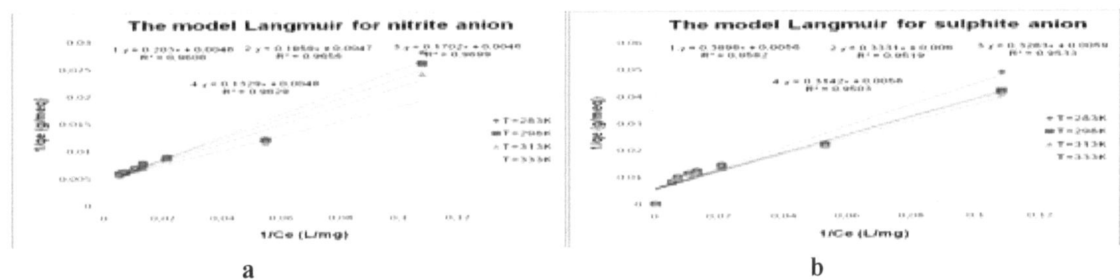

a b

Fgure 2. Linear representation of Langmuir isotherm for adsorption of nitrite (a), sulphite (b) on composite 10 μm membrane

According to the findings, the shape of the Langmuir adsorption isotherms of the nitrite anion suggests a favorable and spontaneous adsorption fig. 2(a). The Langmuir isotherms fit very well with the experimental results ($R^2 > 0.96$) presented in figure 2. The Langmuir adsorption isotherms of the sulphite are presented in fig. 2(b) and fit well with the experimental results $R^2 > 0.95$.

Freundlich Adsorption Isotherm:

The Freundlich isotherm is an empirical equation which estimates the adsorption intensity of the adsorbent towards the adsorbate. Freundlich equation is suitable for a highly heterogeneous surface and an adsorption isotherm lacking a plateau indicates a multilayer adsorption [5,13]. The model is represented by the equation:

$$q_e = K_F \cdot C_e^{1/n} \quad (5)$$

where:

- K_F is a constant indicative of the relative adsorption capacity of the adsorbent
- n is a constant indicative of the intensity of the adsorption.

The Freundlich adsorption isotherms for nitrite and sulphite adsorption are obtained by plotting the graph of lnqe versus lnCe, as in figure 3, from which the constants of Freundlich adsorption model have been determined.

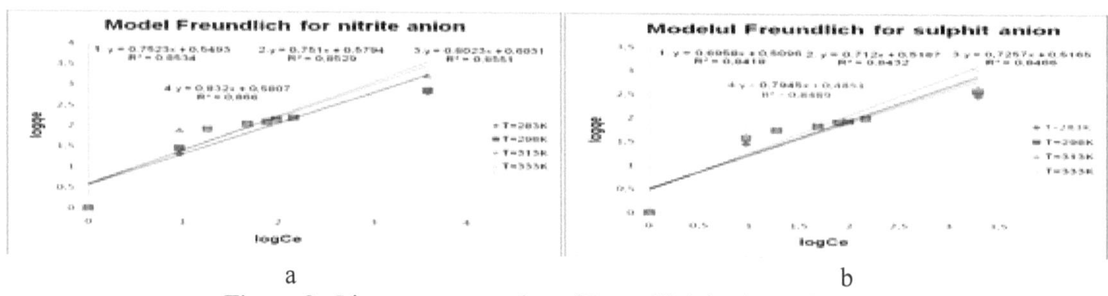

a b

Figure 3. Linear representation of Freundlich isotherm for adsorption of nitrite (a), sulphite (b) on composite 10 μm

The Freundlich expression is an exponential equation and therefore, assumes that as the adsorbate concentration increases, the concentration of adsorbate on the adsorbent surface also increases fig.3 (a and b). To determine the maximum adsorption capacity, it is necessary to operate with constant initial concentration C_0 and variable weights of adsorbent; thus ln q_e is the extrapolated value of ln q for $C=C_0$. Among the adsorption isotherm models studied, the Freundlich model had not satisfactory correlation coefficients ($R^2 < 0.870$) for describing the adsorption of anions studied onto adsorbent.

Dubinin–Radushkevich isotherm model

Dubinin–Radushkevich isotherm is generally applied to express the adsorption mechanism with a Gaussian energy distribution onto a heterogeneous surface. The model has often successfully fitted high solute activities and the intermediate range of concentrations data well.

$$q_e = q_s \cdot \exp(-K_{ad} \cdot \varepsilon^2) \qquad (6)$$

where:

- q_e, q_s, K_{ad}, are q_e = amount of adsorbate in the adsorbent at equilibrium (mg/g);
  q_s = theoretical isotherm saturation capacity (mg/g);
  K_{ad} = Dubinin–Radushkevich isotherm constant (mol^2/kJ^2) ;
  ε = Dubinin–Radushkevich isotherm constant.

The parameter ε can be calculated as:

$$\varepsilon = RT \ln(1 + 1/C_e) \qquad (7)$$

The linear regression of the Dubinin-Radushkevich isotherm plot for the adsorption of the two anions are presented in figure 4.

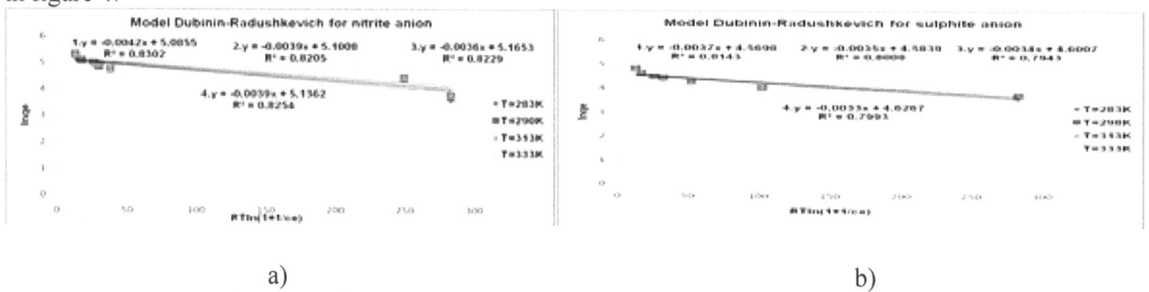

a) b)

Figure 4. Linear representation of Dubinin-Radushkevich isotherm for adsorption of nitrite (a) and sulphite (b) on composite 10 μm

The values decreased with increases in temperature, thus confirming the suitability of low temperatures for adsorption. The decrease in adsorption capacity of the membrane at higher temperatures may be attributed to inactivation of the membrane surface (Dizge et al., 2009). In the range 283-333K, the experimental data fit better to the Temkin model ($0.9779<R^2<0.9922$), Langmuir model ($0.9503 < R^2 <0.9629$) than to the Freundlich ($0.719 < R^2 < 0.985$) or D-R ($0.787 < R^2 < 0.945$).

Adsorption analysis results obtained at various temperatures showed that the adsorption pattern on the composite membrane followed Temkin isotherms rather than those from Langmuir, Freundlich, and Dubinin-Radushkevich models

Kinetic adsorption

The kinetic studies predict the progress of adsorption, however, the determination of the adsorption mechanism is also important for design purposes. In order to investigate the adsorption kinetics of the anions on the adsorbent, pseudo-first order, pseudo-second order, and intra-particle diffusion models were used.

Pseudo-first and second order models

The kinetics of anions adsorption was studied to verify the adsorption mechanism. The first-order rate expression of Lagergren [4,12] and pseudo-second order rate [4,14] expressions were applied in this study. Lagergren suggested that a pseudo first-order rate equation for sorption of solutes from a liquid solution is represented as follows:

$$\log(q_e - q_t) = \log q_e - 1/2.303 k_1 t \qquad (8)$$

where:

- q_e is the adsorption capacity at the equilibrium;
- q_t is the individual capacity in a given time;
- k_1 and k_2 are the pseudo-first and pseudo-second order rate constants, respectively;
- t is the time [min].

The value of the adsorption rate constant (k_1) for nitrate adsorption by adsorbent was determined from the plot of $\log(q_e - q_t)$ against t. Although the correlation coefficient value is 0.91 for nitrate anion and 0.867 for the sulphite anion, the experimental qe value do not agree with the calculated one, obtained from the linear plot. This result indicated that the anions adsorption system do not follow a first-order reaction.

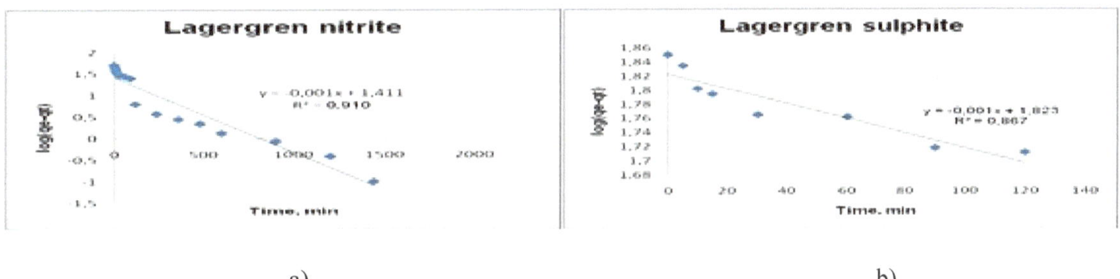

a) b)

Figure 5. Linear representation of Lagergren model for adsorption of nitrite (a), sulphite (b) on composite 10 μm

Another model for the analysis of adsorption kinetics is pseudo second-order. This model proposed by **Ho&McKay**[12,14] can be used to explain the sorption kinetics figure 5 (a and b). The model is based on the assumption that the adsorption follows a second order chemisorption. The pseudo-second order model can be expressed as:

$$\frac{dq_t}{dt} = k_2 (q_e - q_t)^2 \qquad (9)$$

where:

- k_2 is the rate constant of Ho &McKay equation (g/mg/min).

The plot of t/q versus t gives a straight line with slope of 1/q and intercept of $\frac{1}{k_2 q_e^2}$. There is no need to know any parameter before hand and the grams of solute adsorbed per gram of sorbent at equilibrium (q_e) and sorption rate constant (k_2) can be evaluated from the slope and intercept, respectively.

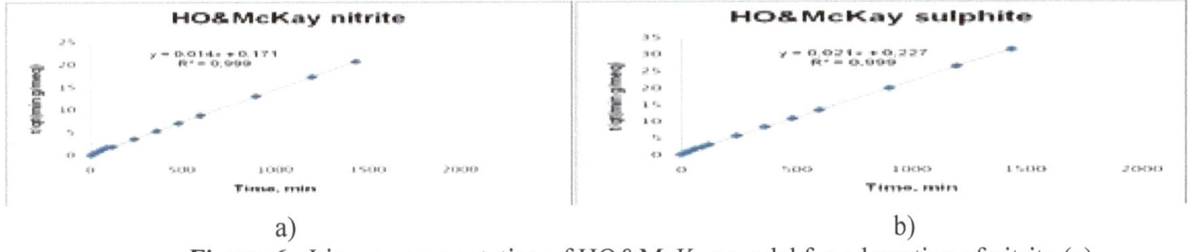

a) b)

Figure 6. Linear representation of HO&McKay model for adsorption of nitrite (a), sulphite (b) on composite 10 μm

Figure 6 shows the adsorption kinetics for pseudo second-order rate equation of nitrite, sulphite on the adsorbent. The theoretical q_t value also agreed very well with the experimental q_e value, indicating the pseudo second-order kinetics. In addition, the correlation coefficient for the second-order kinetic model was higher than 0.99 for the both

anions. This suggest that the applicability of this kinetic equation and the second-order nature of the adsorption process of anions on the adsorbent are very good.
Experimental q_e values were compared with q_e value determined by pseudo-first and second-order rate kinetics model, which suggests that adsorption is not a pseudo first-order reaction and that a pseudo-second-order model can be considered.
Morris-Weber model is one of the most important models of the intra-particle diffusion. The intra-particle diffusion model is considered in two aspects: on the one hand by the membrane pore size, on the other hand, by the surface diffusion which occurs aqueous solutions of the two anions studied.

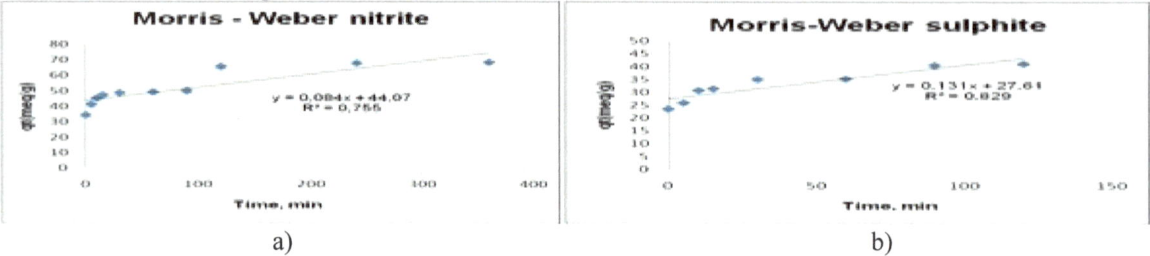

a) b)

Figure 7. Linear representation of Morris – Weber model for adsorption of nitrite (a) and sulphite (b) on composite 40 μm

With Morris-Weber model can be carried out and the two anion adsorption and it is the basic equation can be written as:

$$q_t = k_{id} t^{1/2} \tag{10}$$

where:

- k_{id} is the intra-particle diffusion rateconstant (mg g−1 min$^{1/2}$)

As shown in fig.7, the linear line do not pass through the origin and this deviation from the origin near saturation might be due to the difference in the mass transfer rate in the initial and final stages of adsorption. This means that the pore diffusion is not the only rate limiting mechanism in the adsorption process. It has been reported that if the intra-particle diffusion is the sole rate-limiting step, it is essential for the q_t versus $t^{1/2}$ plots to pass through the origin, which is not the case in this study. It may be concluded that surface adsorption and intra-particle diffusion are concurrently operating during the anions-adsorbent interactions.

4. CONCLUSIONS

In this work, the ability of composite membrane to remove nitrite and sulphite anions from aqueous solution was investigated. The best results obtained at various temperatures revealed that the adsorption isotherms fitted the data in the following order: Temkin > Langmuir > Freundlich > Dubinin-Radushkevich models. The composite membrane exhibited potential application for removal of nitrite and sulphite anions from aqueous solutions, at four different temperature. The most important models for intra-particle diffusion are Morris-Weber model.
The maximum adsorption capacity corresponds to a saturated monolayer of solute molecules on the adsorbent surface, with no lateral interaction between the adsorbed molecules is assumption by the Langmuir model. Adsorption seems to be a more attractive method for the removal of nitrite and sulphite in terms of cost, simplicity of design and operation.
The indirect sorbate/adsorbate interactions on adsorption isotherms has revealed that heat of adsorption of all the molecules in the layer would decrease linearly with coverage and was described by the Temkin and Pyzhev models.

Acknowledgements
The work has been funded by the Sectoral Operational Programme Human Resources Development 2007-2013 of the Ministry of European Funds through the Financial Agreement POSDRU/159/1.5/S/132395 and POSDRU/159/1.5/S/134398.
Faculty of Applied Chemistry and Materials Sciences, Politehnica University of Bucharest, support is also gratefully acknowledged.

References

[1] Segarceanu M., Pascu D.E., Traistaru G.A., Pascu (Neagu) M., Teodorescu S., Orbeci C., Optimization of Membrane Processes with Polysulfone/Polyaniline Composite Membranes, Revista de Chimie, 65(1), p. 8-14, 2014.

[2] Bhatnagar A., Kumar E., Sillanpää M., Fluoride removal from water by adsorption—A review, Chemical Engineering Journal, 171, p. 811–840, 2011.
[3] Dron J., Dodi A., Comparison of Adsorption Equilibrium Models for the Study of Cl^-, NO_3^- and SO_4^{2-}Removal from Aqueous Solutions by an Anion Exchange Resin, Journal of Hazardous Materials,190(1-3), p. 300–307, 2011.
[4] A. de Lima A. C., Nascimento R.F., F. de Sousa F.,Filho J. M.,Oliveira A.C., Modified coconut shell fibers: A green and economical sorbent for the removal of anions from aqueous solutions, Chemical Engineering Journal , 185– 186, p. 274– 284, 2012.
[5] Hannachi C., Guesmi F., Missaoui K., Hamrouni B., Application of adsorption models for fluoride, nitrate, and sulfate ion removal by amx membrane, International Journal of Technology 1, p.60-69, 2014.
[6] Labbez C., Fievet P., Szymczyk A.,Aoubiza B.,Vidonne A., Pagetti J., Theoretical study of the electrokinetic and electrochemical behaviors of two-layer composite membranes, Journal of Membrane Science, 184, p.79–95, 2001.
[7] Li Q, Ito K., Wu Z., Lowry C.S.,Loheide S.P., COMSOL Multiphysics: A Novel Approach to Ground Water Modeling, Ground Water, 47(4), p. 480–487, 2009.
[8] Gholipour M., Hashemipour H., Evaluation of multi-walled carbon nanotubes performance in adsorption and desorption of hexavalent chromium, Chemical Industry&Chemical Engineering Quarterly, 18(4),p.509–523, 2012.
[9] Aikpokpodion P.E., Osobamiro T., Atewolara-Odule O. C., Oduwole O. O. and Ademola S. M., Studies an adsorption mechanism and kinetics of magnesium in selected cocoa growing soilsin Nigeria, Journal of Chemical and Pharmaceutical Research, 5(6), p.128-139, 2013.
[10] Voicu S.I., Nechifor A.C., Serban B., **Nechifor G.**, Miculescu M., Formylated polysulfone membranes for cell immobilization, Journal of Optoelectronics and Advanced Materials, 9 (11), p. 3423, 2007.
[11]Nechifor Gh., Pascu D. E., Pascu (Neagu) M., Foamete (Panait) V.-I., Membrane optimization separation process: "Study of adsorption kinetics of sulphate and nitrate ions", U.P.B. Sci. Bull., Series B, Vol. 75, p. 191-197, Iss. 4, 2013 ISSN 1454 – 2331.
[12] Ho Y. S. and McKay G., A Comparison of Chemisorption Kinetic Models Applied to Pollutant Removal on Various Sorbents, Institution of Chemical Engineers Trans IChemE, 76(4), p.332-340, 1998.
[13] Benmaamar Z. and Bengueddach A., Correlation with different models for adsorption isotherms of m-xylene and toluene on zeolites, Journal of Applied Sciences in Environmental Sanitation, 2 (2),p. 43-56, 2007.
[14] Sharma I., Goyal D., Kinetic modeling:Chromium(III) removal from aqueous solution by microbial waste biomass, Journal of Scientific&Industrial Research, 68, p.640-646, 2009.
[15] Moreno L.,Crawford J. and Neretnieks I, Modelling radionuclide transport for time varying flow in a channel network, Journal of Contaminant Hydrology, 86(3–4), p.215–238, 2006.

TREND, ISSUE AND POTENTIAL ON OPTIMIZING PATH OF COMPOSITE MATERIALS

Catalin Iulian Pruncu

[1)] School of Mechanical Engineering, University of Birmingham, Edgbaston, B15 2TT, United Kingdom
[*]Corresponding author:c.i.pruncu@bham.ac.uk

The indicators governing the powerful of technological development envelop the rating of use the composite materials in nature components. So, composites are designed andmanufactured to be applied in many different areas, taking the place of materialsregarded as typical, such as steel and aluminium[1]. Such, the growing use of composite materials has arisen from their high specific strength and stiffness, when compared to the more conventional materials, and the ability to tailor their structure to produce more efficient structural configurations[2].The route of composite materials covers the automotive, aerospace, civil, marine, and sports areas. In key industry as aeronautics application it is considered that in the futurethe composites materials can contribute more than 50% of structural mass.

Nevertheless very good development of composites application, these materials entails complex mechanical mechanism when are submitted to loading condition (i.e. pressure, thermal load, hard environments and so on). The prediction of mechanical behaviour leads to a multi-objective approachsettled within a robust strategy. This strategy settled on analytical, numerical simulation, and/or experimental tests can bring important solutions for damage mechanisms (i.e.raised from static strength, delamination failures, interlaminar fracture mechanics, matrix cracking, porosity growth and other manufacturingrelated defects that also can introduce nucleation sites for failure). Typically solutions cover the simple situations to complex modelsfrom first initiation to final failure. For example, the mechanical loading applied in the axial tension, on the composite material displays progressive failurethrough several damage mechanisms that willtake place sequentially. A cross-ply laminate, Figure1, present the action of axial load applied in the longitudinal (L) direction, consequently the 0^0 ply is loaded along its reinforcing fibers, whereas the90^0 ply is loaded across the fibers. In this complex conditiona multi-damage mechanisms (delamination, matrix cracking, splitting of L lamina, interface cracking) is activated emphasizing an early failure database.

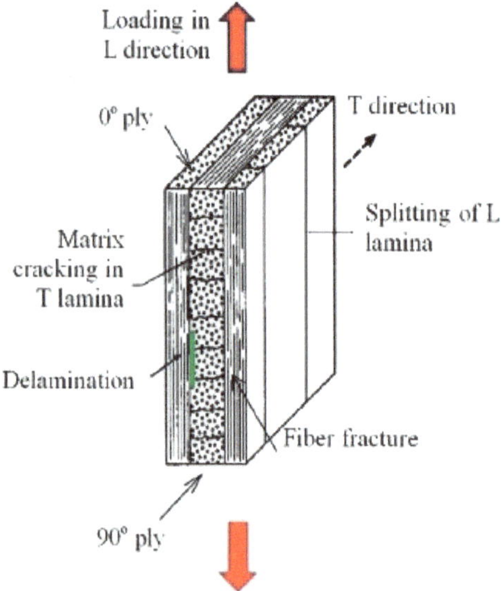

Figure1.Longitudinal tension of a 0/90 composite laminates; Highlight of several damage modes: matrix cracking in transverse (T) lamina, splitting of longitudinal (L) laminae, and delamination between T and L [3]

In relation to these phenomena, the selection of proper set upcircumstances for predictions and explanation of the mechanistic damage approaches candictate to suitable prediction models. Besides, in-depth understandings of individual damage mechanics may pave the road towards further material optimization with respect to fatigue, failure and durability [4].

Tserpes and Koumpias [5] have developed a numerical algorithm to optimizing the geometry of composite structural parts to reduce the failure and obtain a superior product. The algorithms combine the optimization module of the ANSYS FE code and a progressive damage modelling module.The algorithm of the numerical optimization methodology is described by means of the flowchart shown in Figure 2. The methodology proposedwas divided into three basic packages: PDM of initial geometry, optimization and verification.

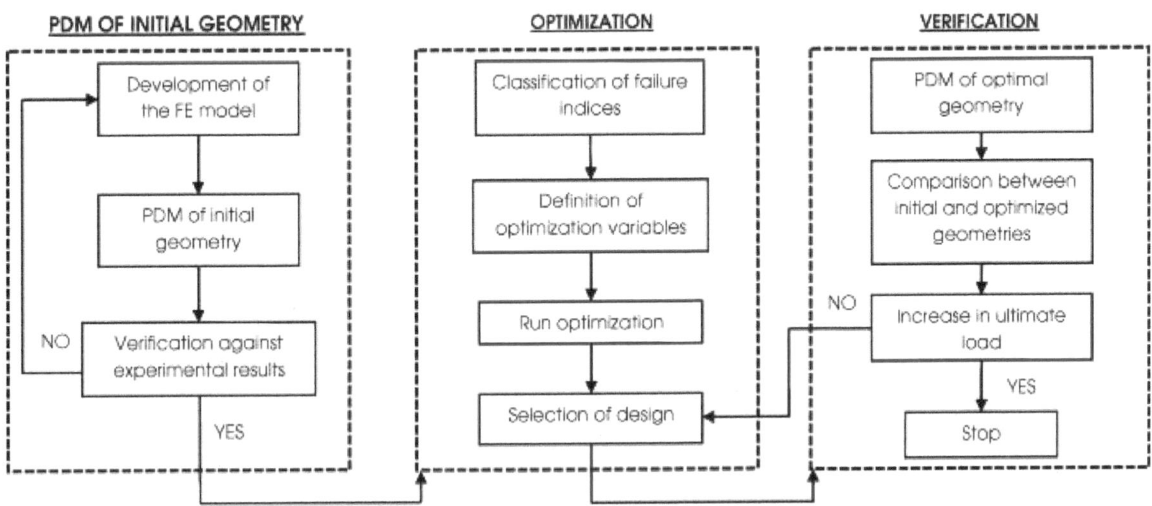

Figure 2.Flowchart of the numerical optimization methodology

Obviously, taken into account this proper model of optimizationthe processes we may obtain effective solutions thatcan led to a considerable increase in joint's strength for composites materials [5].

This paper provides a strong outline of principal application of composite materials, bring in attention typical damage mechanism that occur in composites and finally show an approach to optimize the mechanical properties of these materials.

REFERENCES

1. **Marcio Loos.** Composites. Marcio Loos. *Carbon Nanotube Reinforced Composites, CNR Polymer Science and Technology.* ISBN: 978-1-4557-3195-4, 2015.

2. **C. Soutis.** Introduction: Engineering requirements for aerospace composite materials. P. Irving and C. Soutis. *Polymer Composites in the Aerospace Industry.* 978-0-85709-523-7, 2015.

3. **R.H. Bossi, V. Giurgiutiu.** Nondestructive testing of damage in aerospace composites. P. Irving and C. Soutis. *Polymer Composites in the Aerospace Industry.* 978-0-85709-523-7, 2015.

4. **R.C. Alderliesten.** Critical review on the assessment of fatigue and fracture in composite materials and structures. Engineering Failure Analysis, Volume 35, 15 December 2013, Pages 370–379.

5. **K.I. Tserpes and A.S. Koumpias.** A numerical methodology for optimizing the geometry of composite structural parts with regard to strength. Composites Part B: Engineering, Volume 68, January 2015, Pages 176–184.

PLANE CIRCULAR PLATES OF SMALL THICKNESS. LARGE DISPLACEMENTS AND EIGEN VALUES-BENDING AND MEMBRANE LOADS

Gheorghe N. RADU
TRANSILVANIA University of Brasov, Dept. of Strength of Materials,
B-dul Eroilor 29, tel.0268474761, e-mail: rngh@unitbv.ro

Abstract: *The present paper deals with the state of stress and deformation at thin circular plates subjected to symmetrical axial loading. The loading is a bending owing to the uniform distributed loads which act perpendicular to the mean surface of the plate, simultaneous with a membrane load (loads acting in the mean plane of the plate). This type of problem is solved by help of equations which result from the equilibrium of a plate's element and from the boundary and continuity conditions of the mean surface of the plate.*
One considers the following two cases:
- the membrane stresses are small comparatively to the bending stresses; in this case the calculus is precise enough if we take into account only the mean plane extensions which will be superposed over the effects given by the transversal bending stress q;
- the membrane stresses are considerable and can not be neglected; in this case, second order calculus is required.
Keywords: *deformation, plate, stress, bending, membrane stress*

1. GENERAL CONSIDERATIONS

The present paper deals with the state of stress and deformation at thin circular plates under symmetrical axial loading. The loading is a bending owing to the uniform distributed loads, which act perpendicular to the mean surface of the plate, simultaneous with a membrane load (loads which are acting in the mean plane of the plate). This type of problem can be described by equations which result from the equilibrium of an attached plate element and from boundary conditions. In order to solve these problems, different methods have been suggested and used: Love's displacement function, the use of the working lengthening tube of Popkowitsch, etc. Uses of the finite element method and of the finite difference method have shown a great usefulness, leading to very good results. The behavior of the structures is described by help of a stiffness matrix, in the case of the displacements method, by a suppleness matrix, in case of the forces method, or by a stiffness matrix, in the case of the joint method. The above mentioned matrix are established using the finite element method, which is a method of structures division and which requires the similitude of the model's behavior to the real structure. The displacements method is adequate for the symmetric structures and symmetric loaded structures. The displacements are defined in the interior of an element by polynomials containing a number of parameters equal to the number of the unknown displacements number of the element nodes. One gets the basis equations of this method by help of the energetic method, which is based on the principle of the constant value of the elastic potential. In calculus one may consider that the material remains in the limits of Hooke's law validity, when external loads are acting upon it; a linear condition between stresses and deformations is required. At symmetrical axial state of stress correspond symmetrical axial states of deformation and the calculus can be made regarding the mean plane. Under these circumstances, the general 3-D problem can be reduced to a 2-D problem, where the stiffness is expressed by help of a mean section displacements. One divides the continuous structure into a system of symmetrical axial elements, whose unit element is defined as a plane finite element which rotates around the symmetry axis of the structure.

2. GREAT DISPLACEMENTS. EQUATIONS OF EQUILIBRIUM

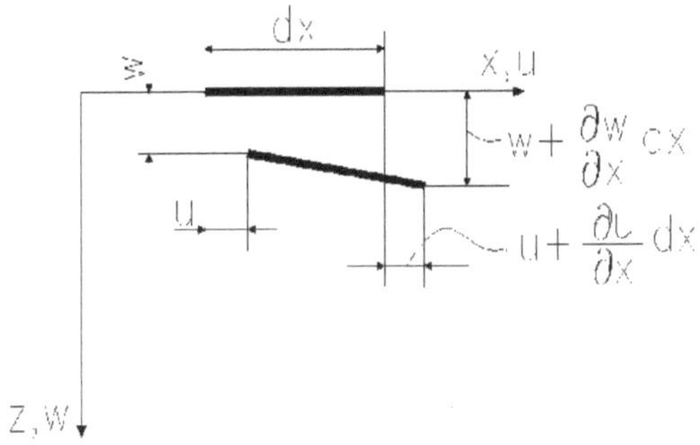

Fig. 1. Displacements

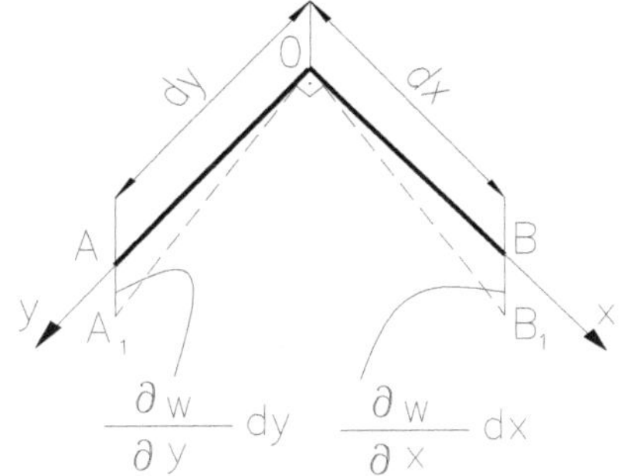

Fig. 2. Displacements

On basis of the notations from Fig. 1, one calculates the length after deformation of an element whose initial length is *dx*:

$$\sqrt{\left(dx+\frac{\partial u}{\partial x}dx\right)^2+\left(\frac{\partial w}{\partial x}dx\right)^2} \cong dx\left[1+\frac{\partial u}{\partial x}+\frac{1}{2}\left(\frac{\partial u}{\partial x}\right)^2+\frac{1}{2}\left(\frac{\partial w}{\partial x}\right)^2\right] \tag{1}$$

where the approximation of the square root does not influence the calculus or the final results accuracy. One mention the fact that the mean plane deformations allow to neglect the term $\frac{1}{2}\left(\frac{\partial u}{\partial x}\right)^2$; equation (1) shows that the element's length *dx* is changing with the quantity:

$$\varepsilon_x=\frac{\partial u}{\partial x}+\frac{1}{2}\left(\frac{\partial w}{\partial x}\right)^2; \qquad \varepsilon_y=\frac{\partial v}{\partial y}+\frac{1}{2}\left(\frac{\partial w}{\partial y}\right)^2 \tag{2}$$

In order to get the angular strain corresponding to the deformation *w*, normal to the mean plane, one study the case of the perpendicular elements of length *dx* and *dy* ($\angle AOB=\pi/2$), Fig. 2. The wanted angular strain is given by the subtraction of the final value of the angle $\angle AOB$ (after deformation), $\angle A_1OB_1$ from the initial angle $\angle AOB$, plus the share of angular deformation brought by the linear displacements u and v (displacements in the mean plane of the plate). One writes the cosine of the angle $\angle A_1OB_1$ as a dot product, getting:

$$\cos\angle A_1OB_1=\frac{\frac{\partial w}{\partial x}dx\cdot\frac{\partial w}{\partial y}dy}{dx\cdot dy}, \quad \text{or} \quad \sin\angle A_1OB_1=\cos(90-A_1OB)=\frac{\partial w}{\partial x}\cdot\frac{\partial w}{\partial y}$$

One takes into account the share of the linear displacements *u* and *v*; one finally get the wanted angular strain:

$$\gamma_{xy}=\frac{\partial u}{\partial y}+\frac{\partial v}{\partial x}+\frac{\partial w}{\partial x}\cdot\frac{\partial w}{\partial y} \tag{3}$$

From the specific literature (elastic theory), we have the expression of the mean surface curvatures:

$$\chi_x=\frac{\partial^2 w}{\partial x^2}; \qquad \chi_y=\frac{\partial^2 w}{\partial z^2}; \qquad \chi_{xy}=\frac{\partial^2 w}{\partial x\partial y}. \tag{4}$$

If one works in plane coordinates, Fig. 3, the equations of the strains and curvatures are easier. Equations (2) and (4) become:

$$\varepsilon_r=\frac{du}{dr}+\frac{1}{2}\left(\frac{dw}{dr}\right)^2; \quad \varepsilon_\theta=\frac{(r+u)d\theta-rd\theta}{rd\theta}=\frac{u}{r} \tag{5}$$

$$\chi_r=-\frac{d^2w}{dr^2}; \qquad \chi_\theta=-\frac{1}{r}\cdot\frac{dw}{dr} \tag{6}$$

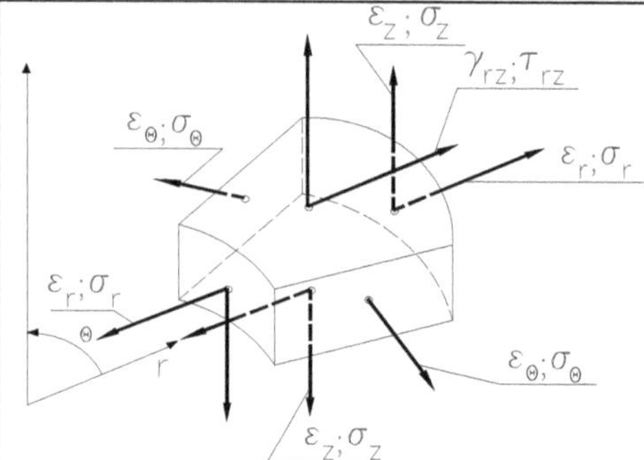

Fig. 3. Strains and curvatures in plane coordinate

The load is performed in the field of linear elastic plastic deformations, so that between stresses and deformations following equations are valid (Hooke's law):

$$\varepsilon_x = \frac{1}{E}\left[\sigma_x - \nu\sigma_y\right];\ \varepsilon_y = \frac{1}{E}\left[\sigma_y - \nu\sigma_x\right];\ \gamma_{xy} = \frac{2(1+\nu)}{E}\tau_{xy} \qquad (7)$$

One analyses equations (2) and (3). From these equations we get the deformations from the mean plane, these deformations being membrane deformations. One calculates the second derivative of the equations (2) and (3) with respect to *y* and to *x* and one neglects the high order infinitesimal; one derives equation (4) with respect to *x* and then with respect to *y*. We get the following expressions:

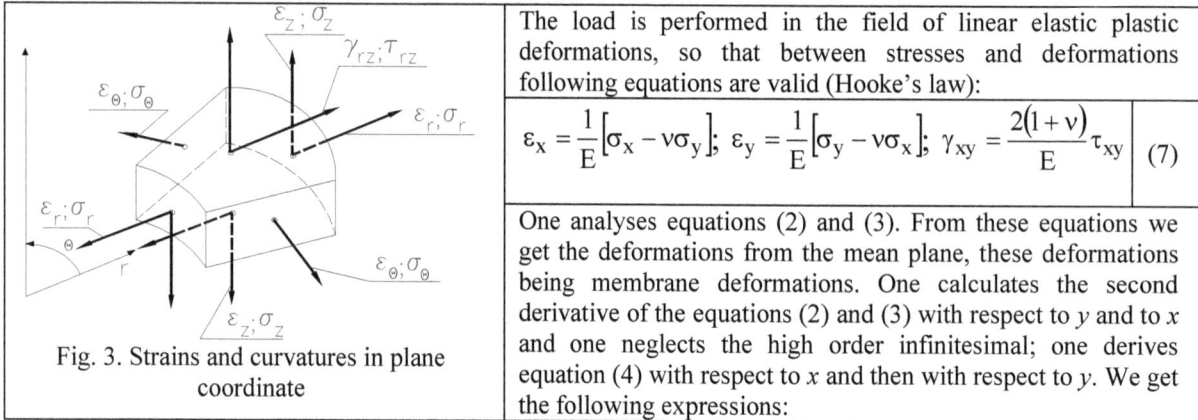

Substituting equation (7) in relation (9) it results a relation between stresses and deformations corresponding to the membrane state of stress :

$$E\left[\left(\frac{\partial^2 w}{\partial x \partial y}\right)^2 - \frac{\partial^2 w}{\partial x^2}\frac{\partial^2 w}{\partial y^2}\right] = \frac{\partial \sigma_x}{\partial y^2} + \frac{\partial \sigma_y}{\partial x^2} - \nu\left(\frac{\partial^2 \sigma_x}{\partial x^2} + \frac{\partial^2 \sigma_y}{\partial y^2}\right) - 2(1+\nu)\frac{\partial^2 \tau_{xy}}{\partial x \partial y} \qquad (10)$$

3. GENERAL EQUATIONS AT BENDING WITH CONSTANT q AND LOADING IN THE MEAN PLANE OF THE PLATE (FIG. 4)

Further on we want to establish a calculus relation between stresses and deformations for the loading of the circular plate, in Fig. 4. In this case, besides the bending due to the constant load *q*, the plate is subjected to tension because of the loads N_o (membrane load).

One specifies the existence of the following two calculus methods [1]:

a) the membrane stresses are small in comparison to those due to bending; in this case the calculus is precise enough if we consider only the extensions of the mean plane which are superposed on the effects caused by the transversal load *q*;

b) the membrane stresses due to the loads *p* are big and can not be neglected. That is the reason why the calculus must be continued.

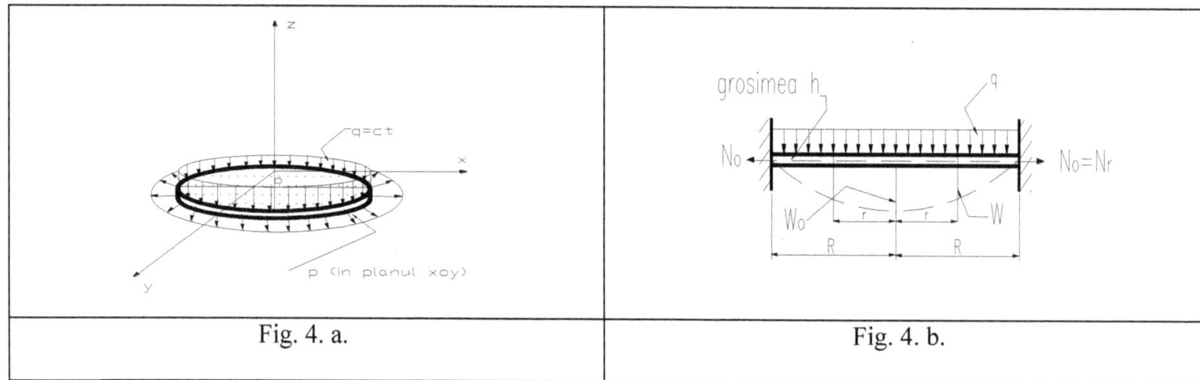

Fig. 4. a.   Fig. 4. b.

In order to take into account both the membrane stresses and the state stress caused by the bending loads, normal to the mean plane, one analyses the plate's element of dimensions $dx \cdot dy$, in deformed state of the mean plane, Fig. 5.

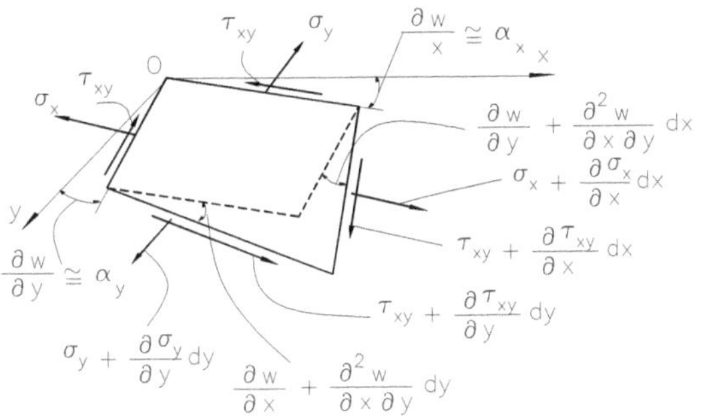

Fig. 5. Equilibrium of the plate's element

The plate's element in Fig. 5 is in equilibrium under the action of the shown state of stresses and deformations. On basis of the projections upon axes x and y, taking into account the angles in the deformed state, we get the following two equilibrium equations:

$$\left(\sigma_x + \frac{\partial \sigma_x}{\partial x} dx\right) \cdot \cos\left(\alpha_x + \frac{\partial \alpha_x}{\partial x} dx\right) - \sigma_x \cdot \cos\alpha_x + \left(\tau_{xy} + \frac{\partial \tau_{xy}}{\partial y} dy\right) \cdot \cos\left(\alpha_x + \frac{\partial \alpha_x}{\partial y} dy\right) - \tau_{xy} \cdot \cos\alpha_x = 0 \tag{11}$$

$$\left(\sigma_y + \frac{\partial \sigma_y}{\partial y} dy\right) \cdot \cos\left(\alpha_y + \frac{\partial \alpha_y}{\partial y} dy\right) - \sigma_y \cdot \cos\alpha_y + \left(\tau_{xy} + \frac{\partial \tau_{xy}}{\partial x} dx\right) \cdot \cos\left(\alpha_y + \frac{\partial \alpha_y}{\partial x} dx\right) - \tau_{xy} \cdot \cos\alpha_y = 0 \tag{12}$$

Equations (11) and (12) take simple forms, easy to use after the approximations below :

$$\left\{ \cos\alpha_x \cong \sqrt{1 - \left(\frac{\partial w}{\partial x}\right)^2} \cong 1 - \frac{1}{2}\left(\frac{\partial w}{\partial x}\right)^2; \quad \cos\left(\alpha_x + \frac{\partial \alpha_x}{\partial x} dx\right) \cong 1 - \frac{1}{2}\left(\frac{\partial w}{\partial x}\right)^2 - \frac{\partial^3 w}{\partial x^2 \partial y} \right. \tag{13}$$

The equations of equilibrium for a plane plate's element are :

$$\left\{ \frac{\partial \sigma_x}{\partial x} + \frac{\partial \tau_{xy}}{\partial y} = R_x; \quad \frac{\partial \tau_{xy}}{\partial x} + \frac{\partial \sigma_y}{\partial y} = R_y \right. \tag{14}$$

One derives them with respect to x and to y ; by adding the yielded relations, we get :

$$2\frac{\partial^2 \tau_{xy}}{\partial x \partial y} = -\frac{\partial^2 \sigma_x}{\partial x^2} - \frac{\partial^2 \sigma_y}{\partial y^2} + \frac{\partial R_x}{\partial x} + \frac{\partial R_y}{\partial y} \tag{15}$$

One replaces equation (15) in Equation (10), which was obtained from the relation of compatibility (9). One gets finally :

$$E\left[\left(\frac{\partial^2 w}{\partial x \partial y}\right)^2 - \frac{\partial^2 w}{\partial x^2} \cdot \frac{\partial^2 w}{\partial y^2}\right] = \Delta(\sigma_x + \sigma_y) - (1 - \nu)\left(\frac{\partial R_x}{\partial x} + \frac{\partial R_y}{\partial y}\right) \tag{16}$$

In Equation (17) one takes into account the relations (11) and (12), the projections upon axis z, the curvatures given by relations (4) as well as the independent relations (14) and one yields the equation below [1]:

$$D\Delta^2 w = q(x, y) + h(\sigma_x \gamma_x + 2\tau_{xy}\gamma_{xy} + \sigma_y \gamma_y) - R_x \frac{\partial w}{\partial x} - R_y \frac{\partial w}{\partial y} \tag{17}$$

Equations (16) and (17) show the state of stress and deformation of the plate, under the following circumstances : bending under transversal loading q, taking into account both the membrane stresses and the massic forces (calculated on surface unit of the plate's mean plane).

The solution yields by help of numerical methods (FEM or FDM), taking into account each effective loading of the plate and the way the plate is supported.
We further on analyze the plate in Fig. 4, subjected to the transversal loading q and to the forces in the mean plane, forces denoted by N_r=p.

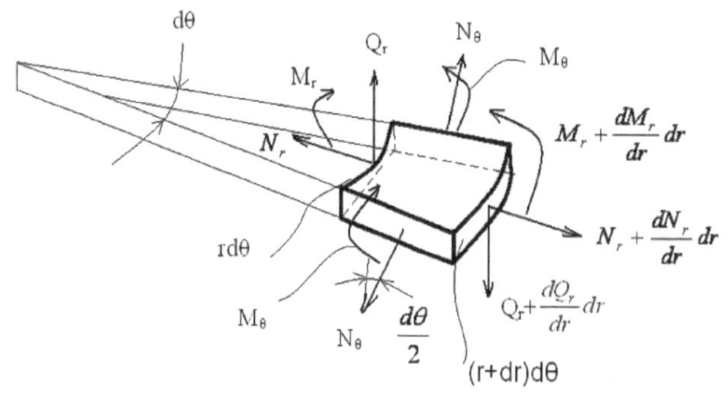

Fig. 7.

The plate's element presented in Fig. 7. is in equilibrium.

$$N_r \cdot rd\theta - (N_r + dN_r)(r + dr)d\theta + 2N_\theta dr \sin\frac{d\theta}{2} = 0$$

$$\left(M_r + \frac{dM_r}{dr}dr\right)(r + dr)d\theta - M_r \cdot rd\theta - 2M_\theta \cdot dr\frac{d\theta}{2} - Q_r \cdot rd\theta \cdot dr = 0 \quad (18)$$

One cancels the analogous terms, we neglect the infinitesimal terms and it successively yields:

$$\frac{N_r - N_\theta}{r} + \frac{dN_r}{dr} = 0 \quad (19)$$

where N_r and N_θ, on basis of Hooke's law and referring to the plate's unit of length, take the following expressions:

$$N_r = \frac{E \cdot h}{1 - \nu^2}(\varepsilon_r + \nu \cdot \varepsilon_\theta); \qquad N_\theta = \frac{E \cdot h}{1 - \nu^2}(\varepsilon_\theta + \varepsilon_r \cdot \nu) \quad (20)$$

By replacing equations (5) into (20), yields:

$$N_r = \frac{E \cdot h}{1 - \nu^2}\left[\frac{du}{dr} + \frac{1}{2}\left(\frac{dw}{dr}\right)^2 + \nu\frac{u}{r}\right]; \qquad N_\theta = \frac{E \cdot h}{1 - \nu^2}\left[\frac{u}{r} + \nu\frac{du}{dr} + \frac{1}{2}\nu\left(\frac{dw}{dr}\right)^2\right] \quad (21)$$

For the equations of moments one neglects the infinitesimal terms, one divides them to $r \cdot d\theta \cdot dr$ and the following expression yields:

$$\frac{M_r - M_\theta}{r} + \frac{dM_r}{dr} - Q_r = 0 \quad (22)$$

A sum of projections on vertical direction, taking into account both the curvature of the deformed mean surface of the plate and the loading q leads to a third equation of equilibrium:

$$\frac{d}{dr}(rQ_r) + \frac{d}{dr}\left(rN_r\frac{dw}{dr}\right) + rq = 0 \quad (23)$$

Equations (19), (22) and (23) are equations of equilibrium of the plate, Fig. 4. N_r and N_θ are radial loading and circumference loading, respectively, Q_r is the shear force, M_r and M_θ being the bending moments upon the radius and the circumference.

If one attaches a diametric strip of the plate, subjected to bending, and one considers the loading towards a single direction then, the equation of equivalence between stresses and external loading is (using the notation in Fig. 8):

$$M_i = \int_{-h/2}^{h/2} \sigma_x \cdot z \cdot dz = -\int_{-h/2}^{h/2} \frac{Ez^2}{1 - \nu^2} \cdot \frac{d^2w}{dx^2}dz = -D\frac{d^2w}{dx^2} \quad (24)$$

where $D = \dfrac{Eh^3}{12(1 - \nu^2)}$ represents the cylindrical stiffness modulus at bending.

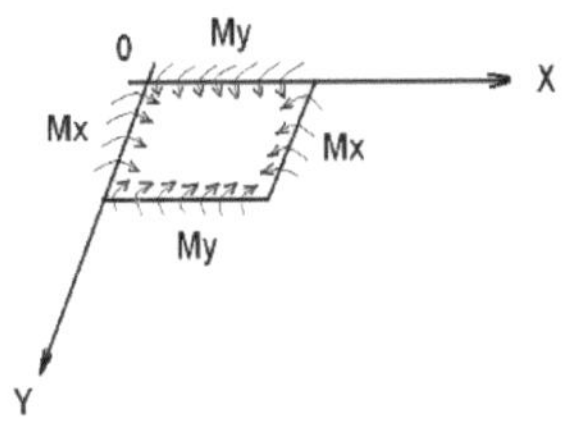

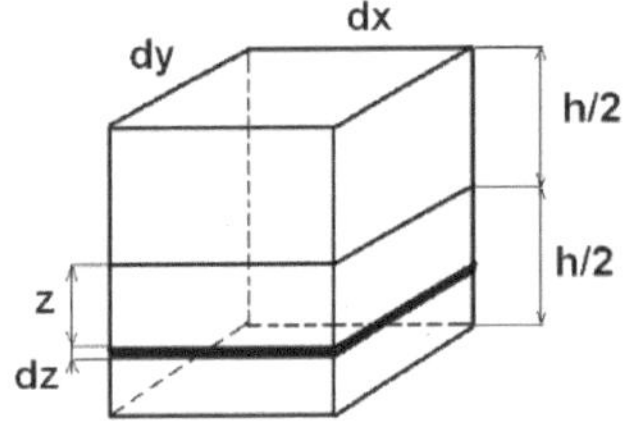

Fig. 8.

The well-known equations have been used too:

$$\varepsilon_x = -z\frac{d^2w}{dx^2}; \varepsilon_x = \frac{\sigma_x}{E} - \nu\frac{\sigma_y}{E}; \varepsilon_y = \frac{\sigma_y}{E} - \nu\frac{\sigma_x}{E} = 0 \text{ yielding: } \sigma_x = \frac{E\varepsilon_x}{1-\nu^2}$$

On basis of the deduction method of relation (24) and of the plate's element shown in Fig. 8, the moment's expressions yield:

$$M_x = D\left(\frac{1}{r_x} + \nu\frac{1}{r_y}\right) = -D\left(\frac{\partial^2 w}{\partial x^2} + \nu\frac{\partial^2 w}{\partial y^2}\right); \quad M_y = D\left(\frac{1}{r_y} + \nu\frac{1}{r_x}\right) = -D\left(\frac{\partial^2 w}{\partial x^2} + \nu\frac{\partial^2 w}{\partial y^2}\right) \tag{25}$$

Depending on the curvatures γ_r and γ_θ of the deformed mean surface of the plate, the relations of the moments M_r and M_θ have the expressions below, analogous to Equations (25):

$$M_r = D(\gamma_r + \nu\gamma_\theta); \quad M_\theta = D(\gamma_\theta + \nu\gamma_r) \tag{26}$$

By replacing equations (6) in relations (26) and then in relation (22), the calculus relation of the stresses yields:

$$M_r = -D\left(\frac{d^2w}{dr^2} + \frac{\nu}{r}\frac{dw}{dr}\right); \quad M_\theta = -D\left(\frac{1}{r}\frac{dw}{dr} + \nu\frac{d^2w}{dr^2}\right); \quad Q_r = -D\left(\frac{d^3w}{dr^3} + \frac{1}{r}\frac{d^2w}{dr^2} - \frac{1}{r^2}\frac{dw}{dr}\right) \tag{27}$$

The analyzed plates in the present paper respect the condition $\frac{h}{a} < \frac{1}{25}$, being thin plates; in this case one neglects the shear deformations [1].

By integration of the equation of equilibrium (23), we get:

$$Q_r + N_r\frac{dw}{dr} + \frac{1}{2}qr = 0 \tag{28}$$

On basis of relation (5), one establishes a compatibility relation. One follows the succession below:

$$\varepsilon_\theta - \varepsilon_r = \frac{u}{r} - \frac{du}{dr} - \frac{1}{2}\left(\frac{dw}{dr}\right)^2; \quad \frac{d}{dr}(\varepsilon_\theta) = \frac{d}{dr}\left(\frac{u}{r}\right) = \frac{1}{r}\left(\frac{du}{dr} - \frac{u}{r}\right) \text{sau} \frac{du}{dr} - \frac{u}{r} = r\frac{d\varepsilon_\theta}{dr};$$

Finally, the relation of compatibility yields:

$$r\frac{d\varepsilon_\theta}{dr} + \varepsilon_\theta - \varepsilon_r + \frac{1}{2}\left(\frac{dw}{dr}\right)^2 = 0 \tag{29}$$

By replacing the third relation (27) in relation (28), yield:

$$\frac{d^3w}{dr^3} + \frac{1}{r}\frac{d^2w}{dr^2} - \frac{1}{r^2}\frac{dw}{dr} - \frac{1}{D}N_r\frac{dw}{dr} = \frac{q}{2D}r \tag{30}$$

Relations (20) express the strains as function of forces, yielding the following expressions:

$$\varepsilon_r = \frac{1}{Eh}(N_r - \nu N_\theta) \quad \text{and} \quad \varepsilon_\theta = \frac{1}{Eh}(N_\theta - \nu N_r) \tag{31}$$

Based on relations (31), the compatibility relation yields:

$$\frac{dN_\theta}{dr} + \frac{dN_r}{dr} + \frac{Eh}{2r}\left(\frac{dw}{dr}\right)^2 = 0 \tag{32}$$

One replaces the equation of equilibrium (19) in relation (32), yielding:

$$\frac{dN_r}{dr} = -\frac{N_r - N_\theta}{r}; \quad \frac{dN_\theta}{dr} - \frac{N_r - N_\theta}{r} + \frac{Eh}{2r}\left(\frac{dw}{dr}\right)^2 = 0 \tag{33}$$

Equations (19), (30) and (33) are three nonlinear equations with three unknowns: dw/dr, N_r and N_θ. The nonlinearity appears in the term N_r, dw/dr (in Eq. 30) and $(dw/dr)^2$ in Eq. 33.

For the problem in Fig. 4, the circular plate is first subjected to stretching by a load of plane stress N_o, around its circumference and then it is subjected to bending under a constant load q. One gets the solution concerning the initial plane stress from the general equations (19)-(27), putting the condition $w=0$ and $q=0$. One yields:

$$N_r = N_\theta = N_0 \quad \text{and} \quad u = \frac{N_0}{Eh}(1-\nu)r \tag{34}$$

N_o results from the constant load p.

Equations (34) fulfill the boundary condition $N_r = N_o$, at $r=a$. If $N_r = \sigma_r h$ is the radial load, the number $N_0/Eh = \sigma_r/E = \varepsilon_0$ can be interpreted as the uniaxial load. The plate is first subjected to a stretching under the load N_o and then it is subjected to a vertical load q. The solution yields for each case partly, using the boundary conditions of the plate (conditions for continuity and for relation of the deformed mean surface). For instance, the initial radial load N_o will be affected, $\overline{N}_r$ and $\overline{N}_\theta$ due to the action of the load q.

Depending on the current radius r, in the mean surface the following loads yield:

$$N_r = N_o + \overline{N}_r; \quad N_\theta = N_o + \overline{N}_\theta$$

Equations (19), (30) and (33) are changing accordingly. In the new relations, one may use the following notations:

$$\zeta = \frac{r}{a}; \quad W = \frac{w}{h}; \quad U = \frac{u}{h}; \quad \theta = \frac{dW}{d\zeta} = \frac{a}{h} \cdot \frac{dW}{dr} \tag{35}$$

yielding dimensionless quantities. These are more accessible in engineering. We have to emphasize the fact that such an approach is intricate enough. In general, the analytical solutions are very difficult.

In the case of Fig. 4, the expressions of the stresses will have compulsory components due to the initial, membrane and bending loading, respectively.

The equations presented in the present paper are enough for a complex approach of the circular thin plates bending, in the case of the initial loading in the mean surface. One can also take into account the temperature's effects, the calculus is reported on the linear theory (small deformations), but particularly on the nonlinear theory (large deformations). The solutions are presented in other papers.

The complexity of the approach of different loading consists in the fact that the solution can not be given dimensionless for each case; that is because of the different dependence of the membrane stress fields on the dimensional parameters. The FEM and FDM present important advantages in the approach of the above problems.

We present further on an example solved by help of FEM (specialized program).

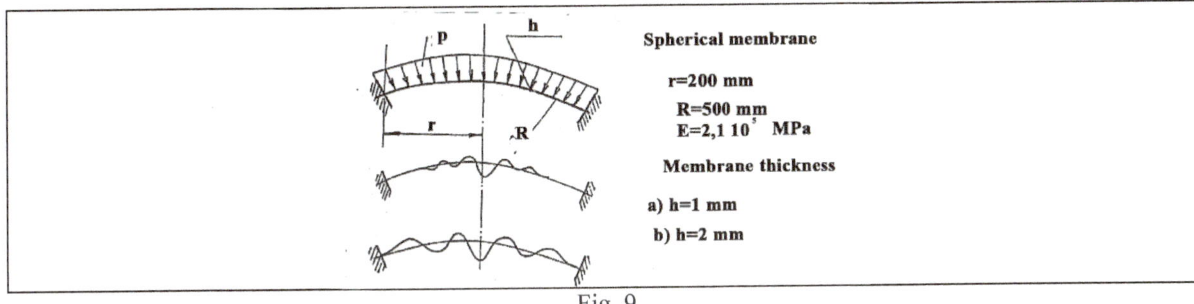

Fig. 9

Fig. 9. presents a fixed spherical membrane subjected to a constant pressure q. One divide in 16 finite elements, using the theory presented in the first part of the paper. We have obtained a critical pressure of 1.047 MPa, for h = 1 mm and 4.359 MPa for h=2 mm, respectively. The yielded values are very precise. They have been compared to the values yielded from the equation given by the differential equation made up by help of Bessel's funcons (Ponomariov). The equation is:

$$f_w = \frac{Eh^2}{R^2} \cdot \frac{2}{\sqrt{3(1-v^2)}}$$

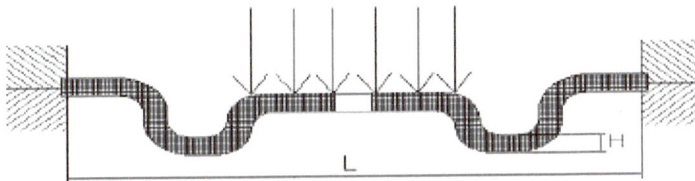

On basis of the above equation the critical values of the pressure yield: 1.017 MPa and 4.070 MPa, respectively. The eigen ways of loosing stability are shown in the figure above (due to the forking).

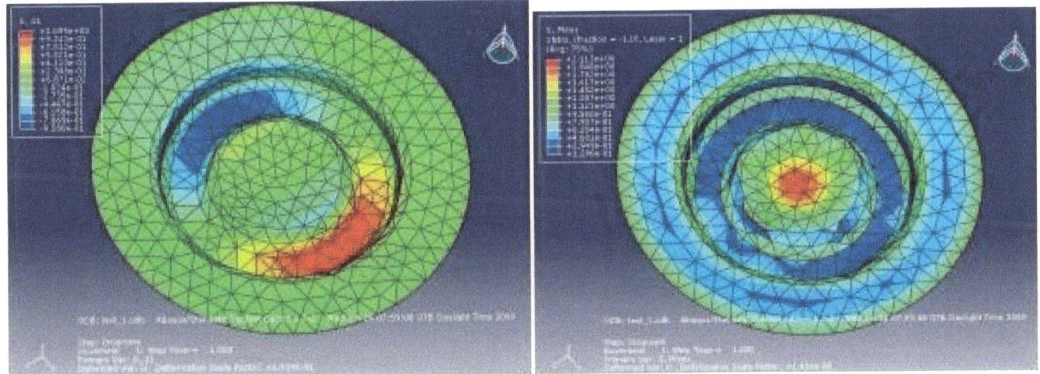

Figure 7. Rubber and sectioned textile membrane.

Figure11.Displacement on X direction.

Figure 12.The tensions distribution for the rubber textile membrane.

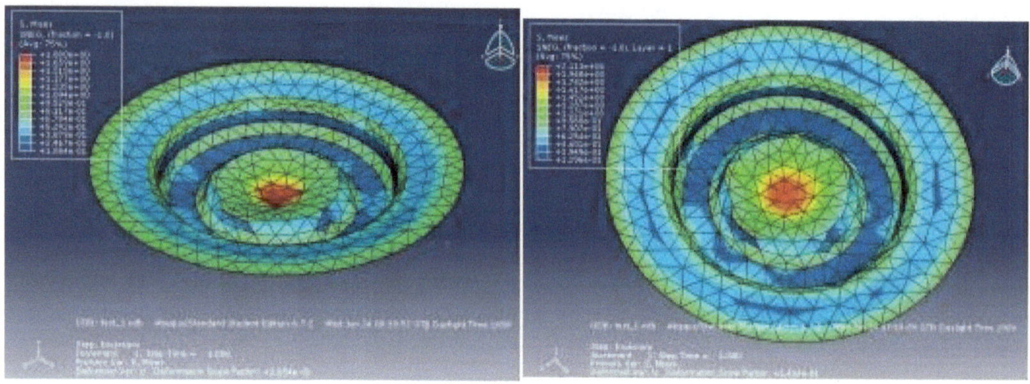

Figure 13.Displacement on X direction(ruber membrane). Figure 14.The tensions distribution(rubber membr).

REFERENCES

[1] TIMOSHENKO, S.P., WOINOWSKY-KRIEGER, S., *Teoria plăcilor plane şi curbe,* Editura Tehnică, Bucureşti, 1968.

[2] RADU, Gh. N., Curtu, I., *Elastic Stability of the Thin disks due to the Temperature and Rotation Stress*, 14th Symposium DANUBIA –ADRIA, Porec, Croatia, 1997.

[3] SHEPLAK, M., *Large Deflections of Clamped Circular Plates Under Initial Tension and Transitions to Membrane Behavior,* ASME, 1998.

[4] RADU, Gh. N., *Contribuţii la studiul discurilor subţiri în mişcare de rotaţie şi în regim termic axial simetric* – Teză de doctorat, Braşov, 1992.

OPTIMIZATION OF A "SANDWICH" STRUCTURE FOR THE INSULATION OF A PREFABRICATED WOODEN HOUSE

Scutaru Maria Luminita[1], Baba Marius[1]

[1] Transilvania University of Brasov, Department of Mechanical Engineering, 29 Eroilor Blvd, 500036, Brasov, Romania, lscutaru@unitbv.ro

Abstract: This paper aims to determine an optimal solution in terms of cost to the problem of determining a structure composed of layers of different material which ensures a required heat transfer from inside a building to outside them (a suitable insulation at a minimum price) . The problem is placed in a zone of interference between the heat exchange and construction research domains. A theoretical study on the method of optimizing the objective function cost price considering heat transfer by laminated flat surfaces was carried out, and experimental measurements were made in order to validate the results. The literature reveals the existence of a corrective coefficient for each insulation material but not for structures of the type we studied. Knowledge of these factors allows the design of dimensionally and thermally constructions having similar composition as those in the study and allows optimizing the performance in terms of costs.
Keywords: insulation material, structures, heat transfer.

1. INTRODUCTION

In order to achieve resistance elements and structures with outstanding performance, industry experts are increasingly concerned about the many possibilities of combining wood with other materials and products made from wood or other materials, with different properties. The paper proposes the use of an optimization method for the determination of a solution to combine several types of boards so as to obtain a material with good thermal insulating properties. [1]-[6]

There have been theoretical and experimental research on a large number of "sandwich" structures that can be widely used in the construction of prefabricated wooden houses. The method we adopted was following on the possibility of combining aspects of common insulation materials, cheap, domestic (mineral wool, polystyrene, PAL) to form a sandwich structure thermophysical properties that are suitable for use as wall panels for prefabricated houses.

In practice, this kind of panels are developed without scientifically motivating a number of issues such as: the possibility of combining different materials (with various physical and mechanical, elastic and strength characteristics) of large diversity, the influence the dimensions (width, length thickness, etc.) on the characteristics of rigidity and stability of panels of thermal and sound absorbing properties. Therefore, a consistent methodology according to objective criteria in conformity with the related problems of choosing materials to achieve, in terms of minimum cost, of insulating panels for construction is required,.

These panels can be classified as composite materials with a special structure - designed to capitalize on the higher characteristics of each material that is included in them, in order to be able to cope with a variety of operating conditions characterized by variations in humidity, temperature, mechanical, static and dynamic stresses.

2. OPTIMIZATION METHOD

The general formulation of a single-criterion linear programming problem is the following:

minimize $f(x)$ with $x = (x_1, x_2, \ldots, x_N)$

subjected to: $g_j(x) \leq 0$ where $j=1,2,\ldots,P$

$h_i(x) = 0$ where $i=P+1,\ldots,P+M$

f(x) is a linear function, $g_j(x)$ and $h_i(x)$ are inequality and equality constraints, respectively.

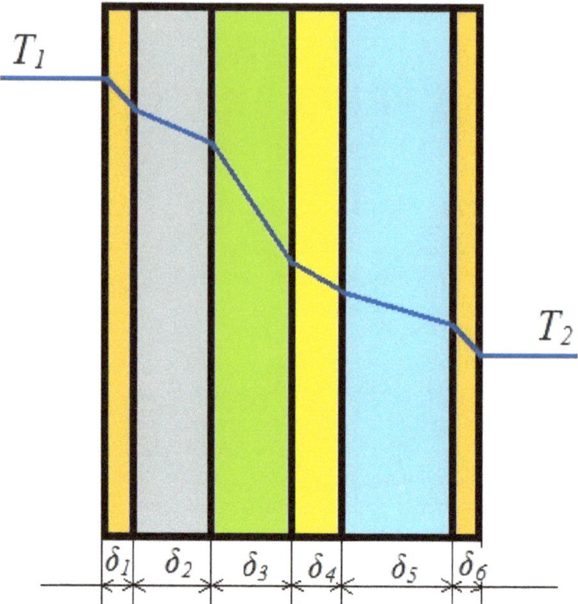

Figure 1. Multilayered construction material

The cost function f(x) can be written using a linear expression:

$$f(x) = \alpha_1\delta_1 + \alpha_2\delta_2 + ... \quad . + \alpha_N\delta_N$$

where α_1, α_2 ... ., α_N are cost coefficients and $\delta_1, \delta_2, ...$, δ_N represent, respectively, the width of the layers 1, 2,... ., N.

The cost calculation is founded on material costs and fabrication costs, which have direct effect on the dimensions and geometry of the structure. Generally, the cost function includes the cost of material, assembly, painting, cutting, forming the shell but for our purposes we will consider the costs of the materials used considering, for the begining, the other costs being the same for all solutions. Informations considering different fabrication costs can be found in [7]-[12].

3. EXPERIMENTAL DATA

Based on the principle of minimizing the cost price by imposing conditions on the total wall resistance and maximum wall thickness, we are considering solutions resulting from the calculation and measure the heat transfer coefficient variation for assemblies in the various combinations of these materials.The conditions that were considered in choosing the types of structures made were:

- to be common (most commonly used in the construction of wooden houses) ;
- to allow drawing conclusions by comparing the values obtained from tests that are practical recommendations for users.Other conditions were imposed in the development, not only in the choice of these structures :
- the insulating layer to be formed only of mineral wool and polystyrene , so the outside is chipboard ;
- for multilayered structures, the polystyrene layer to be on the warm side
- for three-layer structures, the insulating layer to be formed by combining two symmetrical insulation materials.

For ease of tracking, processing and interpretation of experimental results we have developed a code for each specimen, using the notation: P - PA;L p - polystyrene; v - wool, followed by figures representing the layer thickness. We have carried out experiments on three types of structures:

I.monolayer
a) mineral wool insulation layer
b) polystyrene insulating layer

a) b)

Figure 2. Monolayer

II.dual-layer
a) the insulating materials are mineral wool (varies) and polystyrene (constant)
b) the insulating materials are mineral wool (constant) and polystyrene (varies)

a) b)

Figure 3. Dual-layer

We determined the thermal conductivity coefficient equivalent to the material resulting from the composition of different types of materials. Knowledge of these factors determines a correction factor (denoted with $c = \frac{\lambda_e}{\lambda_t}$ where λ_e is the coefficient of thermal conductivity experimentally determined and λ_t is the coefficient of thermal conductivity determined by calculation).

Based on the Fourier's law the heat flux is proportional to the local temperature gradient. For a three layer the heat transfer rate through the first layer is:

$$q_1 = \frac{\lambda_1 A}{t_1}(T_0 - T_1) = \frac{(T_0 - T_1)}{R_1} \qquad (1)$$

while the heat rates through the second and the third layer are:

$$q_2 = \frac{\lambda_2 A}{t_2}(T_1 - T_2) = \frac{(T_1 - T_2)}{R_2} \qquad (2)$$

$$q_3 = \frac{\lambda_3 A}{t_3}(T_2 - T_3) = \frac{(T_2 - T_3)}{R_3} \qquad (3)$$

As a steady-state analysis is considered in this paper, the heat flows passing each layer are equal while no internal heat is generated:

$$q_1 = q_2 = q_3 = q \qquad (4)$$

Therefore by substituting equations (1), (2) and (3), the heat transfer rate thorough the layered composite is:

$$q = \frac{T_0 - T_3}{R_1 + R_2 + R_3} = \frac{T_0 - T_3}{\Sigma R_i} \qquad (5)$$

where R_i is the conductive thermal resistance of layer *i*:

$$R_i = \frac{t_i}{\lambda_i A} \qquad (6)$$

The global heat transfer coefficient is:

$$\lambda_e = \frac{\delta}{\Sigma \frac{\delta_i}{\lambda_i}}.$$

The results of measurements and theoretical calculations are summarized in tabelar form corresponding to each category of structures in order to be analyzed and interpreted graphically. Table 2 is a complete data on monolayer structures in two variants:

a) mineral wool insulation layer
b) polystyrene insulating layer

Table 2. Table of cumulative data for monolayer samples

Sample	Total width (mm)	The coefficient of thermal conductivity experimentally determined λ_e (W/mK)	The coefficient of thermal conductivity theoretically determined λ_t (W/mK)	Theoretically calculated thermal resistance R (m^2K/W)	The correction coefficient $c = \frac{\lambda_e}{\lambda_t}$
PvvP 16,20,20,16	72	0,051	0,065	1,104	0,784
PvvP 16,35,35,16	102	0,049	0,055	1,843	0,890
PvvP 16,35,50,16	117	0,047	0,053	2,209	0,886
PvvP 16,50,50,16	132	0,040	0,051	2,575	0,784
PppP 16,40,30,16	102	0,048	0,058	1,764	0,827
PppP 16,40,50,16	122	0,047	0,055	2,229	0,854
PpppP 16,50,20,50,16	152	0,043	0,052	2,927	0,826
PpppP 16,50,40,50,16	172	0,040	0,051	3,392	0,784

Having all the necessary data, a graphic was made for the dependence between the coefficient of thermal conductivity experimentally determined (λ_e) and the theoretically determined one (λ_t), taking into considerration the mineral wool insulation thickness (fig 1.)

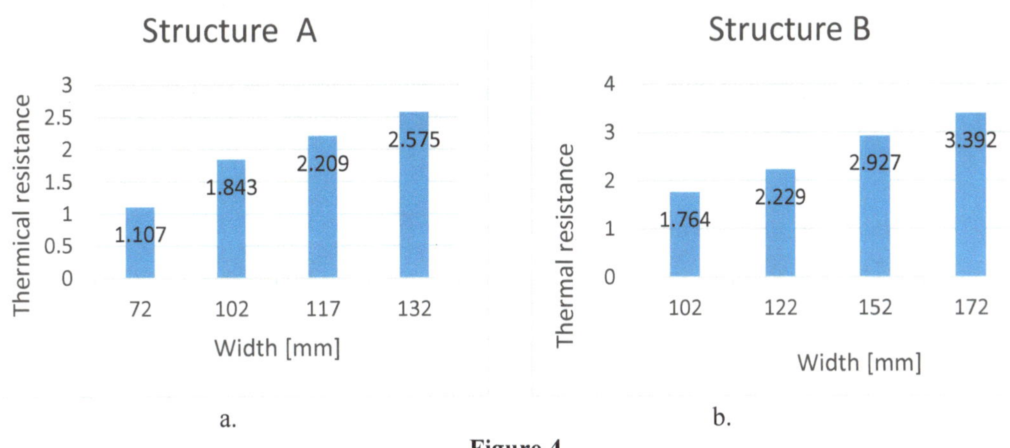

a. b.

Figure 4

a. thermal resistance variation for the (PvvP) variant

b. thermal resistance variation for the (PppP) variant

Fig. 5 Shows the temperature variation through the composite layups considered.

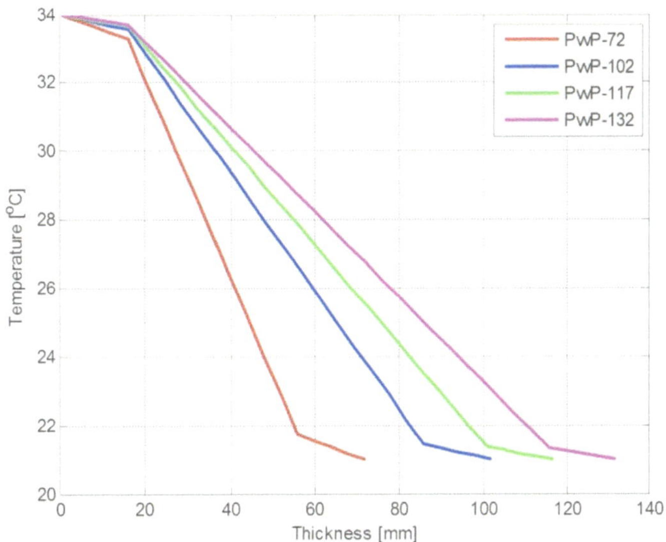

Figure 5. Temperature variation through the layered composite PvvP

An almost identical decrease for the two heat transfer coefficients can be very well seen and, therefore, an increased equivalent thermal resistance compared to using mineral wool, reinforcing further the findings from the literature regarding the influence on the heat transfer of moisture and uniformity of the material.

A planar steady state heat transfer finite element analysis has been performed for all the composite layups presented in the previous section. Fig. 6 shows the 2D finite element model and the corresponding primary output data represented by the temperature variation through the layered composite PvvP-72.

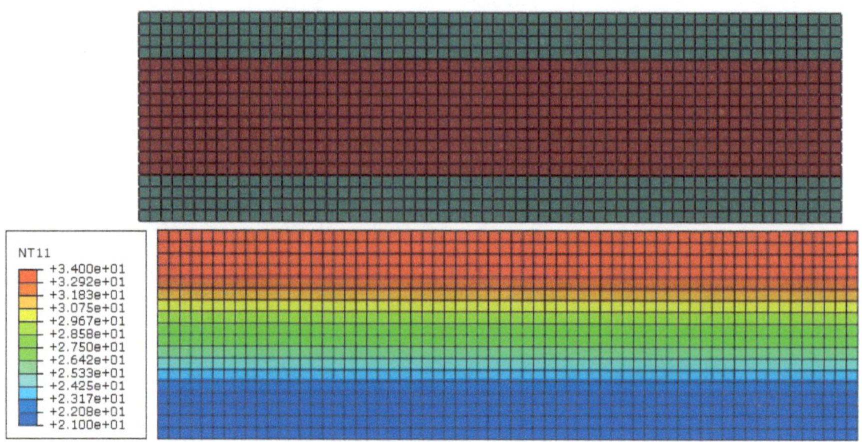

Figure 6. Temperature variation through the layered composite PvvP-72

Fig. 7 shows the temperature variation through the composite layups considered.

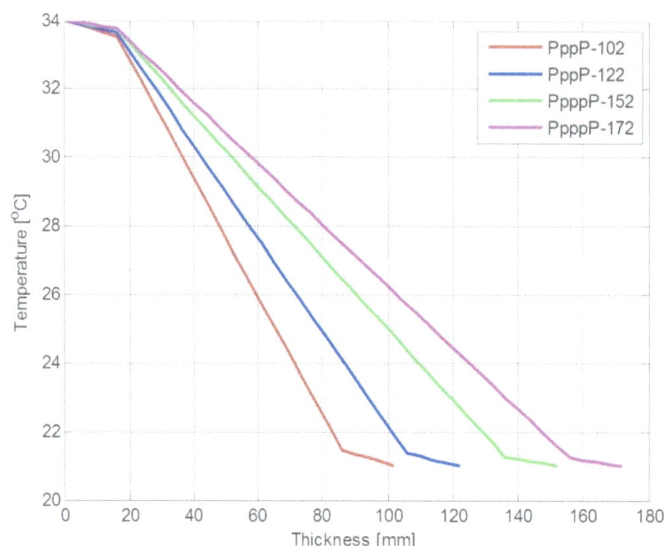

Figure 7. Temperature variation through the layered composite PppP

From the study of this representation , an almost linear decrease of the two heat transfer coefficient can be seen, depending on the insulation thickness. The somewhat larger differences that occur in specimens of thickness 72 mm and 132 mm are explained by the fact that both specimens have some moisture at the surface and in the interior and the materials used were not the same for each sample. No material was changed from one sample to another, using commercially available materials that vary even from batch to batch and, also, the mineral wool is not as compact as polystyrene.

The decrease transfer coefficient theory should involve a linear increase of thermal resistance, which is well highlighted in Figure 8.

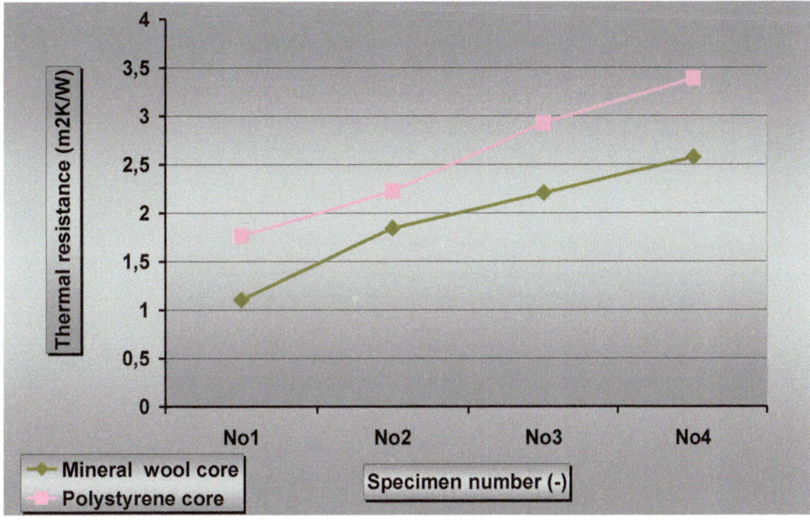

Mineral wool core - POLYSTYRENE CORE

Figure 8. Thermal resistance variation for PvvP , PppP

Analyzing the chart above we can draw the conclusion that, with the increasing thickness of both thread and polyester wool, an increase of the value of thermal resistance of the sample can be observed and, at the same time, involving a decrease in the value of heat transfer coefficient (both the experimentally and the theoretically determined) . This linear increase in the thermal resistance happens in concordance with the linear decrease of the value of the heat transfer coefficient, being inversely proportional to the thickness of the insulating layer. Experimental and theoretical research data for dual-layered structures are summarized in Table 3 in order to be able to be interpreted.

Table 3. Table of cumulative data for dual-layer samples

Epruveta	Total width (mm)	The coefficient of thermal conductivity experimentally determined λ_e (W/mK)	The coefficient of thermal conductivity theoretically determined λ_t (W/mK)	Theoretically calculated thermal resistance R (m^2K/W)	The correction coefficient $c = \frac{\lambda_e}{\lambda_t}$
PpvP 16,20,20,16	72	0,060	0,066	1,089	0,909
PpvP 16,20,30,16	82	0,058	0,062	1,333	0,935
PpvP 16,20,50,16	102	0,050	0,056	1,821	0,892
PpvP 16,20,80,16	132	0,044	0,052	2,552	0,846
PpvP 16,20,20,16	72	0,060	0,066	1,089	0,909
PpvP 16,30,20,16	82	0,057	0,062	1,322	0,919
PpvP 16,50,20,16	102	0,050	0,057	1,787	0,877
PpvP 16,80,20,16	132	0,046	0,053	2,484	0,867

For these types of structures, in the case when the mineral wool insulation layer varies and the polystyrene remains constant, based on data in Table 3, we were able to represent the variation of the two heat transfer coefficients, λ_e and λ_t (Fig.7) and also we were able to show the change in resistance heating of the specimen (Fig. 8). Both graphs are based on the thickness of the specimen, thus the thickness of mineral wool.

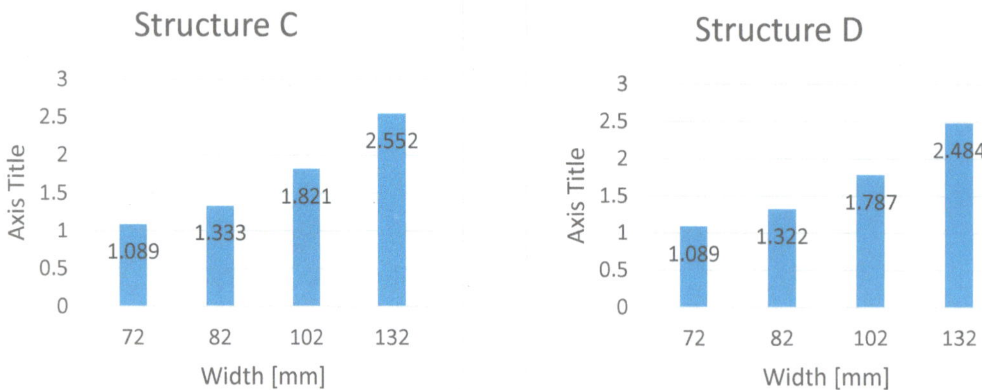

Figure 9.
a. thermal resistance variation for the (PpvP) variant
b. thermal resistance variation for the (PvpP) variant

A quite close variation of the two factors can be seen, the differences being due to the fact that the mineral wool insulation layer is not compact and has air gaps in the structure, factors which have influence on the value of the heat transfer coefficient.

This time, considering only the increase in the polystyrene layer, which is more compact than the wool, the most important aspect is the almost constant difference between the results obtained experimentally and those which were theoretically determined.

For these types of dual-layered structures, with a simultaneous increase of both the layer of mineral wool and the polystyrene layer so that the overall insulation thickness remains constant, we started to study on the same graph of the variation of heat transfer coefficient determined experimentally (Fig. 8) in the two situations.

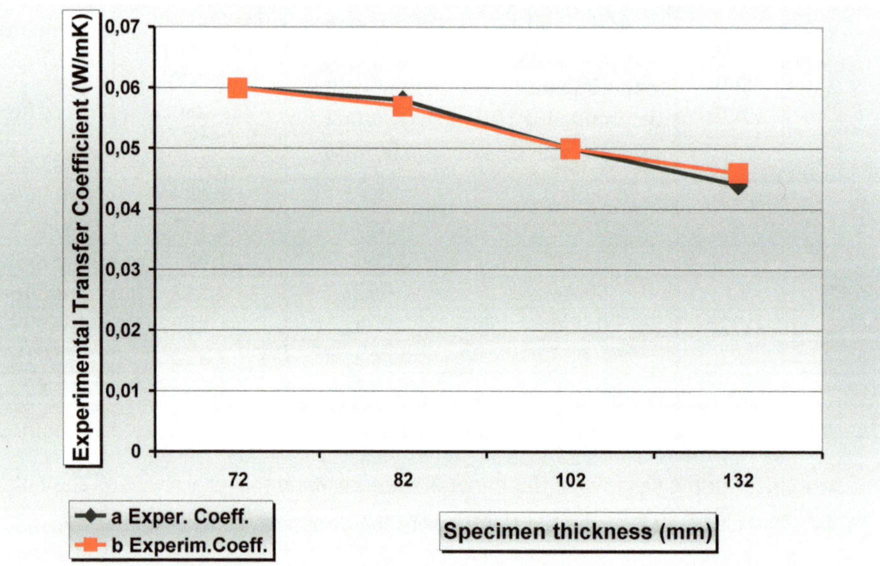

Figure 10. The heat transfer coefficient in the dual-layered structure options (a- the polystyrene-layer remains constant, b- the wool layer remains constant)

It can be seen from the superposition of the two graphs that the two solutions provides almost identical results. The order of layers in achieving the final solution is not important.

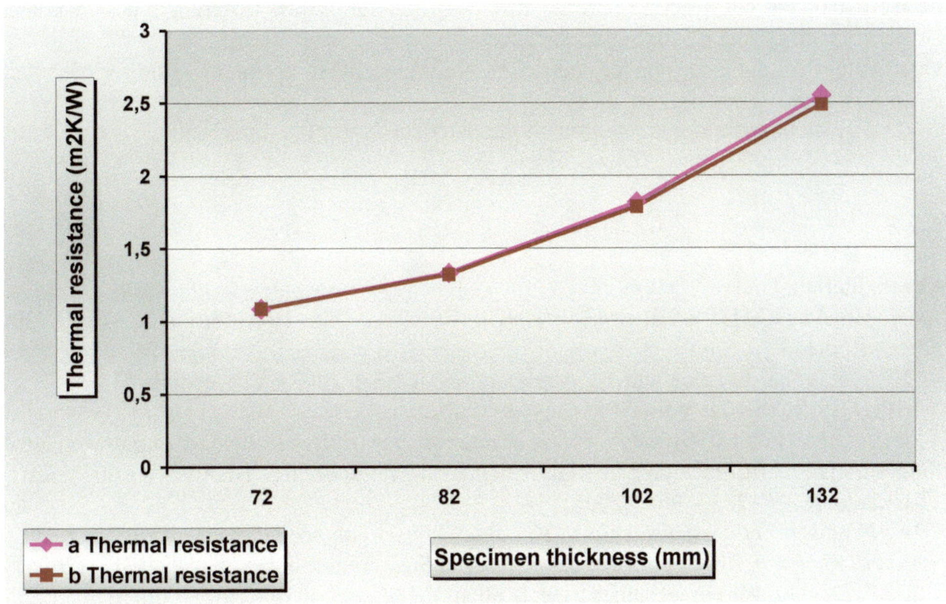

Figure 11. Thermal resistances for the two cases of dual-layered structures (a - the polystyrene layer remains constant, b - the wool layer remains constant)

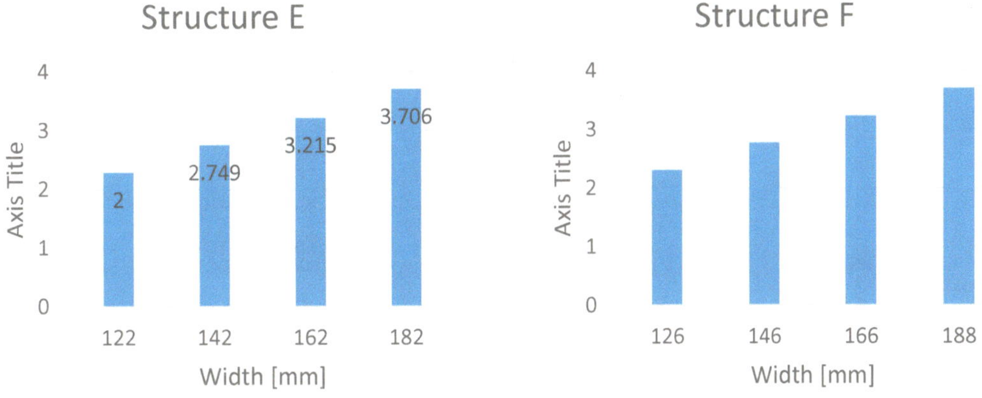

Figure 12. The thermal resistances for the two cases of triple-layered structures for two studied variants

4. CONCLUSIONS

The diagrams resulted from the experiments on the presented structures show the change of layout for the analyzed characteristics and can eventually be used as guidance for thicknesses greater than those analyzed in the paper , with relatively small errors by extrapolation. This observation is supported by the small differences that occur between the values of theoretical and experimental data

Analyzing the results obtained, we could appreciate against the influence of types of insulation material, and the thickness thereof, for the "sandwich " structure . We can draw the following conclusions:

• a good insulating material - with a low thermal conductivity - has a certain porosity , pores containing gases that contribute to the reduction of thermal conductivity and volume density.

• the lowering of the heat transfer coefficient with increasing insulation thickness - but this also depends on the sequence of layers . A large increase in the insulating layer is not recommended , it's better to follow a very good correlation between the insulation thickness and heat transfer coefficient in order to simultaneously solve the problem of the weight of the construction and the low heat transfer A few different types of constraints can be formulated in order to optimize the performance, such as:

• determined values for the heat transfer coefficient are imposed. An optimum value for the total heat transfer coefficient 0.3...0.5 W/m2K ;

• a given temperature is required on the outer surface or the inner wall ;

REFERENCES

1. Sterian, Andreea; Sterian, Paul - *Mathematical Models of Dissipative Systems in Quantum Engineering*, MATHEMATICAL PROBLEMS IN ENGINEERING, DOI: 10.1155/ 2012/ 347674 (2012).
2. Sterian, Andreea Rodica - *Computer modeling of the coherent optical amplifier and laser systems,* International Conference on Computational Science and Its Applications (ICCSA 2007). Proceedings Book Series: LECTURE NOTES IN COMPUTER SCIENCE Vol. 4705, 436-449 (2007).
3. C. Itu, F, Dogaru, M. Baba, *Dynamic con-rod analysis for different type of materials based on virtual simulation.*, "International Conference on Materials Science and Engineering "BRAMAT 2007", 22 – 24 February 2007, Braşov, Romania, ISSN 1223 – 9631(2007).
4. Marius BABA, *Fenomene microstructurale in mecanica ruperi compozitelor lignocelulozice din structurile de mobilier si constructii din lemn*, Revista de Politica Stiintei si Scientometrie, ISSN-1582-1218, (2006).
5. M. Baba, C. Itu, F. Dogaru, Micromechanics aspects about the prediction of elastic properties for a lignocellulosic composite lamina., "International Conference on Materials Science and Engineering BRAMAT 2007"., 22 – 24 February 2007, Brasov, Romania, ISSN 1223 – 9631 (2007).
6. Panouri termoizolante din lemn si materiale lemnoase utilizate in constructia caselor. Teza de doctorat, Universitatea TRANSILVANIA din Brasov (2006).
7. Farkas, J., Jármai, K., Optimum Design of Steel Structures. Springer (2013).
8. Klansek, U., Kravanja,S., Cost estimation, optimization and competitiveness of different composite floor systems- Part I. Self manufacturing cost estimation of composite and steel structures. Journal of Constructional Steel Research 62(5), 434-448 (2006).
9. Jalkanen,J., Tubular truss optimization using heuristic algorithms, PhD. Thesis, 104 p. Tampere University of Thechnology, Finland (2007).
10. Timár, I., Horváth,P., Borbély,T., Optimierung von profilierten Sandwichbalken. Stahlbau 72(2), 109-113 (2003).
11. Farkas, J., Jármai, K., Design and optimization of metal structures, p.328, Horwood Publishers, Chichester (2008).
12. Bader,M.G., Selection of composite materials and manufacturing routes for cost effective performance. Composite: Part A 33, 913-934 (2002).

THE INFLUENCE OF THE MASSES' VALUE AND OF THE MOMENTS OF INERTIA'S VALUE REGARDING THE MONOWHEEL VEHICLE' STABILITY

Dan Botezatu
Transilvania University of Brasov, Romania, dan.botezatu@unitbv.ro

Abstract: Monowheel vehicles present a number of challenges to the designer and several compromises have to be made to get everything come together into a functional machine. The first problem is stability; because monowheel depends on gyroscopic effect to keep it upright.
Keywords: monowheel, vehicle simulation, stability

1. INTRODUCTION

The monowheel [1] consists of an inner frame (1) and a wheel (2). The inner frame (1) has three small wheels (4) that make contact with the wheel (2).
The wheel (2) is the actual rotating wheel and has a solid rubber tire.
The rider sits inside the inner frame that also contains the driving roller (3), the engine, the clutch, the propulsion mechanism and the petrol tank.
Let us consider the following representation:

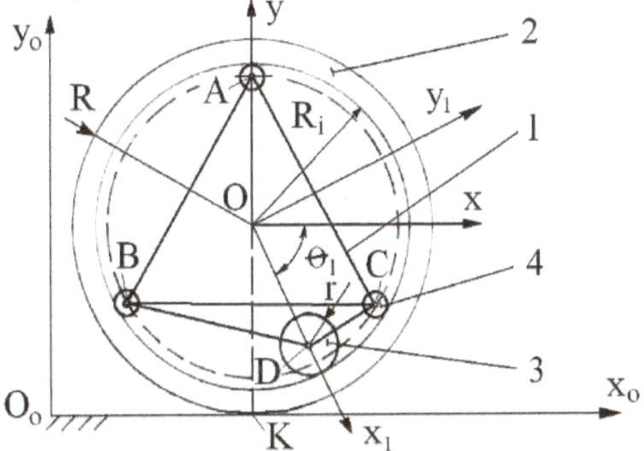

Figure 1. Simplified model

2. ADAMS SIMULATION

Vehicle handling and stability are significantly affected by inertial properties including moments of inertia and center of gravity location. Therefore let's generally analyze how vehicle inertial properties (i.e., weight, moment of inertia and center of gravity location) relate to typical dimensions (length, width and height) and how these properties affect vehicle dynamics.

The first contribution of this study is the creation of a realistic physical model which includes diverse kinematic and dynamic effects that are often difficult to foresee. The monowheel dynamic model was developed using the Lagrangian approach. In the first step the free Lagrangian is computed from the total kinetic energy.
After having implemented it, the dynamics model must be validated before being used for control analysis. To this purpose, several tests have been made, encompassing different situations. Namely, the spectrum of simulation comprehends diverse torques applied by the driver, starting conditions and external forces.

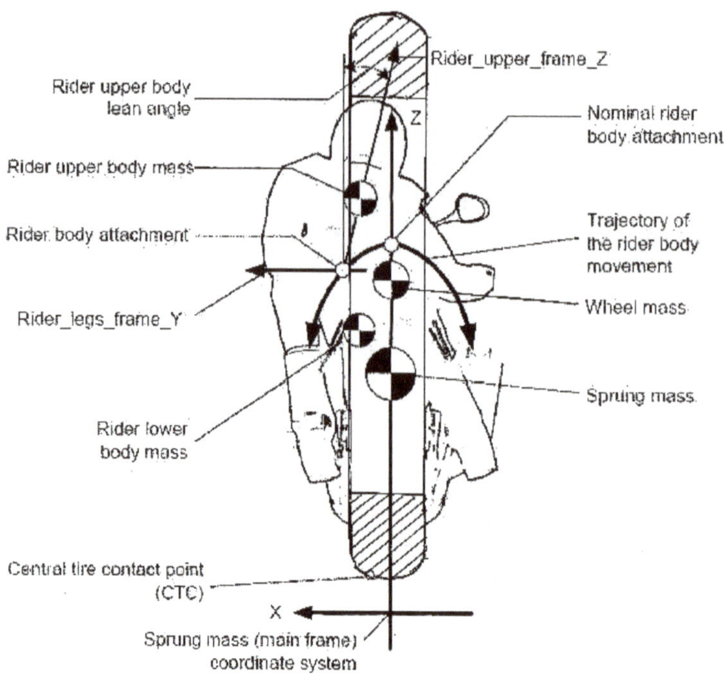

Figure 2: Monowheel cross-section

Usually it is useful to have a general appreciation of the effects of vehicle properties on stability and handling. This information is helpful in the preliminary phases of vehicle design. In the same time useful relationships between typical dimensions and inertial properties are provide through the use of regression analysis.

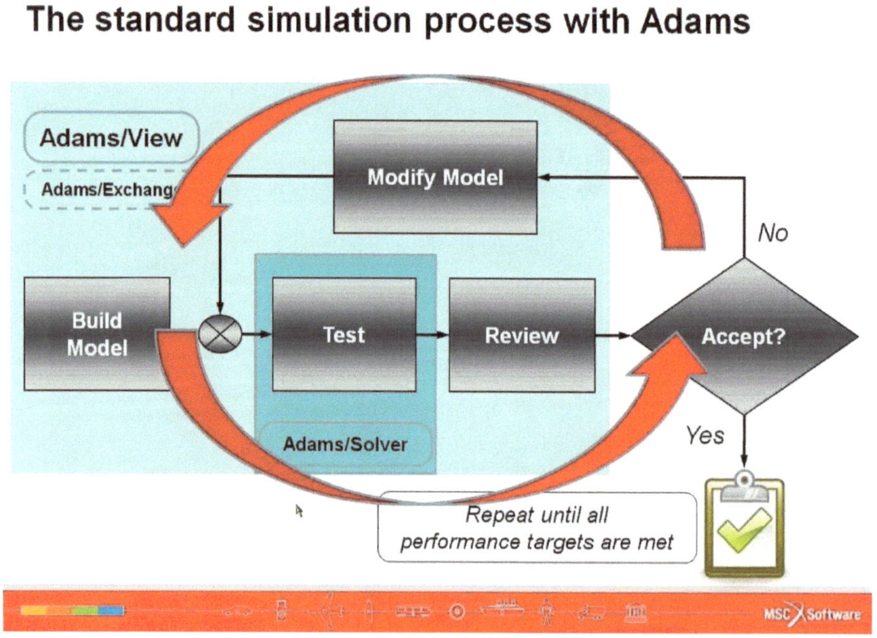

The standard simulation process with Adams

Adams/View
Adams/Exchange
Modify Model
No
Build Model
Test
Review
Accept?
Adams/Solver
Yes
Repeat until all performance targets are met
MSC Software

Figure 3: Simulation steps

Inertial properties have known effects on vehicle dynamic response, and key inertial variables relate to directional and roll mode stability issues. A validated nonlinear simulation is used to demonstrate stability problems related to inertial properties. Then we will use regression analysis to reveal the relationships between vehicle inertial properties and basic size dimensions. Finally, we carry out some nonlinear computer simulation analysis with detailed vehicle models to show how size and speed interact to create stability problems.

3. LATERAL/DIRECTIONAL STABILITY

Lateral/directional vehicle dynamics models give a general feeling for the effect of vehicle inertial properties on vehicle handling and stability. The following matrix expresses vehicle lateral velocity (v) and yaw rate (r) as a function.

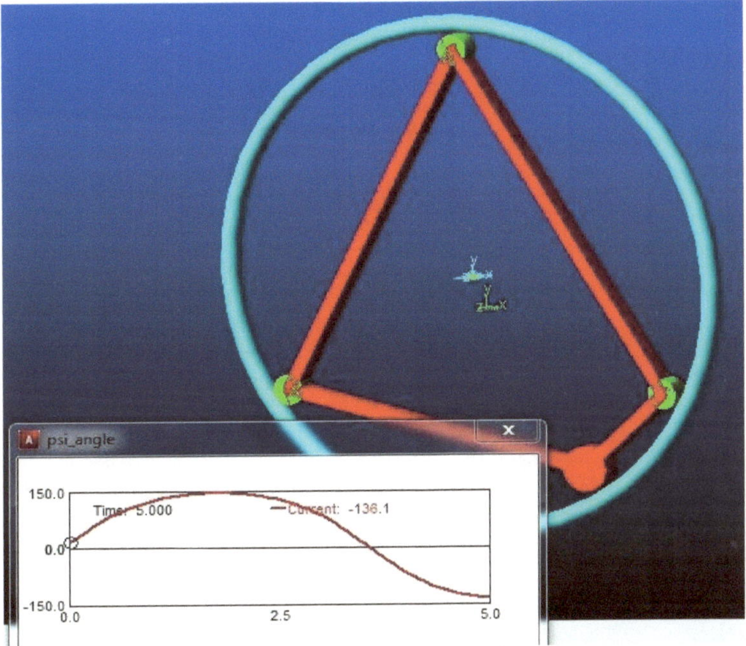

Figure 4: Yaw angle caption

Basic vehicle dynamics have been traditionally subdivided into lateral/directional dynamic modes including yawing and rolling motions. The basic input for these dynamics is steering, and speed is a key operating point. Lateral/directional dynamics are affected significantly by inertial properties including mass, moments of inertia and the location of center of gravity as will be analyzed subsequently. Inertial properties affect the time constants of various response modes, and also the influence of control inputs.

4. CONCLUSION

The analysis in this paper shows that vehicle inertial properties are strongly correlated with standard measures of length, width and height. It has also been shown that these inertial properties are related to lateral/directional handling and stability. In particular, specific inertial parameters are related to specific dynamic response properties.
The challenging factor of this stage was connected to the minimal influence of the effect of sustaining vertical position of the monowheel resulting from the gyroscopic effect (low angular velocity of the wheel) or its lack.

REFERENCES

1. Deliu, G. (2002) *Mechanics for Engineering Students.* Editura Albastră, ISBN 973-650-082-9, Cluj-Napoca, 2002
2. http://web.mscsoftware.com/Academia/Student-Center
3. https://github.com/hazelnusse/OBD

ACKNOWLEDGEMENT:

This paper is supported by the Sectoral Operational Programme Human Resources Development (SOP HRD), ID134378 financed from the European Social Fund and by the Romanian Government.

THE USE OF PERMEABLE REACTIVE BARRIERS FOR REMOVAL OF PHOSPHATE FROM GROUND WATERS

A. R. Miron[1*], D. E. Pascu[1], O. D. Orbuleţ[1], C. Modrogan[1], M. Neagu (Pascu)[1,2], G.Al. Popa[1]

[1] Politehnica University, Bucharest, ROMANIA, e-mail *andra3005@yahoo.com
[2] S.C. Hofigal S.A, Bucharest, ROMANIA

Abstract:*Permeable reactive barriers represent a promising remediation method for surface and groundwater containing a variety of inorganic and organic pollutants. Phosphate is a nutrient which leads to an extensive water eutrophication. It can be removed from surface and ground waters by means of permeable reactive barriers after using precipitation or adsorption.*
The aim of this study was to evaluate the permeable reactive barriers efficiency in the case of phosphate removal from groundwaters.
In this paper, clay mold soil type was used as a reactive permeable matrix for ground water phosphate removal. Also, the influence of the main operational parameters, namely: pH, initial solution concentration and contact time on the thermodynamic system behavior was investigated. The experimental results obtained were interpreted with the help of Temkin and Flory-Huggins adsorption isotherms.
Keywords : *phosphate removal, permeable reactive barriers, groundwater remediation, adsorption isotherms*

1. INTRODUCTION

Between 1982 and 1997 for groundwater treatment only surface techniques were applied. The groundwater surface treatment technology is a major energy consumption source and can not remove the contaminants adsorbed on the soil. As a result of the researches, an in situ passive treatment method was discovered named "barrier/permeable and reactive treatment area". [1,2]
In 1998, USEPA defined it as being:" a site of the underground containing a reactive material, designed to intercept the pollution plume, to ensure the flow of the plume through the reactive medium and for transforming the pollutants into less harmful compounds, from the environmental point of view."[3]
Currently, permeable reactive barriers are some of the most promising technologies for passive treatment of the groundwater due to high retention efficiency of different contaminants and low price compared to other *in situ* technologies. [4]
A typical configuration of the permeable reactive barriers is presented in figure 1. The chemical compounds which are retained in the permeable reactive barriers are either broken down into other compounds less hazardous or are effectively fixed on the reactive material. [5]
A variety of pollutants are degraded, precipitated, adsorbed, or change in the reactive area, including chlorinated solvents, heavy metals, radionuclides and other inorganic or organic species.
In general, the mechanism by which the groundwater is treated consists into its direct passage through the reactive area of the permeable reactive barriers where contaminants are either immobilized or chemically transformed. [6]
The permeable reactive barrier (PRB) acts as a barrier, not for water, but for the compounds which are generating contamination. The pollutants removal takes place in two ways:

- by introducing an adsorption medium for the contaminants;
- by introducing a medium which will alter the contaminants and reduce their bioavailability.[7]

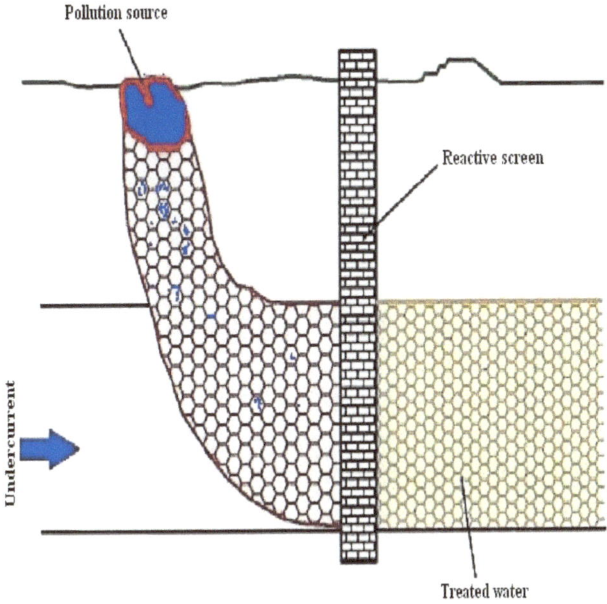

Figure 1: Permeable reactive screen configuration [8]

This fact can be achieved both by changing the speciation and by provision of a ligand with which to form a solid. The second version was more often used simply because the adsorbent surface quickly saturates, leaving the contaminants unchanged and thus, the barriers permeability is reduced. The contaminant is transformed into a harmless form by changing its oxidation state. The main configurations of reactive barrier systems are: [9]

- continues barriers;
- "funnel-gate" type systems;
- walls filled with reactive material systems;
- injection systems.

For the construction of a permeable reactive barrier several types of reactive materials can be used:

- granular metals: Al, Zn, Cu, stainless steel, but especially iron are the most used materials in laboratory facilities, at pilot or large scale;
- granular iron with amendments;
- bimetallic materials.

The choice of material for reactive barriers is governed by the following considerations:

- reactivity: are preferred the materials which provide lower half-times (greater degradation rate);
- stability: the time period during which the maintains its reactivity;
- availability and cost: cheap materials are preferred as against the expensive ones, especially when the performances differ only slightly;
- compatibility with the environment: reactive material must not introduce dangerous byproducts into the environment;
- security: the material must not endanger the safety and health of workers. [10]

2. MATERIALS AND METHODS

2.1. Working procedure

All the reagents used for the present study were analytical grade and purchased from Merck. The experiments were performed on clay mold soil type samples, at 20 ° and a soil: solution phase mass ratio of 1: 5. The mobile phosphorus amount (expressed as phosphate) was determined according to the procedure described in SR 11411-2:

1998. During the thermodynamic study, the soil samples were left in contact with a KH_2PO_4 solution having the following concentrations: 10, 20, 40, 60, 80, 100 mg/L for 24 hours in order to achieve equilibrium. The water-soil suspension was vacuum filtered. From the resulting aqueous extract 10 mL were pipetted in a volumetric flask in which were added 12.5 ml of vanadomolybdenum reactive and brought to 50 mL with distilled water. The samples obtained were spectrophotometrically analyzed at a wavelength of 470 nm.

2.2. Soil sample characterization

The clay mold soil type (argillic) used in this study comes from the Urziceni city. An important characteristic of the county is that on an area not very large are succeeding: carbonate chernozem, chernozem itself, cambic chernozem, argillic chernozem and typical reddish brown soil, the last one present in small areas in the extreme west of the county.
These soils have high fertility thus allowing the widespread practice of agriculture, predominant being the grain crop production. The main physico-chemical properties of clay mold soil type are presented in table 1.

Table 1: The physico-chemical properties of clay mold soil type [4,11]

Property	Value	Analysis method
pH	7,85	SR ISO 1039:1999
Humus (%)	2,00	STAS 7184/21-82
$CaCO_3$ (%)	0,00	STAS 7184/16-80
Macroelements		
N_{total} (%)	0,12	SR ISO 11261:2000
P_{AL} (mg/kg)	20,00	STAS 7184/19-82
P_{total} (%)	0,065	STAS 7184/14-79
K_{AL} (mg/kg)	12,00	STAS 7184/18-80
Microelements		
Zn (mg/kg)	80,20	SR ISO 11047:1999
Cu (mg/kg)	27,40	
Pb (mg/kg)	21,70	
Co (mg/kg)	9,70	
Ni (mg/kg)	55,80	
Cr (mg/kg)	13,20	
Cd (mg/kg)	0,01	
Fe (mg/kg)	256,71	SR ISO 6332:1996
Al (mg/kg)	295,00	STAS 6326-90

Note: The soil characterization was done and offered by Agrochemistry and Pedology Research Institute (ICPA) Bucharest

3. RESULTS AND DISCUSSIONS

Phosphate transformations in the soil reactive permeable barrier must be done analytically on components and overall synthetic, because this is the only way that allows us to study the intimate mechanisms and interactions that occur and regulates soil retention processes.
Soil, as a poly-dispersed, heterogeneous system with three phases (solid, liquid, gas) tends to achieve a state of dynamic equilibrium that changes depending on the environmental factors mode of action (climate, plants, crop technique).
The relationships that exist between the solid and liquid phase reflects phosphate dynamics both through physical and chemical mechanisms. [1,12]

3.1. Temkin model

The thermodynamic study of the phosphate removal from groundwaters by means of permeable reactive barriers use aimed to determine the maximum soil adsorption capacity for this pollutant based on the experimental data obtained. The data were interpreted using classical adsorption models, namely Temkin and Flory-Huggins. [13]
In the case of Temkin isotherm equation assumes that the heat of adsorption of all the molecules in layer decreases linearly with coverage due to adsorbent-adsorbate interactions, and the adsorption is characterized by a uniform distribution of the bonding energies, up to some maximum binding energy. [14,15]

The equation used for describing Temkin isotherm is:

$$q_e = \frac{RT}{b} Ln(K_T C_e) \quad (1)$$

By linearization, equation (1) becomes:

$$q_e = B_T \ln K_T + B_T Ln C_e \quad (2)$$

where: T = the absolute temperature (K), R = universal gas constant (8.314 J/molK), K_T = equilibrium binding constant (L/mg), b_T = variation of the adsorption energy (kJ/mol), B_T = Temkin constant related to the heat of adsorption (kJ/mol).

The Temkin adsorption isotherm model was chosen to evaluate the adsorption potentials of the adsorbent for adsorbates.

In figure 2 are presented the adsorption isotherms achieved for the experimental data obtained for two working solution initial pH values (5 and 7), as well the theoretical adsorption isotherms obtained based on the Temkin model.

From the data presented it can be observed that the initial pH value has an influence on the phosphate removal through adsorption mechanisms, the adsorption capacity (q_e) at pH = 5 being with 10% higher than the one obtained for pH = 7.

Also, the graphical representation reveals the fact that experimental data are well fitted by the curves obtained using Temkin model.

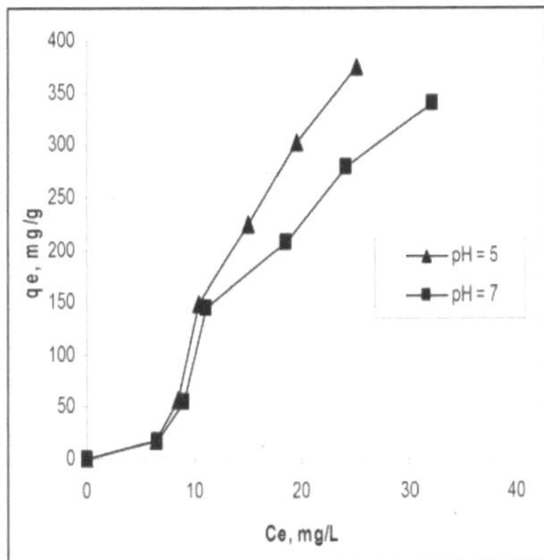

Figure 2: Adsorption isotherms for Temkin model

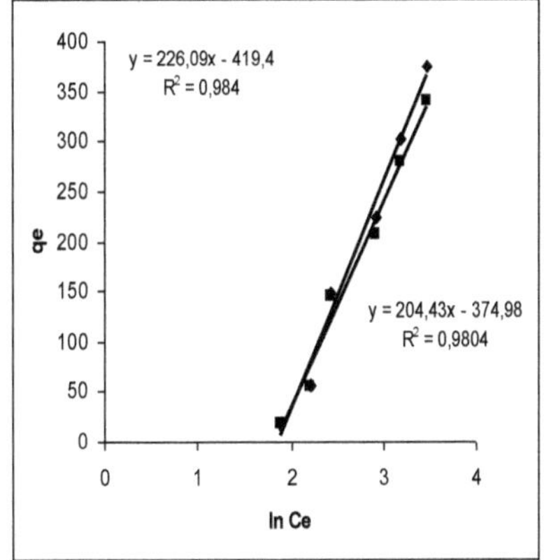

Figure 3: Graphical representation of the linear form for Temkin equation, pH 5 (▲), pH 7(■)

The results presented in figure 3, revealed that the correlation coefficients (R^2) obtained for two data sets have values higher than 0.97, fact which suggests that the phosphate adsorption equilibrium on the permeable reactive barrier can be described by means of Temkin model. [15]

Table 2: Temkin parameters

Initial pH value	B_T	K_T
5	10,04	0,156
7	11,10	0,159

From the data presented in Table 2, resulted that Temkin adsorption potential K_T, had a very low value (0.15) fact that clearly reveals a very low potential of the reactive barrier for the phosphate ion. Regarding Temkin constant, b_T, related to heat of phosphate sorption the values were higher than 10 $kJmol^{-1}$. According to the literature, the typical range of bonding energy for ion-exchange mechanism is 8-16 kJ mol^{-1}. The values relatively low obtained for this parameter indicates a relatively weak interaction between sorbate and sorbent, supporting an ion-exchange mechanism for the present study. [16]

3.2. Flory-Huggins model

This mathematical model was chosen in order to determine the surface coverage degree characteristics of the adsorbate on the adsorbent. The linear form of the Flory-Huggins equation can be expressed as follows:

$$ln\left(\frac{\theta}{C_0}\right) = ln K_{FH} + n ln(1-\theta) \quad (3)$$

where: $\theta = (1\text{-}C_e/C_0)$, the degree of surface coverage, n = number of phosphate ion occupying adsorption sites, K_{FH} = the equilibrium constant (L/mol). [17, 18]

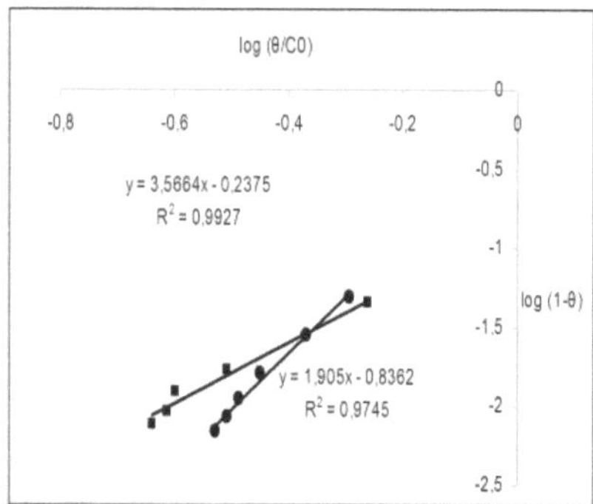

Figure 4: The linear form of Flory-Huggins equation representation, pH 5 (●), pH 7(■)

In figure 4 is represented the linear form of Flory-Huggins equation for experimental data obtained at pH=5 and pH =7.

From graphical representation analysis we can affirm that better results are obtained in the case of pH = 5, corresponding to a correlation coefficient (R^2) higher than 0.99, compared to 0.97 obtained in the case of initial solution pH = 7.

A value of R^2 higher than 0.97 suggest that the phosphate adsorption during the interaction with the permeable reactive barrier can be also described using Flory-Huggins model.

4. CONCLUSION

The aim of this study was to evaluate the permeable reactive barriers efficiency in the case of phosphate removal from groundwaters. Different permeable reactive mixtures can be used for phosphate removal. Mixtures can be placed in situ, as horizontal or vertical reactive barriers in sediments receiving wastewaters, or can be used in a single pass, as self-contained treatment modules in the alternative treatment systems. The thermodynamic system behavior was investigated. The experimental results obtained were interpreted with the help of Temkin and Flory-Huggins adsorption isotherms. The first conclusion that resulted after the study was that the use of a permeable reactive barrier made from a mold clay soil containing metal oxide responsible for adsorption (e.g. iron oxides, alumina), a calcium source for precipitation (limestone, $CaCO_3$) for ground water phosphate removal lead to satisfactory results. The second conclusion was that the values relatively low obtained for b_T indicates a relatively weak interaction between sorbate and sorbent, supporting an ion-exchange mechanism for the present study.

Acknowledgements

The work has been funded by the Sectoral Operational Programme Human Resources Development 2007-2013 of the Ministry of European Funds through the Financial Agreement POSDRU/159/1.5/S/134398 and POSDRU/159/1.5/S/132395.

REFERENCES

[1] Sposito G., The surface chemistry of soils, Oxford University Press, Oxford, England.

[2] Starr R.C., Cherry J.A., In situ remediation of contaminated ground water: the funnel-and-gate system, Ground water, vol.32, no.3, p.465-476, 1994.

[3] http://www.epa.gov/

[4] McMurty D.C., Elton R.O., New approach to in-situ treatment of contaminated groundwaters, Environmental Progress, vol.4, no.3, p.168-170, 1985.

[5] Ijoor G.C, Modeling of a permeable reactive barrier ,PhD Thesis, 1999.

[6] Modrogan C., Diaconu E., Orbuleţ O. D., Miron A. R., Forecasting study for nitrate ion removal using reactive barriers, Revista de Chimie, vol. 61(6):580-584, 2010.

[7] Powell R.M., Puls R.W., Blowes D.W., Gillham R.W., Vogan J.L., Schultz D., Powell P.D., Sivavec T. Landis R., Permeable Reactive Barrier, Technologies for Contaminant Remediation, USEPA, 1998.
[8] Bica I., Dimache Al., Iancu I., Vraciu S., Constantinoiu C., Stefanescu M., Voicu A., Dumitrescu C., 9, 2008.
[9] Navarro A., Chimenos J. M., Muntaner D., Fernández A. I., Permeable Reactive Barriers for the Removal of Heavy Metals: Lab-Scale Experiments with Low-Grade Magnesium Oxide, Ground Water Monitoring & Remediation, vol.26, no.4, p.142-152, 2006.
[10] Golab A.N., Peterson M.A., Indraratna B., Selection of potential reactive materials for a permeable reactive barrier for remediating acidic groundwater in acid sulphate soil terrains, Quarterly Journal of Engineering Geology and Hydrogeology, vol.39, p.209–223, 2006.
[11] Theivarasu C., Mylsamy S., Removal of Malachite Green from Aqueous Solution by Activated Carbon Developed from Cocoa *(Theobroma Cacao)* Shell - A Kinetic and Equilibrium Studies, E-Journal of Chemistry, vol.8, no.S1, p.363-371, 2011.
[12] Diaconu E., Orbuleţ O. D., Miron A. R., Modrogan C., Forecasting the sorption of phosphates in soil with artificial neural networks, U.P.B. Scientific Bulletin, Series B, vol.72, no.3, p.175-182, 2010.
[13] Saratha R., Priya S.V., Thilagavathy P., Investigation of *Citrus aurantiifolia* leaves extract as corrosion inhibitor for mild steel in 1M HCl, E-Journal of Chemistry, vol.6, no.3, p.785-795, 2009.
[14] Shahmohammadi-Kalalagh Sh., Babazadeh H., Nazemi A. H., Manshouri M., Isotherm and kinetic studies of adsorption of Pb, Zn, Cu by kaolinite, Caspian Journal of Environmental Sciences, vol.9, no.2, p.243-255, 2011.
[15] Temkin M.J., Pyzhev V., Kinetics of ammonia synthesis on promoted iron catalysts, Acta Physiochimica Urss, vol.12, p.217-222, 1940.
[16] Ho Y.S., Wase D.A.J., Forster C.F., Removal of lead ions from aqueous solution using sphagnum moss peat as adsorbent, Water SA, vol.23, no.3, p.219-224, 1996.
[17] Blowes D.W., Ptacek C.J., Geochemical remediation of groundwater by permeable reaction walls: removal of chromate by reaction with iron - bearing solids. Subsurface Restoration Conference, Third International Conference on Groundwater Quality Research, Dallas, 1992.
[18] Jnr M. H. Spiff A. I., Equilibrium Sorption Study of Al^{3+}, Co^{2+} and Ag^{+} in Aqueous Solutions by Fluted Pumpkin (*TelfairiaOccidentalis* HOOK f) Waste Biomass, Acta Chimica Slovacia, vol.52, p.174-181, 2005

RESEARCH ON LIQUID METAL HEAT PIPE STARTUP AND OPERATION

Virgil B. Ungureanu
Transilvania University, Braşov, ROMANIA, virbung@unitbv.ro

Abstract: *The paper presents experimental investigations on liquid metal (sodium and potassium) heat pipe operating and startup. Thus, is studied the wickless gravity assisted heat pipes, that presents a high non-uniformity of the temperature profile along the entire pipe. Because the temperature increasing of the evaporator lower end, it is compulsory to provide wick in the liquid metal heat pipes. There are recommended composite wicks with grooves and screen, or some screen layers.*
Keywords: *heat pipe, gravity assisted heat pipe, wickless heat pipe, screen wick, heat pipe startup.*

1. INTRODUCTION

Because of the diversity of field utilization, the heat pipes there are very extended ranges of the operating temperature. The heat pipes that operate in the middle and low temperature ranges and also the effects of working fluid and wick configuration on the startup characteristics of heat pipes have studied by many authors. Thus, it was demonstrated that in the case of the gravity assisted wickless heat pipes, the startup is suddenly, by reaching a quasi-metastable state. Further on, it is establish an accurate operating, with pulsatory liquid boiling that assure the constant temperature within the entire heat pipe [1].

El-Genk, Huang and Tournier [2], [3] investigate the operation and design constrains pertinent to uses of water and liquid metal heat pipes in space reactor systems, and the modelling capabilities of the startup from a frozen state. A free molecular, transition and continuum vapor flow model based on the dusty gas model is developed and a two dimensional heat pipe transient analysis model, to analyses the startup of a radiatively-cooled sodium heat pipe from a frozen state.

Chang [4] studied the feasibility of employing heat pipes to cool the hot sections of the Army's ground-to-ground missile fins.

Faghri [5] studied the heat pipe startup from the frozen state. For a safe startup Ponnapan [6] studied a diffusion controlled startup of a liquid metal heat pipe.

Gravity assisted wickless heat pipes have studied by Fetcu, [7], in a view to utilize for heat pipe heat recovery systems.

Dickinson [8] analyses the performance of liquid metal heat pipes and characterize the frozen startup and restart behavior of liquid metal heat pipes in a microgravity environment.

This paper investigates the startup of sodium and potassium heat pipes.

2. TYPICAL TRANSIENT PHENOMENA OF METAL LIQUID HEAT PIPES

The startup behavior of heat pipes is difficult to provide because it depends of many factors. In the startup period, the vapor flows with a high speed from the evaporator to the condenser zone because the low density of vapor [5]. Thus, the pressure drop and implicit of temperature along the heat pipe is great. Because the axial temperature gradient in a heat pipe is determined by the vapor pressure drop, the initial evaporator temperature shall be greater than the evaporator one. The temperature level realized in the evaporator zone is dependent by the working fluid. If the heat flow rate transferred by the heat pipe is great, it is observed a temperature front that moves towards the condenser section. During a normal startup of a heat pipe, the evaporator temperature increase with some degrees until the heat pipe becomes quasi-isothermal. If the vapor speed is high during the liquid metal startup, the liquid returning towards evaporator zone can be braked.

3. EXPERIMENTAL RESULTS

3.1. The experimental stand

Figure 1 presents the schematic stand having the main components: a radiation oven (1) composed by a stainless steel covered by an insulation (2); the electric installation of the radiation oven composed by a voltage control device, a transformer (4), two flexible cables (5 and 6) coupled to the radiation oven by two clamps (7 and 8) cooled by an water circuit; the oven temperature measuring installation composed by two K thermocouples (9 and 10) connected successively to the measuring apparatus; the temperature measuring installation along the heat pipe composed by a K thermocouple connected to the correspondent measuring apparatus; chassis for the oven-heat pipe assembly, having the possibility to be oriented at various tilt inclination angles towards the horizontal between 0 to 90°.
Figure 1 presents a section through heat pipes too. Thus, for an accurate temperature measuring in the heat pipe it is introduced a pipe Φ6×1.2 from stainless steel fixed by welding at both caps of the heat pipe.
The evaporator zone cooling is realized by radiation in the ambient.

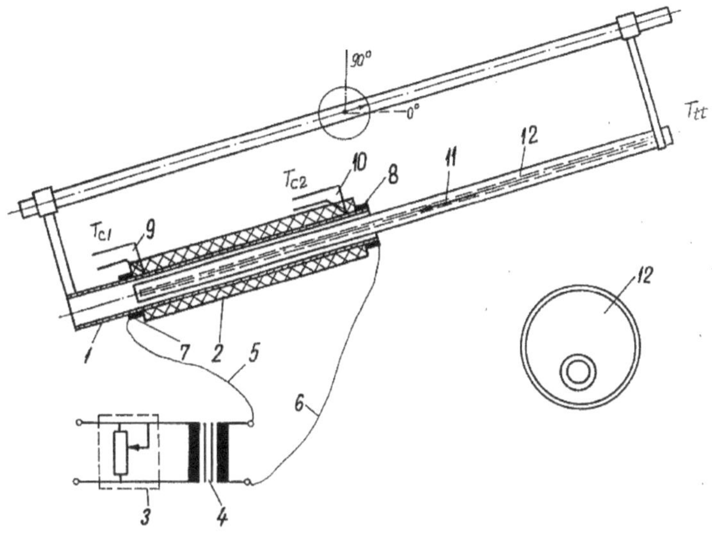

Figure 1: The experimental stand

There are manufactured three wickless gravity assisted heat pipes having the dimensions: outer diameter 28 mm, wall thickness 2.4 mm, overall length of the pipe 1000 mm. Volumes of working fluid are: 1st heat pipe with sodium: 90 cm^3 (the filling coefficient 0.228); the 2nd heat pipe with sodium, too: 53 cm^3 (the filling coefficient: 0.134); and respectively the 3rd heat pipe with potassium: 102 cm^3 (the filling coefficient 0.259). The filling coefficient is the working fluid volume (liquid) to the entire void of the heat pipe ratio.
However, there are manufactured two heat pipes: 1st heat pipe with sodium: outer diameter: 32 mm, wall thickness: 2.5mm, overall length of the pipe 1480 mm, and two wrapping screens: the first composed by three layers of 50 mesh per inch and the second adjoined at the wall, composed by 1.25 layers of 25 mesh per inch to create a channel with high permeability for liquid flow, both being spot welded, overall volume of working fluid being 61.6 cm^3, (corresponding to a filling coefficient 0.072); the 2nd heat pipe with sodium, too: outer diameter 27.5 mm, wall thickness 2.55 mm, overall length 980 mm, inner diameter of the wick: 20 mm, the wick being composed by three layers with 50 mesh per inch.

3.2. The gravity assisted wickless heat pipe testing

First, it was tested the 1st heat pipe for an evaporator length of 480mm and condenser zone of 530mm, at the oven temperature of about 700°C. It is observed that the heat pipe operates very instable. The next tests are realized by using a cylindrical insulation of 250mm length over the condenser zone of the heat pipes.
Figure 2 presents the temperature profile along the wickless heat pipe in this configuration.

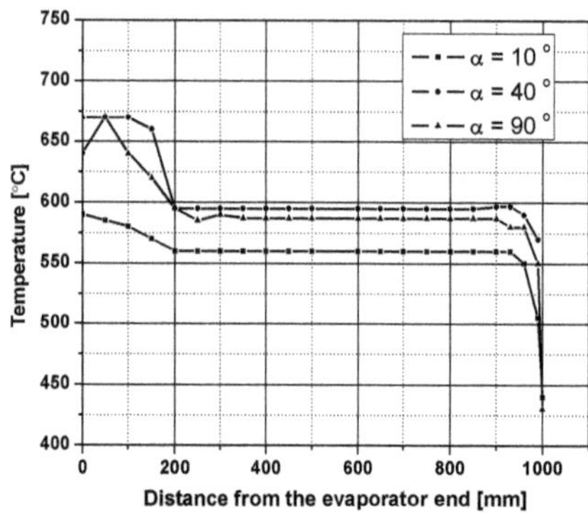

Figure 2: The longitudinal temperature profile of the 1st wickless heat pipe

The 2nd heat pipe was tested in the configuration: evaporator zone: 630 mm; condenser zone: 370 mm; insulated zone: 180mm. The oven temperature was about 760°C, experimental results being presented in figure 3.

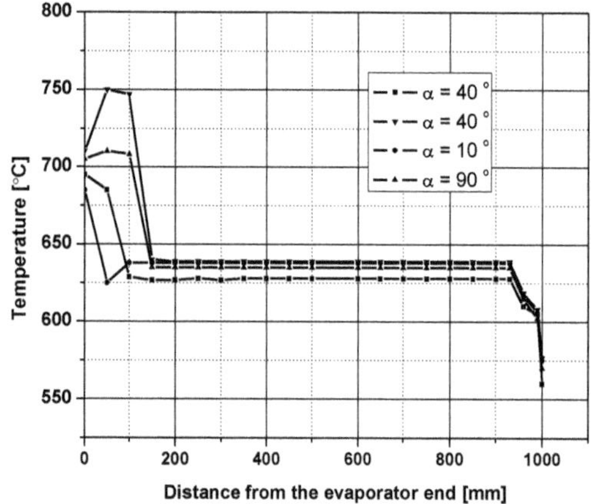

Figure 3: The longitudinal temperature profile of the 2nd wickless heat pipe

It is observed a quasi-isothermal operation on a greater length of the condenser zone, due, probably the higher filling coefficient and/or an accurately manufacturing of the heat pipe (higher degasing of the container and working fluid). The potassium heat pipe was tested with the geometry: evaporator zone length: 640mm; condenser zone length placed in the ambient: 360mm. There are no used the insulation on the condenser zone. Like the sodium heat pipes this one has an instable operation. Figure 4 presents temperature profiles. However, there are remarked a lower overheating of the evaporator zone, probably because of the higher filling coefficient.

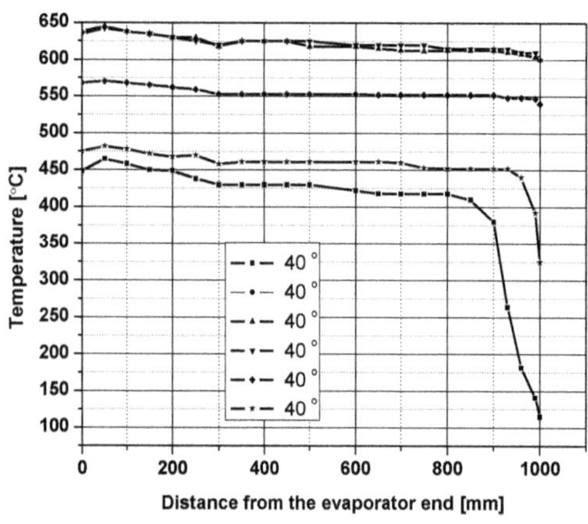

Figure 4: The longitudinal temperature profile of the 3rd wickless heat pipe

3.3. The gravity assisted wicked heat pipe testing

The 1st heat pipe was tested with the geometry: evaporator zone length: 680mm; condenser zone length 800mm; insulated zone: 600 mm. Test results are presented in figure 5.

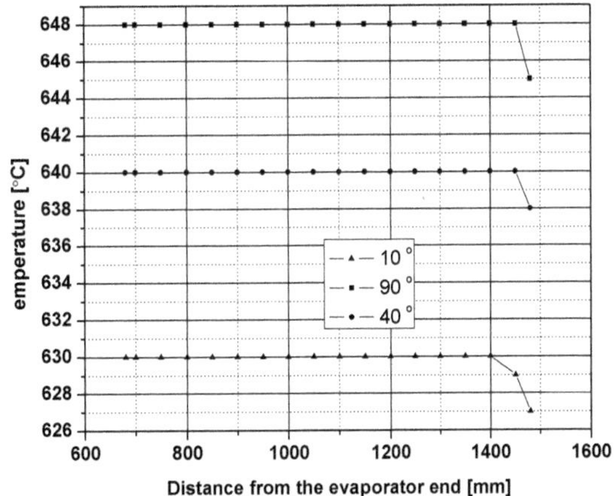

Figure 5: The longitudinal temperature profile of the 1st wicked heat pipe

It is remarked that the insulation was used because the source hasn't sufficient power. The heat pipe was a stable operation. By using a high power oven, the heat pipe was a uniform temperature on the overall condenser zone (about 600 mm).

The 2nd heat pipe was tested by using the geometry: evaporator zone length: 630 mm, condenser zone length, free in ambient: 350 mm. The oven temperature was between 775 to 800°C and there wasn't utilizing thermal insulation on the condenser zone.

Figure 6 presents the temperature profile. It is remarked that the startup is very good like the 1st heat pipe, the operation is proper and the temperature profile is very good.

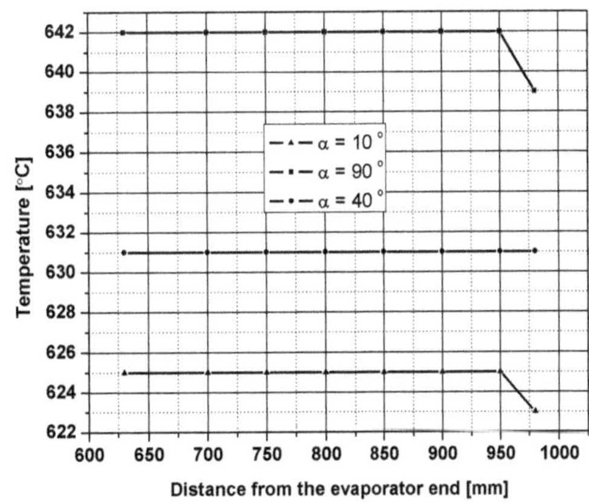

Figure 6: The longitudinal temperature profile of the 2nd wicked heat pipe

It is remarked that the operation, startup and stability are not influenced by the tilt angle.

4. CONCLUSION

While the startup and operation of the wickless heat pipes that operates in the middle and low temperature ranges is very accurate, the liquid metal gravity assisted wickless heat pipes operates very faulty, with overheating of the lower end caps of the evaporator zone and pulsating temperature. Thus, it is necessitating a wick, even a summary wick having a high permeability for providing the liquid return. There are recommended composite wick, with grooves and screen, or some screen layers.

REFERENCES

1. Fetcu D. Hoffmann V., Ungureanu V.B., Theoretic and experimental researches on the gravity assisted wickless heat pipes operation. Heat transfer and optimization of thermal energy utilization, Brasov, Oct. 30-31. 1986, pp I-55.
2. El-Genk M.S., Huang L., Experimental investigation of low temperature heat pipe response to transient heating. Proc. of the 8 th International Heat Pipe Conference, Beijing, Sept. 14-18, 1992.
3. El-Genk M.S., Tournier J.M.P. Uses of liquid-metal and water heat pipes in space reactor power systems. Frontier in Heat Pipes (FHP), 2, 2011.
4. Chang W.S., Startup of the liquid-metal heat pipe in aerodynamic heating environments Final report. Aero Propulsion and Power Directorate, Wright Laboratory, Ohio.. Feb. 1996.
5. Faghri A., Bean J., Heat pipe startup from the frozen state, Proc. of the 8-th International Heat Pipe Conference, Beijing, Sept. 14-18, 1992.
6. Ponnapan R., Boehman L.I., Mahefkey E.T., Diffusion controlled start-up of a bas-loaded liquid metal heat pipe. Journal of Thermophysics and Heat Transfer, vol. 4. No. 3, 1990, pp 332-340.
7. Fetcu D., Ungureanu V.B. Bacanu G., Gravity assisted wickless heat pipe heat exchangers. Proc. of the 8-th International Heat Pipe Conference, Beijing, Sept. 14-18, 1992.
8. Dickinson T.J., Performance analysis of a liquid metal heat pipe space shuttle experiment. Thesis. Air Force Institute of Technology. Wright-Patterson Air Force Base, Ohio, Dec. 1996.

TABLE OF CONTENTS